Modern
Food
Microbiology
Second Edition

Modern
Food
Microbiology
Second Edition

James M. Jay
Wayne State University

D. Van Nostrand Company
New York
Cincinnati
Toronto
London
Melbourne

D. Van Nostrand Company Regional Offices:
New York   Cincinnati

D. Van Nostrand Company International Offices:
London   Toronto   Melbourne

Published by D. Van Nostrand Company
450 West 33rd Street, New York, N. Y. 10001

10 9 8 7 6 5 4 3 2 1

Preface

MODERN FOOD MICROBIOLOGY, SECOND EDITION is designed primarily for a subsequent course in microbiology in the liberal arts, food science, nutrition, or related course programs. Organic chemistry and biochemistry are desirable prerequisites, but students with a good grasp of biology and chemistry from the secondary level should be equipped to handle much of the material.

The second edition contains two new chapters—"Determining Microorganisms and their Products in Foods" (Chapter 4) and "Fermented Foods and Related Products of Fermentation" (Chapter 14), both of which add valuable information to the book and increase its scope. Classification of microorganisms in the new edition is in conformity with the 8th Edition of *Bergey's Manual of Determinative Bacteriology*. The new edition also cites numerous recent references that have been consulted in its preparation.

The book begins with a brief history of man's awareness of the presence and role of microorganisms in nature and in foods. Foodborne microorganisms are presented in the context of their continual recycling of the elements of organic matter. Frequently encountered genera of these organisms are described in an early chapter, along with the intrinsic and extrinsic parameters of foods that affect their growth and activity. A synopsis of the traditional methods of examining foods for microorganisms and/or their products is presented in Chapter 4, along with an introduction to recently developed and promising rapid methods of analysis, including fluorescent antibody and radioimmunoassay techniques. Chapters on food spoilage present the general chemical composition of various foods in order to emphasize the ecologic relationships between these foods and their microorganisms and to develop a general knowledge of the chemical and physical properties of foods—indispensable to a proper understanding of their microbiology. The chapters on food spoilage, food preservation, and food poisoning emphasize the interplay of ecologic parameters that affect the entry of microorganisms into foods and their subsequent fate. Chapter 14 presents an up-to-date treatment of the fermentation process and related topics. It provides background information on the microorganisms important in fermentation products, and contains brief synopses of some of these products. The brief descriptions given for the preparation and production of the products discussed are intended as a guide to their microbiology and other references should be sought for details. The last three chapters develop the molecular biology of some important foodborne microorganisms.

I am indebted to a large number of individuals for their kind assistance in reviewing various parts of this book in the previous and current editions. For reading the entire first-edition manuscript, I thank Dr. B. W. Koft of Rutgers University and Dr. L. A. Shelef of Wayne State University; and Dr. W. Litsky of the University of Massachusetts for reading the second edition manuscript. For their assistance on various parts of the first or second-edition manuscripts, I thank the following colleagues at Wayne State University: Drs. R. S. Berk, D. L. DeGiusti, L. H. Mattman, H. W. Rossmoore, and R. Teodoro. I am grateful to the following for helpful comments and suggestions on individual chapters of the book: Drs. O. R. Collins of the University of California at Berkeley; M. G. Johnson of Clemson University; R. V. Lechowich of Virginia Polytechnic and State University; Z. J. Ordal of the University of Illinois; D. F. Splittstoesser of Cornell University; and W. I. Taylor of West Suburban Hospital, Oak Park, Illinois. I am indebted to the many publishers from whose materials I have drawn for granting me this privilege, and to the many researchers whose findings I have cited liberally.

Detroit, Michigan   James M. Jay

Contents

1
History of Microorganisms in Food

Although it is extremely difficult to pinpoint the precise beginnings of man's awareness of the presence and role of microorganisms in foods, the available evidence indicates that this knowledge preceded the establishment of bacteriology or microbiology as a science. The era prior to the establishment of bacteriology as a science may be designated the pre-scientific era. This era may be further divided into what has been called man's **food gathering period** and his **food producing period.** The former covers the time from man's origin over 1 million years ago up to 8,000 to 10,000 years ago. During this period, man was presumably carnivorous in his eating habits with plant foods coming into his diet later in this period. It is also during this period that man learned to cook his foods.

The food producing period dates from about 8,000 to 10,000 years ago and, of course, includes the present time. It is presumed that man first encountered the problems of spoilage and food poisoning early in this period. With the advent of prepared foods, the problems of disease transmission by foods and faster spoilage due to improper storage both made their appearance. Spoilage of prepared foods apparently dates from around 6,000 B.C. The practice of making pottery was brought to Western Europe about 5,000 B.C. from the Near East. The first boiler pots are thought to have originated in the Near East sometime between 6,000–8,000 years ago. The art of cereal cookery, brewing, and the storage of foods, was either started at about this time or given boosts by this new development, which also included basket making. The first evidence of beer manufacture has been traced to ancient Babylonia, possibly dating as far back as 5,000 to 7,000 B.C. (4). The Sumarians of about 3,000 B.C. are believed to have been the first great livestock breeders and dairymen, and were among the first to make butter around 3,100 B.C. Salted meats, fish, fat, dried skins, wheat, barley, etc., are also known to have been associated with this culture at about this time. Milk, butter, and cheese were used by the Egyptians as early as 3,000 B.C. Between 3,000 B.C. and 1,200 B.C., the Jews employed salt from the

1

Dead Sea in the preservation of various foods. The Chinese and Greeks used salted fish in their diets around this time and are credited with passing this practice on to the Romans, who included pickled meats in their diets. Mummification and preservation of foods seem to have influenced each other. Wines are known to have been prepared by the Assyrians by 3,500 B.C. Fermented sausages were prepared and consumed by the ancient Babylonians and the people of ancient China as far back as 1,500 B.C. *(4)*.

Another method of food preservation that apparently arose during this time was the use of oils such as olive, and sesame. Jensen *(3)* has pointed out that the use of oils leads to high incidences of staphylococcal food poisoning. The Romans excelled in the preservation of meats other than beef about 1,000 B.C. and are known to have used snow to pack prawns and other perishables, according to Seneca. The practice of smoking meats as a form of preservation is presumed to have emerged sometime during this period as did the making of cheese and wines. It is doubtful whether man at this time understood the nature of these newly found preservation techniques. It is also doubtful whether man of this period understood the role of foods in the transmission of disease, or the danger of eating meat from infected animals.

Little was apparently contributed towards man's understanding of the nature of food poisoning and food spoilage between the time of the birth of Christ and 1,100 A.D. Ergot poisoning (caused by *Claviceps purpurea*, a fungus which grows on rye and other grains) caused many deaths during the Middle Ages. Over 40,000 deaths due to ergot poisoning were recorded in France in 943 A.D., but it was not known that the toxin of this disease was produced by a fungus. Meat butchers are mentioned for the first time in 1156, and by 1248 the Swiss were concerned with marketable and nonmarketable meats. In 1276 a compulsory slaughter and inspection order was issued for public abbatoirs in Augsburg. Although man was aware of quality attributes in meats by the turn of the 13th Century, it is doubtful if he had any knowledge of the possible causal relationship between meat quality and microorganisms.

Perhaps the first man to suggest the role of microorganisms in spoiling foods was A. Kircher, a monk, who as early as 1658, examined decaying bodies, meat, milk, and other substances and saw what he referred to as "worms" invisible to the naked eye. Kircher's descriptions lacked precision, however, and his observations did not receive wide acceptance. In 1765, L. Spallanzani showed that beef broth which had been boiled for an hour and sealed remained sterile and did not spoil. Spallanzani performed this experiment to disprove the doctrine of the spontaneous generation of life. However, he did not convince the proponents of the theory since they believed that his treatment excluded oxygen, which they felt was vital to spontaneous generation. In 1837 Schwann showed that heated infusions remained sterile in the presence of air which he supplied by passing it through heated coils into the infusion. While both of these men demonstrated the idea of the heat preservation of foods, neither took ad-

vantage of his findings with respect to application. The same may be said of
D. Papin and G. Leibniz who hinted at the heat preservation of foods at the
turn of the 18th century.

The event that led to the discovery of canning had its beginnings in 1795,
when the French government offered a prize of 12,000 fr. for the discovery
of a practical method of food preservation. In 1809, a Parisian confectioner,
François (Nicholas) Appert, succeeded in preserving meats in glass bottles
that had been kept in boiling water for varying periods of time. This discovery was made public in 1810, when Appert was issued a patent for his
process. Not being a scientist, Appert was probably unaware of the long-
range significance of his discovery or why it worked. This, of course, was
the beginning of canning as it is known and used today. This event occurred
some 50 years before L. Pasteur demonstrated the role of microorganisms in
the spoilage of French wines, a development which gave rise to the redis-
covery of bacteria. A. Leeuwenhoek in the Netherlands had examined bacteria through a microscope and described them in 1683, but it is unlikely that
Appert was aware of this development since he was not a scientist, and
Leeuwenhoek's report was not available in French.

The first man to appreciate and understand the presence and role of
microorganisms in food was L. Pasteur. In 1837 he showed that the souring
of milk was caused by microorganisms, and about 1860 he employed for the
first time the use of heat to destroy undesirable organisms in wine and beer
(pasteurization).

LIST OF HISTORICAL DEVELOPMENTS

Some of the more significant dates and events in the history of food preservation, food spoilage, food poisoning, and food legislation are tabulated
below.

Food Preservation

1782—Canning of vinegar was introduced by a Swedish chemist.

1810—Preservation of food by canning was patented by Appert.

1813—Use of SO_2 as a meat preservative is thought to have originated
around this time.

1820—The commercial production of canned foods was begun in the United
States by W. Underwood and T. Kensett.

1835—A patent was granted to Newton in England for making condensed
milk.

1837—Winslow was the first to can corn on the cob.

1839—Tin cans came into wide use in the U.S.

1840—Fish and fruit were first canned.

1842—A patent was issued to H. Benjamin in England for freezing foods by
immersion in an ice and salt brine.

1843—Sterilization by steam was first attempted by I. Winslow in Maine.

1854—Pasteur began wine investigations. Heating to remove undesirable organisms was introduced commercially in 1867–68.

1855—Grimwade in England was the first to produce powdered milk.

1856—A patent for the manufacture of unsweetened condensed milk was granted to Gail Borden in the U.S.

1865—The artificial freezing of fish on a commercial scale was begun in the United States. Eggs followed in 1889.

1874—The first extensive use of ice in transporting meat at sea was begun.

1874—Steam pressure cookers or retorts were introduced.

1878—The first successful cargo of frozen meat went from Australia to England. The first from New Zealand to England was sent in 1882.

1880—The pasteurization of milk was begun in Germany.

1882—Krukowitsch was the first to note the destructive effects of ozone on spoilage bacteria.

1886—A mechanical process of drying fruits and vegetables was carried out by an American, A. F. Spawn.

1887—Malted milk first appeared.

1890—The commercial pasteurization of milk was begun around this time in the United States.

1890—Mechanical refrigeration for fruit storage was begun about this time in Chicago.

1893—The Certified Milk movement was begun by H. L. Coit in N.J.

1895—The first bacteriological study of canning was made by Russell.

1907—E. Metchnikoff & coworkers isolated and named one of the yogurt bacteria, *Lactobacillus bulgaricus*.

1907—The role of acetic acid bacteria in cider production was noted by B. T. P. Barker.

1908—Sodium benzoate was given official sanction as a preservative in certain U.S. foods.

1916—The quick freezing of foods was achieved in Germany by R. Plank, E. Ehrenbaum, and K. Reuter.

1917—Clarence Birdseye in the U.S. began work on the freezing of foods for the retail trade.

1917—Franks was issued a patent for preserving fruits and vegetables under CO_2.

1923–28—Heat-process calculations were introduced into the canning industry.

1928—The first commercial use of controlled-atmosphere storage of apples was made in Europe (first in N.Y. state in 1940).

1929—A patent issued in France proposed the use of high-energy radiation for the processing of foods.

1929—Birdseye frozen foods were placed in retail markets.

1943—B. E. Proctor in the United States was the first to employ the use of ionizing radiation to preserve hamburger meat.

1954—The antibiotic nisin was patented in England for use in certain processed cheese to control clostridial defects.

1955—Sorbic acid was approved for use as a food preservative.

1955—The antibiotic chlortetracycline was approved for use in fresh poultry (oxytetracycline followed a year later).

1967—The first commercial facility designed to irradiate foods was planned and designed in the United States.

Food Spoilage

1659—Kircher demonstrated the occurrence of bacteria in milk; Bondeau did same in 1847.

1780—Scheele identified lactic acid as the principal acid in sour milk.

1836—Latour discovered the existence of yeasts.

1839—Kircher examined slimy beet juice and found organisms which formed slime when grown in sucrose solutions.

1857—Pasteur showed that the souring of milk was due to the growth of organisms in it.

1866—L. Pasteur's *Étude sur le Vin* was published.

1867—Martin advanced the theory that cheese ripening was similar to alcoholic, lactic, and butyric fermentations.

1873—The first reported study on the microbial deterioration of eggs was carried out by Gayon.

1873—Lister was first to isolate *Streptococcus lactis* in pure culture.

1876—Tyndall observed that bacteria in decomposing substances were always traceable to air, substances or containers.

1878—Cienkowski reported the first microbiological study of sugar slimes and isolated *Leuconostoc mesenteroides* therefrom.

1887—Forster was the first to demonstrate the ability of pure cultures of bacteria to grow at 0°C.

1888—Miquel was the first to study thermophilic bacteria.

1895—The first records on the determination of numbers of bacteria in milk were those of Von Geuns in Amsterdam.

1895—S. C. Prescott and W. Underwood traced the spoilage of canned corn to improper heat processing for the first time.

1902—The term psychrophile was first used by Schmidt-Nielsen for microorganisms that grow at 0°C.

1912—The term osmophilic was coined by Richter to describe yeasts that grow well in an environment of high osmotic pressure.

1915—*Bacillus coagulans* was first isolated from coagulated milk by B. W. Hammer.

1917—*Bacillus stearothermophilus* was first isolated from cream-style corn by P. J. Donk.

Food Poisoning

1820—The German poet Justinus Kerner described "sausage poisoning" and its high fatality rate (which in all probability was botulism).

1857—Milk was incriminated as transmitter of typhoid fever by W. Taylor of Penrith, England.

1870—Francesco Selmi advanced his theory of ptomaine poisoning to explain illness contracted by eating certain foods.

1888—Gaërtner first isolated *Salmonella enteriditis* from meat which had caused 57 cases of food poisoning.

1894—T. Denys was the first to associate staphylococci with food poisoning.

1896—Van Ermengen first discovered *Clostridium botulinum*.

1926—The first report of food poisoning by streptococci was made by Linden, Turner, and Thom.

1945—McClung was the first to prove the etiologic status of *Clostridium perfringens (welchii)* in food poisoning.

1951—*Vibrio parahaemolyticus* was shown to be an agent of food poisoning by T. Fujino of Japan.

1960—The production of aflatoxins by *Aspergillus flavus* was first reported.

Food Legislation

1890—The first national meat inspection law was enacted. It required the inspection of meats for export only.

1895—The previous meat inspection act was amended to strengthen its provisions.

1906—The U.S. Federal Food and Drug Act was passed by Congress.

1910—New York City Board of Health issued order requiring the pasteurization of milk.

1939—The New Food, Drug, and Cosmetic Act became law.

1954—The Miller Pesticide Chemicals Amendment to the Food, Drug, and Cosmetic Act was passed by Congress.

1957—The U.S. Compulsory Poultry and Poultry Products law was enacted.

1958—The Food Additives Amendment to the Food, Drug, and Cosmetics Act was passed.

1962—The Talmadge-Aiken Act (allowing for federal meat inspection by states) was enacted into law.

1963—The U.S. Food & Drug Administration approved the use of irradiation for the preservation of bacon.

1967—The U.S. Wholesome Meat Act was passed by Congress and enacted into law on December 15.

1968—The Food & Drug Administration withdrew its 1963 approval of irradiated bacon.

1968—The Poultry Inspection Bill was signed into law by President Johnson on August 19.

1973—The state of Oregon adopted microbial standards for fresh and processed retail meats.

REFERENCES

1. **Brandly, P. J., G. Migaki, and K. E. Taylor.** 1966. *Meat Hygiene,* 3rd Edition, Chapter 1. Lea & Febiger, Philadelphia, Penna.

2. **Goldblith, S. A., M. A. Joslyn, and J. T. R. Nickerson.** 1961. *Introduction to Thermal Processing of Foods.* Vol. 1. Avi Publishing Company, Westport, Conn.

3. **Jensen, L. B.** 1953. *Man's Foods.* Chapters 1, 4, and 12. The Garrard press, Champaign, Illinois.

4. **Pederson, C. S.** 1971. *Microbiology of Food Fermentations.* (Avi Publishing: Westport, Conn.).

5. **Schormüller, J.** 1966. *Die Erhaltung der Lebensmittel.* Ferdinand Enke Verlag, Stuttgart, Germany.

6. **Stewart, G. F. and M. A. Amerine.** 1973. *Introduction to Food Science and Technology,* Chap. 1 (Academic Press: N.Y.).

7. **Tanner, F. W.** 1944. *The Microbiology of Foods,* 2nd Edition. The Garrard Press, Champaign, Illinois.

8. **Tanner, F. W. and L. P. Tanner.** 1953. *Food-Borne Infections and Intoxications,* 2nd Edition. The Garrard Press, Champaign, Illinois.

2

The Role
and Significance
of Microorganisms
in Nature
and in Foods

Man's sources of food are of plant and animal origin. Therefore, it is important to understand the biological principles of the microbial flora associated with plants and animals in their natural habitats and their respective roles. While it sometimes appears that microorganisms are trying to ruin our food sources by infecting and destroying plants and animals, including man, this is by no means their primary role in nature. From our present concepts of life on this planet, the primary role of microorganisms in nature is self-perpetuation. During this process, the heterotrophs carry out the following general reaction:

All organic matter (carbohydrates, proteins, lipids, etc.) $\rightarrow$
$$\text{Energy} + \text{Inorganic compounds (nitrates, sulfates, etc.)}$$

This, of course, is essentially nothing more than the operation of the nitrogen cycle and the cycle of other elements (see Figure 2-1). As will be discussed in a later chapter, the microbial spoilage of foods may be viewed simply as an attempt by the food flora to carry out what appears to be their primary role in nature. This should not be taken in the teleological sense. In spite of their simplicity when compared to higher forms, microorganisms are capable of carrying out many complex chemical reactions essential to their perpetuation. To do this, they must obtain nutrients from organic matter, some of which constitutes our food supply.

If one considers the types of microorganisms associated with plant and animal foods in their natural states, one can then predict the general types of microorganisms to be expected on this particular food product at some later stage in its history. Results from many laboratories show that untreated foods may be expected to contain varying numbers of bacteria, molds, or yeasts, and the question often arises as to the safety of a given food product based upon total microbial numbers. The question should be twofold: (1) What is the total number of microorganisms present per g or ml? (2) What

9

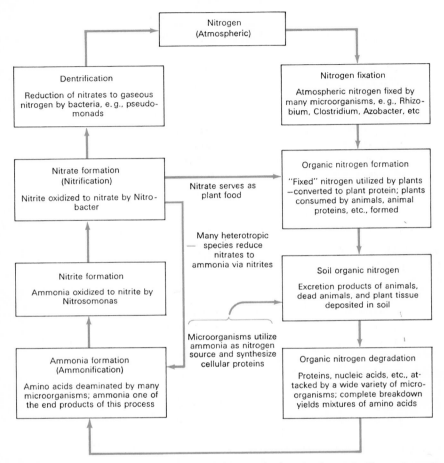

FIGURE 2-1. Nitrogen cycle in nature is here depicted schematically to show the role of microorganisms. (From *Microbiology* by M. J. Pelczar and R. Reid *(3)*, copyright 1965 by McGraw-Hill Book Company. Used with permission of the publisher.)

types of organisms are represented in this number? It is necessary to know which organisms are associated with a particular food in its natural state, and which of the organisms present are not normal for that particular food. It is, therefore, of value to know the general distribution of bacteria in nature and the general types of organisms normally present under given conditions where foods are grown and handled.

Bacteria. Twenty-five of the most important genera of bacteria known to cause food spoilage and food poisoning are listed on page 11 in alphabetical order. Some of these are highly desirable in certain foods. Eight of the genera belong to the family *Enterobacteriaceae*.

Acetobacter   *Lactobacillus*
Acinetobacter   *Leuconostoc*
Alcaligenes   *Micrococcus*
Bacillus   *Pediococcus*
Bacteroides   *Proteus*
Citrobacter   *Pseudomonas*
Clostridium   *Salmonella*
Corynebacterium   *Serratia*
Enterobacter   *Shigella*
Erwinia   *Staphylococcus*
Escherichia   *Streptococcus*
Flavobacterium   *Streptomyces*
Kurthia

For a synopsis of these 25 genera, see pp. 14–19 in this chapter.

Molds.   Eighteen of the most common genera of molds associated with foods are as follows:

Alternaria   *Gleosporium*
Aspergillus   *Helminthosporium*
Botrytis   *Monilia (Neurospora)*
Byssochlamys   *Mucor*
Cephalosporium   *Penicillium*
Cladosporium   *Rhizopus*
Colletotrichum   *Sporotrichum*
Fusarium   *Thamnidium*
Geotrichum   *Trichothecium*

These organisms are further described on pp. 20–24 in this chapter.

Yeasts.   Twelve of the most common genera of yeasts encountered in and on foods are as follows:

Brettanomyces   *Mycoderma*
Candida   *Rhodotorula*
Debaromyces   *Saccharomyces*
Endomycopsis   *Schizosaccharomyces*
Hansenula   *Torulopsis (Torula)*
Kloeckera   *Trichosporon*

These genera are further described on pp. 24–26 in this chapter.

PRIMARY SOURCES OF MICROORGANISMS TO FOODS

The genera listed above represent perhaps the most important organisms normally found in foods. Each genus has its own particular functions of nutrition and consequent degradative processes. These organisms may be generally associated with the particular aspects of the food environment presented below.

1. Soil and Water.   It may be assumed that at one time, all microorganisms existed in water. The drying of surface soils gives rise to dust which, when disseminated by winds, carries adhering microorganisms to many places including other areas of the soil, rivers, oceans, etc. The formation of clouds over large bodies of water and the subsequent rainfall over land as well as other waters has the same effect. It is not surprising, then, that soil and water microorganisms are often one and the same.

The following genera of food-borne bacteria generally found in soils and waters may be expected in foods: *Alcaligenes, Bacillus, Citrobacter, Clostridium, Corynebacterium, Enterobacter, Micrococcus, Proteus, Pseudomonas, Serratia,* and *Streptomyces,* among others.

Many of the molds listed above occur in soils and also in waters. Molds are in general very widespread in nature, where they participate in the degradation of both plant and animal matter as well as cause many diseases of plants and animals (see Chapter 6). Among those that are nearly always present in soils are *Aspergillus, Rhizopus, Penicillium, Trichothecium, Botrytis, Fusarium,* and others.

A large number of yeast genera are associated with plants and may, therefore, be expected to be found in soils. Their numbers in water are generally low.

2. Plants and Plant Products.   Most of the organisms discussed above for soil and water are found also on plants, since soil and water constitute the primary sources of microorganisms to plants. On the other hand, there are some bacteria that are associated more with plants than with soil. Among these genera are *Acetobacter, Erwinia, Flavobacterium, Kurthia, Lactobacillus, Leuconostoc, Pediococcus,* and *Streptococcus.* Any and all of the other 25 genera may at times be found on plants and in plant products along with other genera not listed.

Among the molds, the most important plant-borne genera are those that cause the spoilage of vegetables and fruits (so-called market diseases). These are presented in more detail in Chapter 6. The genus *Saccharomyces* is the most notable of the yeasts that may be found on many plant products, especially fruits. Others include the genera *Rhodotorula* and *Torula.*

3. Food Utensils.   The genera of microorganisms to be found on food utensils depend upon the types of foods handled, the care of these utensils, their storage, and other factors. If vegetables are handled in a given set of utensils, one would, of course, expect to find some or all of the organisms associated with vegetables. When utensils are cleaned with hot or boiling water, the remaining flora would normally be those best able to withstand the effects of this treatment. Utensils that are stored in the open where dust might collect should be expected to have air-borne bacteria, yeasts, and molds.

4. Intestinal Tract of Man and Animals. There are several genera of bacteria that are more commonly found in this environment than in soils, water, or other places. Among these are: *Bacteroides, Escherichia, Lactobacillus, Proteus, Salmonella, Shigella, Staphylococcus,* and *Streptococcus.* The most notable of these is the genus *Escherichia,* which has as its natural habitat the intestinal tract of man and other mammals. Species of other genera common to the intestinal tract include *Clostridium, Citrobacter, Enterobacter,* and *Pseudomonas.* From the intestinal tracts of animals, intestinal microorganisms find their way directly to the soil and water. And from the soil they may find their way onto plants, in dust, to utensils, etc. Molds are not thought to be transmitted by fecal sources though the yeast genus *Candida* is very often found in the intestinal tract of man. (See Chapter 5 for discussion of the enteroviruses).

5. Food Handlers. The microflora on the hands and outer garments of food handlers generally reflects the environment and habits of the individuals. This flora would normally consist of organisms found on any object handled by the individual as well as some of those picked up from dust, water, soil, and the like. In addition, there are several genera of bacteria that are specifically associated with the hands, nasal cavities, and mouth. Among these are the genera *Micrococcus* and *Staphylococcus,* the most notable of which are the staphylococci which are found on hands, arms, in nasal cavities, the mouth, and other parts of the body. While the genera *Salmonella* and *Shigella* are basically intestinal forms, they may be deposited onto foods and utensils by food handlers if sanitary practices are not followed by each individual. Any number of molds and yeasts may be found on the hands and garments of food handlers depending upon the immediate history of each individual.

6. Animal Feeds. Any one or all of the genera of bacteria, yeasts, and molds cited earlier in this chapter may be found in animal feeds. The types of organisms to be found would depend, of course, on the source of the feeds, the treatment given them to destroy microorganisms, the containers in which they are stored, and the like. As discussed further in Chapter 18, animal feeds are of great importance in the spread of food poisoning *Salmonella.* Organisms from this source have been shown to be rapidly disseminated throughout processing plants where feeds are handled.

7. Animal Hides. Just about any or all of the microorganisms associated with soils, water, animal feeds, dust, and fecal matter may be found on the hides of animals. From animal hides, these organisms may again be deposited in the air, onto the hands of workers, and directly into foods. Some members of the hide flora find their way into the lymphatic system of slaughter animals from which they migrate after slaughter into the muscle tissue proper.

8. Air and Dust.   The types of organisms to be found in air and dust, with the exception of some of the pathogens, include the 25 genera of bacteria, the 16 genera of molds, and many of the yeasts. Although *Staphylococcus* and *Salmonella* spp. may at times be found in air and dust, these are not the major sources of these organisms to foods. Notable among the bacterial genera in air and dust are *Bacillus* and *Micrococcus* spp., all of which are able to endure dryness to varying degrees. Notable among the yeasts is the genus *Torulopsis*, and many mold genera may be found from time to time.

9. The Primary Sources of Food-Poisoning Bacteria to Foods.   The most important food-poisoning bacteria belong to the following genera: *Staphylococcus, Salmonella, Streptococcus,* and *Clostridium.* The staphylococci are associated with the nasal cavities of man and animals as well as with other parts of the body. Salmonellae are indigenous to the intestinal tract of man and animals but may enter foods from other sources contaminated from fecal matter. The streptococci exist both in man and animals as well as on plants, while the clostridia are basically soil forms. In addition to the above, *Bacillus cereus* and *Vibrio parahaemolyticus* cause food poisoning. The food-poisoning syndromes caused by all of these organisms are discussed in Chapters 16, 17, and 18. Other biological hazards associated with the consumption of foods are presented in Chapter 19.

SYNOPSIS OF 25 GENERA OF COMMON FOOD-BORNE BACTERIA

A brief description of each of the genera of bacteria presented earlier in this chapter is given below. References to number of species of the genera are based upon *Bergey's Manual,* 8th Edition. This synopsis is not meant to be employed alone in the identification of these organisms but is designed to characterize these important genera with respect to their roles in foods. For the purpose of generic and species identifications, *Bergey's Manual* and the Appendix section should be consulted. Information on the more general and basic aspects of bacteria may be obtained from a standard textbook on general microbiology.

1. Acetobacter.   This genus contains 7 species. These are gram negative, rod-shaped cells that are strict aerobes. They are commonly found in fermented grain mash, mother of vinegar, beer, wines, and souring fruits and vegetables. Some species such as *A. aceti* oxidize ethanol to acetic acid and thereby give rise to vinegar, perhaps their greatest industrial use. The G + C content of DNA = 55–64 moles %.

2. Acinetobacter.   This genus of gram negative rods shows some affinity to the family *Neisseriaceae* and only one species is recognized (*A. calcoaceticus*). Some organisms formerly classified as *Achromobacter* have been placed in this genus. These are strict aerobes that do not reduce nitrates. While they are related to the genus *Moraxella,* they differ in being oxidase negative. Although rod-shaped cells are formed in young cultures, old cultures contain many coccoid-shaped cells. They are common in soils and water. The G + C content of DNA = 39–47 moles %.

3. Alcaligenes.   This genus consists of 4 species. They are gram negative rods which sometimes stain variable or gram positive. They do not ferment sugars, but produce alkaline reactions which are also produced in litmus milk. They do not produce pigments. They are widely distributed in nature in decomposing matter of all types, in raw milk, poultry products, the intestinal tract, etc. The G + C content of DNA = 58–70 moles %.

4. Bacillus.   This genus belongs to the family *Bacillaceae.* Some 48 species are recognized. Most are aerobic, gram positive rods that produce endospores. They often exist in long chains on cultural media. Most are mesophiles with some being psychrophilic and some thermophilic in nature. The thermophilic members are of great importance in the canning industry due to the extreme heat resistance of their spores. This genus contains one pathogen for man and other vertebrates—*B. anthracis,* which causes anthrax and one food poisoning species, *B. cereus.* Some species are insect pathogens. The nonpathogens are widely distributed and may be found in air, dust, soil, water, on utensils, and in various foods. They are important in the spoilage of many foods held above refrigerator temperatures. The G + C content of DNA = 32–62 moles %.

5. Bacteroides.   This group of anaerobes belongs to the family *Bacteroidaceae.* Some 22 species of these gram negative, mesophilic, nonsporing rods are recognized. They are found in the intestinal tract of man and animals from which they find their way to meats where they play a role in spoilage. Due to their anaerobic and nonsporeforming habits, these organisms are often overlooked in routine analyses of foods. The G + C content of DNA = 40–55 moles %.

6. Citrobacter.   This genus belongs to the family *Enterobacteriaceae,* and two species are recognized. The slow, lactose-fermenting, gram negative rods previously designated paracolons and the Bethesda-Ballerup group have been placed in this genus. All members can use citrate as their sole carbon source. They are common in fecal matter, water, and in some foods.

7. Clostridium.   This is an important group of anaerobic bacteria that belong to the family *Bacillaceae.* Some 61 species are now recognized. They are gram positive, sporeforming rods. The genus contains some thermophilic species which are of great importance in the canning industry due to the extreme heat resistance of their endospores. The etiologic agents of tetanus, gas gangrene, perfringens food poisoning, and botulism are members of this genus. Some species are of commercial importance in the production of certain solvents. They are very widely distributed in nature, in soils, water, the intestinal tract of man and animals, and other places. They may be found in many foods where they may or may not grow. Many species are strongly proteolytic as are many of the *Bacillus* spp. in the same family. The G + C content of DNA = 23–43 moles %.

8. Corynebacterium.   This genus belongs to the coryneform group and contains some 30 species. They are aerobic, gram positive rods that often show granules and club-shaped swellings. They are nonsporeformers. This group contains the etiologic agent of diphtheria and other species common on the body. Most are mesophiles with some being psychrophiles. These organisms are widely distributed among certain plants such as wheat, beans, tomatoes, and others. They are also found in the intestinal tract of man and animals and have been isolated from spoiling foods of various types. The G + C content of DNA = 57–60 moles %.

9. Enterobacter.   These organisms belong to the family *Enterobacteriaceae* and are represented by 2 species. They are short, gram negative, nonpigment-forming rods that grow well on many cultural media and ferment glucose and lactose with the production of acid and gas. In their sugar fermentations, these organisms produce two or more times as much CO_2 as hydrogen gas. The most distinctive taxonomic features are: MR − and VP + (MR = methyl red; VP = Voges-Proskauer reaction). They are widely distributed in nature, being especially present on plants, grain, in water, and the intestinal tract. This genus is one of the coliform genera along with *Escherichia*. Formerly *Aerobacter*. The G + C content of DNA = 52–59 moles %.

10. Erwinia.   This genus belongs to the family *Enterobacteriaceae*. Some 13 species are now recognized. They are gram negative rods that are motile. Many species grow within the psychrophilic range and many produce pigments of various shades of red when growing on culture media and certain foods. They are characteristically associated with plants where they cause dry necroses, galls, wilts, and soft rots. These are the most important bacteria in the cause of market diseases of fruits and vegetables (see Chapter 6). The G + C content of DNA = 53.6–54.1 moles %.

11. Escherichia.   As stated above, this group belongs to the family *Enterobacteriaceae* and contains one specie. They are short, gram negative rods that are indistinguishable from *Enterobacter* on culture media and under the microscope. They differ from *Enterobacter* spp. in being MR + and VP −. Their main habitat appears to be the intestinal tract of man and other animals from which they may be found in soils, waters, and many other places in nature. *E. coli* is the most important member of the coliform group along with *E. aerogenes*. Their presence in large numbers in foods is generally taken to indicate fecal contamination. The G + C content of DNA = 50–51 moles %.

12. *Flavobacterium*. This genus contains 12 species. They are gram negative rods that generally produce yellow to red pigments on agar. They are widely distributed in soils, water, on fish, and plants. They may be isolated from decaying plant materials. Many are psychrophilic while most are mesophiles. For motile species, the G + C content of DNA = 63–70 moles % while for some nonmotile species 30–42 moles %.

13. *Kurthia*. There is only one specie of this genus. These are long, gram positive, nonsporing mesophilic rods. They are motile and do not attack carbohydrates. They are found on plant and animal matter and in decomposing organic matter in general.

14. *Lactobacillus*. This group belongs to the family *Lactobacillaceae* and contains some 27 species. They are long, gram positive, nonsporing rods that are catalase negative. They often occur in long chains when viewed under the microscope. Most are microaerophilic or anaerobic while both homofermentative and heterofermentative species exist among them. They are widely distributed among plants and in dairy products. Some are employed in the production of fermented milks such as Acidophilus and Bulgaricus milks. Some are important in cheese making. Many are used in the microbiological assay of B-vitamins and amino acids due to their exacting growth requirements. One species, *L. thermophilus*, survives milk pasteurization temperatures. They are common in and on cured and processed meat products. G + C content = 35–53%.

15. *Leuconostoc*. This genus belongs to the family *Streptococcaceae*. It contains at least 6 species. They are gram positive, spherical to oval, catalase negative, heterofermentative organisms. They are widely distributed among plants from which they find their way into milk and dairy products. Some cause problems in sugar refineries where they form slime in sugar lines. Some species are employed in dairy starter cultures while others are often found in cured meat products. Some synthesize the medically important polymer dextran. The G + C content of DNA of most = 38–42 moles %.

16. *Micrococcus*. This group belongs to the family *Micrococcaceae* and contains at least 3 species. They are gram positive cocci that are catalase positive in contrast to the streptococci. Some produce nonpigmented colonies while others produce pink to orange-red to red pigments on culture media. Most or all can tolerate high levels of salt. They are widely distributed in nature, on the skin of man, and the hides of animals, as well as in dust, soil, water, and in many foods. Most are mesophiles while some are capable of growth in the psychrophilic range. Several species are associated with dairy products through which they enter processed meats such as frankfurters. The formerly recognized genera *Gaffkya* and *Sarcina* have been reduced to species of this genus. The G + C content of DNA = 66–75 moles %.

17. Pediococcus. These homofermentative cocci belong to the family *Streptococcaceae* and differ microscopically from streptococci and leuconostocs by their arrangement in pairs and tetrads resulting from cell division in two planes. Like the other lactic acid genera, they are widespread in nature, especially on plants. At least 5 species are recognized. Most are important as starters in certain fermented foods. The G + C content of DNA = 34–44 moles %.

18. Proteus. These organisms belong to the family *Enterobacteriaceae* and consist of 5 species. Like the other members of this family, they are gram negative rods that are aerobic and often show pleomorphism. All are motile while all but one species hydrolyze urea. They are found in the intestinal tract of man and animals and on decaying materials in general. They may be isolated from spoiled eggs and meats, especially those allowed to spoil above refrigerator temperatures. For most species, G + C content is 38–42 moles %.

19. Pseudomonas. These important bacteria belong to the family *Pseudomonadaceae* and while only 29 species are recognized, over 200 others have been reported. They are short, gram negative, aerobic rods that usually produce a single polar flagellum. Many psychrophilic species and strains as well as mesophiles exist in this genus. They are widely distributed in nature in soils, water, on plants, and in the intestinal canal of man and other animals. These are by far the most important bacteria in the low-temperature spoilage of foods such as meats, poultry, eggs, and seafoods. Some strains produce fluorescent pigments (pyocyanin and/or fluorescin). Most are capable of oxidizing glucose to gluconic acid, 2-keto-gluconic acid or other intermediates. Many are plant pathogens, causing leaf spot, leaf stripe, and related diseases. Many of those that cause food spoilage do not produce the water-soluble pigment but may fluoresce under ultraviolet light. The G + C content of DNA ranges from 58 to 70 moles %.

20. Salmonella. This important genus of bacteria belongs to the family *Enterobacteriaceae* and contains approximately 1,800 serotypes with new ones being added yearly. They are gram negative, short rods that are aerobic and do not produce pigments on culture media. Most ferment glucose and other simple sugars with the production of acid and gas. They generally do not ferment lactose although some do. These are basically intestinal forms, as are most of the *Enterobacteriaceae* genera. They may be found widely distributed in nature. The etiologic agents of typhoid and paratyphoid fevers belong to this group as well as those that cause food-borne salmonellosis in man. All species and strains of this genus are undesirable in foods. The G + C content of DNA = 50–53 moles %.

21. Serratia. This genus belongs to the family *Enterobacteriaceae* and contains one specie. They are gram negative, aerobic, proteolytic, and mesophilic rods that generally produce red pigments on culture media and on certain foods. Most species are rather widespread in nature in water and soil, and in decaying plant and animal matter. The G + C content of DNA ranges from 53 to 59 moles %.

22. Shigella. These organisms also belong to the family *Enterobacteriaceae* and contain at least 4 species. They are short, gram negative rods that are nonmotile. They are aerobic and mesophilic in nature and occur in polluted waters and in the intestinal canal of man where they cause bacillary dysentery and other intestinal disorders. Their primary sources to foods are polluted water and human carriers. These organisms are undesirable in foods.

23. Staphylococcus. These organisms belong to the family *Micrococcaceae* and consist of only 3 species. They are gram positive cocci that divide irregularly as do the micrococci. They are catalase positive in contrast to the streptococci. Most *S. aureus* strains produce a golden pigment and coagulate blood plasma, while *S. epidermidis* is nonpigmented and does not produce coagulase. Both are common in the nasal cavities of man and certain other animals as well as on the skin and other parts of the body. *S. aureus* produces boils, carbuncles, and an important food-poisoning syndrome in man. Their presence in foods in large numbers is undesirable. The G + C content of DNA ranges from 30 to 40 moles %.

24. Streptococcus. This genus belongs to the family *Streptococcaceae* and consists of 21 species. They are gram positive, catalase negative cocci that often appear as spherical to ovoid forms. All produce small colonies when growing on culture media as do the lactobacilli. They are nonpigmented and microaerophilic in nature. Some are associated with the upper respiratory tract of man and other animals where they may cause diseases such as scarlet fever, septic sore throat, etc. Others are found in the intestinal tract of man and animals and tend to be rather widespread on plants and plant parts, and in dairy products. While most are mesophilic, some grow within the psychrophilic range. Some cause mastitis in cattle while others are important in dairy starter cultures. One species, *S. lactis,* is the most common cause of sour milk, and is important in the manufacture of cheese. Some produce a food-poisoning syndrome in man. The presence of some species in foods in large numbers may indicate fecal contamination. The G + C content of DNA = 33–42 moles %.

25. Streptomyces. These organisms are among the so-called higher bacteria and belong to the family *Streptomycetaceae.* Some 463 species are recognized in the 8th Edition of *Bergey's Manual,* with the actual number probably being higher. Unlike the groups presented above, these organisms grow in the form of a highly branched mycelium. They find their way into many foods such as vegetables from the soil, their native habitat. They are found in the oral cavity of man. They are active in the destruction of both plant and animal matter at temperatures above the refrigerator range. When growing on culture media, they produce characteristic soil-type odors. They are perhaps best known for the antibiotic-producing abilities of some species.

SYNOPSIS OF 18 GENERA OF MOLDS COMMON IN AND ON FOODS

Unlike the true bacteria and most yeasts, molds grow in the form of a tangled mass which spreads rapidly and may cover several inches of area in 2–3 days. The total of the mass or any large single portion of it is referred to as **mycelium.** The mycelium is composed of branches or filaments referred to as **hyphae.** At the time of asexual reproduction, **sporangiophores** or **conidiophores** are sent up which bear, usually at their tips, sporangia or conidia (see Figures 2-2K and 2-3A). In those molds that produce sporangia, the spores are borne by these structures and are responsible for the various colors displayed by molds. Conidia represent unprotected spores. In addition to these characteristic and common asexual spores, some molds produce other asexual spores. **Chlamydospores** result when a thick wall develops around any cell of the mycelium. These structures are somewhat resistant to adverse environmental conditions. **Arthrospores** or **oidia** result from fragmentation by some molds that produce a septate mycelium. Chlamydospores and arthrospores are a bit more difficult to destroy than other parts of the mold mycelium, and they sometimes cause concern in the food industry. Molds also reproduce by sexual means where they form either ascospores, oospores, or zygospores.

Most molds of importance in foods are now placed in the *Fungi imperfecti* group. These are molds whose sex cycles are not known. Most or all are thought to be related to the *Ascomycetes* group.

The descriptions below are not meant to be detailed enough to permit generic identifications of these organisms except for the most common ones. For this purpose, the reader should consult a more detailed reference such as Barnett *(1)*.

1. Alternaria.   These molds produce septate mycelia with dark conidiophores and dark conidia. The conidia have both cross and longitudinal septa and are variously shaped. They are active in the spoilage of many plant products (Fig. 2-2A).

2. Aspergillus.   These molds produce upright conidiophores that are simple and terminate in a globose or clavate swelling. The conidia are 1-celled, globose, and variously colored in mass. These molds appear yellow to green to black on a large number of foods. They produce septate mycelia. Some species of this genus produce carcinogenic aflatoxins while others are employed as commercial sources of proteases and citric acid. They are widespread and may be found on cakes, fruits, vegetables, meats, and other foods (Figure 2-2B).

3. Botrytis.   These organisms produce long, slender, and often pigmented conidiophores. The mycelium is septate and the conidia are borne on apical cells. The conidia are gray in mass and 1-celled. Black irregular sclerotia are frequently produced. They cause a "gray mold" condition on many plants

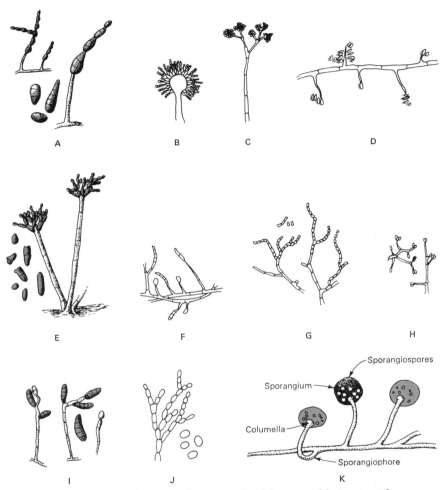

FIGURE 2-2. Illustrated genera of common food-borne molds (see text for identification).

and plant foods. They are important causes of market diseases of fruits and vegetables. *B cinerea* is presented in Figure 2-2C showing conidiophores and conidia.

4. Cephalosporium.   These molds produce a septate mycelium with simple and slender or swollen conidiophores. The microspores of certain species of *Fusarium* are similar in many ways to these. A water-mount of *Cephalosporium* sp. is presented in Figure 2-2D.

5. Cladosporium.   This genus is characterized by the production of septate mycelia with dark conidiophores variously branched near the apex of the middle portion. Conidia are dark, 1- or 2-celled, and some are lemon shaped. One species, *C. herbarum,* produces black spots on beef (Figure 2-2E).

6. Fusarium.   These molds produce an extensive mycelium which is cottony in culture with tinges of pink, purple, or yellow. The conidia are canoe-shaped, borne singly or in chains. These fungi are important in the spoilage of many fruits and vegetables. They have been implicated in the "neck rot" of bananas (Figure 2-2F).

7. Geotrichum (Oidium).   These are yeastlike fungi that produce various colors, but generally white. The mycelium is septate and reproduction occurs by fragmentation of mycelium into arthrospores. These organisms are sometimes referred to as the "dairy mold," since they impart flavor and aroma to many types of cheese. They are referred to also as "machinery molds", since they build up on food-contact equipment in food processing plants, especially tomato canning plants. *G. albidum (Oospora* or *Oidium lactis)* is shown in Figure 2-2G.

8. Gloeosporium.   These molds produce simple and variable-length conidiophores. The conidia are hyaline, 1-celled, and sometimes curved. They cause anthracnoses in plants. *G. fructigenum* is shown in Figure 2-2H.

9. Helminthosporium.   These molds produce a light to dark mycelium in culture. Their conidiophores are short or long, septate, simple or branched. The conidia are borne successively on new growing tips; they are dark and typically contain more than 3 cells. This genus contains both plant pathogens and saprophytes. *H. satiuum* is shown in Figure 2-2I.

10. Monilia.   These molds produce white or gray mycelia that bear branched conidiophores. The conidia are pink or tan in mass. Some species are the imperfect stages of *Neurospora.* Some of those whose perfect states are *Monilinia or Sclerotinia* spp. cause brown rots of fruits. *N. sitophila* is often referred to as the "red bread mold." *M. americana* is shown in Figure 2-2J.

11. Mucor.   These molds produce nonseptate mycelia that give rise to conidiophores that bear columella and a sporangium at the apex. The spores are smooth, regular, and borne within the sporangium. They may be found growing on a large number of foods (Figure 2-2K).

12. Penicillium.   These organisms produce septate mycelia that bear conidiophores arising singularly or sometimes in synnemata with branches near the apex to form a brush-like, conidia-bearing apparatus. The conidia form by pinching off from phialides. Typical colors on foods are blue to blue-green. These molds are important in the making of some cheeses. Some are important in the production of antibiotics (penicillin, e.g.). They are widespread in the soil, air, dust, and in many other places and may be found on foods such as breads, cakes, fruits, and preserves. Some cause soft rots of fruits (Figure 2-3A).

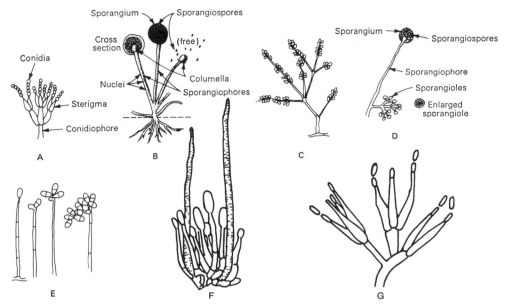

FIGURE 2-3.   Illustrated genera of common food-borne molds
(see text for identification).

13. Rhizopus.   These molds produce nonseptate mycelia that give rise to
stolons and rhizoids. The sporangiophores arise at nodes and bear columella
and sporangia at the apex. The spores are borne within the sporangia and are
usually black in color. Like the penicillia, they are very widespread in nature
and may be found growing on foods such as fruits, cakes, preserves, and
bread. One species, *R. stolonifer,* is often called the "bread mold." Some
are employed in the fermentation of starch to alcohol (Figure 2-3B).

14. Sporotrichum.   These molds produce septate mycelia that give rise to
conidiophores that bear spores near the apex. The conidia are hyaline, 1-
celled, globose, or ovoid, attached apically and laterally. They have been
reported to grow at and below 0°C. Some produce "white spot" on refrigerated beef (Figure 2-3C).

15. Thamnidium.   These organisms produce nonseptate mycelia that bear
sporangiophores with large sporangia at the tip and lateral sporangioles near
the base. They are sometimes found on refrigerated meats, especially
hindquarters held for long periods of time, where they cause a condition
often referred to as "whiskers." They may be found in a large number of decaying foods such as eggs (Figure 2-3D).

16. Trichothecium (Cephalothecium).   These forms produce septate
mycelia that bear long, slender, and simple conidiophores. The conidia appear singly, apically, and sometimes in groups of chains. Some are pink on
foods such as fruits and vegetables. *T. roseum* is shown in Figure 2-3E.

17. Byssochlamys. Members of this genus of *Ascomycetes* produce clusters of asci each of which contains eight ascospores. The asci appear to be without a covering wall or ascocarp. The ascospores of these organisms are heat-resistant leading to the spoilage of some high-acid canned foods. They can grow under conditions of low Eh. They exist in soils and can be recovered from ripening fruits. The conidial structures of *B. fulva* are shown in Figure 2-3F.

18. Colletotrichum. This genus belongs to the order *Melanconiales*. They produce simple but elongate conidiophores and hyaline conidia that are one-celled, ovoid, or oblong. The acervuli produced are disc- or cushion-shaped, waxy, and generally dark in color. They are common contaminants on fruits and vegetables. The conidiophores, conidia, and spines of *C. lindemutheanum* are illustrated in Figure 2-3G.

SYNOPSIS OF 12 GENERA OF YEASTS

Yeasts are microscopic organisms that may be differentiated from the common bacteria by their larger cell size, their oval, elongate, elliptical, or spherical cell shapes, and by their production of buds during the process of division. The sizes of yeast cells vary, some ranging from 5–8 μm in diameter, while others may be as large as 100 μm in length. In general, older yeast cells tend to be smaller in size than young growing cells. Yeasts can grow over rather wide ranges of pH, alcohol, and sugar concentrations. Some have been reported to grow at a pH as low as 1.5 and in up to 18% ethanol. Many grow in the presence of 55% or more sucrose. These organisms produce pigments of many colors, with red and black pigment producers being common. A microscopic slide of growing yeast cells generally reveals cells in varying stages of budding with some showing several buds. The true yeasts or **ascosporogenous** yeasts reproduce by sexual reproduction involving copulation to varying degrees. True yeasts also produce asexual spores and **chlamydospores.** The latter are durable and tend to be produced when a culture encounters unfavorable conditions of growth. The ascosporogenous yeasts presented below all belong to the family *Endomycetaceae*.

The **asporogenous** (false yeasts, wild yeasts) yeasts do not display sexual reproduction and are all placed in the family *Crytococcaceae*. Some of these are yeast-like organisms which are sometimes placed in the *Fungi imperfecti* group along with some of the more common molds. The wild yeasts are difficult to distinguish morphologically and culturally but may be identified serologically. While some are strains of *S. cerevisiae* or *S. carlsbergensis*, some are representatives of many other genera including *Candida, Debaryomyces, Pichia*, etc.

Yeasts are often placed into groupings based upon some particular function or activity. For example, **film yeasts** are those that grow at the surface of certain acid products such as sauerkraut and pickles. Some film yeasts are species and strains of the genera *Candida* and *Hansenula*. These organisms are capable of oxidizing acids and alcohols as sources of energy. **Top yeasts** are those that carry out the conversion of sugars to alcohol at the top of a vessel, while **bottom yeasts** are those that can do the same but from the bottom of the vessel. **Apiculate** or lemon-shaped yeasts are undesirable in wine fermentations where they produce off-flavors.

1. Brettanomyces. These are acid-producing asporogenous yeasts that produce oval, elongate, spherical, or ogive-shaped cells. Reproduction is by multipolar budding which often leads to the formation of chains of cells. Some species carry out an after-fermentation in certain European beers and ales while others have been isolated from spoiled pickles.

2. Candida. These are yeast-like organisms that are sometimes placed among *Fungi imperfeci* in the family *Moniliaceae* along with the genera *Trichothecium* and *Geotrichum*. They reproduce by either fragmentation of mycelium into blastospores, or by budding. These are asporogenous yeasts that produce a pseudomycelium. A few species are of both industrial and medical importance. They are common on many foods such as fresh and cured meats. One species causes rancidity of margarine.

3. Debaromyces. These are ascosporogenous yeasts that sometimes produce a pseudomycelium. They generally reproduce by multipolar budding and also by sexual means. This genus is often found on the surface of spoiling foods such as fresh and cured meats, sausage, pickle brine, wines, etc.

4. Endomycopsis. These are ascospore forming or true yeasts that are oxidative. They are common on stored cereal grains and commonly form films on fermenting sauerkraut and cucumbers. At least one species, *E. fibuliger,* possesses strong alpha and beta-amylase activities.

5. Hansenula. These are ascosporogenous yeasts that produce spherical, elongate, or oval cells and often a pseudomycelium. They reproduce by multipolar budding and by sexual means. When the latter occurs, hat-shaped spores are produced inside of the asci. They are common on citrus fruits, grapes, grape products, and in olive brine and fruit juice concentrates.

6. Kloeckera. These are non-sporeforming yeasts common on fruits where they are disseminated and also consumed by fruit flies. They possess both fermentative and oxidative abilities. Some cause off-flavor and turbidity in wines.

7. Mycoderma. These are asporogenous yeasts that usually grow on the surface of beers, pickle brines, fruit juices, vinegar, and other related products and produce a heavy film or pellicle. One species, *M. vini,* is associated with the ''wine flower'' condition of wines, vinegar, and related products.

8. Rhodotorula.   These are asporogenous yeasts that sometimes produce a primitive type pseudomycelium. Reproduction is by multipolar budding. Many produce red pigments both on cultural media and on various foods. They are widespread in nature and are often found in the air and dust.

9. Saccharomyces.   These are ascosporogenous yeasts that produce ovoid, spherical, or elongate cells. Reproduction is by multipolar budding and ascus formation where from 1–4 spores are formed. This group represents the yeasts of greatest industrial importance with *S. cerevisiae* being employed in the brewing, baking, and distilling industries. These organisms are very widespread on fruits (especially grapes) and vegetables where they cause a fermentation of sugars leading to CO_2 and ethanol. Most osmophilic yeasts belong to this genus including those formerly classified as *Zygosaccharomyces*.

10. Schizosaccharomyces.   These are ascosporogenous yeasts which reproduce either by fission, arthrospores, or by sexual means. When the latter occurs, the asci contain 4–8 spores that are oval, spherical or kidney-shaped. They occur in sugar and other related products.

11. Torulopsis (Torula).   These are asporogenous yeasts that produce spherical or oval cells. They reproduce by multipolar budding and sometimes produce a primitive type pseudomycelium. They are widespread in nature and may be seen growing on refrigerated foods of many types.

12. Trichosporon.   These are non-ascosporeforming oxidative yeasts found in a variety of foods including fermenting maple sap, meats, and beers. At least one species is lipolytic *(T. pullulans)*. They belong to the family *Cryptococcaceae*.

SUMMARY

Microorganisms are quite ubiquitous, having been found everywhere in the earth's atmosphere except in volcanic lava and on the inside of healthy plant and animal tissues. Their primary role in nature is self-perpetuation, the result of which is the conversion of organic matter to inorganic compounds through which process they obtain the necessary energy for growth. All microorganisms such as bacteria, molds, and yeasts, are basically soil and water creatures with the parasitic state being relatively recent in the existence of these forms. From soil and water, these organisms find their way onto plants, into the air, dust, the intestinal tract of man and other animals, and back onto soil and into water where the cycle continues. Each genus appears to occupy a more-or-less specific niche in nature, and knowledge of this fact will allow one to make certain general predictions about the existence of these organisms in and on foods.

REFERENCES

1. **Barnett, R. L.** 1960. *Illustrated Genera of Imperfect Fungi,* 2nd Edition. Burgess Publishing Co., Minneapolis.

2. **Buchanan, R. E. and N. E. Gibbons, Editors.** 1974. *Bergey's Manual of Determinative Bacteriology,* 8th Edition. Williams & Wilkins Co., Baltimore.

3. **Pelczar, M. J., Jr. and R. D. Reid.** 1965. *Microbiology,* 2nd Edition. McGraw-Hill, N.Y. Chapter 36 and Appendix B.

3
Intrinsic and Extrinsic
Parameters of Foods
that Affect
Microbial Growth

Since our foods are of plant and animal origin, it is worthwhile to consider those characteristics of plant and animal tissues that affect the growth of microorganisms therein. The plants and animals that serve as food sources have all evolved mechanisms of defense against the invasion and proliferation of microorganisms, and some of these remain in effect in fresh foods. By taking these natural phenomena into account, one can make effective use of each or all in preventing or retarding the microbial spoilage of the products which are derived from them.

INTRINSIC PARAMETERS

Those parameters of plant and animal tissues that are an inherent part of the tissues are referred to as **intrinsic parameters.** These parameters are:

1. pH
2. Moisture content
3. Oxidation-reduction potential (Eh)
4. Nutrient content
5. Antimicrobial constituents
6. Biological structures

Each one of these is discussed below with emphasis placed upon their effect on microorganisms in foods.

1. pH. It has been well established that most microorganisms grow best at pH values around 7.0 (6.6–7.5) while few grow below 4.0 (see Table 3-1). Bacteria tend to be more fastidious in their relationships to pH than molds and yeasts, with the pathogenic bacteria being the most fastidious. With respect to pH minima and maxima of microorganisms, those given in Table 3-1 should not be taken to be precise boundaries, since the actual values are

TABLE 3-1. Approximate minimum and maximum pH values for the growth of some microorganisms.

Organism	Minimum	Maximum
Escherichia coli	4.4	9.0
Salmonella typhi	4.5	8.0
Streptococcus lactis	4.3–4.8	
Lactobacillus spp.	Ca. 3.0	7.2
*Thiobacillus thiooxidans**	<1.0	9.8
Molds	1.5–2.0	11.0
Yeasts	1.5	8.0–8.5
Acontium velatum (fungus)	0.2–0.7	7.0

*Optimum between 2–3.

known to be dependent upon certain other growth parameters. For example, the pH minima of certain lactobacilli have been shown to be dependent upon the type of acid used with citric, HC1, phosphoric, and tartaric acids, permitting growth at lower pH than acetic or lactic acids *(12)*. Of the foods presented in Table 3-2, it can be seen that fruits, soft drinks, vinegar, and wines all fall below the point at which bacteria normally grow. The excellent keeping quality of these products is due in great part to pH. It is a common observation that fruits generally undergo mold and yeast spoilage, and this is due to the capacity of these organisms to grow at pH values below 3.5, which is considerably below the minima for most food spoilage and all food poisoning bacteria. It may be further noted from Table 3-3 that most of the meats and seafoods have a final ultimate pH of about 5.6 and above. This makes these products susceptible to bacterial as well as to mold and yeast spoilage. Likewise, most vegetables have higher pH values than fruits and consequently vegetables are more subject to bacterial than fungal spoilage (Table 3-3). An inspection of Table 3-4 reveals that most common market diseases of fruits are fungal, with the exception of pears, which are sometimes spoiled by bacteria of the genus *Erwinia*. On the other hand, the most common market diseases of vegetables are caused by bacteria.

With respect to the keeping quality of meats, it is well known that meat from fatigued animals spoils faster than that from rested animals, and that this is a direct consequence of final pH attained upon completion of rigor mortis. Upon the death of a well-rested meat animal, the usual 1 percent glycogen is converted into lactic acid, which directly causes a depression in pH values from about 7.4 to about 5.6, depending upon the type of animal. Callow *(6)* found the lowest pH values for beef to be 5.1 and the highest 6.2 after rigor mortis. The usual pH value attained upon completion of rigor mortis of beef is around 5.6 *(3)*. The lowest and highest values for lamb and pork, respectively, were found by Callow to be 5.4 and 6.7, and 5.3 and 6.9. More recently, Briskey *(4)* reported that the ultimate pH of pork may be as low as approximately 5.0 under certain conditions. The effect of pH of this

TABLE 3-2.   Approximate pH values of some fresh fruits and vegetables.

Vegetables	pH
Asparagus (buds and stalks)	5.7–6.1
Beans (string and Lima)	4.6 & 6.5
Beets (sugar)	4.2–4.4
Broccoli	6.5
Brussel sprouts	6.3
Cabbage (green)	5.4–6.0
Carrots	4.9–5.2; 6.0
Cauliflower	5.6
Celery	5.7–6.0
Corn (sweet)	7.3
Eggplant	4.5
Lettuce	6.0
Olives	3.6–3.8
Onions (red)	5.3–5.8
Parsley	5.7–6.0
Parsnip	5.3
Potatoes (tubers & sweet)	5.3–5.6
Pumpkin	4.8–5.2
Rhubarb	3.1–3.4
Spinach	5.5–6.0
Squash	5.0–5.4
Tomatoes (whole)	4.2–4.3
Turnips	5.2–5.5
Fruits	
Apples	2.9–3.3
Bananas	4.5–4.7
Figs	4.6
Grapefruit (juice)	3.0
Limes	1.8–2.0
Melons (honey dew)	6.3–6.7
Oranges (juice)	3.6–4.3
Plums	2.8–4.6
Watermelons	5.2–5.6
Grapes	3.4–4.5

magnitude upon microorganisms, especially bacteria, is obvious. With respect to fish, it has been known for some time that halibut, which usually attains an ultimate pH of about 5.6, has better keeping qualities than most other fish, whose ultimate pH values range between 6.2 and 6.6 *(15)*.

While some foods are characterized by what Weiser *(20)* refers to as inherent acidity, some foods owe their acidity or pH to the actions of certain microorganisms. Weiser refers to the latter type of acidity as biological acidity, which is displayed by products such as fermented milks, sauerkraut,

TABLE 3-3.   Approximate pH values of dairy, meat, poultry, and fish products.

Product	pH
Meat & Poultry	
Beef (ground)	5.1–6.2
Ham	5.9–6.1
Veal	6.0
Chicken	6.2–6.4
Fish & Shellfish	
Fish (most species)*	6.6–6.8
Clams	6.5
Crabs	7.0
Oysters	4.8–6.3
Tuna fish	5.2–6.1
Shrimp	6.8–7.0
Salmon	6.1–6.3
White fish	5.5
Dairy Products	
Butter	6.1–6.4
Buttermilk	4.5
Milk	6.3–6.5
Cream	6.5
Cheese (American mild & Cheddar)	4.9, 5.9

*Just after death.

TABLE 3-4.   Some microorganisms associated with the spoilage of fruits and vegetables in California *(19)*.

Organisms	Fruits and Vegetables Affected
	Fruits
Penicillium spp.	Apples, pears, apricots, cherries, citrus fruits.
Rhizopus spp.	Cranberries, apricots, cherries, peaches, strawberries
Cladosporium spp.	Apples, figs, cherries
Alternaria spp.	Figs, apricots, cherries, peaches
Botrytis spp.	Cherries, grapes, strawberries
Erwinia spp.	Pears
	Vegetables
Rhizopus spp.	Carrots, snap beans, sweet potatoes, tomatoes
Alternaria spp.	Crucifers, tomatoes
Botrytis spp.	Artichokes, lettuce, onions, crucifers, snap beans, tomatoes
Erwinia spp.	Carrots, lettuce, asparagus, celery, onions, potatoes

and pickles. Regardless of the source of acidity, the effect upon keeping quality appears to be the same.

Some foods are better able to resist changes in pH than others. Those that tend to resist changes in pH are said to be **buffered.** In general, meats are more highly buffered than vegetables. Contributing to the buffering capacity of meats are their various proteins. Vegetables are generally low in proteins and consequently lack the buffering capacity to resist changes in their pH by the growth of microorganisms (see Table 6-1 for the general chemical composition of vegetables).

The natural or inherent acidity of foods may be thought of as nature's way of protecting the respective plant or animal tissues from destruction by microorganisms. It is of interest that fruits should have pH values below those required by many spoilage organisms. The biological function of the fruit is the protection of the plant's reproductive body, the seed. This one fact alone has no doubt been quite important in the evolution of present-day fruits. While the pH of a living animal favors the growth of most spoilage organisms, other intrinsic parameters come into play to permit the survival and growth of the animal organism.

2. Moisture Content.   One of man's oldest methods of preserving foods is drying or desiccation, and precisely how this method came to be used is not known. The preservation of foods by drying is a direct consequence of removal or binding of moisture without which microorganisms do not grow. It is now generally accepted that the water requirements of microorganisms should be defined in terms of the **water activity (a_w)** in the environment. This parameter is defined by the ratio of the water vapor pressure of food substrate to the vapor pressure of pure water at the same temperature, i.e., $a_w = p/p_o$, where p = vapor pressure of solution and p_o = vapor pressure of solvent (usually water). This concept is related to relative humidity (R.H.) in the following way: R.H. $= 100 \times a_w$ (7).

The a_w of most fresh foods is above 0.99. The minimum a_w values reported for the growth of some microorganisms in foods are presented in Table 3-5 (see also Table 13-1). Most spoilage bacteria do not grow below a_w of 0.91 while spoilage molds can grow as low as 0.80. With respect to food-poisoning bacteria, *Staphylococcus aureus* has been found to grow as low as 0.86 while *Clostridium botulinum* does not grow below 0.95. Just as yeasts and molds grow over a wider pH range than bacteria, the same is true for a_w. The lowest reported values for bacteria of any type is 0.75 for halophilic (literally, "salt loving") bacteria while xerophilic (literally, "dry loving") molds and osmophilic (prefering high osmotic pressures) yeasts have been reported to grow at a_w values of 0.65 and 0.60, respectively. When salt is employed to control a_w, it can be seen from Table 3-6 that an extremely high level is necessary to achieve a_w values below 0.80.

Certain relationships have been shown to exist between a_w, temperature, and nutrition. First, at any temperature, the ability of microorganisms to grow is reduced as the a_w is lowered. Second, the range of a_w over which growth occurs is greatest at the optimum temperature for growth; and third, the presence of nutrients increases the range of a_w over which the organisms

TABLE 3-5.   Approximate minimum a_w values for the growth of microorganisms of importance in foods.

Organisms	Minimum a_w
Groups	
Most spoilage bacteria	0.91
Most spoilage yeasts	0.88
Most spoilage molds	0.80
Halophilic bacteria	0.75
Xerophilic molds	0.65
Osmophilic yeasts	0.60
Specific organisms	
Acinetobacter	0.96
Enterobacter aerogenes	0.95
Bacillus subtilis	0.95
Clostridium botulinum	0.95
Escherichia coli	0.96
Pseudomonas	0.97
Staphylococcus aureus	0.86
Saccharomyces rouxii	0.62

TABLE 3-6.   Relationship between water activity and concentration of salt solutions*

Water Activity	Sodium Chloride Concentration	
	Molal	Percent, w/v
0.995	0.15	0.9
0.99	0.30	1.7
0.98	0.61	3.5
0.96	1.20	7
0.94	1.77	10
0.92	2.31	13
0.90	2.83	16
0.88	3.33	19
0.86	3.81	22

*From *The Science of Meat and Meat Products,* by the American Meat Institute Foundation. W. H. Freeman and Company, San Francisco, Copyright © 1960 *(8).*

can survive *(13).* The specific values given above, then, should be taken only as reference points, since a change in temperature or nutrient content might permit growth at lower values of a_w.

3. Oxidation-Reduction Potential (O/R, Eh).   It has been known for many years that microorganisms display varying degrees of sensitivity to the oxidation-reduction potential of their growth medium *(10).* The O/R potential of a substrate may be defined generally as the ease with which the substrate loses or gains electrons. When an element or compound loses electrons, the substrate is said to be oxidized while a substrate that gains electrons becomes reduced:

$$\text{Cu} \underset{\text{reduction}}{\overset{\text{oxidation}}{\rightleftharpoons}} \text{Cu} + e.$$

Oxidation may also be achieved by the addition of oxygen as illustrated in the following reaction:

$$2\,\text{Cu} + \text{O}_2 \rightarrow 2\,\text{CuO}$$

Therefore, a substance that readily gives up electrons is a good reducing agent, while one that readily takes up electrons is a good oxidizing agent. When electrons are transferred from one compound to another, a potential difference is created between the two compounds. This difference may be measured by use of an appropriate instrument and expressed as millivolts (mv). The more highly oxidized a substance, the more positive will be its electrical potential, and the more highly reduced a substance, the more negative will be its electrical potential. When the concentration of oxidant and reductant is equal, a zero electrical potential exists. The O/R potential of a system is expressed by the symbol Eh. Aerobic microorganisms require positive Eh values (oxidized) for growth while anaerobes require negative Eh values (reduced). Among the substances in foods that help to maintain reducing conditions are —SH groups in meats and ascorbic acid and reducing sugars in fruits and vegetables.

According to Frazier *(9),* the O/R potential of a food is determined by: (1) the characteristic O/R potential of the original food, (2) the **poising capacity,** that is, the resistance to change in potential of the food, (3) the oxygen tension of the atmosphere about the food, and (4) the access which the atmosphere has to the food.

With respect to Eh requirements of microorganisms, some bacteria require reduced conditions for growth initiation (Eh of about −200 mv) while others require a positive Eh for growth. In the former category are the anaerobic bacteria such as the genus *Clostridium,* while in the latter belong aerobic bacteria such as the genus *Bacillus.* Some aerobic bacteria actually grow better under slightly reduced conditions and these organisms are often referred to as **microaerophiles.** Examples of microaerophilic bacteria are lactobacilli and streptococci. Some bacteria have the capacity to grow under either aerobic or anaerobic conditions. Such types are referred to as **facultative anaerobes.** Most molds and yeasts encountered in and on foods are aerobic though a few tend to be facultative anaerobes.

In regard to the Eh of foods, plant foods, especially plant juices, tend to have Eh values of from +300 to +400. It is not surprising to find that aerobic bacteria and molds are the common cause of spoilage of products of this type. Solid meats have Eh values of around −200 mv while in minced meats the Eh is generally around +200 mv. Cheeses of various types have been reported to have Eh values on the negative side from −20 to around −200 mv.

With respect to the Eh of prerigor as opposed to postrigor muscles, Barnes and Ingram *(1, 2)* undertook a study of the measurement of Eh in muscle over periods of up to 30 hr postmortem and its effect upon the growth of anaerobic bacteria. These authors found that the Eh of the sternoce-phalicus muscle of the horse immediately after death was +250 mv, at which time clostridia failed to multiply. At 30 hr postmortem, the Eh had fallen to about −130 mv in the absence of bacterial growth. When bacterial growth was allowed to occur, the Eh fell to about −250 mv. Growth of clostridia was observed at Eh values of −36 mv and below. These authors confirmed for horse meat the finding of Robinson *et al. (16)* for whale meat: that anaerobic bacteria do not multiply until the onset of rigor mortis, because of the high Eh in prerigor meat. The same is undoubtedly true for beef, pork, and other meats of this type.

4. Nutrient Content. In order to grow and function normally, the microorganisms of importance in foods require the following:

1. Water
2. Source of energy
3. Source of nitrogen
4. Vitamins and related growth factors
5. Minerals

The importance of water to the growth and welfare of microorganisms was presented earlier in this chapter. With respect to the other four groups of substances, molds have the lowest requirement, followed by yeasts, gram negative bacteria, and gram positive bacteria.

As sources of energy, food-borne microorganisms may utilize sugars, al-cohols, and amino acids. Some few microorganisms are able to utilize com-plex carbohydrates such as starches and cellulose as sources of energy by first degrading these compounds to simple sugars. Fats are used also by microorganisms as sources of energy but these compounds are attacked by a relatively small number of microbes in foods.

The primary nitrogen sources utilized by heterotrophic microorganisms are amino acids. A large number of other nitrogenous compounds may serve this function for various types of organisms. Some microbes, for example, are able to utilize nucleotides and free amino acids while others are able to utilize peptides and proteins. In general, simple compounds such as amino acids will be utilized by most all organisms before any attack is made upon the more complex compounds such as high molecular weight proteins. The same is true of polysaccharides and fats.

Microorganisms may require B-vitamins in low quantities and most all natural foods tend to have an abundant quantity for those organisms that are unable to synthesize their essential requirements. In general, gram positive bacteria are the least synthetic and must, therefore, be supplied with one or more of these compounds before they will grow. The gram negative bacteria and molds are able to synthesize most or all of their requirements. Consequently, these two groups of organisms may be found growing on foods low in B-vitamins. Fruits tend to be lower in B-vitamins than meats and this fact along with the usual low pH and positive Eh of fruits all help to explain the usual spoilage of these products by molds rather than bacteria.

5. Antimicrobial Constituents.   The stability of some foods against attack by microorganisms is due to the presence in these foods of certain naturally occurring substances which have been shown to have antimicrobial activities. For example, fresh milk contains **lactenin** and a substance which has been designated **anticoliform** factor, both of which are antimicrobial. The **lactoperoxidase** complex in raw milk is effective against some streptococci. **Lysozyme** is present in egg white while **benzoic acid** is present in cranberries. Lipids and essential oils, especially **eugenol** in cloves and **cinnimic aldehyde** in cinnamon, all possess antimicrobial properties.

6. Biological Structures.   The natural covering of some foods provides excellent protection against the entry and subsequent damage by spoilage organisms. In this category are such structures as the testa of seeds, the outer covering of fruits, the shell of nuts, the hide of animals, and the shells of eggs. In the case of nuts such as pecans and walnuts, the shell or covering is sufficient to prevent the entry of all organisms. Once cracked, of course, nutmeats are subject to spoilage by molds. The outer shell and membranes of eggs, if intact, prevent the entry of nearly all microorganisms when stored under the proper conditions of humidity and temperature. Fruits and vegetables with damaged covering undergo spoilage much faster than those not damaged. The skin covering of fish and meats such as beef and pork prevents the contamination and spoilage of these foods partly because it tends to dry out faster than freshly cut surfaces.

Taken together, these six intrinsic parameters represent nature's way of preserving plant and animal tissues from microorganisms. By determining the extent to which each exists in a given food, one can predict the general types of microorganisms that are likely to grow and consequently the overall stability of this particular food. Their determination may also aid one in determining age and possibly the handling history of a given food.

EXTRINSIC PARAMETERS

The extrinsic parameters of foods are those properties of the storage environment that affect both the foods and their microorganisms. Those of greatest importance to the welfare of food-borne organisms are: (1) temperature of storage, (2) relative humidity of environment, and (3) presence and concentration of gases in the environment.

1. *Temperature of Storage.* Microorganisms grow over a very wide range of temperatures. Therefore, it would be well to consider at this point the temperature growth ranges for organisms of importance in foods as an aid in selecting the proper temperature for the storage of different types of foods (see Figure 3-1).

The lowest temperature at which a microorganism has been reported to grow is −34 C while the highest is somewhere in excess of 90 C. It is customary to place microorganisms into three groups based upon their temperature requirements for growth. Those organisms that grow well below 20 C and have their optimum between 20 and 30 C are referred to as **psychrophiles** or **psychrotrophs** (see Chapter 11). Those that grow well between 20 and 45 C with optima between 30 and 40 C are referred to as **mesophiles,** while those that grow well at and above 45 C with optima between 55–65 C are referred to as **thermophiles.** Physiological properties of these groups are treated in Chapters 20 and 21.

In regard to bacteria, psychrotrophic species and strains are found among the following genera of those presented in the previous chapter: *Alcaligenes, Corynebacterium, Flavobacterium, Lactobacillus, Micrococcus, Pseudomonas, Streptococcus, Streptomyces,* and others. The psychrotrophs found most commonly on foods are those that belong to the genera *Alcaligenes, Pseudomonas,* and *Streptococcus.* These organisms grow well at refrigerator temperatures and cause spoilage of meats, fish, poultry, eggs,

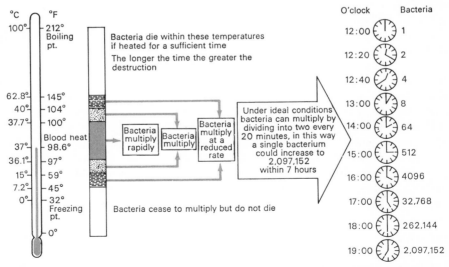

FIGURE 3-1.   Effect of temperature and time on the growth of bacteria. Safe and dangerous temperatures for foodstuffs (from Hobbs, *11,* reproduced with permission of the publisher).

and other foods normally held at this temperature. Standard plate counts of viable organisms on such foods are generally higher when the plates are incubated at about 7 C for at least 7 days than when incubated at 30 C and above. Mesophilic species and strains are known among all 25 genera presented in the previous chapter and may be found on foods held at refrigerator temperature. They apparently do not grow at this temperature but do grow at temperatures within the mesophilic range if other conditions are suitable. It should be pointed out that some organisms can grow over a range from 0 and 30 C or above. One such organism is *Streptococcus faecalis*.

Most thermophilic bacteria of importance in foods belong to the genera *Bacillus* and *Clostridium*. While only a few species of these genera are thermophilic, they are of great interest to the food microbiologist and food technologist in the canning industry.

Just as molds are able to grow over wider ranges of pH, osmotic pressure, and nutrient content, they are also able to grow over wider ranges of temperature than bacteria. Many molds are able to grow at refrigerator temperatures, notably some strains of *Aspergillus, Cladosporium,* and *Thamnidium,* which may be found growing on eggs, sides of beef, and fruits. Yeasts grow over the psychrophilic and mesophilic temperature ranges but generally not within the thermophilic range.

The quality of the food product must also be taken into account in selecting a storage temperature. While it would seem desirable to store all foods at refrigerator temperatures or below, this is not always best for the maintenace of desirable quality in some foods. For example, bananas keep better if stored at 13–17°C than at 5–7°C. A large number of vegetables are favored by temperatures of about 10°C, including potatoes, celery, cabbage, and many others. In every case, the success of storage temperature depends to a great extent upon the R.H. of the storage environment and the presence or absence of gases such as CO_2 and O_3.

2. Relative Humidity of Environment.   The relative humidity (R.H.) of the storage environment is important both from the standpoint of a_w within foods and the growth of microorganisms at the surfaces. When the a_w of a food is set at 0.60, it is important that this food be stored under conditions of R.H. that do not allow the food to pick up moisture from the air and thereby increase its own surface and subsurface a_w to a point where microbial growth can occur. When foods with low a_w values are placed in environments of high R.H., the foods pick up moisture until equilibrium has been established. Likewise, foods with a high a_w lose moisture when placed in an environment of low R.H. There is a relationship between R.H. and temperature which should be borne in mind in selecting proper storage environments for the storage of foods. In general, the higher the temperature, the lower the R.H., and vice versa.

Foods that undergo surface spoilage from molds, yeasts, and certain bacteria should be stored under conditions of low R.H. Improperly wrapped meats such as whole chickens and beef cuts tend to suffer surface spoilage in the refrigerator much before deep spoilage occurs, due to the generally high R.H. of the refrigerator and the fact that the meat spoilage flora is essentially aerobic in nature. While it is possible to lessen the chances of surface spoilage in certain foods by storing under low conditions of R.H., it should be remembered that the food itself will lose moisture to the atmosphere under such conditions and thereby become undesirable. In selecting the proper environmental conditions of R.H., considerations must be given to both the possibility of surface growth and the desirable quality to be maintained in the foods in question. By altering the gaseous atmosphere, it is possible to retard surface spoilage without lowering R.H.

3. Presence and Concentration of Gases in the Environment.   The storage of food in atmospheres containing increased contents of CO_2 up to about 10 percent is referred to as "controlled atmosphere" or c-a storage. The effect of c-a storage on plant organs has been known since 1917 *(18)*, and was first put into commercial use in 1928. The use of c-a storage for fruits is employed in a number of countries, with apples and pears being the fruits most commonly treated. The concentration of CO_2 generally does not exceed 10 percent and is applied either from mechanical sources or by use of dry ice (solid CO_2). Carbon dioxide has been shown to retard fungal rotting of fruits caused by a large number of fungi. While the precise mechanism of action of CO_2 in retarding fruit spoilage is not known, it is probable that it acts as a competitive inhibitor of ethylene action. Ethylene seems to act as a senescence factor in fruits *(17)*, and its inhibition would have the effect of maintaining a fruit in a better state of natural resistance to fungal invasion. For a more detailed treatment of the c-a storage of fruits and vegetables, the review by Smith *(18)* should be consulted.

It has also been known for many years that ozone added to food storage environments has a preservative effect upon certain foods. At levels of several ppm, this gas has been tried with several foods and found to be effective against spoilage microorganisms. It is effective against a variety of microorganisms *(5)*. Since it is a strong oxidizing agent, it should not be used on high lipid-content foods, since it would cause an increase in rancidity. Both CO_2 and O_3 are effective in retarding the surface spoilage of beef quarters under long-term storage.

SUMMARY

Each food product is characterized by a set of intrinsic parameters consisting of pH, moisture content, O/R potential, nutrient content, antimicrobial substances, and/or a biological structure which is capable of preventing the entry of many microorganisms to certain foods. When a given food is characterized with respect to these parameters, it then becomes possible to make predictions about the perishability of that food by taking into consideration the overall capacity of certain flora to grow within these parameters. The microorganisms of importance in foods all have limitations relative to these parameters and may be inhibited by one or more of them. The extrinsic parameters of temperature of storage, relative humidity of the storage environment, and the presence or absence of gases such as CO_2 and O_3 may also be employed to inhibit the growth of some organisms. The storing of dried foods in areas of high R.H. may offset the intrinsic parameter of water content and allow for growth in an otherwise stable product. The operation of these as well as other less definable parameters leads to the establishment of certain organisms as the dominant types in given foods, as Mossel and Ingram (14) and others before them have called attention to. For example, citrus fruits are largely spoiled by certain molds; refrigerated ground beef is spoiled mainly by certain gram negative bacteria; bread is spoiled mainly by certain molds; etc.

REFERENCES

1. **Barnes, E. M. and M. Ingram.** 1955. Changes in the oxidation-reduction potential of the sterno-cephalicus muscle of the horse after death in relation to the development of bacteria. J. Sci. Food Agr. *6:* 448–455.

2. **Barnes, E. M. and M. Ingram.** 1956. The effect of redox potential on the growth of *Clostridium welchii* strains isolated from horse muscle. J. Appl. Bacteriol. *19:* 117–128.

3. **Bate-Smith, E. C.** 1948. The physiology and chemistry of rigor mortis, with special reference to the aging of beef. Adv. Food Res. *1:* 1–38.

4. **Briskey, E. J.** 1964. Etiological status and associated studies of pale, soft, exudative porcine musculature. Adv. Food Res. *13:* 89–178.

5. **Burleson, G. R., T. M. Murray, and M. Pollard.** 1975. Inactivation of viruses and bacteria by ozone, with and without sonication. Appl. Microbiol. *29:* 340–344.

6. **Callow, E. H.** 1949. Science in the imported meat industry. J. Roy. Sanitary Inst. *69:* 35–39.

7. **Christian, J. H. B.** 1963. Water activity and the growth of microorganisms. In: *Recent Advances in Food Science,* Vol. 3: 248–255. Edited by Leitch, J. M. and Rhodes, D. N., Butterworths, London.

8. **Evans, J. B. and C. F. Niven, Jr.** 1960. Microbiology of meat: Bacteriology. In: *The Science of Meat and Meat Products.* W. H. Freeman & Co., San Francisco.

9. **Frazier, W. C.** 1968. *Food Microbiology,* 2nd Edition. McGraw-Hill, N.Y., p. 171.

10. **Hewitt, L. F.** 1950. *Oxidation-Reduction Potentials in Bacteriology and Biochemistry,* 6th Ed., E & Livingston, Ltd., Edinburgh.

11. **Hobbs, B. C.** 1968. *Food Poisoning & Food Hygiene,* 2nd Edition. Edward Arnold Publishers, Ltd., London.

12. **Juven, B. J.** 1976. Bacterial spoilage of citrus products at pH lower than 3.5. J. Milk Food Technol. *39:* 819–822.

13. **Morris, E. O.** 1962. Effect of environment on micro-organisms. In: *Recent Advances in Food Science,* Vol. 1: 24–36. Hawthorn, J. and Leitch, J. M., Editors, Butterworths, London.

14. **Mossel, D. A. A. and M. Ingram.** 1955. The physiology of the microbial spoilage of foods. J. Appl. Bacteriol. *18:* 232–268.

15. **Reay, G. A. and J. M. Shewan.** 1949. The spoilage of fish and its preservation by chilling. Adv. Food Res. *2:* 343–398.

16. **Robinson, R. H. M., M. Ingram, R. A. M. Case, and J. G. Benstead.** 1952. Whale-meat: bacteriology and hygiene. Spec. Rep. Serv. Food Invest. Board (London), No. 59.

17. **Salisbury, F. B. and C. Ross.** 1969. *Plant Physiology,* p. 467 (Wadsworth Publ.: Belmont, Ca.).

18. **Smith, W. H.** 1963. The use of carbon dioxide in the transport and storage of fruits and vegetables. Adv. in Food Res. *12:* 95–146.

19. **Vaughn, R. H.** 1963. Microbial spoilage problems of fresh and refrigerated foods. In: *Microbiological Quality of Foods.* L. W. Slanetz *et al.,* Editors. Academic Press, N.Y., pp. 193–197.

20. **Weiser, H. H.** 1962. *Practical Food Microbiology and Technology.* Avi Publishing Co., Conn., Chapter 8.

4

Determining Microorganisms and Their Products in Foods

The examination of foods for the presence, types, and numbers of microorganisms and/or their products is basic to food microbiology. In spite of the importance of this, none of the methods in common use permit the determination of exact numbers of microorganisms in a food product. Although some methods of analysis are better than others, every method has certain inherent limitations associated with its use.

The four general methods that are employed for "total" numbers are: (1) direct microscopic counts (DMC) for both viable and nonviable cells, (2) standard plate counts (SPC) for viable cells, (3) the most probable numbers (MPN) method for a statistical determination of viable cells, and (4) dye reductions for viable cells that possess reducing capacities (see Figure 4-1). The SPC is by far the most widely used of these methods for determining the numbers of viable cells or colony-forming units (cfu) in a food product.

When total viable counts are reported for a food product, the counts should be looked upon as being a function of at least some of the following factors.

1. The sampling methods employed.
2. Distribution of the organisms in the food sample.
3. Nature of the food flora.
4. Nature of the food material.
5. The pre-examination history of the food product.
6. Nutritional adequacy of the plating medium employed.
7. Incubation temperature and time used.
8. The pH, a_w, and Eh of the plating medium.
9. Type of diluent used.
10. Relative number of organisms in food sample.
11. The existence of other competing or antagonistic organisms, etc.

Plate count procedures are employed for the enumeration of viable numbers of various groups of microorganisms as well as for food-borne pathogens as illustrated in Figure 4-1. In addition to the limitations noted above for SPC, plating procedures for selected groups are further limited by the degree of inhibition and effectiveness of the selective and differential agents employed. A synopsis of all of the microbiological methods from Figure 4-1 is given below.

SYNOPSIS OF PROCEDURES FOR MICROBIOLOGICAL EXAMINATION OF FOODS

A. "Total" Numbers

Direct Microscopic Count (DMC).   Smears of food specimen (or low dilutions thereof) are prepared on microscope or special slides, stained with appropriate dye preparation, microbial cells (individual or clumps) are counted in a given number of microscopic fields, and the number of organisms/g are determined by use of the microscope factor.

Standard Plate Count (SPC).   Aliquots of food samples are blended, serially diluted in an appropriate diluent, plated in or onto a suitable agar medium, incubated at an appropriate temperature for a given time, and all visible colonies are counted using a Quebec counter. **Thermoduric** organisms are determined by above procedure following exposure of food product to specified temperature and time. **Thermophilic** organisms are determined as for SPC with use of appropriate media and incubation of plates at or above 55 C.; **psychrotrophic** organisms as for SPC but with incubation of plates at 5 to 7 C.

Most Probable Numbers (MPN).   Food samples are handled as for SPC with three serial aliquots or dilutions planted into 9 or 15 tubes of appropriate medium for the 3- or 5-tube method, respectively. Numbers of organisms in original sample are determined by use of standard MPN tables. This method is statistical in nature and MPN results are generally higher than SPC.

Dye Reductions.   Properly prepared supernatants of foods are added to standard solutions of methylene blue or resazurin and observed for reduction of dyes (from blue to white for methylene blue; from slate blue to pink or white for resazurin). The time for dye reductions to occur is referable to number of organisms in sample.

B. Selected Groups

Yeasts and Molds.   The culture method is similar to SPC method with plating onto media at pH 3.5 (such as acidified potato dextrose agar), or onto

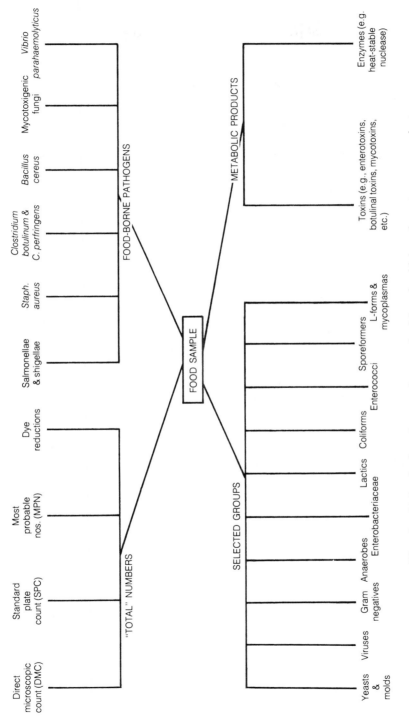

FIGURE 4-1.   Ways in which a food product may be examined for microbial numbers, types, or products.

media containing antibacterial agents. Mold mycelia may be determined by impression of food specimen directly onto microscope slide followed by staining and viewing.

Viruses.   A slurry of food sample is prepared and clarified with appropriate agents; the clarified effluent is concentrated by use of polyethylene glycol or other suitable agents; viruses are concentrated by ultracentrifugation and enumerated by plaque count following inoculation and incubation of suitable tissue culture medium.

Gram Negative Bacteria.   Same as SPC method but with plating in or on selective media such as violet red bile, eosin methylene blue, or MacConkey agars.

Anaerobes.   Same as SPC method but with special attention to diluents and media pre-reduced to an appropriate negative Eh. Differentiation between facultative anaerobes and strict anaerobes may be necessary.

Enterobacteriaceae.   Food homogenates are enriched in buffered glucose, brilliant green medium (EE-broth) followed by plating onto selective media such as VRBA.

Lactic Acid Bacteria.   Same as SPC method but with plating on selective media such as MRS, Eugon, or Elliker's lactic agars followed by biochemical characterizations.

Coliforms.   Presumptive and confirmed tests are carried out by inoculation of food dilutions into lauryl sulfate tryptose (LST) broth and then gas-positive tube of LST to brilliant green lactose bile broth (BGLB), both at 35 C. Positive BGLB tube streaked onto EMB agar at 35 C. Or by inoculation of LST with incubation at 44 C followed by streaking onto EMB agar. For fecal coliforms, inoculate EC broth from positive LST tube and incubate at 45.5 C. Enteropathogenic *E. coli* by use of appropriate antisera to positive tubes of above.

Enterococci.   Same as SPC or MPN methods with direct plating onto selective medium such as KF agar, or by use of broths such as KF or azide dextrose for MPN determinations with confirmation of the latter medium on Pfizer selective enterococcus (PSE) agar.

Sporeformers.   Food samples or diluents are heat-treated at 70–80 C for 10 min followed by plating and incubation under aerobic and/or anaerobic conditions. **Flat-sour** spores are determined by steaming or boiling aliquots of food for around 5 min, plating with medium containing glucose and acid-base indicator, and incubating aerobically in the mesophilic range. **Sulfide-spoilers,** same as flat-sour spores with culturing in sulfide-containing media.

L-forms and Mycoplasmas.   Food homogenates or diluents are plated on L-form media such as brain heart infusion or tryptic soy with 1% agar, 10–20%

horse serum, 10–20% sucrose and antibiotics such as methicillin or penicillin with incubation at around 30 C for up to several weeks. Periodic examinations are made for the presence of "fried egg" or diffuse-spreading colonies characteristic of bacterial L-forms or mycoplasmas. Mycoplasmas have G + C values of 23–40% while the values for bacterial L-forms are higher.

C. Food-borne Pathogens

Salmonellae & Shigellae.   Food aliquots or dilutions thereof are pre-enriched and either plated directly onto selective media (such as brilliant green, XLD, and/or Hektoen agars), or planted into enrichment media such as selenite cystine or tetrathionate broths. Enrichment tubes streaked onto selective media plates such as those above. Suspect colonies characterized biochemically, by use of appropriate O and H antisera, and/or by fluorescent antibody (FA) procedure in the case of salmonellae.

Staphylococcus aureus.   Ditto SPC or MPN methods with direct plating onto selective medium such as Baird-Parker agar, or onto similar media following enrichment in broth containing 7.5–10% NaCl. Determine coagulase reaction, lysostaphin sensitivity, and/or heat-stable nuclease production.

Clostridium botulinum.   Food aliquots are enriched in reduced media such as cooked meat or other nonselective media containing trypsin followed by isolation on media such as liver veal-egg yolk agar. Suspects are characterized biochemically and for toxigenesis by injection into protected and unprotected mice. For presence of toxin in food, mice may be challenged directly with trypsin-treated food extracts.

Clostridium perfringens.   Same as SPC methods with plating onto selective media such as tryptose-sulfite-cycloserine egg yolk or sulfite-polymyxin-sulfadiazine agars followed by enumeration of suspect colonies and biochemical characterizations. Or, by determining presence of *C. perfringens* alpha-toxin in food product.

Bacillus cereus.   Same as SPC method with plating onto selective media such as KG or phenol red-egg yolk-polymyxin agars followed by enumeration of suspect colonies and biochemical confirmations.

D. Metabolic Products

Microbial Toxins.   Prepare extracts of foods, concentrate extracts and assay by immunoelectrophoresis (for staphylococcal enterotoxins), or by inoculation of food homogenates or diluents into the appropriate experimental animal (e.g., mice for botulinal toxins; rhesus monkeys for staphylococcal enterotoxins), or by UV fluoresence and/or silica gel chromatography of foods or food extracts for aflatoxins, or by appropriate extraction of foods for endotoxins and subsequent assay by use of the *Limulus* lysate test.

Microbial Enzymes.   Foods are extracted with appropriate extractants, the extracts are concentrated (if necessary), and subjected to the relevant substrate—DNA for heat-stable DNase of *S. aureus,* and so on.

Microbiological methods for examining foods may be divided into two groups: traditional and recent methods. The traditional methods include DMC, SPC, dye reductions, MPN, animal inoculations, and some serologic methods for toxins. For further information on the background, theory, application, and use of these widely used methods, the reader is referred to one or more of the standard references noted in Table 4-1.

TABLE 4-1.   Standard references for methods of microbiological analyses of foods.

Reference	Direct Microscopic Counts	Standard Plate Counts	Most Probable Numbers	Dye Reductions	Food Poisoning Organisms	Coliforms & E. coli	Yeasts & Molds	Fluorescent Antibody	Heat-Stable Nuclease	Animal Inoculations	Canned Foods	Sampling Methods
(6)	X	X	X		X	X	X	X				
(4)		X			X	X					X	
(134)	X	X	X		X	X	X	X	X	X		X
(70)	X											X
(100)		X			X	X						X
(123)	X	X	X		X	X		X	X	X	X	X
(3)	X	X	X	X		X						
(131)		X	X		X	X						X

RECENT DEVELOPMENTS—CULTURING

In an effort to improve the efficiency and speed of quantitating and recovering food-borne microorganisms and their products, a large number of techniques have been proposed in recent years and some show excellent potential for future widespread use. One of these is designed to improve both the qualitative and quantitative aspects of culture methods. This involves the recovery of viable but metabolically injured cells which do not grow on selective media by conventional plating procedures.

Metabolically Injured Organisms

When microorganisms are subjected to certain environmental stresses such as sublethal heat and freezing, many of the individual cells undergo metabolic injury resulting in their inability to form colonies on selective media that uninjured cells can tolerate. Whether or not a culture has suffered metabolic injury can be determined by plating aliquots separately on a nonselective and a selective medium and enumerating the colonies that develop after suitable incubation. The colonies that develop on the nonselective medium represent both injured and uninjured cells; only the uninjured cells develop on the selective medium, and the difference between the number of colonies on the two media is a measure of the number of injured cells in the original culture or population. This principle is illustrated in Figure 4-2 by data from Tomlins *et al. (133)* on sublethal heat injury of *S. aureus.* These investigators subjected *S. aureus* to 52 C for 15 min in a phosphate buffer at pH 7.2 to inflict cell injury. The plating of cells at zero time and up to 15 min of heating on nonselective trypticase soy agar (TSA) and selective TSA + 7.0% salt (stress medium) reveals only a slight reduction in numbers on TSA, while the numbers on TSAS were reduced considerably indicating a high degree of injury relative to a level of salt which uninjured *S. aureus* can withstand. To allow the heat-injured cells to repair, the cells were placed in nutrient broth (recovery medium) followed by incubation at 37 C for 4 hr. With hourly plating of aliquots from the recovery medium onto TSAS, it can be seen that the injured cells regained their capacity to withstand the 7.0% NaCl in TSAS after the 4-hr incubation. The existence of metabolically injured cells in foods and their recovery during culturing procedures is obviously of great importance not only from the standpoint of pathogenic organisms but for spoilage organisms as well. The data cited above suggest that if a high-salt medium had been employed to examine a heat-pasteurized product for *S. aureus,* the number of viable cells found would have been lower than the actual number by a factor of 3 log cycles. Injury of food-borne microorganisms has been shown by a large number of investigators to be induced not only by sublethal heat and freezing but also by freeze-drying, drying, irradiation, aerosolization, dyes, sodium azide, salts, antibiotics, and a large number of other chemicals such as EDTA and sanitizing compounds.

The recognition of sublethal stresses on food-borne microorganisms and their effect upon growth under varying conditions dates back to the turn of the century. However, a full appreciation of this phenomenon did not come until the late 1960s. The increased nutritional requirement of bacteria which had undergone heat treatment was noted by Nelson *(101)* in 1943 (who also reviewed the work of others up to that time). Gunderson and Rose *(59)* noted the progressive decrease in the numbers of coliforms from frozen chicken products that grew on violet red bile agar with increasing storage time of products. Hartsell *(65)* inoculated foods with salmonellae, froze the inoculated foods, and then studied the fate of the organisms during freezer

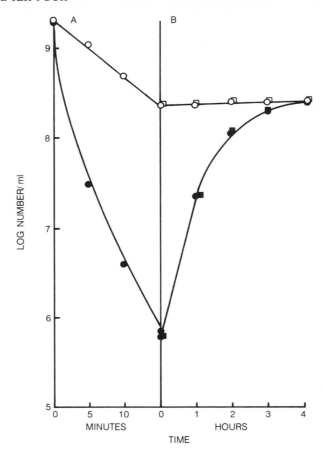

FIGURE 4-2.   Tomlins *et al. (133)*. Survival and recovery curve for *S. aureus* MF-31.
A. Heat injury at 52C for 15 min in 100 m*M* potassium phosphate buffer, pH 7.2. B.
Recovery from heat injury in NB at 37C. Symbols: ○, samples plated on TSA to give
a total viable count; ●, samples plated on TSAS to give an estimate of the uninjured
population. Cells recovered in NB containing 100 μg/ml of chloramphenicol; □,
samples plated on TSA; ■, samples plated on TSAS. "Reproduced by permission of
Nat'l. Res. Coun. of Canada from Can. J. Microbiol. *17:* 759–765, 1971"

storage. This investigator found that larger numbers of organisms could be
recovered on highly nutritive nonselective media than on selective media
such as MacConkey, deoxycholate, or VRB agars. The importance of the
isolation medium in recovering stressed cells was noted also by Postgate and
Hunter *(105)* and by Harris *(63)*. In addition to the more exacting nutritional
nature of food-borne organisms which undergo environmental stresses,
these organisms may be expected to manifest their injury via increased lag
phases of growth, increased sensitivity to a variety of selective media
agents, damage to cell membranes and TCA-cycle enzymes, and breakdown
of ribosomes *(27, 28, 58, 68, 71, 74, 75, 96, 108, 109, 110, 125, 133)*. While
damage to ribosomes and cell membranes appear to be common conse-

quences of sublethal heat injury, not all harmful agents exert identical injuries.

With respect to the recovery of metabolically injured cells, the process can occur at least in *S. aureus* in no-growth media *(67),* and at a temperature of 15 C but not 10 C *(48).* In some instances at least, the recovery process is not instantaneous, for it has been shown that stressed coliforms do not all recover to the same degree but that the process takes place in a step-wise manner *(90).* Not all cells in a given population suffer the same degree of injury. Hurst *et al. (69)* have found dry-injured *S. aureus* cells which failed to develop on the nonselective recovery medium (TSA) but did recover when pyruvate was added to this medium. These cells were said to be severely injured in contrast to injured and uninjured cells. It has been found that sublethally heated *S. aureus* cells may recover their NaCl tolerance before certain membrane functions are restored *(66).* It is well established that injury repair occurs in the general absence of cell wall and protein synthesis. It can be seen from Figure 4-2 that the presence of chloramphenicol in the recovery medium had no effect upon the recovery of *S. aureus* from sublethal heat injury. The repair of cell ribosomes and membrane appears to be essential for recovery at least from sublethal heat, freezing, drying, and irradiation injuries.

The protection of cells from heat and freeze injury is favored by complex media and menstra or certain specific components thereof. Milk provides more protection than saline or mixtures of amino acids *(94, 99),* and the milk components that are most influential appear to be phosphate, lactose, and casein *(75).* Sucrose appears to be protective against heat injury *(2, 86),* while glucose has been reported to decrease heat protection for *S. aureus* *(94).*

The recovery of injured microbes from foods is of primary concern here. The need to provide for a recovery procedure is well established and some of the consequences of not doing so have been reviewed by Busta *(20).* In spite of this need, many methods currently in wide use do not include recovery steps for injured organisms. A summary of some of the many recovery or resuscitation methods employed to induce recovery of injured cells is presented in Table 4-2. While there appears to be no consensus method of recovery at this time, the use of TSB with incubations ranging from 1 to 24 hr at temperatures from 20 to 37 C is widely used for various organisms. The enumeration of sublethally heated *S. aureus* strains on various media has been studied *(31, 69).* In one of these studies, seven staphylococcal media were compared on their capacity to recover 19 strains of sublethally heated *S. aureus,* and the Barid-Parker medium was found to be clearly the best of those studied including nonselective TSA. Similar findings by others led to the adoption of this medium in the official methods of AOAC for the direct determination of *S. aureus* in foods that contain ≥ 10 cells/g *(9).* The greater efficacy of the Baird-Parker medium has been shown to be due to the presence of pyruvate in it. The use of the Baird-Parker medium following

TABLE 4-2.   Summary of some methods employed to recover foodborne microorganisms subjected to various stresses.

Organisms	Method of Injury	Stress Medium	Recovery Medium & Method	Reference
Clostridium perfringens	0.1 N NaOH	TSN agar	TSN agar + lysozyme	(39)
C. perfringens spores	Ultra-high temp.	Polymyxin & neomycin	TSN agar + lysozyme; no antibiotics	(11)
C. botulinum Type E	Sublethal heat	PYE + 0.07% bile salts	PY agar	(104)
C. botulinum spores	Irradiation	TYT + 5% NaCl	TYT thioglycolate medium	(26)
Coliforms	Freezing	VRBA	TSA surface plate; VRBA overlay	(122)
Enterobacteriaceae	Dried foods	E. E. medium	TSPB, 1–6 hr at 19–25 C	(97)
Escherichia coli	Freeze-drying	Antibiotics & Na-deoxycholate	TSYA and minimal salts	(120)
E. coli	Organic & inorganic acids	VRBA	Nutrient agar at 37 C	(112)
E. coli	Frozen foods	VRBA and DCA	TSB for 1 hr at 25 C	(121)
Pseudomonas spp.	Freezing	Minimal agar	Trypticase soy agar	(126)
Salmonella typhimurium	Sublethal heat	EMBA + 2% NaCl	TSA at 37 C for 24 hr	(28)
S. anatum	Freeze-drying	XLPA	XLPA + 0.25% Nadeoxycholate	(108)
S. anatum	Freezing	XLPA	XLPA + 0.2% Nadeoxycholate	(75)
S. typhimurium	Sublethal heat	TSA + yeast extract	M-9 agar at 37 C	(57)
S. senftenberg	Sublethal heat	TSYA	M-9 medium at 37 C for 2 hr	(140)
Salmonella spp.	Irradiation	SS and DCA	BGA; NB + TT + BG (MPN)	(88)
Shigella sonnei	Freezing	Synthetic medium	Blood heart inf., NA plates	(99)
Staphylococcus aureus	Sublethal heat	SM #110	Plate count agar	(19)
S. aureus	Sublethal heat	TSA + 7.5% NaCl	5% glucose or galactose	(125)
S. aureus	Sublethal heat	SM #110	Sodium pyruvate medium	(10)
S. aureus	Sublethal heat	TSA + 7.5% NaCl	Trypticase soy broth, 4 hr	(71)
S. aureus	Freeze-drying	PCA + 7.5% NaCl	Plate count agar at 20–50 C	(48)
S. aureus	Sublethal heat	TSA + 7.0% NaCl	Catalase-treated media	(89)
Streptococcus faecalis	Sublethal heat	High salts in APT	All purpose tween broth	(15)
S. faecalis	Sublethal heat	TSA + 6% NaCl	Synthetic nongrowth	(27)
S. faecium	Sublethal heat	TYGA + 2.5% NaCl	TYG medium at 37 C for 6 hr	(40)

Vibrio parahaemolyticus	Sublethal heat	TCBS + 6% NaCl	TCBS medium	(16)
Aspergillus parasiticus conidia	Sublethal heat	Reduced a$_w$ medium; YEA + 10% NaCl	High a$_w$ media; YEA	(1)
Yeasts	Sublethal heat	PDA at pH 3.5	Potato dextrose agar at pH ca. 8.0	(101)

BGA = brilliant green agar; DCA = deoxycholate agar; EE = enterobacteriaceae medium; EMBA eosin methylene blue agar; M-9 = a glucose salts medium; NA = nutrient agar; NB = nutrient broth; PDA = potato dextrose agar; PYE = peptone-yeast extract; SM #110 = Staphylococcus medium #110; SS = Salmonellae-Shigella medium; TCBS = thiosulphate-citrate-bile salts-sucrose medium; TSA = trypticase soy agar; TSB = trypticase soy broth; TSN = tryptone-sulfite-neomycin; TSPB = tryptone-soya-peptone broth; TSYA = tryptone-soya-yeast extract agar; TT = tetrathionate broth; TYGA = tryptone-yeast extract-glucose-phosphate medium; TYT = thiotone-yeast extract-trypticase-sodium thioglycollate; VRBA = violet red bile agar; XLPA = xylose-lysine-peptone agar; YEA = yeast extract agar.

recovery in an antibiotic-containing, nonselective medium has been suggested *(69)*. While this approach may be suitable for *S. aureus* recovery,
some problems may be expected to occur with the widespread use of antibiotics in recovery media to prevent cell growth. It has been shown that
heat-injured spores of *C. perfringens* are actually sensitized to polymyxin
and neomycin *(11)*, and it is well established that the antibiotics which affect
cell wall synthesis are known to induce L-phase variations in many bacteria.
More recently, it has been found that stressed cells recover even on selective media if catalase is added to the media surfaces prior to plating *(89)*.
Catalase was shown to be effective in recovering sublethally heated *S.
aureus* and *P. fluorescens* and injured *S. typhimurium* and *E. coli*. Catalase
and pyruvate both act to degrade peroxides, and the effectiveness of these
agents in allowing for the recovery of injured cells suggests that the normal
abilities of uninjured cells to destroy peroxides are affected during injury.
The inability of heat-damaged *E. coli* cells to grow as well when surface-
plated as they did when pour-plated with the same medium *(62)* may be
explained by the loss of capacity to destroy peroxides.

Special plating procedures have been found by Speck *et al. (122)* and
Hartman *et al. (64)* to allow for recovery from injury and subsequent
enumeration in essentially one step. The procedures consist of using the agar
overlay plating technique with one layer consisting of TSA onto which are
plated the stressed organisms. Following a 1 to 2 hr incubation at 25 C for
recovery, the TSA layer is overlaid with VRBA followed by an additional incubation at 35 C for 24 hr. The overlay method of Hartman *et al.* involved
the use of a modified VRBA. The principle involved in the overlay technique
could be extended to other selective media, of course. An overlay technique
has been recommended for the recovery of coliforms. By this method, coliforms are plated with TSA and incubated at 35 C for 2 hr followed by an
overlay of VRBA *(103)*.

In his comparison of 18 plating media and 7 enrichment broths to recover
heat-stressed *Vibrio parahaemolyticus,* Beuchat *(16)* found the two most
efficient plating media to be water blue-alizarin yellow agar and arabinose-
ammonium-sulphate-cholate agar, while arabinose-ethyl violet broth was
found to be the most suitable enrichment broth.

RECENT DEVELOPMENTS—SEROLOGIC

The immunodiffusion test was the first immunologic technique to be employed to detect the presence of microbial products in foods on a routine
basis. This technique is employed either as a single or double-diffusion
method to detect the presence of staphylococcal enterotoxins. The detection
of botulinal and *C. perfringens* toxins can be achieved also by use of immunodiffusion methods. Of more recent use in food microbiology are the
fluorescent antibody (FA) and radioimmunoassay (RIA) techniques. The
relative sensitivity of FA and RIA is compared in Table 4-3 to other
serologic techniques including reversed passive hemagglutination and

TABLE 4-3. Comparison of serologic methods for detecting in foods toxins of food poisoning bacteria.

Methods	Toxins	Incubation time for test	Sensitivity	Reference
Single gel diffusion	S. aureus enterotoxin B	20–24 hr	2 μg/ml	(138)
Single gel diffusion	C. perfringens enterotoxin	24 hr	0.9 μg/ml	(51)
Micro capillary agar-gel diffusion	C. botulinum Types A, B, E	3–24 hr	100 mouse MLD/ml	(92)
Capillary tube	S. aureus enterotoxins	<1 to 24 hr	1–2 μg/ml	(47,49)
Microslide double diffusion	S. aureus enterotoxins A & B	24–72 hr	0.1 μg/ml	(22)
Microslide double diffusion	C. perfringens Type A toxin	24 hr	0.5 μg/ml	(51)
Hemagglutination-inhibition	S. aureus enterotoxin B	3–4 hr	0.0013 μg/0.1 ml	(79)
Hemagglutination-inhibition	C. botulinum Types A & B toxins	4–18 hr	0.15–0.46 mouse LD$_{50}$/0.1 ml	(78)
Reversed passive hemagglutination	S. aureus enterotoxin B	2 hr	0.0015 μg/ml	(119)
Reversed passive hemagglutination	C. perfringens Type A toxin	2 hr	0.001 μg/ml	(51)
Electroimmunodiffusion	C. perfringens enterotoxin	24–36 hr	0.01 μg	(38)
Electroimmunodiffusion	C. botulinum Type A toxin	2 hr	14 mouse LD$_{50}$/0.1 ml	(93)
Electroimmunodiffusion	Botulinal toxins	24 hr	3.7–5.6 mouse LD$_{50}$/0.1 ml	(136)
Fluorescent antibody	S. aureus enterotoxin B	4–5 hr	ca. 0.05 μg/ml	(50)
Radioimmunoassay	S. aureus enterotoxin A	3–4 hr	0.01 μg	(30)
Radioimmunoassay	S. aureus enterotoxin B	4 hr	1–5 ng/ml	(80)

hemagglutination-inhibition in their detection of some food-borne bacterial toxins. The FA and RIA techniques are discussed further below.

Fluorescent Antibody

This technique has had extensive use in clinical microbiology since its development in 1942, and from all indications it will continue to find greater application in food microbiology. By this technique, an antibody to a given antigen is made fluorescent by coupling it to a fluorescent compound, and when the antibody reacts with its antigen, the antigen-antibody complex emits fluorescence and can be detected by the use of a fluorescent microscope. The fluorescent markers used are rhodamine B, fluorescein isocyanate, and fluorescein isothiocynate, with the latter being used most widely. The FA technique can be carried out by use of either of two basic methods. The direct method employs antigen and specific antibody to which is coupled the fluorescent compound (antigen coated by specific antibody with fluorescent label). With the indirect method, the homologous antibody is not coupled with the fluorescent label, but instead an antibody to this antibody is prepared and coupled (antigen coated by homologous antibody which is in turn coated by antibody to the homologous antibody bearing the fluorescent label). In the indirect method, the labeled compound detects the presence of the homologous antibody, while in the direct method it detects the presence of the antigen. The use of the indirect method eliminates the need to prepare FA for each organism of interest. The FA technique obviates the necessity of pure culture isolations of salmonellae if H antisera are employed. A commonly employed conjugate is polyvalent salmonellae OH globulin labeled with fluorescein isothiocyanate with somatic groups A to Z represented. Because of the cross-reactivity of salmonellae antisera with other closely related organisms (e.g., *Arizona, Citrobacter, E. coli*), false positive results are to be expected when naturally contaminated foods are examined. The early history and development of the FA technique for clinical microbiology has been reviewed by Cherry and Moody *(24)* and for food applications by Ayres *(7)* and Goepfert and Insalata *(55)*.

The application of FA to the examination of foods for organisms such as salmonellae is made after cultural enrichment procedures have been carried out. The general outline of a slide method is as follows. Smears are made on glass slides, air dried, fixed with an appropriate fixative such as Kirkpatrick's, rinsed with ethanol, and air dried. To the dried slides is added the conjugate (labeled antibody) followed by incubation in a moist chamber to allow antigen-antibody reaction to occur. The unbound conjugate is washed off, the slide rinsed in distilled water, air dried, and mounted with buffered saline (pH 7.5–9.5) containing glycerol. The smears are next examined under a fluorescence microscope equipped with appropriate exciter and barrier filters. Positive cells display a yellow-green fluorescence, and the degree of fluorescence is scored from 1+ to 4+ when manually done.

The first successful use of the FA technique for the detection of food-borne organisms was made by Russian workers who employed the technique to detect salmonellae in milk *(5)*. The technique has been employed successfully to detect the presence of salmonellae in a large number of different types of foods and related products (Table 4-4). The procedure has been used also to monitor food processing plant utensils and equipment. While most investigators have obtained some false positive results with FA, the incidence of false negatives has been lower. Of the false negatives obtained by Laramore and Moritz, 68% were from meatmeal samples. Eleven percent of meatmeal samples examined by Thomason *et al. (132)* yielded false negative results. Just why this product is so conspicuous in this regard is not entirely clear at this time. With the exception of this product, the false negative findings have been extremely low, thus, making FA an excellent screening test for salmonellae in foods. The adoption of the direct FA method in 1975 by AOAC as an official method is limited to its use as a screening method. The generally low percentage of false positive results is not significant in a screening test as long as false negative results can be avoided. The most obvious reason for false positive results is cross-reaction of salmonellae antibodies with related organisms as noted above.

More recently, the FA direct slide technique first described by Insalata *et al. (73)* has been adapted to an automated method *(98, 132)*. By the automated method, slides are prepared by machine at the rate of 120/hr and machine-read at the rate of 360/hr. Both groups who have evaluated this procedure to date find that at least 10^6 salmonellae/ml of culture are necessary to obtain fluorescence readings within useful ranges. The au-

TABLE 4-4.   The application of the fluorescent antibody (FA) technique to the detection of salmonellae in foods and related products.

Foods examined	No. of samples	FA Method Used	% False Negatives	% False Positives	Reference
Meats	286	Indirect	0	14.7	*(52)*
Egg products	20	Indirect	0	0	*(60)*
Various foods (656 raw beef)	706	Direct	6.7	7	*(53)*
Dried foods & products	420	Indirect	0	0	*(118)*
Various inoculated foods	48	Direct	0	0	*(72)*
Various foods	3,991	Direct	<0.1	<0.2	*(116)*
Animal feed & ingredients	1,013	Direct	2.2	5.7	*(85)*
Various foods (39)	894	Direct	0	0	*(43)*
Various foods (7)	422	Indirect	0	7.3	*(56)*
Variety of human foods & animal feeds	65	Direct slide	0	4.6–14.8	*(73)*
Food products	*ca.*4,000	Direct	<1	7	*(44)*
Meatmeal	100	Auto Direct	11.1	1.1	*(132)*
Powdered eggs & candy	201	Auto Direct	0	5.3–6.6	*(132)*
Frog legs & dried products	283	Auto Direct	0	4	*(98)*

tomated procedure was found to correlate well with cultural methods on samples of milk powder, dried yeasts, and frog legs *(98)*. While the method gave excellent results on samples of powdered eggs and chocolate candy, false negative results were obtained on meatmeal and a high rate of false positives on samples of poultry products and sausage *(132)*. At least one group of workers believe that the automated system shows potential for screening of samples for salmonellae and that all positives should be confirmed by manual methods *(98)*.

Although the FA technique has had its widest food application in the detection of salmonellae, it has been used to trace bacteria associated with plants and to identify clostridia including *C. perfringens* and *C. botulinum*. It can be applied to any organism to which an antibody can be prepared. Among the many advantages of the procedure for detecting salmonellae in foods are the following: (1) FA is more sensitive than cultural methods, (2) FA is more rapid; time for results can be reduced from around 5 days to 18–24 hr, (3) reduction in time for negative tests on food products makes it possible to free foods earlier for distribution, (4) larger numbers of samples can be analyzed making for increased sampling of food products, and (5) the qualitative determination of specific salmonellae serotypes is possible for FA if desired for epidemiologic reasons.

Radioimmunoassay

This technique consists of adding a radioactive label to an antigen, allowing the labeled antigen to react with its specific antibody, and measuring the amount of antigen that combined with the antibody by use of a counter to measure radioactivity. Solid-phase RIA refers to methods that employ solid materials or surfaces onto which a monolayer of antibody molecules binds electrostatically. The solid materials used include polypropylene, polystyrene, bromacetylcellulose, and so on. The ability of antibody-coated polymers to bind specifically with radioactive tracer antigens is essential to the basic principle of solid-phase RIA *(23)*. When the free-labeled antigen is washed out, the radioactivity measurements are quantitative. The label used by many workers is ^{125}I.

Johnson *et al. (80)* developed a solid-phase RIA procedure for the determination of *S. aureus* enterotoxin B and found the procedure to be 5 to 20 times more sensitive than the immunodiffusion technique. These workers found the sensitivity of the test to be in the 1–5 ng range employing polystyrene and counting radioactivity with an integral counter. Collins *et al. (29)* employed RIA for enterotoxin B with the concentrated antibody coupled with bromacetylcellulose. Their findings indicated the procedure to be 100-fold more sensitive than immunodiffusion and to be reliable at an enterotoxin level of 0.01 μg/ml. Staphylococcal enterotoxin A was extracted from a variety of foods including ham, milk products, crab meat, etc., by Collins *et al. (30)* and measured by RIA all within 3–4 hr. These investigators agreed with earlier workers that the method was highly sensitive and useful to 0.001 μg/ml and quantitatively reliable to 0.01 μg/ml of enterotoxin A (see Table 4-3 and Crowther and Holbrook (33) for comparison of RIA sensitivity to other serologic methods).

The RIA technique lends itself to the examination of foods for biological hazards other than staphylococcal enterotoxins. It may find application in the measurement of a large number of toxic substances of microbial origin including gram negative endotoxins, fungal toxins, and the toxins of other gram positive bacteria. The detection and identification of bacterial cells within 8 to 10 min was achieved by Strange et al. (127) by use of ^{125}I-labeled homologous antibody filtered and washed on a millipore membrane. By this method, these workers were able to detect as few as 500 to 1,000 vegetative cells or endospores of pure culture. Multibacterial species were detected in one operation when mixtures of homologous antibodies were used (128). The detection and identification of low numbers of bacterial cells in an hr was achieved by Benbough and Martin (14) employing either the direct or indirect radiolabeling technique. Use of the indirect technique required only one labeled globulin to select bacterial species from a mixed sample, while with the direct technique a labeled antibody to each organism is needed. These investigators were able to detect $10^3 - 10^4$ cells/ml of B. subtilis spores, Serratia marcescens, and Francisella by use of an indirect technique.

RECENT DEVELOPMENTS—CHEMICAL/PHYSICAL METHODS

Heat-Stable Nuclease

The growth of S. aureus to significant numbers in a food can be determined by examining the food for the presence of heat-stable nuclease (DNase). This is possible because of the high correlation between the production of coagulase and heat-stable nuclease by S. aureus strains, especially enterotoxin producers. In one study, 232 of 250 (93%) enterotoxigenic strains produced coagulase while 242 or 95% produced heat-stable nuclease (137).

The examination of foods for heat-stable nuclease as an indirect test for S. aureus was first carried out by Chesbro and Auborn (25) employing a spectrophotometric method for nuclease determination. These investigators showed that as the numbers of S. aureus increased in ham sandwiches there was also an increase in the amount of extractable heat-stable nuclease of staphylococcal origin. They suggested that the presence of 0.34 unit of nuclease indicated certain staphylococcal growth and that at this nuclease level it was unlikely that enough enterotoxin was present to cause staphylococcal food poisoning symtoms. The 0.34 unit was shown to correspond to 9.5 $\times$ 10^{-3} μg of enterotoxin by S. aureus 234. The reliability of the heat-stable nuclease assessment as an indicator of S. aureus growth has been shown by others. While *S. epidermidis* and some micrococci produce DNase, it is not as stable to heating as is that produced by *S. aureus (84)*. The heat-stable nuclease will withstand boiling for 15 min. It has been shown to have a D value (D_{130}) of 16.6 in BHI broth at pH 8.2, and a z value of 51 *(41)*. The enzyme is produced under all conditions that permit cell growth *(25)* and the greatest amount is present in culture supernatants.

The fastest way to detect heat-stable nuclease in a food product is by use of the microslide method described by Lachica *et al. (82, 83)*. This method consists of combining DNA and toluidine blue 0 in a buffered salts solution with 1 percent agar. To a microscope slide, 3 ml of the molten DNA-dye preparation are layered. Small wells are cut in the agar layer and particles of food (*ca.* 5 mg) are added to the wells. The inoculated wells are covered, incubated at 37 C for 3 hr, and read for the appearance of a bright pink halo around the food particles indicating reaction of DNase with DNA. Heat-stable nuclease is detected by heating food samples at 97 C for 15 min before adding them to slide wells. These investigators found the rapid technique to be reliable on inoculated beef and pork samples, and Tatini *et al. (129)* found the technique to be reliable on naturally contaminated products. The latter authors assessed heat-stable nuclease and *S. aureus* growth and enterotoxin production in broth, milk products, ground beef, and bologna. They also followed the production of nuclease in Genoa sausage and during the curing and smoking of sausage relative to the efficacy of heat-stable nuclease to assess the safety of these products and found the indirect test to be a reliable product indicator. A high statistical correlation has been found between heat-stable nuclease and *S. aureus* growth in Cheddar, Colby, and brick cheeses *(32)*.

Among the advantages of testing for heat-stable nuclease as an indicator of *S. aureus* growth and activity are the following: (1) because of its heat-stable nature, the enzyme will persist even if the bacterial cells are destroyed by heat, chemicals, bacteriophage, or if they are induced to L-forms; (2) the heat-stable nuclease can be detected faster than enterotoxin (about 3 hr versus several days); (3) the nuclease appears to be produced by enterotoxigenic cells before enterotoxins appear (see Figure 4-3), (4) the nuclease is detectable in unconcentrated cultures of food specimen while enterotoxin detection requires concentrated samples, and (5) the nuclease of concern is stable to heat as are the enterotoxins.

Gram Negative Endotoxins

Gram negative bacteria are characterized by their production of endotoxins which consist of a lipopolysaccharide (LPS) layer of the cell envelope. The LPS material is pyrogenic and responsible for many of the symptoms that accompany infections caused by gram negative bacteria. While the gram negative flora of foods can be determined by plating procedures employing bile salts-containing media, this method depends

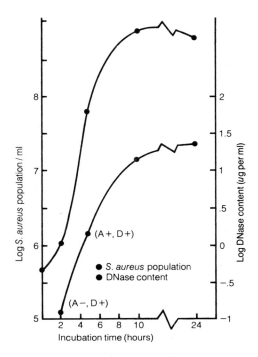

FIGURE 4-3.   Growth of *S. aureus* (196E) and production of DNase and enterotoxins in Brain Heart Infusion Broth at 37°C. DNase and enterotoxin D were detectable within 2 hr at a population of 2×10^6 whereas enterotoxin A was detected after 4 hr at higher cell populations. DNase was detectable in unconcentrated cultures, enterotoxins in 50-fold concentrates. [Tatini *et al.*, *129*.] Copyright © 1975, Institute of Food Technologists.

upon the cfu capacity of the cells under the cultural and incubation conditions employed. A 24-hr incubation period is necessary for results, with longer times being required for psychrotrophic types. The use of the *Limulus* lysate test to measure the endotoxins of gram negative bacteria in foods can be carried out in a shorter period of time with only a 1-hr incubation required. The use of this method is applicable to foods where large numbers of gram negative bacteria may exist.

The *Limulus* lysate test employs a lysate protein obtained from the blood (actually haemolymph) cells (amoebocytes) of the horseshoe crab *(Limulus polyphemous)*. This lysate protein is the most sensitive substance known for endotoxins. The lysate test for endotoxins consists of adding aliquots of food suspensions or other test materials to small quantities of a *Limulus* lysate preparation followed by incubation at 37 C for 1 hr. The presence of endotoxins causes gel formation of the lysate material. While most applications of this method have been made in clinical microbiology, some food applications have been made. The first food application was the use of this method for detecting the microbial spoilage of ground beef. It was found that endotoxin titers increased in proportion to the viable counts of gram negative bacteria. Since the normal spoilage of refrigerated fresh meats is caused by gram negative bacteria, the lysate test was found to be a good rapid indicator of the total numbers of gram negative bacteria (76, 77). The *Limulus* test was found by Terplan *et al.* (130) to be suitable for the rapid evaluation of the hygienic quality of milk relative to the detection of coliforms before and after pasteurization. These investigators found the *Limulus* test to be a good and rapid method for measuring the amount of endotoxin in raw and pasteurized liquid milk and, consequently, a good method for determining the history of a milk product relative to its content of gram negative bacteria. Since both viable and nonviable gram negative bacteria are detected by the *Limulus* test, a simultaneous plating is necessary in order to determine the number of viable or cfu's. The value of the *Limulus* test lies in the speed at which results can be obtained. Foods that have high *Limulus* titers may be candidates for further testing by other methods, while those that have low titers may be placed immediately into categories of lower risk relative to numbers of gram negative bacteria.

Radiometry

The radiometric detection of microorganisms is based upon the incorporation of a ^{14}C-labeled metabolite in a growth medium so that when the organisms utilize this metabolite, $^{14}CO_2$ is released and measured by use of a radioactivity counter. For organisms that utilize glucose, ^{14}C-glucose is usually employed. For those that cannot utilize this compound, others such as ^{14}C-formate or ^{14}C-glutamate are used. The overall procedure consists of using capped 50 ml serum vials to which are added anywhere from 12 to 36 ml of medium containing the labeled metabolite. The vials are made either aerobic or anaerobic by sparging with appropriate gases, and are then inoculated. Following incubation, the headspace is tested periodically for the presence of $^{14}CO_2$.

The use of radiometry to detect the presence of microorganisms was first suggested by Levin *et al.* (89). The technique has been largely confined to clinical microbiology, but some applications have been made to foods and water. One of the earliest nonclinical uses was the detection of coliforms in water and sewage (117). These investigators showed that a direct relationship existed between coliform numbers and the amount of $^{14}CO_2$ produced by the organisms. The experimental detection of *S. aureus, S. typhimurium,* and spores of P.A. 3679 and *C. botulinum* in beef loaf was studied by Previte (106). The inocula employed ranged from about 10^4 to 10^6/ml of medium and the detection time ranged from 2 hr for *S. typhimurium* to 5–6 hr for *C. botulinum* spores. For these studies, 0.0139 μCi of ^{14}C-glucose/ml of tryptic soy broth was employed. In another study, Lampi *et al.* (84) found that 1 cell/ml of *S. typhimurium* or *S. aureus* could be detected by a radiometric method in 9 hr. For 10^4 cells, 3 to 4 hr were required. With respect to spores, a level of 90 of P.A. 3679 was detected in 11 hr while 10^4 were detectable within 7 hr. These and other investigators have shown that spores required 3–4 hr longer for detection than vegetative cells. From the findings of Lampi *et al.*, the radiometric detection procedure could be employed as a screening procedure for foods containing high numbers of organisms, for such foods produced results by this method within 5–6 hr while those with lower numbers required longer times.

The detection of nonfermenters of glucose by this method is possible when metabolites such as labeled formate and/or glutamate are used *(107)*. These investigators further showed that a large number of food-borne organisms can be detected by this method in 1 to 6 hr. The radiometric detection of 1 to 10 coliforms in water within 6 hr was achieved by Bachrach and Bachrach *(8)* by employing ^{14}C-lactose with incubation at 37 C in a liquid medium. It is conceivable that a differentiation can be made between fecal *E. coli* and total coliforms by employing 45.5 C incubation along with 37 C incubation.

More recently, an automated radiometric system has been developed and used, largely in clinical laboratories. The automated system can be set up to make periodic sampling of inoculated vials with recordings of the results. Some workers have found the automated method to be comparable to conventional methods for culturing blood *(36)*, while others have found it to be a bit less effective *(111)*. Heat-stressed spores of *C. botulinum* were detected in an automated system in 14 hr compared to 34 hr for first detection by the MPN method *(42)*. Unstressed *C. botulinum* spores required 7 to 15 hr while spores of *C. sporogenes* required only 3–5 hr by this method. The potential use of radiometry for field and surface samples is suggested by a procedure employed by Schrot *et al. (115)*. By this method, membrane filter samples are collected and moistened with a small quantity of labeled medium in a closed container. The evolved $^{14}CO_2$ is trapped by $Ba(OH)_2$-moistened filter pads which are assayed later by a radioactivity counter.

Microcalorimetry

Microcalorimetry is the study of small heat changes, the measurement of the enthalpy change involved in the breakdown of growth substrates. The heat production that is measured is closely related to the cell's catabolic activities *(46)*. It has been applied to the study of spoilage in canned foods *(114)*, to differentiate between the *Enterobacteriaceae*, and to detect the presence of *S. aureus*. As a means of identifying bacteria, Boling *et al. (17)* were able to differentiate between 17 species of 10 genera of *Enterobacteriaceae* by inoculating approximately 500 cells into BHI broth and recording the temperature changes over an 8 to 14 hr incubation period. The thermograms produced were distinctive for each organism. In detecting *S. aureus*, Lampi *et al. (84)* achieved results in 2 hr using an initial number of $10^7 - 10^8$ cells/ml, and in 12–13 hr when only two cells/ml were used. Russell *et al. (113)* examined over 250 cultures representing 24 genera and found that most of the organisms typically produced maximum outputs of 40–60 μcal/(sec)(ml) and returned to baseline in 5–7 hr thereafter. Some profiles developed within 3 hr while others required up to 14 hr. A monitoring use of microcalorimetry is suggested by the work of Beezer *et al. (12)* who used flow microcalorimetry to determine the viability of recovered frozen cells of *S. cerevisiae* within 3 hr after thawing.

The determination of microcalorimetric data requires the use of specialized calorimeters. The two general types most commonly used are adiabatic calorimeters and thermal fluxmeters. These apparati are so designed that the small amounts of heat produced by a respiring culture are detected by strip-chart recorders. One of the most widely used microcalorimeters for microbiological work is the Calvet instrument. This instrument is sensitive to a heat flow of 0.01 cal/hr from a 10 ml sample *(46)*. Considerably more work needs to be done before this technique can be reliably applied to the detection of microorganisms in foods on a routine basis, for it has been shown that the thermograms produced by bacteria are affected by various growth parameters such as nutrients, Eh, and so forth *(13, 95)*.

Electrical Impedance

Impedance is the apparent resistance in an electric circuit to the flow of alternating current corresponding to the actual electrical resistance to a direct current. When microorganisms grow in culture media, they metabolize substrates of low conductivity into products of higher conductivity and thereby increase the impedance of the media. When the impedance of broth cultures is measured, the curves that result resemble bacterial growth curves. The signals or curves are reproducible for species and strains, and mixed cultures can be identified by use of specific growth inhibitors. The technique has been shown capable of detecting as few as several hundred cells *(21)*, or 10^3 to 10^4 bacterial cells/ml within 2 hr *(135)*. It has been used to detect as few as 10^3 bacteria/ml in monitoring urine specimens for suspected bacteriuria *(139)*. The technique has been employed to monitor bacteria in frozen foods *(61)*. These authors assessed 200 pureed vegetable samples and found a 90–95% agreement between impedance measurements and plate count results relative to the presence of unacceptable levels of bacteria. The impedance analyses required 5 hr, and the method is reported to be applicable to other foods such as cream pies, ground meat, and so on.

RECENT DEVELOPMENTS—*IN VIVO* TECHNIQUES

Ligated Loop Techniques

These techniques are based on the fact that certain enterotoxins elicit fluid accumulation in the small intestines of susceptible animals. While they may be performed with a variety of animals, rabbits are most often employed.

Young rabbits 7 to 20 weeks old and weighing 1.2–2.0 kg are kept off food and water for a period of 24 hr, or off food for 48–72 hr with water ad libitum prior to surgery. Under local anaesthesia, a midline incision about 2 in. long is made just below the middle of the abdomen through the muscles and peritoneum in order to expose the small intestines (34). A section of the intestine midway between its upper and lower ends or just above the appendix is tied with silk or other suitable ligatures in 8–12 cm segments with intervening sections of at least 1 cm or more. Up to 6 sections may be prepared by single or double ties.

Meanwhile, the specimen or culture to be tested is prepared, suspended in sterile saline, and injected intraluminally into the ligated segments. A common inoculum size is 1 ml, although smaller and larger doses may be used. Different doses of test material may be injected into adjacent loops or into loops separated by a blank loop or by a sham (inoculated with saline). Following injection, the abdomen is closed with surgical thread and the animal is allowed to recover from anaesthesia. The recovered animal may be kept off of food and water for an additional 18–24 hr period, or water or feed or both may be allowed. With ligatures intact, the animals may not survive beyond 30–36 hr (18).

To assess the effect of the materials previously injected into ligated loops, the animal is sacrificed and the loops are examined and measured for fluid accumulation. The fluid may be aspirated and measured. The reaction can be quantitated by measuring loop fluid volume to loop length ratios (18), or by determining the ratio of fluid volume secreted/mg dry weight intestine (91). The appearance of a ligated rabbit ileum 24 hr after injection of a C. perfringens culture is presented in Figure 4-4. The minimum amount of C. perfrin-

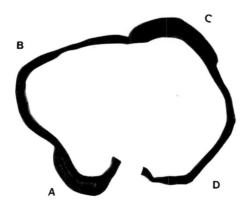

FIGURE 4-4. Gross appearance of the ligated rabbit ileum 24 hr after injection of 2 ml of cultures of Clostridium perfringens grown 4 hr at 37 C in Skim Milk. Loop A, strain NCTC 8798, 8 ml of fluid in loop; loop B, 2 ml of sterile milk, negative loop; loop C, strain T-65, 10 ml of fluid; loop D, strain 6867, negative loop (Redrawn from Duncan *et al.*, 37.) Copyright © 1968, American Society for Microbiology.

gens enterotoxin necessary to produce a loop reaction has been reported variously to be 28 to 40 μg and as high as 125 μg of toxin by the standard loop technique. The 90 min loop technique has been found to respond to as little as 6.25 μg and the standard technique to 29 μg of toxin *(51)*.

This technique was developed originally to study the mode of action of the cholera organism in producing the disease *(34)*. Recently, it has been employed widely in studies on the virulence and pathogenesis of food-borne pathogens including *B. cereus, C. perfringens, E. coli,* and *V. parahaemolyticus*.

Suckling Mouse Test and Others

While this technique has been used largely as an assay method for *E. coli* enterotoxins, it is based upon the same principle as the ligated loop technique and can be employed perhaps for the same enterotoxins. It was developed by Dean *et al. (35)* and has been used by a large number of investigators who have found it to be reliable and reproducible.

Newborn suckling mice, 1–5 days old, are separated from their mothers and divided into groups of three. Each is inoculated intragastrically with 0.1 ml of a toxin preparation containing about 2 drops of a 2% Evans blue solution. The mice are incubated for 3–4 hr at 25 C and sacrificed by surgical dislocation. The abdomen is opened and the entire intestine (not stomach) is removed with forceps. The intestines from 3 mice are pooled and weighed (those without the blue dye are rejected), and the ratio of gut weight to remaining carcass weight is determined. Gianella *(54)* found the following gut weight/carcass weight ratios for *E. coli* enterotoxins: <0.074 = negative test; 0.075–0.082 = intermediate (should be retested); and >0.083 = positive test. This investigator found the day-to-day variability among various *E. coli* strains to range from 10.5–15.7% and *ca.* 9% for replicate tests with the same strain. This test has been found to correlate well with the rabbit ileal loop assay technique for *E. coli* enterotoxins.

Among other *in vivo* tests is the infant rabbit test which has been used as an assay for *E. coli* enterotoxins. By this method, 1–5 ml of culture filtrate is administered intragastrically to 6–10 day old rabbits. A positive test is indicated by diarrhea after 6–8 hr. Among the other *in vivo* techniques used to assay *E. coli* enterotoxins is one employing adrenal cells in tissue culture and another employing hamster ovary cells. The mouse lethality test long used to detect for and to quantitate botulinal toxins has been used more recently as an assay for *C. perfringens* enterotoxin. A MLD for mice of about 3 μg of purified, enterotoxin-administered i.v. has been reported. In one study the mouse lethal dose was found to be 6.8 μg/ml *(51)*. The oral administration of staphylococcal enterotoxins to human volunteers and to monkeys and kittens has been employed for many years as a means of detection and quantitation of emetic doses of these toxins (see Chapter 16 for minimal dose responses). The enterotoxin of *C. perfringens* has been evaluated by use of dogs with doses of toxin >100 μg being necessary for response. Guinea pig and rabbit intradermal skin tests have been employed to assess the toxins of *C. perfringens, B. cereus,* and *V. parahaemolyticus.* Erythemal dose for *C. perfringens* enterotoxin was found by Genigeorgis *et al. (51)* to be 1.25–2.50 μg/ml.

For more details on *in vivo* responses to toxins of food-borne organisms, see later chapters dealing with the food poisoning syndromes of relevant organisms, and references *51, 54, 91,* and *124.*

SUMMARY

The ability to make rapid and precise determinations of microorganisms in food products is critical to food microbiology. Faster methods of processing and distributing foods demand greater speed of microbiological analysis. Most traditional methods rely mainly on the capacity of food-borne microorganisms to form colonies when plated on suitable culture media. In addition to the several days required for results, metabolically injured cells may not develop, and a certain portion of the uninjured population may not develop because of the unfavorable parameters of the cultural conditions employed. The methods noted and discussed in this chapter are summarized in Table 4-5. The first seven have been established for various applications while the latter five are still experimental relative to use in food microbiology.

TABLE 4-5.   Summary of established and experimental methods of microbiological analysis of foods with respect to incubation time and capacity of method to detect cells or metabolic products.

Method	Viable cells	Nonviable cells	Products	Approx. time for results	Comments
Direct microscopic counts	+	+		minutes	
Standard plate counts	+			24–48 hr	
Most probable numbers	+			24–48 hr	
Dye reductions	+			4–8 hr	Use restricted to few products
Microslide double diffusion			+	24–72 hr	
Fluorescent antibody	+	+	+	4–5 hr	
Heat-stable nuclease			+	3 hr	For *S. aureus*
Radioimmunoassay	+	+	+	3–4 hr	
Endotoxins	+	+		1 hr	Only gram negatives detected
Radiometry	+			1–6 hr	
Microcalorimetry	+			5–7 hr	
Electrical impedance	+			2–5 hr	
Animal inoculations	+		+	1.5–24 hr	

Direct microscopic counts on food specimens, although rapid, are wrought with problems relative to the differentiation of cells from particulate matter. Dye reduction times are useful only within narrow limits, and the number of food products that lend themselves to this analysis is limited. In spite of the shortcomings of the traditional cultural methods, a vast amount of data exists relative to the use of these methods over many years in assessing and predicting the microbiological safety of numerous food products. The development of newer methods is designed not only to speed up microbial food analyses but also to improve the efficiency of analyses, both qualitative and quantitative. The adoption of steps to allow stressed cells to recover prior to culturing in or on selective media improves the efficiency of culture methods. The use of fluorescent antibody, radioimmunoassay, and other serologic techniques in the food microbiology laboratory not only increases the speed of identifying microorganisms and/or their products but extends detection to more sensitive levels. The *Limulus* lysate assay shows promise as a rapid method of assessing foods for the presence of living or dead gram negative bacteria. Greater speed and sensitivity of detection of microorganisms are provided by the techniques of radiometry and microcalorimetry, although more research is needed before these methods receive wider applications for foods. The use of electrical impedance measurements also shows promise as a rapid method for quantitating and identifying food-borne microorganisms. The measurement of heat-stable nuclease of *Staphylococcus aureus* provides a model for the rapid, specific, and sensitive detection of other important organisms or products in foods. The use of the ileal loop and related techniques, such as the suckling mouse test, make it possible to relate more accurately laboratory findings to actual human clinical conditions caused by food-borne pathogens.

REFERENCES

1. **Adams, G. H. and Z. J. Ordal.** 1976. Effects of thermal stress and reduced water activity on conidia of *Aspergillus parasiticus*. J. Food Sci. *41:* 547–550.

2. **Allwood, M. C. and A. D. Russell.** 1967. Mechanism of thermal injury in *Staphylococcus aureus*. I. Relationship between viability and leakage. Appl. Microbiol. *15:* 1266–1269.

3. **American Public Health Association.** 1971. *Standard Methods for the Examination of Dairy Products,* 13th Edition (APHA:N.Y.).

4. **APHIS.** 1974. *Microbiology Laboratory Guidebook* (U. S. D. A., D.C.).

5. **Arkhangel'skii, I. I. and V. M. Kartoshova.** 1962. Accelerated methods of detecting *Salmonella* in milk. Veterinariya *9:* 74–78.

6. **Assoc. Official Anal. Chem.** 1975. *Official Methods of Analysis,* 12th Edition.

7. **Ayres, J. C.** 1967. Use of fluorescent antibody for the rapid detection of enteric organisms in egg, poultry & meat products. Food Technol. *21:* 631–640.

8. **Bachrach, U. and Z. Bachrach.** 1974. Radiometric method for the detection of coliform organisms in water. Appl. Microbiol. *28:* 169–171.

9. **Baer, E. F., J. W. Messer, J. E. Leslie, and J. T. Peeler.** 1975. Direct plating method for enumeration of *Staphylococcus aureus:* Collaborative study. J. Assoc. Off. Anal. Chem. *58:* 1154–1158.

10. **Baird-Parker, A. C. and E. Davenport.** 1965. The effect of recovery medium on the isolation of *Staphylococcus aureus* after heat treatment and after the storage of frozen or dried cells. J. Appl. Bacteriol. *28:* 390–402.

11. **Barach, J. T., R. S. Flowers, and D. M. Adams.** 1975. Repair of heat-injured *Clostridium perfringens* spores during outgrowth. Appl. Microbiol. *30:* 873–875.

12. **Beezer, A. E., R. D. Newell, and H. J. V. Tyrrell.** 1976. Application of flow microcalorimetry to analytical problems: The preparation, storage and assay of frozen inocula of *Saccharomyces cerevisiae.* J. Appl. Bacteriol. *41:* 197–207.

13. **Belaich, A. and J. P. Belaich.** 1976. Microcalorimetric study of the anaerobic growth of *Escherichia coli:* Growth thermograms in a synthetic medium. J. Bacteriol. *125:* 14–18.

14. **Benbough, J. E. and K. L. Martin.** 1976. An indirect radio-labelled antibody staining technique for the rapid detection and identification of bacteria. J. Appl. Bacteriol. *41:* 47–58.

15. **Beuchat, L. R. and R. V. Lechowich.** 1968. Effect of salt concentration in the recovery medium on heat-injured *Streptococcus faecalis.* Appl. Microbiol. *16:* 772–776.

16. **Beuchat, L. R.** 1976. Survey of media for the resuscitation of heat-stressed *Vibrio parahaemolyticus.* J. Appl. Bacteriol, *40:* 53–60.

17. **Boling, E. A., G. C. Blanchard, and W. J. Russell.** 1973. Bacterial identifications by microcalorimetry. Nature *241:* 472–473.

18. **Burrows, W. and G. M. Musteikis.** 1966. Cholera infection and toxin in the rabbit ileal loop. J. Infect. Dis. *116:* 183–190.

19. **Busta, F. F. and J. J. Jezeski.** 1963. Effect of sodium chloride concentration in an agar medium on growth of heat-shocked *Staphylococcus aureus.* Appl. Microbiol. *11:* 404–407.

20. **Busta, F. F.** 1976. Practical implications of injured microorganisms in food. J. Milk Food Technol. *39:* 138–145.

21. **Cady, P.** 1975. Rapid automated bacterial identification by impedance measurements, pp. 73–99 IN: *New Approaches to the Identification of Microorganisms,* edited by C.-G. Hedén and T. Illéni (Wiley & Sons: N.Y.).

22. **Casman, E. P. and R. W. Bennett.** 1965. Detection of staphylococcal enterotoxin in food. Appl. Microbiol. *13:* 181–189.

23. **Catt, K. and G. W. Tregear.** 1967. Solid-phase radioimmunoassay in antibody-coated tubes. Science *158:* 1570–1572.

24. **Cherry, W. B. and M. D. Moody.** 1965. Fluorescent-antibody techniques in diagnostic bacteriology. Bacteriol. Rev. *29:* 222–250.

25. **Chesbro, W. R. and K. Auborn.** 1967. Enzymatic detection of the growth of *Staphylococcus aureus* in foods. Appl. Microbiol. *15:* 1150–1159.

26. **Chowdhury, M. S. U., D. B. Rowley, A. Anellis, and H. S. Levinson.** 1976. Influence of postirradiation incubation temperature on recovery of radiation-injured *Clostridium botulinum* 62A spores. Appl. Env. Microbiol. *32:* 172–178.

27. **Clark, C. W., L. D. Witter, and Z. J. Ordal.** 1968. Thermal injury and recovery of *Streptococcus faecalis.* Appl. Microbiol. *16:* 1764–1769.

28. **Clark, C. W. and Z. J. Ordal.** 1969. Thermal injury and recovery of *Salmonella typhimurium* and its effect on enumeration procedures. Appl. Microbiol. *18:* 332–336.

29. **Collins, W. S. II, J. F. Metzger, and A. D. Johnson.** 1972. A rapid solid phase radioimmunoassay for staphylococcal B enterotoxin. J. Immunol. *108:* 852–856.

30. **Collins, W. S., A. D. Johnson, J. F. Metzger, and R. W. Bennett.** 1973. Rapid solid-phase radioimmunoassay for staphylococcal enterotoxin A. Appl. Microbiol. *25:* 774–777.

31. **Collins-Thompson, D. L., A. Hurst, and B. Aris.** 1974. Comparison of selective media for the enumeration of sublethally heated food-poisoning strains of *Staphylococcus aureus.* Can. J. Microbiol. *20:* 1072–1075.

32. **Cords, B. R. and S. R. Tatini.** 1973. Applicability of heat-stable deoxyribonuclease assay for assessment of staphylococcal growth and the likely presence of enterotoxin in cheese. J. Dairy Sci. *56:* 1512–1519.

33. **Crowther, J. S. and R. Holbrook.** 1976. Trends in methods for detecting food-poisoning toxins produced by *Clostridium botulinum* and *Staphylococcus aureus,* pp. 215–225, IN: *Microbiology in Agriculture, Fisheries and Food,* edited by F. A. Skinner and J. G. Carr (Academic Press: N.Y.).

34. **De, S. N. and D. N. Chatterje.** 1953. An experimental study of the mechanism of action of *Vibrio cholerae* on the intestinal mucous membrane. J. Path. Bacteriol. *66:* 559–562.

35. **Dean, A. G., Y.-C. Ching, R. G. Williams, and L. B. Harden.** 1972. Test for *Escherichia coli* enterotoxin using infant mice: Application in a study of diarrhea in children in Honolulu. J. Infect. Dis. *125:* 407–411.

36. **DeBlanc, H. J., Jr., F. DeLand, and H. N. Wagner, Jr.** 1971. Automated radiometric detection of bacteria in 2,967 blood cultures. Appl. Microbiol. *22:* 846–849.

37. **Duncan, C. L., H. Sugiyama, and D. H. Strong.** 1968. Rabbit ileal loop response to strains of *Clostridium perfringens.* J. Bacteriol. *95:* 1560–1566.

38. **Duncan, C. L. and E. B. Somers.** 1972. Quantitation of *Clostridium perfringens* Type A enterotoxin by electroimmunodiffusion. Appl. Microbiol. *24:* 801–804.

39. **Duncan, C. L., R. G. Labbe, and R. R. Reich.** 1972. Germination of heat- and alkali-altered spores of *Clostridium perfringens* Type A by lysozyme and an initiation protein. J. Bacteriol. *109:* 550–559.

40. **Duitschaever, C. L. and D. C. Jordan.** 1974. Development of resistance to heat and sodium chloride in *Streptococcus faecium* recovering from thermal injury. J. Milk Food Technol. *37:* 382–386.

41. **Erickson, A. and R. H. Deibel.** 1973. Turbidimetric assay of staphylococcal nuclease. Appl. Microbiol. *25:* 337–341.

42. **Evancho, G. M., D. H. Ashton, and A. A. Zwarun.** 1974. Use of a radiometric technique for the rapid detection of growth of clostridial species. J. Food Sci. *39:* 77–79.

43. **Fantasia, L. D.** 1969. Accelerated immunofluorescence procedure for the detection of *Salmonella* in foods and animal by-products. Appl. Microbiol. *18:* 708–713.

44. **Fantasia, L. D., J. P. Schrade, J. F. Yager, and D. Debler.** 1975. Fluorescent antibody method for the detection of *Salmonella:* Development, evaluation, and collaborative study. J. Assoc. Off. Anal. Chem. *58:* 828–844.

45. **Food and Drug Administration, U.S.** 1976. *Bacteriological Analytical Manual for Foods* (AOAC: D.C.).

46. **Forrest, W. W.** 1972. Microcalorimetry. Methods in Microbiol. *6B:* 285–318.

47. **Fung, D. Y. C. and J. Wagner.** 1971. Capillary tube assay for staphylococcal enterotoxins A, B, and C. Appl. Microbiol. *21:* 559–561.

48. **Fung, D. Y. C. and L. L. VandenBosch.** 1975. Repair, growth, and enterotoxigenesis of *Staphylococcus aureus* S-6 injured by freeze-drying. J. Milk Food Technol. *38:* 212–218.

49. **Gandhi, N. R. and G. H. Richardson.** 1971. Capillary tube immunological assay for staphylococcal enterotoxins. Appl. Microbiol. *21:* 626–627.

50. **Genigeorgis, C. and W. W. Sadler.** 1966. Immunofluorescent detection of staphylococcal enterotoxin B. II. Detection in foods. J. Food Sci. *31:* 605–609.

51. **Genigeorgis, C., G. Sakaguchi, and H. Riemann.** 1973. Assay methods for *Clostridium perfringens* Type A enterotoxin. Appl. Microbiol. *26:* 111–115.

52. **Georgala, D. L. and M. Boothroyd.** 1964. A rapid immunofluorescence technique for detecting salmonellae in raw meat. J. Hyg. *62:* 319–326.

53. **Georgala, D. L., M. Boothroyd, and P. R. Hayes.** 1965. Further evaluation of a rapid immunofluorescence technique for detecting salmonellae in meat and poultry. J. Appl. Bacteriol. *28:* 421–425.

54. **Giannella, R. A.** 1976. Suckling mouse model for detection of heat-stable *Escherichia coli* enterotoxin: Characteristics of the model. Infect. Immunity *14:* 95–99.

55. **Goepfert, J. M. and N. F. Insalata.** 1969. Salmonellae and the fluorescent-antibody technique: A current evaluation. J. Milk Food Technol. *32:* 465–473.

56. **Goepfert, J. M., M. E. Mann, and R. Hicks.** 1970. One-day fluorescent-antibody procedure for detecting salmonellae in frozen and dried foods. Appl. Microbiol. *20:* 977–983.

57. **Gomez, R. F., A. J. Sinskey, R. Davies, and T. P. Labuza.** 1973. Minimal medium recovery of heated *Salmonella typhimurium* LT 2. J. Gen. Microbiol. *74:* 267–274.

58. **Gray, R. J. H., L. D. Witter, and Z. J. Ordal.** 1973. Characterization of milk thermal stress in *Pseudomonas fluorescens* and its repair. Appl. Microbiol. *26:* 78–85.

59. **Gunderson, M. F. and K. D. Rose.** 1948. Survival of bacteria in a precooked, fresh-frozen food. Food Res. *13:* 254–263.

60. **Haglund, J. R., J. C. Ayres, A. M. Paton, A. A. Kraft, and L. Y. Quinn.** 1964. Detection of *Salmonella* in eggs and egg products with fluorescent antibody. Appl. Microbiol. *12:* 447–450.

61. **Hardy, D., S. W. Dufour, and S. J. Kraeger.** 1975. Rapid detection of frozen food bacteria by automated impedance measurements. Proc. Inst. Food Technol.

62. **Harries, D. and A. D. Russell.** 1966. Revival of heat-damaged *Escherichia coli.* Experientia *22:* 803–804.

63. **Harris, N. D.** 1963. The influence of the recovery medium and the incubation temperature on the survival of damaged bacteria. J. Appl. Bacteriol. *26:* 387–397.

64. **Hartman, P. A., P. S. Hartman and W. W. Lanz.** 1975. Violet red bile 2 agar for stressed coliforms. Appl. Microbiol. *29:* 537–539.

65. **Hartsell, S. E.** 1951. The longevity and behavior of pathogenic bacteria in frozen foods: The influence of plating media. A. J. Pub. Hlth. *41:* 1072–1077.

66. **Hurst, A., A. Hughes, J. L. Beare-Rogers, and D. L. Collins-Thompson.** 1973. Physiological studies on the recovery of salt tolerance by *Staphylococcus aureus* after sublethal heating. J. Bacteriol. *116:* 901–907.

67. **Hurst, A., A. Hughes, D. L. Collins-Thompson, and B. G. Shah.** 1974. Relationship between loss of magnesium and loss of salt tolerance after sublethal heating of *Staphylococcus aureus.* Can. J. Microbiol. *20:* 1153–1158.

68. **Hurst, A., A. Hughes, M. Duckworth, and J. Baddiley.** 1975. Loss of D-alanine during sublethal heating of *Staphylococcus aureus* S-6 and magnesium binding during repair. J. Gen. Microbiol. *89:* 277–284.

69. **Hurst, A., G. S. Hendry, A. Hughes, and B. Paley.** 1976. Enumeration of sublethally heated staphylococci in some dried foods. Can. J. Microbiol. *22:* 677–683.

70. **International Commission on Microbiological Specifications for foods (ICMSF).** 1974. *Microorganisms in Foods* 2 (Univ. of Toronto Press: Canada).

71. **Iandolo, J. J. and Z. J. Ordal.** 1966. Repair of thermal injury of *Staphylococcus aureus.* J. Bacteriol. *91:* 134–142.

72. **Insalata, N. F., S. J. Schulte, and J. H. Berman.** 1967. Immunofluorescence technique for detection of salmonellae in various foods. Appl. Micriobiol. *15:* 1145–1149.

73. **Insalata, N. F., C. W. Mahnke, and W. G. Dunlap.** 1972. Rapid, direct fluorescent-antibody method for the detection of salmonellae in food and feeds. Appl. Microbiol. *24:* 645–649.

74. **Jackson, H. and M. Woodbine.** 1963. The effect of sublethal heat treatment on the growth of *Staphylococcus aureus.* J. Appl. Bacteriol. *26:* 152–158.

75. **Janssen, D. W. and F. F. Busta.** 1973. Influence of milk components on the injury, repair of injury, and death of *Salmonella anatum* cells subjected to freezing and thawing. Appl. Microbiol. *26:* 725–732.

76. **Jay, J. M.** 1974. Use of the *Limulus* lysate endotoxin test to assess the microbial quality of ground beef. Bacteriol. Proc., p. 13.

77. **Jay, J. M.** 1977. The *Limulus* lysate endotoxin assay as a test of microbial quality of ground beef. J. Appl. Bacteriol. *43:* 99–109.

78. **Johnson, H. M., K. Brenner, R. Angelotti, and H. E. Hall.** 1966. Serological studies of Types A, B, and E botulinal toxins by passive hemagglutination and bentonite flocculation. J. Bacteriol. *91:* 967–974.

79. **Johnson, H. M., H. E. Hall, and M. Simon.** 1967. Enterotoxin B: Serological assay in cultures by passive hemagglutination. Appl. Microbiol. *15:* 815–818.

80. **Johnson, H. M., J. A. Bukovic, P. E. Kauffman, and J. T. Peeler.** 1971. Staphylococcal enterotoxin B: Solid-phase radioimmunoassay. Appl. Microbiol. *22:* 837–841.

81. **Lachica, R. V. F., P. D. Hoeprich, and C. Genigeorgis.** 1971a. Nuclease production and lysostaphin susceptibility of *Staphylococcus aureus* and other catalase-positive cocci. Appl. Microbiol. *21:* 823–826.

82. **Lachica, R. V. F., C. Genigeorgis, and P. D. Hoeprich.** 1971b. Metachromatic agar-diffusion methods for detecting staphylococcal nuclease activity. Appl. Microbiol. *21:* 585–587.

83. **Lachica, R. V. F., P. D. Hoeprich, and C. Genigeorgis.** 1972. Metachromatic agar-diffusion microslide technique for detecting staphylococcal nuclease in foods. Appl. Microbiol. *23:* 168–169.

84. **Lampi, R. A., D. A. Mikelson, D. B. Rowley, J. J. Previte, and R. E. Wells.** 1974. Radiometry and microcarolimetry—techniques for the rapid detection of foodborne microorganisms. Food Technol. *28:* No. 10, 52–55.

85. **Laramore, C. R. and C. W. Moritz.** 1969. Fluorescent-antibody technique in detection of salmonellae in animal feed and feed ingredients. Appl. Microbiol. *17:* 352–354.

86. **Lee, A. C. and J. M. Goepfert.** 1975. Influence of selected solutes on thermally induced death and injury of *Salmonella typhimurium*. J. Milk Food Technol. *38:* 195–200.

87. **Levin, G. V., V. R. Harrison, and W. C. Hess.** 1956. Preliminary report on a one-hour presumptive test for coliform organisms. J. Amer. Water Works Assoc. *48:* 75–80.

88. **Licciardello, J. J., J. T. R. Nickerson, and S. A. Goldblith.** 1970. Recovery of salmonellae from irradiated and unirradiated foods. J. Food Sci. *35:* 620–624.

89. **Martin, S. E., R. S. Flowers, and Z. J. Ordal.** 1976. Catalase: Its effect on microbial enumeration. Appl. Env. Microbiol. *32:* 731–734.

90. **Maxcy, R. B.** 1973. Condition of coliform organisms influencing recovery of subcultures on selective media. J. Milk Food Technol. *36:* 414–416.

91. **Mehlman, I. J., M. Fishbein, S. L. Gorbach, A. C. Sanders, E. L. Eide, and J. C. Olson, Jr.** 1976. Pathogenicity of *Escherichia coli* recovered from food. J. Assoc. Off. Anal. Chem. *59:* 67–80.

92. **Mestrandrea, L. W.** 1974. Rapid detection of *Clostridium botulinum* toxin by capillary tube diffusion. Appl. Microbiol. *27:* 1017–1022.

93. **Miller, C. A. and A. W. Anderson.** 1971. Rapid detection and quantitative estimation of Type A botulinum toxin by electroimmunodiffusion. Infec. Immun. *4:* 126–129.

94. **Moats, W. A., R. Dabbah, and V. M. Edwards.** 1971. Survival of *Salmonella anatum* heated in various media. Appl. Microbiol. *21:* 476–481.

95. **Monk, P. and I. Wadso.** 1975. The use of microcalorimetry for bacterial classification. J. Appl. Bacteriol. *38:* 71–74.

96. **Moss, C. W. and M. L. Speck.** 1966. Identification of nutritional components in trypticase responsible for recovery of *Escherichia coli* injured by freezing. J. Bacteriol. *91:* 1098–1104.

97. **Mossel, D. A. A. and M. A. Ratto.** 1970. Rapid detection of sublethally impaired cells of Enterobacteriaceae in dried foods. Appl. Microbiol. *20:* 273–275.

98. **Munson,T. E., J. P. Schrade, N. B. Bisciello, Jr., L. D. Fantasia, W. H. Hartung, and J. J. O'Connor.** 1976. Evaluation of an automated fluorescent antibody procedure for detection of *Salmonella* in foods and feeds. Appl. Env. Microbiol. *31:* 514–521.

99. **Nakamura, M. and D. A. Dawson.** 1962. Role of suspending and recovery media in the survival of frozen *Shigella sonnei*. Appl. Microbiol. *10:* 40–43.

100. **National Academy of Sciences.** 1971, *Reference Methods for the Microbiological Examination of Foods* (Nat'l Acad. Sci.: D.C.).

101. **Nelson, F. E.** 1943. Factors which influence the growth of heat-treated bacteria. I. A comparison of four agar media. J. Bacteriol. *45:* 395–403.

102. **Nelson, F. E.** 1972. Plating medium pH as a factor in apparent survival of sublethally stressed yeasts. Appl. Microbiol. *24:* 236–239.

103. **Ordal, Z. J., J. J. Iandola, B. Ray, and A. G. Sinskey.** 1976. Detection and enumeration of injured microorganisms, pp. 163–169 IN: *Compendium of Methods for the Microbiological examination of foods,* edited by M. L. Speck (Amer. Pub. Hlth. Assoc.: D.C.).

104. **Pierson, M. D., S. L. Payne, and G. L. Ades.** 1974. Heat injury and recovery of vegetative cells of *Clostridium botulinum* Type E. Appl. Microbiol. *27:* 425–426.

105. **Postgate, J. R. and J. R. Hunter.** 1963. Metabolic injury in frozen bacteria. J. Appl. Bacteriol. *26:* 405–414.

106. **Previte, J. J.** 1972. Radiometric detection of some food-borne bacteria. Appl. Microbiol. *24:* 535–539.

107. **Previte, J. J., D. B. Rowley, and R. Wells.** 1975. Improvements in a non-proprietary radiometric medium to allow the detection of some *Pseudomonas* species and *Alcaligenes faecalis*. Appl. Microbiol. *30:* 339–340.

108. **Ray, B., J. J. Jezeski, and F. F. Busta.** 1971. Repair of injury in freeze-dried *Salmonella anatum*. Appl. Microbiol. *22:* 401–407.

109. **Ray, B. and M. L. Speck.** 1972. Repair of injury induced by freezing *Escherichia coli* as influenced by recovery medium. Appl. Microbiol. *24:* 258–263.

110. **Ray, B. and M. L. Speck.** 1973. Freeze-injury in bacteria. CRC Crit. Rev. Clin. Lab. Sci. *4:* 161–213.

111. **Renner, E. D., L. A. Gatheridge, and J. A. Washington, II.** 1973. Evaluation of radiometric system for detecting bacteremia. Appl. Microbiol. *26:* 368–372.

112. **Roth, L. A. and D. Kennan.** 1971. Acid injury of *Escherichia coli*. Can. J. Microbiol. *17:* 1005–1008.

113. **Russell, W. J., J. F. Zettler, G. C. Blanchard, and E. A. Boling.** 1975. Bacterial identification by microcalorimetry, pp. 101–121, IN: *New Approaches to the Identification of Microorganisms,* edited by C.-G. Hedén and T. Illéni (Wiley & Sons: N.Y.).

114. **Sacks, L. E. and E. Menefee.** 1972. Thermal detection of spoilage in canned foods. J. Food Sci. *37:* 928–931.

115. **Schrot, J. R., W. C. Hess, and G. V. Levin.** 1973. Method for radiorespirometric detection of bacteria in pure culture and in blood. Appl. Microbiol. *26:* 867–873.

116. **Schultz, S. J., J. S. Witzeman, and W. M. Hall.** 1968. Immunofluorescent screening for *Salmonella* in foods: Comparison with cultural methods. J. Assoc. Off. Anal. Chem. *51:* 1334–1338.

117. **Scott, R. M., D. Seiz, and H. J. Shaughnessy.** 1964. I. Rapid carbon[14] test for coliform bacteria in water. II. Rapid carbon[14] test for sewage bacteria. A. J. Pub. Hlth. *54:* 827–844.

118. **Silliker, J. H., A. Schmall, and J. Y. Chiu.** 1966. The fluorescent antibody technique as a means of detecting salmonellae in foods. J. Food Sci. *31:* 240–244.

119. **Silverman, S. J., A. R. Knott, and M. Howard.** 1968. Rapid, sensitive assay for staphylococcal enterotoxin and a comparison of serological methods. Appl. Microbiol. *16:* 1019–1023.

120. **Sinskey, T. J. and G. J. Silverman.** 1970. Characterization of injury incurred by *Escherichia coli* upon freeze-drying. J. Bacteriol. *101:* 429–437.

121. **Speck, M. L. and B. Ray.** 1973. Recovery of *Escherichia coli* after injury from freezing. Bull. Inst. Intern. Froid, Annexe *5:* 37–46.

122. **Speck, M. L., B. Ray, and R. B. Read, Jr.** 1975. Repair and enumeration of injured coliforms by a plating procedure. Appl. Microbiol. *29:* 549–550.

123. **Speck, M. L.** 1976. *Compendium of Methods for the Microbiological Examination of Foods* (Amer. Pub. Hlth. Assoc.: D.C.).

124. **Stark, R. L. and C. L. Duncan.** 1971. Biological characteristics of *Clostridium perfringens* type A enterotoxin. Infect. Immunity *4:* 89–96.

125. **Stiles, M. E. and L. D. Witter.** 1965. Thermal inactivation, heat injury, and recovery of *Staphylococcus aureus.* J. Dairy Sci. *48:* 677–681.

126. **Straka, R. P. and J. L. Stokes.** 1959. Metabolic injury to bacteria at low temperatures. J. Bacteriol. *78:* 181–185.

127. **Strange, R. E., E. O. Powell, and T. W. Pearce.** 1971. The rapid detection and determination of sparse bacterial populations with radioactively labelled homologous antibodies. J. Gen. Microbiol. *67:* 349–357.

128. **Strange, R. E. and K. L. Martin.** 1972. Rapid assays for the detection and determination of sparse populations of bacteria and bacteriophage T7 with radioactively labelled homologous antibodies. J. Gen. Microbiol. *72:* 127–141.

129. **Tatini, S. R., H. M. Soo, B. R. Cords, and R. W. Bennett.** 1975. Heat-stable nuclease for assessment of staphylococcal growth and likely presence of enterotoxins in foods. J. Food Sci. *40:* 352–356.

130. **Terplan, V. G., K.-J. Zaadhof, and S. Buchholz-Berchtold.** 1975. Zum nachweis von Endotoxinen gramnegativer Keime in Milch mit dem *Limŭlus*-test. Archiv. Lebensmittelhyg. *26:* 217–221.

131. **Thatcher, F. S. and D. S. Clark.** 1968. *Microorganisms in Foods: Their Significance and Methods of Enumeration* (Univ. of Toronto Press: Canada).

132. **Thomason, B. M., G. A. Hebert, and W. B. Cherry.** 1975. Evaluation of a semiautomated system for direct fluorescent antibody detection of salmonellae. Appl. Microbiol. *30:* 557–564.

133. **Tomlins, R. I., M. D. Pierson, and Z. J. Ordal.** 1971. Effect of thermal injury on the TCA cycle enzymes of *Staphylococcus aureus* MF 31 and *Salmonella typhimurium* 7136. Can. J. Microbiol. *17:* 759–765.

134. **U.S. Food & Drug Administration.** 1976. *Bacteriological Analytical Manual for Foods.*

135. **Ur, A. and D. Brown.** 1975. Monitoring of bacterial activity by impedance measurements, pp. 61–71, IN: *New Approaches to the Identification of Microorganisms,* edited by C.-G. Hedén and T. Illéni (Wiley & Sons: N.Y.).

136. **Vermilyea, B. L., H. D. Walker, and J. C. Ayres.** 1968. Detection of botulinal toxins by immunodiffusion. Appl. Microbiol. *16:* 21–24.

137. **Victor, R., F. Lachica, K. F. Weiss, and R. H. Deibel.** 1969. Relationships among coagulase, enterotoxin, and heat-stable deoxyribonuclease production by *Staphylococcus aureus.* Appl. Microbiol. *18:* 126–127.

138. **Weirether, F. J., E. E. Lewis, A. J. Rosenwald, and R. E. Lincoln.** 1966. Rapid quantitative serological assay of staphylococcal enterotoxin B. Appl. Microbiol. *14:* 284–291.

139. **Wheeler, T. C. and M. C. Goldschmidt.** 1975. Determination of bacterial cell concentrations by electrical measurements. J. Clin. Microbiol. *1:* 25–29.

140. **Wilson, J. M. and R. Davies.** 1976. Minimal medium recovery of thermally injured *Salmonella senftenberg* 4969. J. Appl. Bacteriol. *40:* 365–374.

5

The Incidence
and Types
of Microorganisms
in Food

In general, the numbers and types of microorganisms present in a finished food product are influenced by the following: (1) The general environment from which the food was originally obtained, (2) the microbiological quality of the food in its raw or unprocessed state, (3) the sanitary conditions under which the product is handled and processed, and (4) the adequacy of subsequent packaging, handling, and storage conditions of the product to maintain the flora at a low level. In producing good quality market foods, it is important to keep microorganisms at a low level for reasons of aesthetics, public health, and product shelf-life. Other than those foods that have been made sterile, all foods should be expected to contain a certain number of microorganisms of one type or another. Ideally, the numbers of organisms should be as low as is possible under good conditions of production. Excessively high numbers of microorganisms in fresh foods present cause for alarm. It should be kept in mind that the inner parts of healthy plant and animal tissues are generally sterile and that it is theoretically possible to produce many foods free of microorganisms. This objective becomes impractical, however, when mass production and other economic considerations are realized. The number of microorganisms in a fresh food product, then, may be taken to reflect the overall conditions of raw product quality, processing, handling, storage, and so forth. This is one of the most important uses made of the standard plate count in food microbiology. The question immediately arises as to the attainability of low numbers under the best and most economic production conditions known. With few exceptions, it is difficult to know what is the lowest number of microorganisms attainable under good production conditions due to the many variables that must be considered. Due to advances in modern technology, it has been possible to reduce the microbial load in a large number of foods over what was possible even 10 years ago.

The microbiology of a large number of foods is treated below. With each group of foods, the most important sources of organisms are listed and discussed along with the incidence of organisms as reported by various workers. It should be noted that the numbers of organisms reported for a particular food by various authors do not always reflect the microbiology of that product under ideal production, handling, and storage conditions. The chapters on food spoilage and food poisoning should be consulted for additional information on the types of organisms associated with the foods discussed. The incidence of coliforms, enterococci, and other indicator organisms is discussed in Chapter 15.

The numbers of microorganisms per gram are given below in $\log_{10}$ numbers in order both to conserve space and to indicate the lack of precision associated with plate count results.

MEATS

The reported incidence of microorganisms in meats along with various other foods is presented in Tables 5-1 and 5-2. As a category of meats, comminuted meats such as ground beef invariably have higher numbers of microorganisms than noncomminuted meats such as steaks. This has been reported rather consistently for over 60 years and there are several reasons for the generally higher count on these type products than one finds on whole meats. First, commercial ground meats generally consist of trimmings from various cuts and thus represent pieces that have been handled excessively. Consequently, these pieces normally contain more microbes than meat cuts such as steaks. Second, ground meat provides a greater surface area which itself accounts in part for the increased flora. It should be recalled that as particle size is reduced, its net surface area increases with a consequent increase in surface energy. Third, this greater surface area of ground meat favors the growth of aerobic bacteria, the usual low-temperature spoilage flora. Fourth, in some commercial establishments, the meat grinders, cutting knives, and storage utensils are rarely cleaned as often and as thoroughly as is necessary to prevent the successive build-up of microbial numbers. This may be illustrated by data obtained by the author from a study of the bacteriology of several areas in the meat department of a large grocery store. The blade of the meat saw and the cutting block were swabbed immediately after they were cleaned on 3 different occasions with the following mean results. The saw blade had a total $\log/in^2$ count of 5.28, with 2.3 coliforms, 3.64 enterococci, 1.60 staphylococci, and 3.69 micrococci. The cutting block had a mean $\log/in^2$ count of 5.69, with 2.04 coliforms, 3.77 enterococci, <1.00 staphylococci, and 3.79 micrococci. These are among the sources of the high total bacterial count to comminuted meats. Fifth, one heavily contaminated piece of meat is sufficient to contaminate others as well as the entire lot as they pass through the grinder. This heavily contaminated portion is often in the form of lymph nodes which are generally imbedded in fat. These organs have been shown to contain high numbers of microorganisms (60), and account in part for hamburger meat

TABLE 5-1. Log microbial counts/g or ml reported for various foods.

Product	Number	Total counts	Coliforms	Enterococci	Staphy- lococci	Yeasts/ molds	Lactic acid types	References
Meats								
Ground beef	96	6.20–8.36 (7.52)	1.30–6.04 (4.72)	—	—	—	—	83
Ground beef	56	5.57–8.67 (6.72)	—	—	—	—	—	41
Veal breakfast patty	3	(6.15)	2.74	all +	1 of 3 +	—	—	35
Tenderloin steak	—	(6.42)	1.68	—	—	—	—	41
Pork chops	56	(5.97)	—	—	(2.97)	—	—	40
Round beef steak	56	(5.98)	—	—	(3.77)	—	—	40
Beef liver	56	(8.61)	—	—	(4.85)	—	—	40
Raw boneless ham	6	(4.43)	—	—	—	—	—	71
Fresh beef sausage (British)	—	5.00–7.54	0–4.71	2.3–4.74	—	3.42–6.23[a]	3.08–5.08	16
Fresh pork sausage (British)	—	4.76–7.93	2.11–4.59	1.63–4.54	—	2.00–5.48[a]	2.83–5.46	16
Bologna	22	3.00–6.95 (3.54)	—	—	—	—	—	67
Pickle loaf	16	3.00–7.15 (4.08)	—	—	—	—	—	67
Cooked salami	16	3.00–6.20 (3.48)	—	—	—	—	—	67
Macaroni & Cheese loaf	21	3.00–7.28	—	—	—	—	—	67
Poultry & Seafoods								
Poultry meats/cm²	—	(4.40)	—	—	—	—	—	5
Broiler skin	—	(5.02)	—	—	—	—	—	64
Dressed broiler-fryers (Salmonella-free)	72	(5.20)	(2.61)	(2.83)	—	—	—	110
Dressed broiler-fryers (Salmonella-free)	192	(5.38)	(2.81)	(2.64)	—	—	—	110
Breaded shrimp	—	5.98–7.15	2.18–4.15	—	all +	—	—	28
Frozen breaded shrimp	144	4.36–7.73	68% 2.00	—	—	—	—	45
	564[b]	5.64–8.25	1.49–3.04	—	56% +	—	—	100
	297[c]	4.69–6.85	1.18–2.88	—	43% +	—	—	100
Frozen breaded raw shrimp	56	2.00–6.89	0–3.11	—	some +	—	—	86
Breaded cooked shrimp	4	2.00–3.88	0	—	1 +	—	—	86
Peeled raw shrimp	19	4.85–6.97	0–2.20	—	some +	—	—	86
Peeled cooked shrimp	3	2.30–5.20	0	—	0	—	—	86

TABLE 5-1. Log microbial counts/g or ml reported for various foods (*Continued*).

Product	Number	Total counts	Coliforms	Enterococci	Staphy-lococci	Yeasts/molds	Lactic acid types	References
Poultry & Seafoods (cont.)								
Raw unshelled shrimp	9	3.79–6.04	0–1.48	—	some +	—	—	86
Sole fillets	—	5.56–5.67	—	—	—	—	—	61
Haddock fillets	—	5.61–5.65	—	—	—	—	—	74
Shucked soft-shelled clams	—	6.51–6.73	—	—	—	—	—	74
Frozen Precooked Meals								
Tuna pies	12	3.34–4.15	0–2.06	0–0.97	—	—	—	63
Chicken pies	36	3.79–6.38	0–2.66	0–3.10	—	—	—	63
Beef pies	15	3.80–7.84	1.00–3.38	all +	—	—	—	35
Turkey pies	42	2.30–7.04	0–3.33	0–4.15	—	—	—	35
Frozen foods, misc.	42	3.30–5.95	0–2.81	0–5.06	—	—	—	35
	157	—	41% +	83% +	41% +	—	—	35
Frozen Raw Pies								
Chicken	20	3.57–6.59	0–3.08	1.78–5.20	0–5.59	—	—	9
Chicken	56	3.00–7.08	0–4.08	0–4.54	0–4.48	—	—	50
Turkey	20	4.20–6.52	0–3.32	1.85–4.18	0–4.61	—	—	50
Beef	48	2.00–5.48	0–2.08	0–4.64	0–4.18	—	—	50
Tuna	36	1.60–4.45	0–2.11	0–3.54	0–2.42	—	—	50
Other Frozen Foods								
Fish sticks	78	2.51–6.15	0–1.54	—	2 +	—	—	73
Chocolate cream pies	137[b]	2.69–6.38	0–3.05	—	11 +	—	—	99
Coconut cream pies	41[c]	2.48–4.38	0.48–1.28	—	0	—	—	99
	32[b]	3.60–5.52	1.42–2.89	—	1 +	—	—	99
Lemon cream pies	42[c]	3.48–4.54	0.48–2.86	—	0	—	—	99
	18[b]	3.68–6.43	1.94–3.04	—	0	—	—	99
Banana cream pies	33[c]	2.48–3.00	0	—	0	—	—	99
	30[c]	3.65–4.04	0–0.85	—	0	—	—	99
French fried potatoes	353[b]	2.00–4.30	31% +	—	1 +	—	—	101
	244[c]	2.00–4.15	10% +	—	1 +	—	—	101

Potato patties	52[b]	2.00–5.08	85% +	—	0	—	—	101
	220[c]	3.56–7.78	all +	—	59% +	—	—	101
Hash brown potatoes	372[b]	2.90–5.08	87% +	—	3% +	—	—	101
	388[c]	3.38–6.00	99% +	—	28% +	—	—	101
Vegetables								
Lima beans	5–8	5.10–5.54	1.00–1.48	all +	2/5 +	—	—	35
Green beans	5–8	4.20–5.45	1.00–1.48	all +	1/5 +	—	—	35
Mashed potatoes	3–5	(5.88)	(2.85)	2/3 +	0	—	—	35
Puffed potatoes	2	(4.11)	(2.35)	all +	1 +	—	—	35
Green beans	—	4.72–5.94	0.84–1.90	1.15–3.11	35% +	—	4.34–5.55	89
Peas	—	4.92–5.15	1.23–>3.26	1.54–3.26	39% +	—	4.53–4.76	90
Corn	—	—	1.66–2.54	1.30–3.38	64% +	—	5.81	90
Miscellaneous Foods								
Dried soup mix, vegetable type	14	3.54–4.52 (4.04)	0–2.59 (1.86)	—	—	—	—	48
Dried soup mix, beef type	10	3.79–4.15 (3.94)	0–1.85 (1.00)	—	—	—	—	48
Dried soup mix, onion type	16	3.45–4.79 (4.11)	0–1.90 (1.11)	—	—	—	—	48
Dip prepared with dried onion soup mix and cream cheese base	—	(4.26)	(0.71)	—	1.61	2.20	4.08	20, 21
Ditto . . . with sour cream base	—	(7.94)	(0.34)	—	2.53	3.40	7.81	20, 21
Ditto . . . with sour cream base + crab meat	—	(7.72)	(2.49)	—	3.28	4.15	7.36	20, 21
Chip-dip bases, commercial dehydrated	34	3.00–6.00	1.00–3.00	—	—	1.00–3.00	—	30
Dried fruits (apples, peaches, raisins)	—	3.00–5.28	—	—	—	2.30–3.30	—	42
Finished sauerkraut (nonpasteurized)	—	(7.04)	—	—	—	(5.40)	7.00	42
Flour	—	4.30–6.69	—	—	—	—	—	108
Pecan nutmeats	70	4.00–6.79	all +	all +	—	—	—	36
Orangegg	9	4.00	—	all +	—	—	—	109

a = yeasts; b = poor sanitary quality; c = good sanitary quality; + = positive; — = no data; ( ) = mean values.

TABLE 5-2. Incidence of organisms in foods relative to percentage of samples complying with target numbers or conditions (target numbers given as $\log_{10}$ unless noted otherwise).

Products Examined	Number	Microbial Group: Target Number	% Samples Meeting Target	Reference
Raw beef patties	735	APC: 6.00 or less/g	76	103
"	735	Coliforms: 2.00 or less/g	84	103
"	735	E. coli: 2.00 or less/g	92	103
"	735	S. aureus: 2.00 or less/g	85	103
"	735	Presence of salmonellae	0.4	103
Fresh ground beef[1]	1830	APC: 6.70 or less/g	89	10
"	1830	S. aureus: 3.00 or less/g	92	10
"	1830	E. coli: 1.70 or less/g	84	10
"	1830	Presence of salmonellae	2	10
"	1830	Presence of C. perfringens	20	10
Fresh ground beef	16	APC: 6.00 or less/g	100	19
"	16	Coliforms: 2.00 or less/g	81	19
"	16	E. coli: 2.00 or less/g	94	19
"	16	S. aureus: 2.00 or less/g	100	19
"	16	C. perfringens: 2.00 or less/g	100	19
Fresh ground beef	213	APC: 7.00 or less/g	36	18
"	213	Coliforms: 2.00 or less/g	5.2	18
"	213	Presence of S. aureus	17	18
"	213	Presence of salmonellae	0	18
"	95	C. perfringens: 2.00 or less/g	87	58
Fresh ground beef	1090	APC: ≤7.00/g at 35 C	87.8	77B
"	1090	Fecal coliforms: ≤2.00/g	75.6	77B
"	1090	S. aureus: ≤2.00/g	90.7	77B
Comminuted big game meats	113	Coliforms: 2.00 or less/g	42	87
"	113	E. coli: 2.00 or less/g	75	87
"	113	S. aureus: 2.00 or less/g	96	87

Product	n	Criterion	Value	Ref.
Precooked turkey rolls	6	APC: 3.00/g	100	66
"	6	Coliforms: 2.00 or less/g	67	66
"	6	Enterococci: 2.00 or less/g	83	66
"	48	Presence of salmonellae	4	66
"	48	Presence of *C. perfringens*	0	66
Precooked turkey rolls and sliced turkey	30	APC: <2.00/g	20	111
"	29	Presence of coliforms	21	111
"	29	Pres. of *E. coli* or salmonellae	0	111
Ground fresh turkey meat	74	APC: 7.00 or less/g	51	29
"	75	Presence of coliforms	99	29
"	75	Presence of *E. coli*	41	29
"	75	Presence of fecal streptococci	95	29
"	75	Presence of *S. aureus*	69	29
"	75	Presence of salmonellae	28	29
"	75	Presence of *C. perfringens*	52	29
Luncheon meats	150	APC: 5.00 or less/g	56	33
"	150	APC: 6.00 or less/g	69	33
Processed meats[1]	454	APC: 6.00 or less/g	66	10
"	454	Presence of *S. aureus*	0.8	10
"	454	Presence of *C. perfringens*	3	10
"	454	*E. coli*: >1.00/g	0	10
"	454	Presence of salmonellae	0	10
Fresh pork sausage	67	APC: 5.70 or less/g	75	102
"	67	*E. coli*: 2.00 or less/g	88	102
"	67	*S. aureus*: 2.00 or less/g	75	102
"	560	Presence of salmonellae	28	102
Pork trimmings for above sausage	528	Presence of salmonellae	28	102

TABLE 5-2. *(Continued)*

Products Examined	Number	Microbial Group: Target Number	% Samples Meeting Target	Reference
Vacuum-packed sliced imported canned ham samples[2]	180	APC: <3.00/g	97	104
"	180	APC: 3.30 or less/g	99	104
"	180	Presence of *E. coli*, or *S. aureus* in 0.1 g; or salmonellae in 25 g portions	0	104
"	180	Coliforms: present in 0.1 g portions	2	104
Retail frozen breaded raw shrimp	27	APC: 6.00 or less/g	52	105
"	27	Coliforms: 3.00 or less/g	100	105
"	27	Presence of *E. coli*	4	105
"	27	Presence of *S. aureus*	59	105
Fresh channel catfish	335	APC: ≤7.00/g	93	4
"	335	Fecal coliforms: 2.60/g	70.7	4
"	335	Presence of salmonellae	4.5	4
Frozen channel catfish	342	APC: ≤7.00/g	94.5	4
"	342	Fecal coliforms: 2.60/g	92.4	4
"	342	Presence of salmonellae	1.5	4
Semipreserved meat products[3]	372	Presence of *C. botulinum*	1.3	1
Convenience foods[4]	400	"	0.25	39
Freshly smoked white-fish[5]	858	"	1.1	75
Frozen cream-type pies	580	APC: <4.00/g	85	59
"	580	APC: 4.70 or less/g	98	59
"	580	Coliforms: <1.00/g	75	59

Frozen cooked-peeled shrimp	204	APC: <4.70/g	52	59
"	204	APC: 5.30 or less/g	71	59
"	204	Coliforms: none or <0.3/g	52.4	59
"	204	Coliforms: <3/g	75.2	59
Dry food-grade gelatin	185	APC: 3.00 or less/g	74	59
Delicatessen salads	764	Within AAFES microbial limits[6]	44	23
Delicatessen salads	764	APC: 5.00 or less/g	84	23
"	764	Coliforms: 1.00 or less/g	78	23
"	764	Yeasts and molds: 1.30 or less/g	55	23
"	764	Fecal streptococci: 1.00/g	77	23
"	764	Presence of S. aureus	9	23
"	764	Pres. of C. perfringens; salmonellae	0	23
Delicatessen salads	517	APC: 5.00 or less/g	26–85	76
"	517	Coliforms: 2.00 or less/g	36–79	76
"	517	S. aureus: 2.00 or less/g	96–100	76
Retail trade salads	53	APC: >6.00/g	36	14
"	53	Coliforms: 2.00 or less/g	57	14
"	53	Presence of S. aureus	39	14
Retail trade sandwiches	62	APC: >6.00/g	16	14
"	62	Coliforms: >3.00/g	12	14
"	62	Presence of S. aureus	60	14
Imported spices and herbs	113	APC: 6.00 or less/g	73	44
"	114	Spores: 6.00 or less/g	75	44
"	113	Yeasts and Molds: 5.00 or less/g	97	44
"	114	TA spores: 3.00 or less/g	70	44
"	114	Pres. of E. coli, S. aureus, Salm.	0	44
Processed spices	114	APC: 5.00 or less/g	70	80
"	114	APC: 6.00 or less/g	91	80
"	114	Coliforms: 2.00 or less/g	97	80

TABLE 5-2. *(Continued)*

Products Examined	Number	Microbial Group: Target Number	% Sample Meeting Target	Reference
"	114	Yeasts and Molds: 4.00 or less/g	96	80
"	114	*C. perfringens:* <2.00/g	89	80
"	110	Presence of *B. cereus*	53	81
Dehydrated space foods	129	APC: <4.00/g	93	79
"	129	Coliforms: <1/g	98	79
"	129	*E. coli:* negative in 1 g	99	79
"	102	Fecal streptococci: 1.30/g	88	79
"	104	*S. aureus:* neg in 5 g	100	79
"	104	Salmonellae: negative in 10 g	98	79

[1]Under Oregon law.
[2]Machine sliced and vacuum-packed samples were obtained from 16–21 lb canned refrigerated, imported hams.
[3]Five with Type A and one with Type B strains.
[4]The one isolate was Type B.
[5]Nine Type E and one Type B isolates.
[6]Army and Air Force Exchange Service.

having a generally higher total count than ground beef. In some states, the former may contain up to 30% beef fat while the latter should not contain more than 20% fat.

In addition to the above-named sources of microorganisms to ground meats in general, sausage and frankfurters have additional sources of organisms in the seasoning and formulation ingredients that are usually added in their production. From Tables 5-2 and 5-3 it can be seen that most spices and condiments have high microbial numbers. The lactic acid bacteria and yeasts in some composition products are usually contributed by milk solids. In the case of pork sausage, natural casings have been shown to contain high numbers of bacteria. In their study of salt-packed casings, Riha and Solberg (82) found counts to range from log 4.48 to 7.77, and from 5.26 to 7.36 for wet-packed casings. Over 60% of the isolates from these natural casings consisted of *Bacillus* spp. followed by clostridia and pseudomonads.

TABLE 5-3. **Numbers of microorganisms in ordinary nonsterile spices and seasonings available in commercial and retail channels (48).***

	Log number of organisms/g of dry sample					
Types & brand	Total aerobes	Coliforms	Staphy-lococci	Yeasts & molds	Aerobic spores	Anaerobic spores
Bay leaves	5.72	0	—	3.52	3.96	<0.30
Cloves	3.48	0	—	0.69	0.30	<0.30
Curry	>6.88	0	—	1.85	>5.38	>4.38
	5.60	1.48	—	1.85	>5.04	2.63
Marjoram	5.57	0	—	4.26	4.73	>4.38
	5.20	0	—	3.69	0.95	2.18
Paprika	>6.74	2.78	—	3.36	>5.38	2.79
	6.81	1.90	—	4.48	3.46	1.63
Pepper	>6.30	0	—	1.18	>5.38	>4.38
Sage	3.83	0	—	1.00	0.85	3.23
Thyme	6.28	0	—	4.04	5.20	>4.38
	6.11	2.30	—	3.92	2.18	—
Turmeric	>6.83	0	—	0	0.30	<0.30
	6.11	1.69	—	1.85	>5.04	2.63
Celery flakes	7.00	3.77	2.49	2.30	1.97	2.18
	6.32	2.15	1.60	1.90	4.38	>4.04
Onion flakes	4.67	0	0	0	2.97	2.18
Onion, minced	4.80	2.59	0	1.18	4.38	3.66
Garlic, ground	5.40	1.48	0	1.90	3.97	3.66
Garlic powder	3.43	0	0	1.88	2.63	2.59
Seasoned salt	4.08	0	0	0	3.38	2.97
Seasoned pepper	7.40	0	0	2.40	3.97	>4.04

*"1965 Copyright © by Institute of Food Technologists."

Of the individual ingredients of fresh pork sausage, casings have been shown to contribute the largest number of bacteria (*102*).

Processed meats such as bologna, salami, and others may be expected to reflect the sum of their ingredient makeup in regard to microbial numbers and types. The microflora of frankfurters has been shown (*17*) to consist largely of gram positive organisms with micrococci, bacilli, sarcinae, lactobacilli, microbacteria, streptococci, and leuconostocs along with yeasts. In a study of slime from frankfurters, these investigators found that 275 of 353 isolates were bacteria while 78 of the 353 were yeasts. *Microbacterium thermosphactum* was the most conspicuous single isolate.

Wiltshire bacon has been reported to have a total count generally in the range of log 5–6/g (*37*), while high-salt, vacuum-packed bacon has been reported to have a generally lower count—about log 4.0/g. The flora of vacuum-packed sliced bacon consists mostly of catalase positive cocci such as micrococci and coagulase-negative staphylococci as well as catalase negative bacteria of the lactic acid types such as latobacilli, Group D streptococci, leuconostocs, and pediococci (*2, 12, 52*). Cooked salami has been found to have a flora consisting mostly of lactobacilli.

Organ meats such as beef liver generally contain large numbers of microorganisms due in part to the fact that such organs filter out microorganisms from the circulating blood, and in part because of the generally higher pH of these meats which favors microbial growth at a more rapid rate than in nonorgan meats. The predominant flora of fresh beef liver has been found to consist of lactic acid bacteria with a mean count of log 5.0 (*85*).

Both bacilli and clostridia may be found in meats of all types. In a study of the incidence of putrefactive anaerobe (P.A.) spores in fresh and cured pork trimmings and canned pork luncheon meat, Steinkraus and Ayres (*95*) found these organisms to occur at very low levels, generally less than 1/g. In a study of the incidence of clostridial spores in meats, Greenberg *et al.* (*27*) found a mean P.A. spore count/g of 2.8 from 2,358 meat samples. Of the 19,727 P.A. spores isolated, only 1 was a *C. botulinum* spore, and it was recovered from chicken. The large number of meat samples studied by these investigators consisted of beef, pork, and chicken, obtained from all parts of the United States and Canada. The significance of P.A. spores in meats is due to the problems encountered in the heat destruction of these forms in the canning industry (see Chapter 12). The incidence of *C. perfringens* in a variety of American foods was studied by Strong *et al.* (*96*). These investigators recovered this organism from 16.4% of the raw meats, poultry, and fish tested; from 5 percent of spices; from 3.8 percent of fruits and vegetables; from 2.7 percent of commercially prepared frozen foods; and from 1.8 percent of home-prepared foods. Other investigators have found low numbers of this organism in both fresh and processed meats (Table 5-3). In ground beef, *C. perfringens* at 100 or less/g was found in 87 percent of 95 samples while 45 of the 95 (47 percent) samples contained this organism at levels <1,000/g *(58)*. The significance of this organism in foods is discussed in Chapter 17.

The effect of cooking on the numbers of microorganisms in meats is to generally reduce coliforms, staphylococci, and gram negative bacteria, either to a very low level or to completely destroy them. Sporeformers, enterococci, and micrococci sometimes survive in varying numbers. Huber *et al. (35)* found a log number/g of 4.62 organisms in cooked pot roast of beef while McDivitt *et al. (71)* found a mean of log 2.76 in roast ham with a range of log 1.18–3.61/g.

In the case of vended burgers, Mueller *(69)* examined hamburgers, cheeseburgers, and pizzaburgers over a two-year period and found a mean APC for 13 hamburgers of log 4.23 and 5.49 for each year. For 10 pizzaburgers, mean APC was 3.98 and 4.39, and 4.08 for 3 cheeseburgers all as obtained from the vendors. All burgers were negative for *S. aureus*.

POULTRY

Whole poultry tends to have a lower microbial count than cut-up poultry. Most of the organisms on such products are at the surface so that surface counts/cm^2 are generally more valid than counts on surface and deep tissues. May *(65)* has shown how the surface counts of chickens build up through successive stages of processing. In a study of whole chickens from 6 commercial processing plants, the initial mean total surface count was log 3.30/cm^2. After being cut up, the mean total count increased to log 3.81, and further increased to log 4.08 after packaging. The conveyor over which these birds moved showed a count of log 4.76/cm^2. When the above procedures were repeated for 5 retail grocery stores, May found that the mean count before cutting was log 3.18 which increased to log 4.06 after cutting and packaging. The cutting block was shown to have a total count of log 4.68/cm^2.

The incidence of salmonellae in dressed broiler-fryer chickens was investigated by Woodburn *(110)*. This investigator found that 72 of 264 birds (27 percent) harbored salmonellae representing 13 serotypes. Among the serotypes, *S. infantis, S. reading,* and *S. blockley* were the most common. Bryan *et al. (8)* isolated salmonellae from the surfaces of 24 of 208 (11.5 percent) turkey carcasses before further processing. After processing into uncooked rolls, 90 of 336 (26.8 percent) yielded salmonellae. From the processing plants, 24 percent of processing equipment yielded salmonellae. Almost one third of the workers had these organisms on their hands and gloves. Of the 23 serotypes recovered, *S. sandiego* and *S. anatum* were recovered most frequently. With respect to fresh ground turkey meats, these organisms were found in 28 percent of 75 samples by another group of workers *(29)*. Almost one half of the samples had total counts above log 7.0/g. Ninety-nine percent of the samples harbored coliforms, 41 percent *E. coli,* 52 percent *C. perfringens,* and 69 percent *S. aureus*.

In regard to cooked poultry products, precooked turkey rolls have been found to have considerably lower microbial numbers of all types (see Table 5-2). In an examination of 118 samples of cooked broiler products, Lillard *(62)* found *C. perfringens* in 6 or 2.6 percent.

The microbial flora of fresh poultry consists largely of pseudomonads and other closely related gram negative bacteria as well as coryneforms, yeasts, and other organisms *(57)*. Other microorganisms normally present in poultry products and other meats are further discussed in Chapter 7.

SEAFOODS

The incidence of microorganisms in seafoods such as shrimp, oysters, and clams would depend greatly upon the quality of water from which these animals are harvested. Assuming good quality waters, most of the organisms are picked up during the various stages of processing. In the case of breaded raw shrimp, the breading process would be expected to add organisms if not properly done or if the ingredients are of poor microbiological quality. In their study of 91 samples of shrimp of various types, Silverman *et al. (86)* found that all precooked samples except one had a total count of less than log 4.00/g. Of the raw samples 59 percent had total counts below log 5.88 while 31 percent were below log 5.69/g. In a more recent study of 204 samples of frozen cooked-peeled shrimp, 52 percent of the samples had total counts <log 4.70/g, and 71 percent had counts of log 5.30 or less/g *(59)*.

In a study of haddock fillets, Nickerson and Goldblith *(74)* found that most microbial contamination occurred during filleting and subsequent handling prior to packaging. These investigators showed that the total count increased from log 5.61 in the morning to log 5.65 at noon, and to log 5.94/g in the evening for one particular processor. According to these authors, results obtained in other companies were generally similar if the night-time cleanup was good. In the case of shucked soft-shelled clams, these authors demonstrated the same general pattern of build-up in numbers from morning to evening. The mean clostridial count for both haddock fillets and soft-shelled clams was less than 2/g with clams being slightly higher than haddock fillets for these organisms, although both were low. The total count on fresh perch fillets produced under commercial conditions were found to average log 5.54/g with yeasts and molds being about log 2.69/g *(49)*.

FROZEN MEAT PIES

The microbiological quality of frozen meat pies has steadily improved since these products were first marketed. Any and all of the ingredients added may increase the total number of organisms, and the total count of the finished product may be taken to reflect the overall quality of ingredients, handling, and storage. Many investigators have suggested that these products should be produced with total counts not to exceed log 5.00/g. In a study of 48 meat pies, Litsky *et al. (63)* found that 84 percent had total counts of less than log 5 while Kereluk and Gunderson *(50)* found that 93 percent of the 188 meat pies studied by them had total counts less than log 5.00. Accordingly, a microbiological standard of log 5.00 (100,000) has been suggested for such products (see Chapter 15 for further discussion of microbiological standards).

VEGETABLES

The incidence of microorganisms in vegetables may be expected to reflect the sanitary quality of the processing steps and the microbiological condition of the raw product at the time of processing. In a study of green beans before blanch, Splittstoesser *et al. (88)* showed that the total counts ranged from log 5.60 to over 6.00 in 2 production plants. After blanching, the total numbers were reduced to log 3.00–3.60/g. After passing through the various processing stages and packaging, the counts (after packaging) ranged from log 4.72–5.94/g. In the case of French-style beans, one of the greatest buildups in numbers of organisms occurred immediately after slicing. This same general pattern was shown for peas and corn. Pre-blanch green peas from 3 factories showed total counts/g between log 4.94–5.95. These numbers were reduced by blanching and again increased successively with each processing step. In the case of whole-kernel corn, the post-blanch counts rose both after cutting and at the end of the conveyor belt to the washer. Whereas the immediate post-blanch count was about log 3.48, the product had total counts of about log 5.94/g after packaging. Between 40–75 percent of the bacterial flora of peas, snap beans, and corn was shown to consist of leuconostocs and streptococci while many of the gram positive, catalase positive rods resembled corynebacteria *(91, 92)*.

The association of lactic acid cocci with many raw and processed vegetables has been made by Mundt *et al. (70)*. These cocci have been shown to constitute from 41 to 75 percent of the APC flora of frozen peas, snap beans, and corn *(93)*. Splittstoesser *et al. (90)* showed that fresh peas, green beans, and corn all showed the presence of coagulase-positive staphylococci after processing. Peas were found to have the highest count (log 0.86/g) while 64 percent of the corn samples contained this organism. These authors found that a general buildup of staphylococci occurred as the vegetables went through successive stages of processing, with the main source of organisms coming from the hands of employees. Although staphylococci may be found on vegetables during processing, these organisms are generally unable to proliferate in the presence of the more normal lactic flora. Both coliforms (but not *E. coli*) and enterococci have been found at most stages during raw vegetable processing but appear to present no public health hazard *(93)*.

In a study of the incidence of *C. botulinum* in 100 commercially available frozen vacuum pouch-pack vegetables by Insalata *et al. (38)*, these workers were unable to find this organism in 50 samples of string beans, but did find Types A and B spores in 6 of 50 samples of spinach.

SPICES AND CONDIMENTS

In considering the microbiology of these products, they may be separated into two groups based upon whether or not they contain antimicrobial principals. Cinnamon, cloves, horseradish, garlic, onion, allspice, and mustard have all been shown by various investigators to possess antimicrobial activities to varying degrees. Of these spices, cinnamon and cloves are in general the most effective in inhibiting and killing microorganisms. The active principal in cinnamon is cinnamic aldehyde, while eugenol is the agent in cloves. The bactericidal activity of horseradish is best demonstrated by use of horseradish vapors. The vapors of this substance were shown by Foter and Gorlick *(22)* to be more active than those of garlic and onion. The active principal in horseradish is allyl isothiocyanate, a pungent volatile oil which is irritating to the eyes. The volatile oil of mustard has been reported by some to have a greater preserving power than that of cloves or cinnamon oil. The antimicrobial activity of onion apparently resides in some of its volatile aliphatic disulfide compounds such as *n*-propyl allyl and di-*n*-propyl, both of which have been shown by Johnson and Vaughn *(43)* to possess antimicrobial activity against *S. typhimurium* and *E. coli*.

Most spices that do not contain antimicrobial compounds tend to have higher numbers of microorganisms than those that do (Table 5-3). It can be seen from Tables 5-2 and 5-3 that most of these agents contain fairly large numbers of bacteria and molds, with coliforms being present in relatively low numbers. Of 110 samples of processed spices examined by Powers *et al.* *(81)*, *B. cereus* was found in 53%. In a study of 114 samples representing 7 different processed spices, *C. perfringens* was found in 15 percent with oregano having the highest numbers *(80)*. The flora of spices consists in part of those organisms picked up by these products during harvest, and in part of organisms that are added during processing. The treatment of spices with ethylene oxide reduces the number of gram negative bacteria more effectively than that of the sporeformers. As was stated earlier in this chapter, spices provide important sources of microorganisms to all foods in which they are added. In general, the ground products have higher microbial counts than nonground due both to the additional processing steps involved and to the greater surface area provided by the ground product.

DAIRY PRODUCTS

The microbial flora of raw milk consists of those organisms which may be present from the cow's udder and hide, from milking utensils or lines, and so forth. Under proper handling and storage conditions, the predominant flora is gram positive. While yeasts, molds, and gram negative bacteria may be found along with the lactic acid bacteria, most or all of these types are more heat sensitive than gram positives and are more likely to be destroyed during pasteurization. Studies over the past several years have revealed the presence of psychrophilic or psychrotrophic sporeformers and mycobacteria in raw milks. Shehata and Collins (84) found psychrophilic *Bacillus* spp. in 25 to 35% of 97 raw milk samples. These organisms were shown to grow at or below 7 C. Psychrotrophic clostridia were isolated from 4 of 48 raw milk samples by Bhadsavle et al. (7). Because of the proteolytic abilities of sporeformers, these organisms can survive pasteurization temperatures in their spore state and cause problems in the refrigerated pasteurized products. *Mycobacterium* and *Nocardia* spp. were isolated from c. 69% of 51 raw milks examined by Hosty and McDurmont (34). Of 43 samples of pasteurized milk examined, all were negative for these organisms. The 66 isolates consisted largely of mycobacteria with the *M. terrae* complex and *M. fortuitum* being the most common. Two *Nocardia asteroides* were recovered. The presence of these organisms in raw milk is not surprising due to their abundance in soils.

The incidence of microorganisms in commercial cream cheese was studied by Fanelli et al. (21) who found that the total count/g was about log 2.69. In the case of commercial sour cream, the total counts/g ranged from log 4.79–8.58, with most of the organisms being lactic acid bacteria. Commercial onion dips with sour cream bases had total counts of log 7.76–8.28 and log 4.34/g of yeasts and molds. The majority of bacteria in this product were found also to be lactic acid types. The microbial flora of fermented dairy products is discussed in Chapter 14.

DEHYDRATED FOODS

In a detailed study of the microbiology of dehydrated soups, Fanelli et al. (20) showed that approximately 17 different kinds of dried soups from 9 different processors had total counts of less than log 5.00/g. These soups included chicken noodle, chicken rice, beef noodle, vegetable, mushroom, pea, onion, tomato, and others. Some of these products had total counts as high as log 7.30/g, while some had counts as low as around log 2.00. These investigators further found that reconstituted dehydrated onion soup showed a mean total count of log 5.11/ml, with log 3.00 coliforms, log 4.00 aerobic sporeformers, and log 1.08/ml of yeast and molds. Upon cooking, the total counts were reduced to a mean of log 2.15 while coliforms were reduced to less than log 0.26, sporeformers to log 1.64, and yeasts and molds to less than log 1.00/ml. In a study of dehydrated sauce and gravy mixes, soup mixes, spaghetti-sauce mixes, and cheese-sauce mixes, Nakamura and Kelly (72) isolated *C. perfringens* from 10 of 55 samples. The facultative anaerobe counts ranged from log 3 to over log 6/g.

In a study of 185 samples of food-grade dry gelatin, Leininger et al. (59) found no samples that exceeded an APC of log 3.70/g. Of 129 samples of dehydrated space food examined by Powers *et al.* *(79)*, 93% contained total counts of <log 4.00/g (see Table 5-2).

Powdered eggs and milk often contain high numbers of microorganisms on the order of log 6–8/g. One reason for the generally high numbers in dried products is that the organisms have been concentrated on a per gram basis along with product concentration. The same is generally true for fruit juice concentrates, which tend to have higher numbers of microorganisms than the fresh, nonconcentrated products.

The author has investigated the incidence and types of organisms on raw squash seeds to be roasted for food use. Some 12 samples of this product showed a mean total count of log 7.99 and the presence of log 4.72 coliforms. Most of the latter were of the nonfecal type. By adding flour batter and salt to these seeds and roasting, the total count was reduced to less than log 2.00/g. The microbiology of desiccated foods is discussed further in Chapter 13.

DELICATESSEN AND RELATED FOODS

Delicatessen foods, such as salads and sandwiches, are often involved in food poisoning cases. These foods are prepared by hand and this direct contact may lead to an increased incidence of food poisoning agents such as *S. aureus*. Once organisms such as these enter meat salads or sandwiches, they tend to grow well because of the reduction in numbers of the normal food flora by the prior cooking of salad or sandwich ingredients.

In a study of retail salads and sandwiches, Christiansen and King *(14)* found 36% of 53 salads to have total counts above log 6.00/g, but only 16% of the 60 sandwiches had counts as high. With respect to coliforms, 57% of the sandwiches were found to harbor <log 2.00/g. *S. aureus* was present in 60% of sandwiches and 39% of salads. Yeasts and molds were found in high numbers with 6 samples above log 6.00/g.

In a study of 517 salads from ca. 170 Wisconsin establishments, Pace *(76)* found 71–96% to have APC of <log 5.00/g. Ninety-six to 100% of the salads showed coagulase-positive *S. aureus* counts of <log 2.00/g. The salads included chicken, egg, macaroni, shrimp, etc. *S. aureus* was recovered in low numbers from 6 or 64 salads in another study *(23)*. The 12 different salads examined by these investigators had total counts between log 2.08 to 6.76 with egg, shrimp, and some of the macaroni salads having the highest counts. Neither salmonellae nor *C. perfringens* were found in any product. A study of 42 salads by Harris *et al.* *(31)* revealed the products to be of generally good microbial quality. The mean APC was log 5.54/g and mean coliform counts were log 2.66/g on the 6 different products. Staphylococci were found in some products, especially ham salad.

Fresh green salads (green, mixed green, and coleslaw) were studied by Fowler and Foster *(24)* and found to contain mean total counts of log 6.67 for coleslaw to log 7.28 for green salad. Fecal coliforms were found in 26% of mixed, 28% of green, and 29% of coleslaw, while the respective percentage findings for *S. aureus* were 8, 14, and 3. With respect to parsley, Käferstein *(46)* examined 64 samples and found *E. coli* on 11 fresh and unwashed samples and on over 50% of frozen samples. The mean APC of fresh washed parsley was log 7.28/g. Neither salmonellae nor *S. aureus* were found in any samples.

In a study of the microbiological quality of imitation-cream pies from plants operated under poor sanitary conditions, Surkiewicz *(99)* found that the microbial load increased successively as the products were carried through the various processing steps. For example, in one instance the final mixture of the synthetic pie base contained fewer than log 2.00 bacteria/g after final heating to 160 F. After overnight storage, however, the count rose to log 4.15. The pie topping ingredients to be mixed with pie base had a rather low count—log 2.78/g. After being deposited on the pies, the pie topping showed a total count of log 7.00/g. In a study of the microbiological quality of French fries, Surkiewicz *et al. (101)* demonstrated the same pattern, i.e., the successive buildup of microorganisms as the fries underwent processing. Since these products are cooked late in their processing, the incidence of organisms in the finished state does not properly reflect the actual state of sanitation during processing.

A bacteriological study of 580 frozen cream-type pies (lemon, cocoanut, chocolate, and banana) by Leininger *et al. (59)* revealed these products to be of excellent quality with 98% having an APC of log 4.70 or less/g (see Table 5-2).

OTHER FOODS

Breads and cakes generally contain rather low numbers of microorganisms, with the aerobic sporeformers and molds being most conspicuous when these products undergo spoilage. The baking process destroys most organisms in products of these types and the presence of chemical inhibitors, such as the propionates, aids in preventing the growth of those that survive. While some investigators have reported the presence of organisms on the order of log 4–6/g in flour, good quality bleached flour generally contains fewer than log 4 organisms/g. The flour bleaching agent, chlorine dioxide, aids in reducing the number of viable organisms in this product.

Although the primary objective in food preservation by canning is the destruction of microorganisms, canned foods may contain microbes of one type or another depending upon the type of food and the type of heat treatment given during their processing. For example, canned acid foods may contain viable spores that cannot germinate and neutral products may contain obligate thermophilic spores that do not germinate due to lack of proper temperatures. The incidence of microorganisms in commercially canned foods is quite low, and these products have an excellent record for being devoid of food-poisoning bacteria.

Pecans, almonds, and cocoanuts may be expected to carry a high and varied flora reflective of harvesting and storage conditions. A varied bacterial flora has been reported for pecan meats including coliforms and clostridial species *(13)*. No bacteria were found on aseptically shelled pecans and only two fungi were recovered in the above study. With respect to almonds, mean bacterial counts over a two-year period of 271 samples did not exceed log 3.85 *(51)*. Yeast and mold counts were higher with mean counts for one year being slightly over log 4.67. These investigators found coliforms, *E. coli,* and streptococci and correlated their presence with soil contamination. Cocoanuts from 5 countries were examined by Kajs *et al. (47),* and the shell was found to contain extremely high numbers of bacteria, yeasts, and molds. The predominant bacteria were *Acinetobacter, Flavobacterium, Microbacterium, Micrococcus,* and *Enterobacter.* Aseptically removed meats or water from undamaged cocoanuts contained few or no microorganisms.

An examination of 36 samples of cocoa powder revealed the numbers of microorganisms to be low and to consist largely of *Bacillus* spp. *(26).* The only other bacteria found were micrococci. Eighty-six percent of the samples had counts <log 3.97, and among the species of bacilli, *B. cereus* was the second most abundant.

FUNGI AND VIRUSES IN FOODS

The ubiquity of fungi in nature suggests that these organisms may be found on all fresh food products. It is doubtful as to whether one cannot recover yeasts or molds from any raw or fresh food product in commerce and yet many reports on the microbiology of foods do not include the fungi. Some of the reasons for this are discussed in Chapter 3. Among other reasons is the technique employed to recover fungi from foods. The traditional method of recovery consists of acidifying culture media to a pH of 3.5. Evidence is now available that a pH of 3.5 is inhibitory to some fungi; that the use of antibiotics leads to higher recoveries of yeasts and molds from foods than acidified media *(53).* While fungi are not indicated for some of the foods discussed earlier in this chapter, they may be presumed to be present on all except those which have received heat treatments. The significance of fungi in foods, especially molds, stems not only from their spoilage potential but also from the potential of many to produce a variety of mycotoxins to which man is susceptible. These are discussed further in Chapter 19.

The incidence and types of yeasts and molds in a wide variety of foods and beverages have been reviewed extensively *(3, 6, 77A, 94, 106, 107).* In a study by Koburger and Farhat *(54)* of 31 different foods including vegetables, cheeses, meats, etc., combined yeast and mold counts ranged from a low of log 2.93/g for sausage to log 7.99/g for colby cheese. The products that contained log 6.00/g or more were okra, asparagus, and Swiss cheese. Lamb chops were found to contain log 5.91, pork patties log 5.68, with the other foods showing lower numbers/g. These investigators employed culture media without pH adjustment but with 0.1 mg/ml of chloramphenicol and chlortetracycline. Using similar recovery methods, Mossel *et al. (68)* found mold counts to range from log <1 to 2.60/g for 14 samples of minced meat. On 15 samples of chicken meat, the mean mold count was log 1.71/g.

With respect to viruses, there is a paucity of data on their existence in foods. Evidence does exist on the transmission of the hepatitis and polio viruses through foods and this is discussed further in Chapter 19. Of primary concern here are the enteric viruses which, from several indications, probably occur in a variety of foods to significant levels. For example, viruses are generally more resistant to adverse environmental conditions than vegetative cells of bacteria *(78)*. The general lack of evidence for their existence in foods is due to the requirement for special isolation and culturing techniques, but techniques are now available which have been shown to be effective in recovering experimentally contaminated viruses from foods *(15, 97)*.

With respect to the capacity of certain viruses to persist in foods, it has been shown that enteroviruses persisted in ground beef up to 8 days at 23 or 24 C and were not affected by the growth of spoilage bacteria *(32)*. In a study of 14 vegetable samples for the existence of naturally-occurring viruses, none were found, but coxsackievirus B5 inoculated onto vegetables did survive at 4 C for 5 days *(55)*. In an earlier study, these investigators showed that coxsackievirus B5 showed no loss of activity when added to lettuce and stored at 4 C under moist conditions for 16 days. Several enteric viruses failed to survive on the surfaces of fruits, and no naturally-occurring viruses were found in 9 fruits examined *(56)*. Echovirus 4 and poliovirus 1 were found in one each of 17 samples of raw oysters by Fugate *et al. (25),* and poliovirus 3 was found in one of 24 samples of oysters. With respect to the destruction of enteric viruses in broiled hamburgers, Sullivan *et al. (98)* found that viruses could be recovered from 8 of 24 patties cooked rare (to 60 C internally) if the patties were cooled immediately to 23 C. No viruses were detected if the patties were allowed to cool for 3 min at room temperature before testing. The existence of enteric viruses in certain foods may be as high as the existence of enteric bacteria, but until more studies are carried out, this must remain in the area of conjecture.

SUMMARY

Nearly all foods may be expected to contain microorganisms of one type or another. The existence of microorganisms at levels of between log 4–6/g is common to many foods. Fresh and frozen foods that undergo extensive steps in their processing invariably show high numbers of microorganisms, especially in the absence of good sanitary control. The latter is indispensable if the finished products are to have a low total count as well as be free of food-borne pathogens. Carroll *et al. (11)* have stressed that the storage extension of shrimp is not the result of a single act, but instead is facilitated by each separate step that effects the removal of microorganisms and prevents their growth. This is true for many foods. Comminuted or ground foods tend, in general, to have higher microbial loads than whole foods. The same is true for multi-ingredient foods such as meat pies, frankfurters, sausage, and others. In products of the latter type, the numbers of organisms in the final products generally reflect the overall microbial quality of the various ingredients. Fermented milks and products such as sauerkraut are produced by the activities of certain microorganisms and may be expected to contain high numbers of these organisms in nonpasteurized finished products. One of the more obvious effects of high initial counts in foods is a shortening of product shelf-life. Although the incidence and types of yeasts and molds are not reported in many food analyses, these organisms are invariably present unless destroyed during processing. Viruses, especially the enterics, are undoubtedly as abundant in some foods as are intestinal bacteria, but the difficulty of their recovery and growth has limited our knowledge of their true existence.

The microbial spoilage of foods is treated in the next three chapters.

REFERENCES

1. **Abrahamsson, K. and H. Riemann.** 1971. Prevalance of *Clostridium botulinum* in semipreserved meat products Appl. Microbiol. *21:* 543–544.

2. **Allen, J. R. and E. M. Foster.** 1960. Spoilage of vacuum-packed sliced processed meats during refrigerated storage. Food Res. *25:* 1–7.

3. **Anderson, A. W.** 1977. The significance of yeasts and molds in foods. Food Technol. *31:* 47–51.

4. **Andrews, W. H., C. R. Wilson, P. L. Poelma, and A. Romero.** 1977. Bacteriological survey of the channel catfish *(Ictalurus punctalus)* at the retail level. J. Food Sci. *42:* 359–363.

5. **Ayres, J. C., W. S. Ogilvy, and G. F. Stewart.** 1950. Post-mortem changes in stored meats. I. Microorganisms associated with development of slime on eviscerated cut-up poultry. Food Technol. *4:* 199–205.

6. **Beuchat, L. R., Editor.** 1977. *Food and Beverage Mycology* (Avi Publishing Co.: Westport, Conn.).

7. **Bhadsavle, C. H., T. E. Shehata, and E. B. Collins.** 1972. Isolation and identification of psychrophilic species of *Clostridium* from milk. Appl. Microbiol. *24:* 699–702.

8. **Bryan, F. L., J. C. Ayres, and A. A. Kraft.** 1968. Salmonellae associated with further-processed turkey products. Appl. Microbiol. *16:* 1–9.

9. **Canale-Parola, E. and Z. J. Ordal.** 1957. A survey of the bacteriological quality of frozen poultry pies. Food Technol. *11:* 578–582.

10. **Carl, K. E.** 1975. Oregon's experience with microbiological standards for meat. J. Milk Food Technol. *38:* 483–486.

11. **Carroll, B. J., G. B. Reese, and B. Q. Ward.** 1968. Microbiological study of iced shrimp: Excerpts from the 1965 iced-shrimp symposium. U.S. Dept. of Interior Circular #284.

12. **Cavett, J. J.** 1962. The microbiology of vacuum packed sliced bacon. J. Appl. Bacteriol. *25:* 282–289.

13. **Chipley, J. R. and E. K. Heaton.** 1971. Microbial flora of pecan meat. Appl. Microbiol. *22:* 252–253.

14. **Christiansen, L. N. and N. S. King.** 1971. The microbial content of some salads and sandwiches at retail outlets. J. Milk Food Technol. *34:* 289–293.

15. **Cliver, D. O. and J. Grindrod.** 1969. Surveillance methods for viruses in foods. J. Milk Food Technol. *32:* 421–425.

16. **Dowdell, M. J. and R. G. Board.** 1968. A microbiological survey of British fresh sausage. J. Appl. Bacteriol. *31:* 378–396.

17. **Drake, S. D., J. B. Evans, and C. F. Niven, Jr.** 1958. Microbial flora of packaged frankfurters and their radiation resistance. Food Res. *23:* 291–296.

18. **Duitschaever, C. L., D. R. Arnott, and D. H. Bullock.** 1973. Bacteriological quality of raw refrigerated ground beef. J. Milk Food Technol. *36:* 375–377.

19. **Emswiler, B. S., C. J. Pierson, and A. W. Kotula.** 1976. Bacteriological quality and shelf life of ground beef. Appl. Environ. Microbiol. *31:* 826–830.

20. **Fanelli, M. J., A. C. Peterson, and M. F. Gunderson.** 1965A. Microbiology of dehydrated soups. I. A survey. Food Technol. *19:* 83–86.

21. **Fanelli, M. J., A. C. Peterson, and M. F. Gunderson.** 1965B. Microbiology of dehydrated soups. III. Bacteriological examination of rehydrated dry soup mixes. Food Technol. *19:* 90–94.

22. **Foter, M. J. and A. M. Gorlick.** 1938. Inhibitory properties of horseradish vapors. Food Res. *3:* 609–613.

23. **Fowler, J. L. and W. S. Clark, Jr.** 1975. Microbiology of delicatessen salads. J. Milk Food Technol. *38:* 146–149.

24. **Fowler, J. L. and J. F. Foster.** 1976. A microbiological survey of three fresh green salads: Can guidelines be recommended for these foods? J. Milk Food Technol. *39:* 111–113.

25. **Fugate, K. J., D. O. Cliver, and M. T. Hatch.** 1975. Enteroviruses and potential bacterial indicators in Gulf Coast oysters. J. Milk Food Technol. *38:* 100–104.

26. **Gabis, D. A., B. E. Langlois, and A. W. Rudnick.** 1970. Microbiological examination of cocoa powder. Appl. Microbiol. *20:* 644–645.

27. **Greenberg, R. A., R. B. Tompkin, B. O. Bladel, R. S. Kittaka, and A. Anellis.** 1966. Incidence of mesophilic *Clostridium* spores in raw pork, beef, and chicken in processing plants in the United States and Canada. Appl. Microbiol. *14:* 789–793.

28. **Gunderson, M. F., H. W. McFadden, and T. S. Kyle.** 1954. *The Bacteriology of Commercial Poultry Processing.* Burgess Publishing Co., Minneapolis, Minn.

29. **Guthertz, L. S., J. T. Fruin, D. Spicer, and J. L. Fowler.** 1976. Microbiology of fresh comminuted turkey meat. J. Milk Food Technol. *39:* 823–829.

30. **Harmon, L. G., C. M. Stine, and G. C. Walker.** 1962. Composition, physical properties and microbiological quality of chip-dips. J. Milk Food Technol. *25:* 7–11.

31. **Harris, N. D., S. R. Martin, and L. Ellias.** 1975. Bacteriological quality of selected delicatessen foods. J. Milk Food Technol. *38:* 759–761.

32. **Herrmann, J. E. and D. O. Cliver.** 1973. Enterovirus persistence in sausage and ground beef. J. Milk Food Technol. *36:* 426–428.

33. **Hill, W. M., J. Reaume, and J. C. Wilcox.** 1976. Total plate count and sensory evaluation as measures of luncheon meat shelf life. J. Milk Food Technol. *39:* 759–762.

34. **Hosty, T. S. and C. I. McDurmont.** 1975. Isolation of acid-fast organisms from milk and oysters. Hlth. Lab. Sci. *12:* 16–19.

35. **Huber, D. A., H. Zaborowski, and M. M. Rayman.** 1958. Studies on the microbiological quality of precooked frozen meals. Food Technol. *12:* 190–194.

36. **Hyndman, J. B.** 1963. Comparison of enterococci and coliform microorganisms in commercially produced pecan nut meats. Appl. Microbiol. *11:* 268–272.

37. **Ingram, M.** 1960. Bacterial multiplication in packed Wiltshire bacon. J. Appl. Bacteriol. *23:* 206–215.

38. **Insalata, N. F., J. S. Witzeman, J. H. Berman, and E. Borker.** 1968. A study of the incidence of the spores of *Clostridium botulinum* in frozen vacuum pouch-pack vegetables. Proc., 96th Ann. Meet., Am. Pub. Health. Assoc., p. 124.

39. **Insalata, N. F., S. J. Witzeman, G. J. Fredericks, and F. C. A. Sunga.** 1968. Incidence study of spores of *Clostridium botulinum* in convenience foods. Appl. Microbiol. *17:* 542–544.

40. **Jay, J. M.** 1961. Incidence and properties of coagulase-positive staphylococci in certain market meats as determined on three selective media. Appl. Microbiol. *9:* 228–232.

41. **Jay, J. M.** 1964. Beef microbial quality determined by extract-release volume (ERV). Food Technol. *18:* 1637–1641.

42. **Jay, J. M.** 1965. Unpublished Data.

43. **Johnson, M. G. and R. H. Vaughn.** 1969. Death of *Salmonella typhimurium* and *Escherichia coli* in the presence of freshly reconstituted dehydrated garlic and onion. Appl. Microbiol. *17:* 903–905.

44. **Julseth, R. M. and R. H. Deibel.** 1974. Microbial profile of selected spices and herbs at import. J. Milk Food Technol. *37:* 414–419.

45. **Kachikian, R., C. R. Fellers, and W. Litsky.** 1959. A bacterial survey of commercial frozen breaded shrimp. J. Milk Food Technol. *22:* 310–312.

46. **Käferstein, F. K.** 1976. The microflora of parsley. J. Milk Food Technol. *39:* 837–840.

47. **Kajs, T. M., R. Hagenmaier, C. Vanderzant, and K. F. Mattil.** 1976. Microbiological evaluation of coconut and coconut products. J. Food Sci. *41:* 352–356.

48. **Karlson, K. E., and M. F. Gunderson.** 1965. Microbiology of dehydrated soups. II. "Adding machine" approach. Food Technol. *19:* 86–90.

49. **Kazanas, N., J. A. Emerson, H. L. Seagram, and L. L. Kempe.** 1966. Effect of γ-irradiation on the microflora of fresh-water fish. I. Microbial load, lag period, and rate of growth on yellow perch *(Perca flavescens)* fillets. Appl. Microbiol. *14:* 261–266.

50. **Kereluk, K. and M. F. Gunderson.** 1959. Studies on the bacteriological quality of frozen meat pies. I. Bacteriological survey of some commercially frozen meat pies. Appl. Microbiol. *7:* 320–323.

51. **King, A. D., Jr., M. J. Miller, and L. C. Eldridge.** 1970. Almond harvesting, processing, and microbial flora. Appl. Microbiol. *20:* 208–214.

52. **Kitchell, A. G.** 1962. Micrococci and coagulase negative staphylococci in cured meats and meat products. J. Appl. Bacteriol. *25:* 416–431.

53. **Koburger, J. A.** 1970. Fungi in foods. I. Effect of inhibitor and incubation temperature on enumeration. J. Milk Food Technol. *33:* 433–434.

54. **Koburger, J. A. and B. Y. Farhat.** 1975. Fungi in foods. VI. A comparison of media to enumerate yeasts and molds. J. Milk Food Technol. *38:* 466–468.

55. **Konowalchuk, J. and J. I. Speirs.** 1975a. Survival of enteric viruses on fresh vegetables. J. Milk Food Technol. *38:* 469–472.

56. **Konowalchuk, J. and J. I. Speirs.** 1975b. Survival of enteric viruses on fresh fruit. J. Milk Food Technol. *38:* 598–600.

57. **Kraft, A. A., J. C. Ayres, G. S. Torrey, R. H. Salzer, and G. A. N. DaSilva.** 1966. Coryneform bacteria in poultry, eggs and meat. J. Appl. Bacteriol. *29:* 161–166.

58. **Ladiges, W. C., J. F. Foster, and W. M. Ganz.** 1974. Incidence and viability of *Clostridium perfringens* in ground beef. J. Milk Food Technol. *37:* 622–623.

59. **Leininger, H. V., L. R. Shelton, and K. H. Lewis.** 1971. Microbiology of frozen cream-type pies, frozen cooked-peeled shrimp, and dry food-grade gelatin. Food Technol. *25:* 224–229.

60. **Lepovetsky, B. C., H. H. Weiser, and F. E. Deatherage.** 1953. A microbiological study of lymph nodes, bone marrow, and muscle tissue obtained from slaughtered cattle. Appl. Microbiol. *1:* 57–59.

61. **Lerke, P., R. Adams, and L. Farber.** 1963. Bacteriology of spoilage of fish muscle. I. Sterile press juice as a suitable experimental medium. Appl. Microbiol. *11:* 458–462.

62. **Lillard, H. S.** 1971. Occurrence of *Clostridium perfringens* in broiler processing and further processing operations. J. Food Sci. *36:* 1008–1010.

63. **Litsky, W., I. S. Fagerson, and C. R. Fellers.** 1957. A bacteriological survey of commercially frozen beef, poultry and tuna pies. J. Milk Food Technol. *20:* 216–219.

64. **May, K. N., J. D. Irby, and J. L. Carmon.** 1961. Shelf life and bacterial counts of excised poultry tissue. Food Technol. *16:* 66–68.

65. **May, K. N.** 1962. Bacterial contamination during cutting and packaging chicken in processing plants and retail stores. Food Technol. *16:* 89–91.

66. **Mercuri, A. J., G. J. Banwart, J. A. Kinner, and A. R. Sessoms.** 1970. Bacteriological examination of commercial precooked Eastern-type turkey rolls. Appl. Microbiol. *19:* 768–771.

67. **Miller, W. A.** 1961. The microbiology of some self-service, packaged, luncheon meats. J. Milk Food Technol. *24:* 374–377.

68. **Mossel, D. A. A., C. L. Vega, and H. M. C. Put.** 1975. Further studies on the suitability of various media containing antibacterial antibiotics for the enumeration of moulds in food and food environments. J. Appl. Bacteriol. *39:* 15–22.

69. **Mueller, D. C.** 1975. Microbiological safety and palatability of selected vended burgers. J. Milk Food Technol. *38:* 135–137.

70. **Mundt, J. O., W. F. Graham, and I. E. McCarty.** 1967. Spherical lactic acid-producing bacteria of Southern-grown raw and processed vegetables. Appl. Microbiol. *15:* 1303–1308.

71. **McDivitt, M. E. and D. L. Hussemann.** 1957. Growth of micrococci in cooked ham. J. American Dietetic Assoc. *33:* 238–242.

72. **Nakamura, M. and K. D. Kelly.** 1968. *Clostridium perfringens* in dehydrated soups and sauces. J. Food Sci. *33:* 424–426.

73. **Nickerson, J. T. R., G. J. Silverman, M. Solberg, D. W. Duncan, and M. M. Joselow.** 1962. Microbial analysis of commercial frozen fish sticks. J. Milk Food Technol. *25:* 45–47.

74. **Nickerson, J. T. R. and S. A. Goldblith.** 1964. A study of the microbiological quality of haddock fillets and shucked, soft-shelled clams processed and marketed in the greater Boston area. J. Milk Food Technol. *27:* 7–12.

75. **Pace, P. J., E. R. Krumbiegal, R. Angelotti, and H. J. Wisniewski.** 1967. Demonstration and isolation of *Clostridium botulinum* types from whitefish chubs collected at fish smoking plants of the Milwaukee area. Appl. Microbiol. *15:* 877–884.

76. **Pace, P. J.** 1975. Bacteriological quality of delicatessen foods: Are standards needed? J. Milk Food Technol. *38:* 347–353.

77A. **Peppler, H. J.** 1976. Yeasts, pp. 427–454, IN: *Food Microbiology: Public Health and Spoilage Aspects,* edited by M. P. DeFigueiredo and D. F. Splittstoesser (Avi Publishing: Westport, Conn.).

77B. **Pivnick, H., I. E. Erdman, D. Collins-Thompson, G. Roberts, M. A. Johnston, D. R. Conley, G. Lachapelle, U. T. Purvis, R. Foster, and M. Milling.** 1976. Proposed microbiological standards for ground beef based on a Canadian survey. J. Milk Food Technol. *39:* 408–412.

78. **Potter, N. N.** 1973. Viruses in foods. J. Milk Food Technol. *36:* 307–310.

79. **Powers, E. M., C. Ay, H. M. El-Bisi, and D. B. Rowley.** 1971. Bacteriology of dehydrated space foods. Appl. Microbiol. *22:* 441–445.

80. **Powers, E. M., R. Lawyer, and Y. Masuoka.** 1975. Microbiology of processed spices. J. Milk Food Technol. *38:* 683–687.

81. **Powers, E. M., T. G. Latt, and T. Brown.** 1976. Incidence and levels of *Bacillus cereus* in processed spices. J. Milk Food Technol. *39:* 668–670.

82. **Riha, W. E. and M. Solberg.** 1970. Microflora of fresh pork sausage casings. 2. Natural casings. J. Food Sci. *35:* 860–863.

83. **Rogers, R. E. and C. S. McCleskey.** 1957. Bacteriological quality of ground beef in retail markets. Food Technol. *11:* 318–320.

84. **Shehata, T. E. and E. B. Collins.** 1971. Isolation and identification of psychrophilic species of *Bacillus* from milk. Appl. Microbiol. *21:* 466–469.

85. **Shelef, L. A.** 1975. Microbial spoilage of fresh refrigerated beef liver. J. Appl. Bacteriol. *39:* 273–280.

86. **Silverman, G. J., J. T. R. Nickerson, D. W. Duncan, N. S. Davis, J. S. Schachter, and M. M. Joselow.** 1961. Microbial analysis of frozen raw and cooked shrimp. I. General results. Food Technol. *15:* 455–458.

87. **Smith, F. C., R. A. Field, and J. C. Adams.** 1974. Microbiology of Wyoming big game meat. J. Milk Food Technol. *37:* 129–131.

88. **Splittstoesser, D. F., W. P. Wettergreen, and C. S. Pederson.** 1961A. Control of microorganisms during preparation of vegetables for freezing. I. Green beans. Food Technol. *15:* 329–331.

89. **Splittstoesser, D. F., W. P. Wettergreen, and C. S. Pederson.** 1961B. Control of microorganisms during preparation of vegetables for freezing. II. Peas and corn. Food Technol. *15:* 332–334.

90. **Splittstoesser, D. F., G. E. R. Hervey, II., and W. P. Wettergreen.** 1965. Contamination of frozen vegetables by coagulase-positive staphylococci. J. Milk Food Technol. *28:* 149–151.

91. **Splittstoesser, D. F. and I. Gadjo.** 1966. The groups of micro-organisms composing the "total" count population in frozen vegetables. J. Food Sci. *31:* 234–239.

92. **Splittstoesser, D. F., M. Wexler, J. White, and R. R. Colwell.** 1967. Numerical taxonomy of gram-positive and catalase-positive rods isolated from frozen vegetables. Appl. Microbiol. *15:* 158–162.

93. **Splittstoesser, D. F.** 1973. The microbiology of frozen vegetables. How they get contaminated and which organisms predominate. Food Technol. *27:* 54–56.

94. **Splittstoesser, D. F. and D. B. Prest.** 1976. Molds, pp. 455–480, IN: *Food Microbiology: Public Health and Spoilage Aspects,* edited by M. P. de Figueiredo and D. F. Splittstoesser (Avi Publishing: Westport: Conn.).

95. **Steinkraus, K. H. and J. C. Ayres.** 1964. Incidence of putrefactive anaerobic spores in meat. J. Food Sci. *29:* 87–93.

96. **Strong, D. H., J. C. Canada, and B. B. Griffiths.** 1963. Incidence of *Clostridium perfringens* in American foods. Appl. Microbiol. *11:* 42–44.

97. **Sullivan, R., A. C. Fassolitis, and R. B. Read, Jr.** 1970. Method for isolating viruses from ground beef. J. Food Sci. *35:* 624–626.

98. **Sullivan, R., R. M. Marnell, E. P. Larkin, and R. B. Read, Jr.** 1975. Inactivation of poliovirus 1 and coxsackievirus B-2 in broiled hamburgers. J. Milk Food Technol. *38:* 473–475.

99. **Surkiewicz, B. F.** 1966. Bacteriological survey of the frozen prepared foods industry. Appl. Microbiol. *14:* 21–26.

100. **Surkiewicz, B. F., J. B. Hyndman, and M. V. Yancey.** 1967. Bacteriological survey of the frozen prepared foods industry. II. Frozen breaded raw shrimp. Appl. Microbiol. *15:* 1–9.

101. **Surkiewicz, B. F., R. J. Groomes, and A. P. Padron.** 1967. Bacteriological survey of the frozen prepared foods industry. III. Potato products. Appl. Microbiol. *15:* 1324–1331.

102. **Surkiewicz, B. F., R. W. Johnston, R. P. Elliott, and E. R. Simmons.** 1972. Bacteriological survey of fresh pork sausage produced at establishments under federal inspection. Appl. Microbiol. *23:* 515–520.

103. **Surkiewicz, B. F., M. E. Harris, R. P. Elliott, J. F. Macaluso, and M. M. Strand.** 1975. Bacteriological survey of raw beef patties produced at establishments under federal inspection. Appl. Microbiol. *29:* 331–334.

104. **Surkiewicz, B. F., M. E. Harris, and J. M. Carosella.** 1977. Bacteriological survey and refrigerated storage test of vacuum-packed sliced imported canned ham. J. Food Protec. *40:* 109–111.

105. **Vanderzant, C., A. W. Matthys, and B. F. Cobb, III.** 1973. Microbiological, chemical, and organoleptic characteristics of frozen breaded raw shrimp. J. Milk Food Technol. *36:* 253–261.

106. **Walker, H. W. and J. C. Ayres.** 1970. Yeasts as spoilage organisms, pp. 463–527, IN: *The Yeasts. 3. Yeast Technology,* edited by A. H. Rose and J. S. Harrison (Academic Press: N.Y.).

107. **Walker, H. W.** 1977. Spoilage of food by yeasts. Food Technol. *31:* 57–65.

108. **Weiser, H. H.** 1962. *Practical Food Microbiology and Technology.* Avi Publishing Co., Westport, Conn.

109. **Winter, A. R., M. D. Baughman, and H. H. Weiser.** 1954. Microbiological aspects of orangegg. Food Technol. *8:* 38–39.

110. **Woodburn, M.** 1964. Incidence of Salmonellae in dressed broiler-fryer chickens. Appl. Microbiol. *12:* 492–495.

111. **Zottola, E. A. and F. F. Busta.** 1971. Microbiological quality of further-processed turkey products. J. Food Sci. *36:* 1001–1004.

6
Food Spoilage: Spoilage of Market Fruits and Vegetables

Spoiled food may be defined as food that has been damaged or injured so as to make it undesirable for human use. Food spoilage may, therefore, be caused by insect damage, by physical injury of various kinds such as bruising and freezing, by enzyme activity, or by microorganisms. Only that caused by microorganisms will be treated here.

As pointed out in Chapter 2, the microbial spoilage of foods should not be viewed as a sinister plot on the part of microorganisms to deliberately destroy our foods, but instead, as the normal function of these organisms in the total ecology of all living organisms. While plants and animals have both evolved intrinsic mechanisms that aid them in combating harmful microbes, many microorganisms can and do overcome these forces and bring about the destruction of the plant and animal organic matter by converting it into inorganic compounds. These primary activities of microorganisms on foods bring about the phenomenon we refer to as spoilage.

It has been estimated that 20% of all fruits and vegetables harvested for human consumption are lost through microbial spoilage by one or more of 250 market diseases *(2)*. The primary causative agents of microbial spoilage are the bacteria, yeasts, and molds. While viruses have the capacity to damage both plant and animal tissues, these agents along with the *Rickettsia* are not generally regarded as being important in food spoilage as it is now recognized. With respect to yeasts, molds, and bacteria, the latter two are by far the most important etiologic agents of food spoilage in general. On the basis of their growth requirements previously discussed in Chapter 3, each may be expected to occupy its own niche with respect to the various types of foods and even within the same food.

THE MICROBIAL SPOILAGE OF VEGETABLES

The general composition of higher plants is presented in Table 6-1, and the composition of 21 common vegetables is presented in Table 6-2. It can be

TABLE 6-1.   General chemical composition of higher plant material.

A. CARBOHYDRATES AND RELATED
   1. Polysaccharides—pentosan (araban), hexosans (cellulose, starch, xylans, fructans, mannans, galactans, levans).
   2. Oligosaccharides—tetrasaccharide (stachyose), trisaccharides (robinose, mannotriose, raffinose), disaccharides (maltose, sucrose, cellobiose, melibiose, trehalose).
   3. Monosaccharides—hexoses (mannose, glucose, galactose, fructose, sorbose), pentoses (arabinose, xylose, ribose, L-rhamnose, L-fucose).
   4. Sugar alcohols—glycerol, ribitol, mannitol, sorbitol, inositols.
   5. Sugar acids—uronic acids, ascorbic acid.
   6. Esters—tannins.
   7. Organic acids—citric, shikimic, D-tartaric, oxalic, lactic, glycolic, malonic, etc.
B. PROTEINS—albumins, globulins, glutelins, prolamines, peptides, and amino acids.
C. LIPIDS—fatty acids, fatty acid esters, phospholipids, glycolipids, etc.
D. NUCLEIC ACIDS AND DERIVATIVES—purine and pyrimidine bases, nucleotides, etc.
E. VITAMINS—fat-soluble (A, D, E), water-soluble (thiamine, niacin, riboflavin, etc.).
F. MINERALS—Na, K, Ca, Mg, Mn, Fe, etc.
G. WATER
H. OTHERS—alkaloids, porphyrins, aromatics, etc.

seen that the average water content of vegetables is about 88% with an average carbohydrate content of 8.6%, 1.9% proteins, 0.3% fat, and 0.84% ash. While the percentage composition of vitamins, nucleic acids, and other plant constituents is not given, the total of these compounds is generally less than 1%. From the standpoint of nutrient content, vegetables are capable of supporting the growth of molds, yeasts, and bacteria, and consequently of being spoiled by any or all of these organisms. The higher water content of vegetables favors the growth of spoilage bacteria and the relatively low carbohydrate and fat contents suggest that much of this water is in available form. The pH range of most vegetables is within the growth range of a large number of bacteria and it is not surprising, therefore, that bacteria are common agents of vegetable spoilage. The relatively high O/R potential of vegetables and their lack of high poising capacity suggest that the aerobic and facultative anaerobic types would be more important than the anaerobes. This is precisely the case, since some of the most ubiquitious etiologic agents in the bacterial spoilage of vegetables are species of the genus *Erwinia,* and are associated with plants and vegetables in their natural growth environment. The common spoilage pattern displayed by these organisms is referred to as **bacterial soft rot** and is described below with some of the most commonly affected vegetables.

Bacterial Agents

1. Bacterial Soft Rot.   This type of spoilage is caused by *Erwinia caroto-vora* and pseudomonads such as *Ps. marginalis* with the former being the more important. *Bacillus* and *Clostridium* spp. have been implicated, but their roles are probably secondary.

The causative organisms break down pectins, giving rise to a soft, mushy consistency, sometimes a bad odor, and a water-soaked appearance. Some of the vegetables affected by this disease are: asparagus, onions, garlic, beans (green, lima, wax), carrots, parsnip, celery, parsley, beets, endives, globe artichoke, lettuce, rhubarb, spinach, potatoes, cabbage, Brussels sprouts, cauliflower, broccoli, radishes, rutabagas, turnips, tomatoes, cucumbers, cantaloupes, peppers, and watermelons.

While the precise manner in which *Erwinia* spp. bring about soft rot is not yet well understood, it is very likely that these organisms, present on the susceptible vegetables at the time of harvest, subsist upon vegetable sap until the supply is exhausted. The cementing substance of the vegetable body then induces the formation of pectinases, which act by hydrolyzing pectin, thereby producing the mushy consistency. In potatoes, tissue maceration has been shown to be caused by an endopolygalacturonate trans-

TABLE 6-2.   Approximate percentage chemical composition of 21 vegetable foods *(8)*.

Vegetable	Water	Carbohyd.	Proteins	Fat	Ash
Beans, green	89.9	7.7	2.4	0.2	0.8
Beets	87.6	9.6	1.6	0.1	1.1
Broccoli	89.9	5.5	3.3	0.2	1.1
Brussels sprouts	84.9	8.9	4.4	0.5	1.3
Cabbage	92.4	5.3	1.4	0.2	0.8
Cantaloupe	94.0	4.6	0.2	0.2	0.6
Cauliflower	91.7	4.9	2.4	0.2	0.8
Celery	93.7	3.7	1.3	0.2	1.1
Corn	73.9	20.5	3.7	1.2	0.7
Cucumbers	96.1	2.7	0.7	0.1	0.4
Lettuce	94.8	2.9	1.2	0.2	0.9
Onions	87.5	10.3	1.4	0.2	0.6
Peas	74.3	17.7	6.7	0.4	0.9
Potatoes	77.8	19.1	2.0	0.1	1.0
Pumpkin	90.5	7.3	1.2	0.2	0.8
Radishes	93.6	4.2	1.2	0.1	1.0
Spinach	92.7	3.2	2.3	0.3	1.5
Squash, summer	95.0	3.9	0.6	0.1	0.4
Sweet potatoes	68.5	27.9	1.8	0.7	1.1
Tomatoes	94.1	4.0	1.0	0.3	0.6
Watermelon	92.1	6.9	0.5	0.2	0.3
Mean	88.3	8.6	2.0	0.3	0.8

eliminase of *Erwinia* origin *(6)*. With the early and relatively rapid growth of these organisms, molds tend to be crowded out and of less consequence in the spoilage of these products. Once the outer plant barrier has been destroyed by these pectinase producers, other nonpectinase producers, no doubt, enter the plant tissues and help bring about fermentation of the simple carbohydrates that are present. The quantities of simple nitrogenous compounds present, the vitamins (especially the B-complex group), and minerals are adequate for growth of the invading organisms until the vegetables have been essentially consumed or destroyed. The malodors that are produced are probably the direct result of volatile compounds produced by the flora such as NH_3, volatile acids, and the like. When growing in acid media, microorganisms tend to decarboxylate amino acids, leaving amines which cause an elevation of pH towards the neutral range and beyond. Complex carbohydrates such as cellulose are generally the last to be degraded, and a varied flora consisting of molds and other soil organisms is usually responsible, since cellulose degradation by *Erwinia* spp. is doubtful. Aromatic constituents and porphyrins are probably not attacked until late in the spoilage process, and again by a varied flora of soil types.

The genus *Erwinia* belongs to the family *Enterobacteriaceae*. Of the 13 species listed in *Bergey's Manual*, all are associated with plants where they are known to cause plant diseases of the rot and wilt types. These are gram negative rods that are related to the genera *Proteus, Serratia, Escherichia, Salmonella,* and others. *Erwinia* spp. normally do not require organic nitrogen compounds for growth, and the relatively low levels of proteins in vegetables make them suitable for the task of destroying plant materials of this type. The pectinase produced by these organisms is actually a **protopectinase,** since the cementing substance of plants as it actually exists in the plant is protopectin. Many *Erwinia* spp. such as *E. carotovora* are capable of fermenting many of the sugars and alcohols that exist in certain vegetables such as rhamnose, cellobiose, arabinose, mannitol, and so forth—compounds which are not utilized by many of the more common bacteria. While most *Erwinia* spp. grow well at about 37 C, most are also capable of good growth at refrigerator temperatures with some strains reported to grow at 1 C.

2. Other Bacterial Spoilage Conditions.   "Black leg" of potatoes, especially in Scotland, is caused mainly by *E. carotovora* var *atroseptica*. *E. carotovora* var. *carotovora* also causes black leg, generally at temperatures of around 25 C or above *(4)*. With respect to the source of infecting organisms, some investigators have found soil to be the source while others have been unable to confirm this. A recent study by Jones and Paton *(3)* suggests that the black leg organisms may exist as L-phase varients in soils and infect potatoes in this state. Once healthy tissues have been invaded, the L-phase varients then revert to classical forms.

Although *Erwinia* spp. constitute the most important group of bacteria causing vegetable spoilage, members of the genus *Pseudomonas* are pathogens of edible plants. In addition to *Ps. marginalis* noted above which causes soft rot, *Ps. glycinea* causes a disease of soybeans; *Ps. apii* causes **bacterial blight** of celery; *Ps. cichorii* causes **bacterial zonate spot** of cabbage and lettuce; *Ps. lachrymans* causes **angular leaf spot** of cucumbers; *Ps. maculicola* causes **bacterial leaf spot** of broccoli and cauliflower; *Ps. phaseolicola* causes **halo blight** of beans; *Ps. pisi* causes **bacterial blight** of peas; and *Ps. tomato* causes **bacterial speck** of tomatoes. **Black rot** of cabbage and cauliflower is caused by *Xanthomonas campestris;* **common blight** of beans by *X. phaseoli;* and **bacterial spot** of tomatoes and peppers by *X. vesicatoria.* **Bacterial wilt** of beans is caused by *Corynebacterium flaccumfaciens;* **bacterial canker** by *C. michiganense;* and **ring rot** of potatoes by *C. sepedonicum (4).*

The appearance of some market vegetables undergoing bacterial and fungal spoilage may be seen from Plates 6-1 and 6-2.

While the genera *Erwinia* and *Pseudomonas* are the most important bacteria that cause vegetable spoilage, the molds are by far the most important group of organisms in vegetable spoilage. One of the most versatile groups belongs to the genus *Botrytis,* which causes **gray mold rot** on at least 26 common vegetables.

Fungal Agents

1. Gray Mold Rot. Caused by *Botrytis cinerea* which produces a gray mycelium. This type of spoilage is favored by high humidity and warm temperatures. Among the vegetables affected are: asparagus, onions, garlic, beans (green, lima, wax), carrots, parsnip, celery, tomatoes, endives, globe artichokes, lettuce, rhubarb, cabbage, Brussels sprouts, cauliflower, broccoli, radishes, rutabagas, turnips, cucumbers, pumpkin, squash, peppers, sweet potatoes, and others. In this disease, the causal fungus grows on decayed areas in the form of a prominent gray mold. It can enter fruits and vegetables through the unbroken skin, or through cuts and cracks.

2. Sour Rot (Oospora Rot, Watery Soft Rot). This market disease of vegetables is caused by *Geotrichum candidum* and others. Among the vegetables affected are asparagus, onions, garlic, beans (green, lima, wax), carrots, parsnip, parsley, endives, globe artichoke, lettuce, cabbage, Brussels sprouts, cauliflower, broccoli, radishes, rutabagas, turnips, and tomatoes. The causal fungus of this disease is widely distributed in soils and on decaying fruits and vegetables. *Drosophila melanogaster* (fruit fly) carries spores and mycelial fragments on its body from decaying fruits and vegetables to growth cracks and wounds in healthy fruits and vegetables. Since the fungus cannot enter through the unbroken skin, infections usually start in openings of one type or another *(5).*

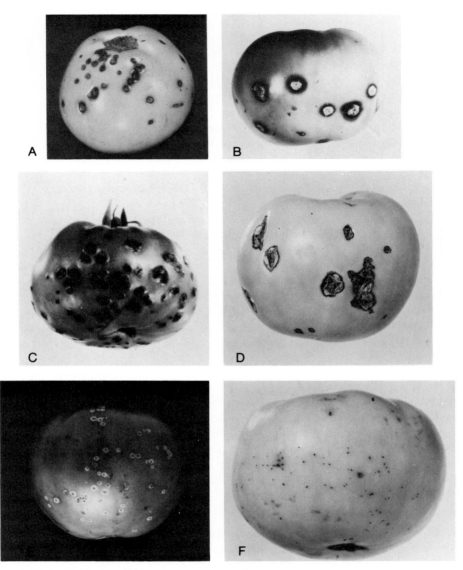

PLATE 6-1.   Tomato Diseases: A and B, Nailhead spot; C and D, bacterial spot; E, bacterial canker; F, bacterial speck. (From Agriculture Handbook 28, USDA, 1968, "Fungus and bacterial diseases of fresh tomatoes.")

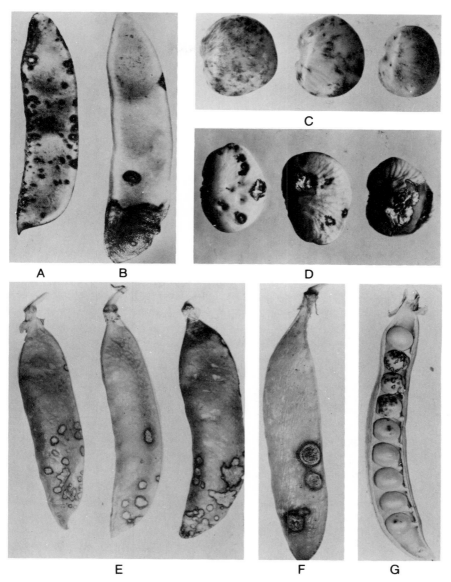

PLATE 6-2.   Lima Bean Diseases: A and B, pod blight; C, seed spotting; D, yeast spot. Pea Diseases: E, pod spot; F, anthracnose; G, scab. (From Agriculture Handbook 303, USDA, 1966, Chapter 5.)

3. Rhizopus Soft Rot.   This market disease is caused by *R. stolonifer* and other species that make vegetables soft and mushy. Cottony growth of the mold with small black dots of sporangia often covers the vegetables. Among those affected are beans (green, lima, wax), carrots, sweet potatoes, potatoes, cabbage, Brussels sprouts, cauliflower, broccoli, radishes, rutabagas, turnips, cucumbers, cantaloupes, pumpkin, squash, watermelons, and tomatoes. This fungus is spread by *D. melanogaster* which lays its eggs in the growth cracks on various fruits and vegetables. The fungus is widespread and is disseminated also by other means. Infections usually occur through wounds and other skin breaks.

4. Phytophora Rot.   This market disease is caused by *Phytophora* spp. It occurs largely in the field as a blight and fruit rot of market vegetables. It appears to be more variable than some other market diseases and affects different plants in different ways. Among the vegetables affected are asparagus, onions, garlic, cantaloupes, watermelons, tomatoes, eggplants, and peppers.

5. Anthracnose.   This plant disease is characterized by spotting of leaves, fruit, or seed pods. It is caused by *Colletotrichum coccodes* and other species. These fungi are considered weak plant pathogens. They live from season to season on plant debris in the soil and on the seed of various plants such as the tomato. Their spread is favored by warm, wet weather. Among the vegetables affected are beans, cucumbers, watermelons, pumpkins, squash, tomatoes, and peppers.

Others.   Other market diseases of vegetables include: *Fusarium* rots, caused by species of this genus; **black rot** and *Alternaria* rot, caused by *Alternaria* spp; **downey mildew,** caused by *Phytophthora, Bremia* spp., and others; **brown rot,** caused by *Sclerotinia* spp.; **blue mold rot,** caused by *Penicillium* spp.; **pink mold rot,** caused by *Trichothecium roseum;* and many others. For further information on market diseases of fruits and vegetables, the monographs issued by the Agricultural Research Service of the U.S. Department of Agriculture should be consulted (See Plate 6-3 for several fungal diseases of onions).

SPOILAGE OF FRUITS

The general composition of 18 common fruits is presented in Table 6-3, which shows that the average water content is about 85 percent and the average carbohydrate content is about 13 percent. The fruits differ from vegetables in having somewhat less water but more carbohydrate. The mean protein, fat, and ash contents of fruits are, respectively, 0.9, 0.5, and 0.5 percent, which are somewhat lower than vegetables except for ash content. Though not shown in the table, fruits contain vitamins and other organic compounds just as do vegetables. On the basis of nutrient content, these

PLATE 6-3.   Onion Diseases: A, white rot; B, black mold rot; C, diplodia stain. (From Agriculture Handbook 303, USDA, 1966, Chapter 5.)

TABLE 6-3. Approximate percentage composition of 18 common fruits *(8)*.

Fruit	Water	Carbohyd.	Protein	Ash	Fat
Apples	84.1	14.9	0.3	0.3	0.4
Apricots	85.4	12.9	1.0	0.6	0.1
Bananas	74.8	23.0	1.2	0.8	0.2
Blackberries	84.8	12.5	1.2	0.5	1.0
Cherries, sweet & sour	83.0	14.8	1.1	0.6	0.5
Figs	78.0	19.6	1.4	0.6	0.4
Grapefruit	88.8	10.1	0.5	0.4	0.2
Grapes, American type	81.9	14.9	1.4	0.4	1.4
Lemons	89.3	8.7	0.9	0.5	0.6
Limes	86.0	12.3	0.8	0.8	0.1
Oranges	87.2	11.2	0.9	0.5	0.2
Peaches	86.9	12.0	0.5	0.5	0.1
Pears	82.7	15.8	0.7	0.4	0.4
Pineapples	85.3	13.7	0.4	0.4	0.2
Plums	85.7	12.9	0.7	0.5	0.2
Raspberries	80.6	15.7	1.5	0.6	1.6
Rhubarb	94.9	3.8	0.5	0.7	0.1
Strawberries	89.9	8.3	0.8	0.5	0.5
Mean	84.9	13.2	0.88	0.53	0.46

products would appear to be capable of supporting the growth of bacteria, yeasts, and molds. However, when the pH of fruits alone is considered, it is found to be below the level that generally favors bacterial growth. This one fact alone would seem to be sufficient to explain the general absence of bacteria in the incipient spoilage of fruits. The wider pH growth range of molds and yeasts suits them as spoilage agents of fruits. With the exception of pears, which sometimes undergo **Erwinia rot,** bacteria are of no known importance in the initiation of fruit spoilage. Just why pears with a reported pH range of 3.8–4.6 should undergo bacterial spoilage is not clear. It is conceivable that *Erwinia* initiates its growth on the surface of this fruit where the pH is presumably higher than on the inside.

Some of the molds important in fruit spoilage along with fruits affected are presented below. Since many of the same organisms discussed above for vegetables also affect fruits, further descriptions are not given here.

1. Blue Mold Rot. Caused by *Penicillium* spp. It affects blackberries, currants, dewberries, grapes, lemons, limes, oranges, grapefruits, cherries, peaches, apricots, plums, prunes, apples, pears and others.

2. Gray Mold Rot. Caused by *Botrytis cinerea.* It affects grapes, blackberries, dewberries, currants, strawberries, oranges, lemons, limes, cherries, peaches, apricots, plums, prunes, apples, pears, and many others.

3. Rhizopus Rot.   Caused by *R. stolonifer*. It infects grapes, strawberries, avocadoes, cherries, peaches, apricots, plums, prunes, pears, apples, blackberries, and others.

4. Black Mold Rot ("smut").   Caused by *Aspergillus niger*. It affects grapes, cherries, peaches, apricots, plums, prunes, and others.

5. Green Mold Rot.   Caused by *Cladosporium, Trichoderma* and other molds. It infects grapes, cherries, peaches, apricots, plums, and prunes among others.

6. Alternaria Rot.   Caused by *A. tenuis*. It infects lemons, limes, oranges, grapefruits, apples, and others.

7. Cladosporium Rot.   Caused by *C. herbarum*. It infects cherries, peaches, apricots, plums, prunes, and others.

Others.   In addition to the above market diseases of fruits, there are many others involving a large number of different fruits and a wide range of fungi. Among them are **anchracnose, brown mold rot, stem-end rot, brown rot,** and **powdery mildew.**

A variety of yeast genera can usually be found on fruits and these organisms often bring about the spoilage of fruit products, especially in the field. Many yeasts are capable of attacking the sugars found in fruits and bringing about fermentation with the production of alcohol and CO_2. Due to their generally faster growth rate than molds, they often precede the latter organisms in the spoilage process of fruits in certain circumstances. It is not clear whether some molds are dependent upon the initial action of yeasts in the process of fruit and vegetable spoilage. The utilization or destruction of the high molecular weight constituents of fruits is brought about more by molds than yeasts. Many molds are capable of utilizing alcohols as sources of energy and when these and other simple compounds have been depleted, these organisms proceed to destroy the remaining parts of fruits such as the structural polysaccharides and rinds.

Of specific mold genera, *Botrytis* and *Rhizopus* apparently affect the largest number of California fruits and vegetables according to data presented by Vaughn *(7)*:

Botrytis rot: Affects *artichoke, lettuce,* cherry, *grape,* onions, *strawberries,* crucifers, snap beans, *tomatoes* (green), and others.

Rhizopus rot: Affects *caneberries,* carrots, *apricots, cherries, peaches,* strawberries, *sweet potatoes, snap beans, tomatoes* (ripe), and others.

The foods in italics are those of which the mold in question is the leading cause of spoilage. *Penicillium* spp. were reported to be the leading cause of spoilage of three fruits while *Alternaria* and *Aspergillus* spp. were reported to be the leading cause of spoilage in one fruit each.

SUMMARY

The composition and properties of vegetables and fruits may be successfully employed to predict the general types of microorganisms that bring about their spoilage or market diseases. The relatively high pH of vegetables along with their high O/R potentials and low poising capacity make them susceptible to bacterial spoilage, while the generally lower pH of fruits is in large part responsible for molds being their primary agents of spoilage. A large number of fruits and vegetables are characterized by a specific market disease which may be tentatively diagnosed upon inspection of its characteristic pattern.

REFERENCES

1. **Agricultural Research Service, U. S. D. A.** Market diseases of fruits and vegetables. (This is a series of publications dating from 1932 to the present. They can be found in most depository libraries. Many of the individual numbers are out of print.

2. **Beraha, L., M. A. Smith, and W. R. Wright.** 1961. Control of decay of fruits and vegetables during marketing. Dev. in Industrial Microbiol. *2:* 73–77.

3. **Jones, S. M. and A. M. Paton.** 1973. The L-phase of *Erwinia carotovora* var. *atroseptica* and its possible association with plant tissue. J. Appl. Bacteriol. *36:* 729–737.

4. **Lund, B. M.** 1971. Bacterial spoilage of vegetables and certain fruits. J. Appl. Bacteriol. *34:* 9–20.

5. **McColloch, L. P., H. T. Cook, and W. R. Wright.** 1968. Market diseases of tomatoes, peppers, and eggplants. Agric. Handbook No. 28, Agr. Res. Serv., U. S. D. A., Washington, D.C.

6. **Mount, M. S., D. F. Bateman, and H. G. Basham.** 1970. Induction of electrolyte loss, tissue maceration, and cellular death of potato tissue by an endopolygalacturonate trans-eliminase. Phytopathol. *60:* 924–000.

7. **Vaughn, R. H.** 1963. Microbial spoilage problems of fresh and refrigerated foods. In: *Microbiological Quality of Foods,* L. W. Slanetz *et al.,* Editors, Academic Press, N.Y., pp. 193–197.

8. **Watt, B. K. and A. L. Merrill.** 1950. Composition of foods—raw, processed, prepared. Agric. Handbook No. 8, U. S. Dept. of Agric., Washington, D.C.

7

Food Spoilage: Spoilage of Fresh and Cured Meats, Poultry, and Seafoods

Meats are the most perishable of all important foods, and the reasons for this may be seen in Table 7-1 where the chemical composition of a typical adult mammalian muscle postmortem is presented. Meats contain an abundance of all nutrients required for the growth of bacteria, yeasts, and molds, and an adequate quantity of these constituents exist in fresh meats in available form. The general chemical composition of a variety of meats is presented in Table 7-2.

The genera of bacteria most often found on fresh and spoiled meats, poultry, and seafoods are listed in Table 7-3. Not all of the genera indicated for a given product would be found at all times, of course. Those that are more often found during spoilage are indicated under the various products. In Table 7-4 are listed the genera of molds most often identified from meats and related products, while the identified yeasts are listed in Table 7-5. When spoiled meat products are examined, only a few of the many genera of bacteria, molds, or yeasts are found, and in almost all cases one or more genus is found to be characteristic of the spoilage of a given type of meat product. The presence of the more varied flora on nonspoiled meats, then, may be taken to represent the organisms that exist in the original environment of the product in question, or contaminants picked up during processing, handling, packaging, and storage.

The question arises, then, as to why only a few types predominate in spoiled meats. It is helpful here to return to the intrinsic and extrinsic parameters that affect the growth of spoilage microorganisms. Fresh meats such as beef, pork, and lamb as well as fresh poultry, seafoods, and processed meats all have pH values within the growth range of most of the organisms listed in Table 7-3. Nutrient and moisture contents as stated above are adequate to support the growth of all organisms listed. While the O/R potential of whole meats is low, O/R conditions at the surfaces tend to be higher so that strict aerobes, facultative anaerobes, as well as strict anaerobes generally find conditions suitable for growth. Antimicrobial

TABLE 7-1.   Chemical composition of typical adult mammalian muscle after rigor mortis but before degradative changes post-mortem (percent wet weight) *(50).* *

WATER				75.5%
PROTEIN			(%)	18.0%

Myofibrillar	{	myosin, tropomyosin, X protein	7.5
	{	actin	2.5
Sarcoplasmic	{	myogen, globulins	5.6
	{	myoglobin	0.36
	{	haemoglobin	0.04
Mitochondrial		cytochrome C	ca. 0.002
		collagen	
Sarcoplasmic reticulum		elastin	2.0
Sarcolemma		"reticulin"	
Connective tissue		insoluble enzymes	

FAT			3.0%
SOLUBLE NON-PROTEIN SUBSTANCES			3.5%

Nitrogenous	creatine	0.55
	inosine monophosphate	0.30
	di- and tri- phosphopyridine nucleotides	0.07
	amino acids	0.35
	carnosine, anserine	0.30
Carbohydrate	lactic acid	0.90
	glucose-6-phosphate	0.17
	glycogen	0.10
	glucose	0.01
Inorganic	total soluble phosphorus	0.20
	potassium	0.35
	sodium	0.05
	magnesium	0.02
	calcium	0.007
	zinc	0.005
Traces of glycolytic intermediates, trace metals, vitamins, etc.		ca. 0.10

TABLE 7-2.   Approximate percentage chemical composition of 9 meats and meat products *(86)*.

Meats	Water	Carbohyd.	Proteins	Fat	Ash
Beef, hamburger	55.0	0	16.0	28.0	0.8
Beef, round	69.0	0	19.5	11.0	1.0
Bologna	62.4	3.6	14.8	15.9	3.3
Chicken (broiler)	71.2	0	20.2	7.2	1.1
Frankfurters	60.0	2.7	14.2	20.5	2.7
Lamb	66.3	0	17.1	14.8	0.9
Liver (beef)	69.7	6.0	19.7	3.2	1.4
Pork, medium	42.0	0	11.9	45.0	0.6
Turkey, medium fat	58.3	0	20.1	20.2	1.0

constituents are not known to occur in products of the type in question. Upon examining the extrinsic parameters, temperature of incubation stands out as being of utmost importance in controlling the types of microorganisms that develop on meats, since these products are normally held at refrigerator temperatures. Essentially all studies on the spoilage of meats, poultry, and seafoods carried out over the past 20 years or so have dealt with low-temperature stored products.

SPOILAGE OF BEEF, PORK, AND RELATED MEATS

Most studies dealing with the spoilage of meats have been done with beef, and most of the discussion in this section is based upon beef studies. It should be pointed out, however, that pork, lamb, veal and similar meats are presumed to spoil in a similar way.

Upon the slaughter of a well-rested beef animal, a series of events take place that lead to the production of meat. Lawrie *(50)* has discussed these events in great detail and they are here presented only in outline form. The events following an animal's slaughter are: (1) Its circulation ceases. The ability to resynthesize ATP (adenosinetriphosphate) is lost. Lack of ATP causes actin and myosin to combine to form actomyosin, which leads to a stiffening of muscles. (2) The oxygen supply falls, resulting in a reduction of the O/R potential. (3) The supply of vitamins and antioxidants ceases, resulting in a slow development of rancidity. (4) Nervous and hormonal regulations cease, thereby causing the temperature of the animal to fall and fat to solidify. (5) Respiration ceases, which stops ATP synthesis. (6) Glycolysis begins, resulting in the conversion of most glycogen to lactic acid, which depresses pH from about 7.4 to its ultimate level of about 5.6. This pH depression also initiates protein denaturation, liberates and activates cathepsins, and completes rigor mortis. Protein denaturation is accompanied by an exchange of divalent and monovalent cations on the muscle proteins. (7) The reticuloendothelial system ceases to scavenge, thus allowing

TABLE 7-3. The genera of bacteria most frequently found on meats, poultry, and seafoods.

Genus	Gram Reaction	Fresh Meats	Processed Meats	Vacuum Packaged Meats	Bacon[1]	Poultry	Fish and Seafoods
Acinetobacter	−	XX	X	X	X	XX	X
Aeromonas	−	XX			X	X	X
Alcaligenes	−	X			X	X	X
Arthrobacter[2]	−/+	X	X	X	X	X	
Bacillus	+	X	X		X	X	X
Bacteroides	−						
Beijerinckia	−						X
Chromobacterium	−					X	X
Citrobacter	−	X				X	
Clostridium	+	X					
Corynebacterium[2]	+	X	X	X	X	XX	X
Cytophaga	−						X
Enterobacter	−	X				X	X
Escherichia	−	X				X	X
Flavobacterium	−	X				XX	X
Halobacterium	−						X
Kurthia[2]	+	X		X			
Lactobacillus	+	X	XX	XX			X
Leuconostoc	+	X	X	X			
Microbacterium[2,3]	+	X	X	X	X	X	X
Micrococcus	+	XX	X	X	X	X	
Moraxella	−	X	X		X	X	X
Neisseria	−				X	X	
Planococcus	+						
Pediococcus	+	X	X	X			
Plesiomonas	−					X	
Proteus	−	X				X	

Pseudomonas	−	XX				X		XX	XX
Salmonella	−	X						X	
Serratia	−	X						X	
Staphylococcus	+	X	X		X	X		X	X
Streptococcus	+	X	XX			X		X	
Streptomyces	+	X					X	X	
Photobacterium	−							X	
Vibrio	−			X				X	X
Yersinia[4]	−			X		X		X	X

[1]Vacuum packaged not included

[2]Belong to the coryneform group. The organisms once classified as *Brevibacterium* belong to this group.

[3]Especially *M. thermosphactum*

[4]Especially *Y. enterocolitica*

X = known to occur; XX = most frequently reported.

121

TABLE 7-4. Genera of molds most often found on meats, poultry, and seafood products (taken from the literature and a previous review, *44*).

Genus	Fresh and Ref. Meats	Poultry	Fish and Shrimp	Processed and Cured Meats
Alternaria	X	X		X
Aspergillus	X	X	X	XX
Botrytis	X			X
Cladosporium	XX	X		X
Fusarium	X			X
Geotrichum	XX	X		X
Monilia	X			X
Monascus	X			
Mortierella	X			
Mucor	XX	X		X
Neurospora	X			
Oidium	X			X
Oospora	X		X	
Penicillium	X	X	X	XX
Rhizopus	XX	X		X
Scopulariopsis			X	X
Sporotrichum	XX			
Thamnidium	XX			X
Wallemia (Sporendonema)			X	
Zygorrhynchus				X

X = known to occur; XX = most frequently found.

TABLE 7-5. Yeast genera most often identified on meats, poultry, and seafood products (taken from the literature and a previous review, *44*).

Genus	Fresh and Ref. Meats	Poultry	Fish and Shrimp	Processed and Cured Meats
Candida	X	XX	XX	X
Cryptococcus			X	
Debaryomyces	X		X	XX
Hansenula			X	
Pichia			X	
Pullularia			X	
Rhodotorula	X	XX	XX	
Saccharomyces		X		
Sporobolomyces			X	
Torula		XX	XX	X
Torulopsis	X	X	X	X
Trichosporon		X	X	X

X = known to occur; XX = most frequently found.

microorganisms to grow unchecked, and (8) various metabolites accumulate which also aid protein denaturation.

These events require between 24–36 hr at the usual temperatures of holding freshly slaughtered beef (35–40 C). Meanwhile, part of the normal flora of this meat has come from the animal's own lymph nodes *(53),* the stick knife used for exsanguination, the hide of the animal, intestinal tract, dust, hands of handlers, cutting knives, storage bins, and the like. Upon prolonged storage at refrigerator temperatures, microbial spoilage begins. In the event that the internal temperatures are not reduced to the refrigerator range, the spoilage that is likely to occur is caused by bacteria of internal sources. Chief among these are *C. perfringens* and *Enterobacteriaceae (34).* On the other hand, bacterial spoilage of refrigerator-stored meats is, by and large, a surface phenomenon reflective of external sources of the spoilage flora *(34).*

With respect to the fungal spoilage of fresh meats, especially beef, the following 6 genera of molds have been recovered from various spoilage conditions of whole beef: *Thamnidium, Mucor,* and *Rhizopus,* all of which produce "whiskers" on beef; *Cladosporium,* which produces "black spot"; *Penicillium,* which produces green patches, and *Sporotrichum,* which produces "white spot". Among the genera of yeasts recovered from refrigerator-spoiled beef with any consistency are *Candida, Torulopsis,* and *Rhodotorula.*

Unlike the spoilage of whole fresh beef, which is sometimes spoiled by molds, ground beef or hamburger meat is spoiled exclusively by bacteria with the following genera the most important: *Pseudomonas, Alcaligenes, Acinetobacter, Moraxella,* and *Aeromonas.* Those generally agreed to be the primary cause of spoilage are *Pseudomonas, Acinetobacter,* and *Moraxella* spp. with others playing relatively minor roles in the process *(4, 5, 11, 42, 46).* A recent study of the aerobic gram negative bacteria recovered from beef, lamb, pork, and fresh sausage revealed all 231 polarly flagellated rods to be pseudomonads and that of 110 nonmotile organisms, 61 were *Moraxella* while 49 were *Acinetobacter (20).*

Beef rounds and quarters are known to undergo deep spoilage, usually near the bone, especially the "aitch" bone. This type of spoilage is often referred to as "bone taint" and also as "sours." Only bacteria have been implicated in its cause, with the genera *Clostridium* and *Streptococcus* reported to be the primary causative organisms *(12).*

Temperature of incubation appears to be the primary reason why only a few genera of bacteria are found in spoiled beef as opposed to fresh beef. In a recent study, only 4 of the 9 genera of bacteria present in fresh ground beef could be found after the meat underwent frank spoilage at refrigerator temperatures *(42).* Ayres *(6)* has pointed out that after processing, more than 80% of the total population of freshly ground beef may be comprised of chromogenic bacteria, molds, yeasts, and sporeforming bacteria, but after spoilage, only nonchromogenic, short gram negative rods are found. While some of the bacteria found in fresh meats can be shown to grow at refrigerator temperatures on culture media, they apparently do not grow fast enough in meats to successfully compete with the *Pseudomonas* and *Acinetobacter-Moraxella* groups.

Beef cuts tend to undergo surface spoilage, and whether or not the spoilage organisms are bacteria or molds depends primarily upon available moisture. Freshly cut meats stored in a refrigerator with high humidity invariably undergo bacterial spoilage preferential to mold spoilage. The essential feature of this spoilage is surface sliminess in which the causative organisms can nearly always be found. The relatively high O/R potential, availability of moisture, and low temperature all favor the pseudomonads. It is sometimes possible to note discrete bacterial colonies on the surface of beef cuts, especially when the level of contamination is low. The slime layer results from the coalescence of surface colonies and is largely responsible for the tacky consistency of spoiled beef. Ayres *(6)* has presented evidence that odors can be detected when the total bacterial number is between log $7.0-7.5/cm^2$ followed by detectable slime with a total count usually about log $7.5-8.0/cm^2$ (see Figure 7-1). Molds tend to predominate in the spoilage of beef cuts when the surface is too dry for bacterial growth, or when beef has

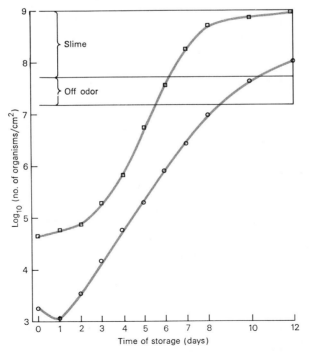

FIGURE 7-1. The development of off-odor and slime on dressed chicken (squares) and packaged beef (circles) during storage at 5° *(6)*.

been treated with antibiotics such as the tetracyclines. Molds virtually never develop on meats when bacteria are allowed to grow freely. The reason for this appears to be that bacteria grow faster than molds, thus consuming available surface oxygen which molds also require for their activities.

Unlike beef cuts or beef quarters, mold growth is quite rare on ground beef except when antibacterial agents have been used as preservatives, or

---

Chemical Methods

a. Measurement of H_2S production.
b. Measurement of mercaptans produced.
c. Determination of noncoagulable nitrogen.
d. Determination of di- and trimethylamines.
e. Determination of tyrosine complexes.
f. Determination of indole and skatol.
g. Determination of amino acids.
h. Determination of volatile reducing substances.
i. Determination of amino nitrogen.
j. Determination of B.O.D. (biochemical oxygen demand).
k. Determination of nitrate reduction.
l. Measurement of total nitrogen.
m. Measurement of catalase.
n. Determination of creatinine content.
o. Determination of dye reducing capacity.
p. Measurement of hypoxanthine.

Physical Methods

a. Measurement of pH changes.
b. Measurement of refractive index of muscle juices.
c. Determination of alteration in electrical conductivity.
d. Measurement of surface tension.
e. Measurement of UV illumination (fluorescence).
f. Determination of surface charges.
g. Determination of cryoscopic properties.

Direct Bacteriological Methods

a. Determination of total aerobes.
b. Determination of total anaerobes.
c. Determination of ratio of total aerobes to anaerobes.
d. Determination of one or more of above at different temperatures.
e. Determination of gram negative endotoxins.

Physio-Chemical Methods

a. Determination of extract-release volume (ERV).
b. Determination of water-holding capacity (WHC).
c. Determination of viscosity.
d. Determination of meat swelling capacity.

---

when the normal bacterial load has been reduced by long-term freezing. Among the early signs of spoilage of ground beef is the development of off-odors followed by tackiness which indicates the presence of bacterial slime. The sliminess is due both to masses of bacterial growth and the softening or loosening of meat structural proteins.

The precise roles played by spoilage microorganisms that result in the spoilage of meats are not well understood at this time. The earlier views on the mechanism of meat spoilage are embodied in the many techniques proposed for its detection, some of which are presented below.

1. Detection and Mechanism of Meat Spoilage.   Approximately 45 different techniques have been proposed for determining meat freshness or spoilage. These techniques (page 125) fall into four categories according to the particular test procedures and product or compound tested for.

It is reasonable to assume that reliable methods of determining meat spoilage should be based upon the cause and mechanism of spoilage. The chemical methods listed above all embody the assumption that as meats spoil, some utilizable substrate is consumed or some new product or products are created by the spoilage flora. It is well established that the spoilage of meats at low temperatures is accompanied by the production of off-odor compounds such as ammonia, H_2S, indole, and others. The drawbacks to the use of these methods are that not all spoilage organisms are equally capable of producing them. Inherent in some of these methods was the incorrect belief that low temperature spoilage was accompanied by a breakdown of primary meat proteins *(38)*. The physical and direct bacteriological methods all tend to show what is obvious, that is, that meat which is obviously spoiled from the standpoint of organoleptic examination (odor, touch, appearance, and taste) is indeed spoiled! They apparently do not allow one to predict spoilage or shelf-life, which a meat freshness test should ideally do.

The extract-release volume (ERV) technique, first described in 1964, has been shown to be of value in determining incipient spoilage in meats as well as in predicting refrigerator shelf-life *(36, 37, 39, 69)*. The ERV technique is based upon the volume of aqueous extract released by an homogenate of beef when allowed to pass through filter paper for a given period of time. By this method, beef of good organoleptic and microbial quality releases large volumes of extract while beef of poor microbial quality releases smaller volumes or none (Figure 7-2). One of the more important aspects of this method is the information which it has provided concerning the mechanism of low-temperature beef spoilage.

The ERV method of detecting meat spoilage reveals two aspects of the spoilage mechanism not previously recognized. First, low-temperature meat spoilage occurs in the absence of any significant breakdown of primary proteins—at least not complete breakdown. Although this fact has been verified by total protein analyses on fresh and spoiled meats, it is also implicit in the operation of the method. That is, as meats undergo microbial spoilage, ERV is *decreased* rather than increased, which would be the case if complete hydrolysis of proteins occurred. The second aspect of meat spoilage revealed by ERV is the increase in hydration capacity of meat proteins by some as yet unknown mechanism, although amino sugar complexes produced by the spoilage flora have been shown to play a role *(77)*. In the absence of complete protein breakdown, the question arises as to how the spoilage flora obtains its nutritional needs for growth.

When fresh meats are placed in storage at refrigerator temperatures, those organisms capable of growth at the particular temperature begin their growth. Their sources of nutrients consist most likely of the soluble nonprotein substances of which there are about 3.5% (Table 7-1), specifically, the carbohydrates in the form of lactic acid, glucose, and others, the free amino acids, and the nucleotides are utilized. The foul odors generally associated with spoiling meats probably owe their origin to free amino acids and related compounds, e.g., H_2S from sulfur-containing acids, NH_4 from many amino acids and related compounds, and indole from tryptophane. Off-odors and off-flavors appear only when amino acids begin to be utilized. Amino acid utilization occurs after glucose is exhausted *(26)*. The primary proteins of meat are probably not attacked until the supply of the simpler constituents has been exhausted. It has been shown, for example, that the antigenicity of salt-soluble beef proteins is not destroyed under the usual conditions of low-temperature spoilage *(56)*. In their studies on fish spoilage, Lerke *et al.* *(54)*

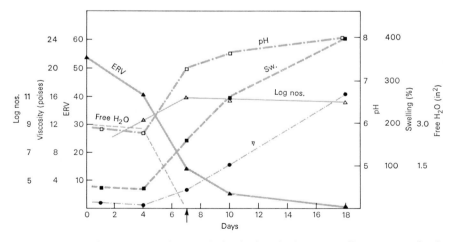

FIGURE 7-2.   The response of several physiochemical meat spoilage tests as fresh ground beef was held at 7 C until definite spoilage had occurred. The arrow indicates the first day off-odors were detected. ERV = extract-release volume; free H_2O = measurement of water-holding capacity (inversely related); Sw = meat swelling; η = viscosity; and log Nos. = total aerobic bacteria/g *(76)*. Copyright © 1969 by Institute of Food Technologists.

have shown that a raw fish press juice displays all apparent aspects of fish spoilage as may be determined by use of whole fish. This finding may be taken to indicate a general lack of attack upon insoluble proteins by the fish spoilage flora, since these proteins are absent from the filtered press juice. The same is apparently true for beef and related meats. Incipient spoilage is accompanied by a rise in pH, increase in bacterial numbers, increase in the hydration capacity of meat proteins, along with other changes. In ground beef, pH may rise as high as 8.5 in putrid meats, although at the time of incipient spoilage mean pH values of ca. 6.5 have been found *(78)*. By plotting the growth curve of the spoilage flora, the usual phases of growth can be observed and the phase of decline may be ascribed to the exhaustion of utilizable nutrients by most of the flora and the accumulation of toxic by-products of bacterial metabolism. Precisely how the primary proteins of meat become destroyed at low temperatures is not well understood, but recent work has helped define the conditions necessary for this event to occur.

Dainty *et al. (19)* inoculated beef slime onto slices of raw beef and incubated them at 5 C. Off-odors and slime were noted after 7 days with counts at $2 \times 10^9/cm^2$. Proteolysis was not detected in either sarcoplasmic or myofibrillar fractions of the beef slices. No changes in the sarcoplasmic fractions could be detected even 2 days later when bacterial numbers reached $10^{10}/cm^2$. The first indication of breakdown of myofibrillar proteins occurred at this time with the appearance of a new band and the weakening of another. All myofibrillar bands disappeared after 11 days with weakening of several bands of the sarcoplasmic fraction. With naturally contaminated beef, odors and slime were first noted after 12 days when the numbers were $4 \times 10^8/cm^2$. Changes in myofibrillar proteins were not noted until 18 days of holding. By the use of pure culture studies, these workers showed that Shewan's group I pseudomonads (see Appendix for the Shewan scheme) were active against myofibrillar proteins while group II organisms were more active against sarcoplasmics. *Aeromonas* spp. were active on both myofibrillar and sarcoplasmic proteins. With pure cultures, protein changes were not detected until counts were above $3.2 \times 10^9/cm^2$. Borton *et al. (10)* showed earlier that *Ps. fragi* (a group II pseudomonad) effected the loss of protein bands from inoculated pork muscle, but no indication is given as to the minimum numbers that were necessary. Other details of the known effects of microorganisms on meats at low temperatures have been reviewed elsewhere *(44)*.

The use of texturized soy proteins (TSP) to extend ground meats is a practice which appears to be increasing. A recent study of ground beef + TSP and ground chicken + TSP mixtures suggests that such mixtures spoil faster than the meat portions alone *(18)*. Higher counts were found in the TSP-added meats held for 8 days at 4 C than in controls. The predominant flora of the spoiled products consisted of *Serratia, Escherichia,* and *Enterobacter* spp. with *Serratia* being the most predominant. Just why the mixtures are more perishable is not clear.

The microbial spoilage of ground beef can be delayed by wrapping or seal-
ing this product in gas-impermeable plastic bags (27, 40, 45, 48). The effect
of this treatment is to reduce the log phase of growth of the spoilage
organisms and to keep down the numbers of spoilage bacteria generally. The
exclusion of oxygen results in a larger number of lactic acid bacteria than
can usually be found in spoiling ground beef. Upon storage of such meat for
several weeks at refrigerator temperatures, the meat is characterized by a
sharp odor and an off-color. The author has held ground beef stored at 5–7 C
in gas-impermeable plastic bags for over 2 yr without dissolution of the
product.

2. Spoilage of Vacuum-Packaged Meats.

As noted above, the packaging
of fresh meats in gas-impermeable bags or wraps increases shelf life. Since
the packaging restricts oxygen supply to the normally occurring aerobic
flora, it may be expected that these types of products will undergo spoilage
different from those packaged in permeable bags. While not using gas-impermeable containers, Clark and Lentz (17) stored prepackaged beef inoculated with 5 strains of meat spoilage bacteria in CO_2-enriched atmospheres
and found that off-odors developed 3 to 10 days later than for control meats.
The most undesirable feature of this method was development of undesirable color which always preceded off-odor development. In a study of
round steaks packaged in oxygen-impermeable bags and held at 38 F for 15
days, Pierson et al. (71), found that 90–95% of the total count consisted of
lactobacilli with no increase in the number of pseudomonads. The latter did
increase rapidly in meats packaged in oxygen permeable films.

On the other hand, Sutherland et al. (82) found that the aerobic gram negative bacteria did increase along with lactic acid bacteria in vacuum-packed
beef stored at 0–2 C for up to 9 weeks. These investigators found that pH
tended to decrease upon storage and because of this, ERV was not reliable
as a spoilage indicator. More recently, studies of vacuum-packaged beef and
lamb by Hanna et al. (28, 29) indicate that initially the dominant flora of both
products consisted of Corynebacterium spp. and M. thermosphactum. In
the case of beef, the dominant flora after 28–35 days consisted of lactobacilli, while after 21 days for lamb, pseudomonads and Moraxella-Acinetobacter spp. were prominent among the dominant flora along with lactobacilli.

If it is correct, as it appears to be, that the most significant microbial effect
of vacuum packaging is the restriction of growth of aerobic bacteria so that
lactic acid bacteria become dominant, recent work by Gill (26) may be of
help in explaining the extended keeping times observed. It was shown by
this investigator that when growing at meat surfaces, pseudomonads utilized
a series of low molecular weight compounds and that their growth was not
limited by a lack of nutrients but by the accumulation of toxic products and
the like. On the other hand, lactobacilli growing on the surface of meat were
limited by the rate of diffusion of fermentable substrates from within the

meat surface. This suggests that the growth of lactobacilli on vacuum-packaged meats is limited considerably more than for aerobes and that once the diffusible fermentable substrate is exhausted, growth of these organisms ceases.

SPOILAGE OF FRANKFURTERS, BOLOGNA, SAUSAGE, AND LUNCHEON MEATS

Unlike other meats covered in this chapter, these are prepared from various ingredients, any one or all of which may contribute microorganisms to the final product. Bacteria, yeasts, and molds may be found in and upon processed meats, but the former two groups are by far the most important in the microbial spoilage of these products.

Spoilage of these meats is generally of three types: sliminess, souring, and greening. Slimy spoilage occurs on the outside of casings, especially of frankfurters, and may be seen in its early stages as discrete colonies which may later coalesce to form a uniform layer of gray slime. From the slimy material may be isolated yeasts, lactic acid bacteria of the genera *Lactobacillus, Streptococcus,* and *Microbacterium, especially M. thermosphactum (21, 22, 58). L. viridescens* has been shown to produce both sliminess and greening. Slime formation is favored by a moist surface and is usually confined to the outer casing. Removal of this material with hot water leaves the product often essentially unchanged.

Souring generally takes place underneath the casing of these meats and results from the growth of lactobacilli, streptococci, and related organisms. The usual sources of these organisms to processed meats are milk solids. The souring results from the utilization of lactose and other sugars by these organisms with the production of acids. Sausage usually contains a more varied flora than most other processed meats due to the different seasoning agents employed, almost all of which contribute their own flora (see Tables 5-2 and 5-3). Of particular importance in sausage is the organism *M. thermosphactum* and related species, which Sulzbacher and McLean *(81)* and more recently Dowdell and Board *(21)* found to be the most predominant in their studies on the microbiology of this product.

Greening occurs more commonly on frankfurters than on other meats in this category, but may be seen on all from time to time. The heterofermentative species of lactobacilli and *Leuconostoc* have been found to be responsible for this condition. These organisms produce peroxides which act upon cured meat pigments and produce the green color. This reaction is made possible due to the inactivation of catalase by heat treatment or by the presence of nitrite. The accumulated H_2O_2 reacts with the meat pigments nitric oxide hemochromogen or nitric oxide myoglobin (see Table 7-6), producing a greenish oxidized porphyrin *(74)*. This condition is caused by growth of the organisms in the interior core where the low O/R potential

TABLE 7-6. Pigments found in fresh, cured or cooked meat (50).*

Pigment	Mode of Formation	State of Iron	State of Haematin Nucleus	State of Globin	Color
(1) Myoglobin	Reduction of metmyoglobin; de-oxygenation of oxymyoglobin	Fe^{++}	Intact	Native	Purplish red
(2) Oxymyoglobin	Oxygenation of myoglobin	Fe^{++}	Intact	Native	Bright red
(3) Metmyoglobin	Oxidation of myoglobin, oxymyoglobin	Fe^{+++}	Intact	Native	Brown
(4) Nitric oxide myoglobin	Combination of myoglobin with nitric oxide	Fe^{++}	Intact	Native	Bright red
(5) Metmyoglobin nitrite	Combination of metmyoglobin with excess nitrite	Fe^{+++}	Intact	Native	Red
(6) Globin haemochromogen	Effect of heat, denaturing agents on myoglobin, oxymyoglobin: irradiation of globin haemichromogen	Fe^{++}	Intact	Denatured	Dull red
(7) Globin haemichromogen	Effect of heat, denaturing agents on myoglobin, oxymyoglobin, metmyoglobin, haemochromogen	Fe^{+++}	Intact	Denatured	Brown
(8) Nitric oxide haemochromogen	Effect of heat, salts on nitric oxide myoglobin	Fe^{++}	Intact	Denatured	Bright red
(9) Sulphmyoglobin	Effect of H_2S and oxygen on myoglobin	Fe^{+++}	Intact but reduced	Denatured	Green
(10) Choleglobin	Effect of hydrogen peroxide on myoglobin or oxymyoglobin: effect of ascorbic or other reducing agent on oxymyoglobin	Fe^{++} or Fe^{+++}	Intact but reduced	Denatured	Green
(11) Verdohaem	Effect of reagents as in 9 in excess	Fe^{+++}	Porphyrin ring opened	Denatured	Green
(12) Bile pigments	Effect of reagents as in 9 in large excess	Fe absent	Porphyrin ring destroyed: chain of porphyrins	Absent	Yellow or colorless

*Reprinted with permission from R. A. Lawrie, *Meat Science,* Copyright 1966, Pergamon Press.

131

allows H_2O_2 to accumulate. A small amount of oxygen favors greening and the green area is very often confined to small parts of the product. The one organism most frequently isolated from green meats is *L. viridescens*, first described by Niven *et al.* *(67).* In spite of the discoloration, the green product is not known to be harmful if eaten.

Although mold spoilage of these meats is not common, it can and does occur under favorable conditions. When the products are moist and stored under conditions of high humidity, they tend to undergo bacterial and yeast spoilage. Mold spoilage is likely to occur only when the surfaces become dry or when the products are stored under other conditions which do not favor bacteria or yeasts.

SPOILAGE OF BACON AND CURED HAMS

The nature of these products and the procedures employed in preparing certain ones such as smoking, brining, etc., make them relatively insusceptible to spoilage by most bacteria. The most common form of bacon spoilage is moldiness, which may be due to *Aspergillus, Alternaria, Fusarium, Mucor, Rhizopus, Botrytis, Penicillium,* and other molds (Table 7-4). The high fat content and low a_w make it somewhat ideal for this type of spoilage. Bacteria of the genera *Streptococcus, Lactobacillus,* and *Micrococcus* are capable of growing well on certain types of bacon such as Wiltshire *(33),* and *S. faecalis* is often present on several types. Vacuum-packed bacon tends to undergo souring due primarily to micrococci and lactobacilli. Vacuum-packed, low-salt bacon stored above 20 C may be spoiled by staphylococci *(84).*

Cured hams undergo a type of spoilage different from that of fresh or smoked hams. This is due primarily to the fact that curing solutions pumped into the hams contain sugars which are fermented by the natural flora of the ham and also by those organisms pumped into the product in the curing solution, such as lactobacilli. The sugars are fermented to produce conditions referred to as "sours" of various types, depending upon their location within the ham. A large number of genera of bacteria have been implicated as the cause of ham "sours," among them *Acinetobacter, Bacillus, Pseudomonas, Lactobacillus, Proteus, Micrococcus, Clostridium,* and others. Gassiness is not unknown to occur in cured hams where members of the genus *Clostridium* have been found. The spoilage of canned hams is treated in Chapter 8.

In their study of vacuum-packed sliced bacon, Cavett *(14)* and Tonge *et al.* *(84)* found that when high-salt bacon was held at 20 C for 22 days, the catalase positive cocci dominated the flora while at 30 C the coagulase negative staphylococci became dominant. In the case of low-salt bacon (5–7% NaCl versus 8–12% in high-salt bacon) held at 20 C, the micrococci as well as *S. faecalis* became dominant; while at 30 C the coagulase negative staphylococci as well as *S. faecalis* and micrococci became dominant. Spoilage of this type product is characterized by "cheesy," scented-sour, and putrid off-odor.

In a study of lean Wiltshire bacon stored aerobically at 5 C for 35 days or 10 C for 21 days, Gardner *(25)* found nitrates were reduced to nitrites when the microbial load reached ca. 10^9/g. The predominant organisms at this stage were micrococci, vibrios, and the yeast genera *Candida* and *Torulopsis*. Upon longer storage, microbial counts reached ca. 10^{10}/g with the disappearance of nitrites. At this stage, *Acinetobacter, Alcaligenes,* and *Arthrobacter-Corynebacterium* spp. became more important. Micrococci were always found while vibrios were found in all bacons with salt contents above 4%.

SPOILAGE OF POULTRY MEATS

Studies on the bacterial flora of fresh poultry meats by several investigators have revealed over 20 genera (Table 7-3). However, when these meats undergo low temperature spoilage, most all workers agree that the primary spoilage organisms belong to the genus *Pseudomonas (4, 8, 66)*. In a recent study of 5,920 isolates from chicken carcasses *(49)*, pseudomonads were found to constitute 30.5%, *Acinetobacter* 22.7%, *Flavobacterium* 13.9, *Corynebacterium* 12.7, with yeasts, *Enterobacteriaceae,* and others in lower numbers. Of the pseudomonads, these investigators found that 61.8% were fluorescent on King's medium and that 95.2% of all pseudomonads oxidized glucose. A previous characterization of pseudomonads on poultry undergoing spoilage was made by Barnes and Impey *(8)* who showed that the pigmented pseudomonads (Shewan's group I) decreased from 34 to 16% from initial storage to the development of strong off-odors, while the nonpigmented actually increased from 11 to 58% (see section below on fish spoilage). *Acinetobacter* and other species of bacteria decreased along with the type I pseudomonads similar to what occurs in spoiling fish.

With respect to the fungi, they are of considerably less importance in poultry spoilage except when antibiotics are employed to suppress bacterial growth. When antibiotics are employed, however, molds become the primary agents of spoilage *(68)*. The genera *Candida, Rhodotorula,* and *Torula* are the most important yeasts found on poultry (Table 7-5). The essential feature of poultry spoilage is sliminess at the outer surfaces of the carcass or cuts. The visceral cavity very often displays sour odors or what is commonly called visceral taint. This is especially true of the spoilage of **New York dressed poultry** where the viscera are left inside. The causative organisms here are also bacteria of the type noted above in addition to streptococci.

The primary reasons why poultry spoilage is mainly restricted to the surfaces are as follows. The inner portions of poultry tissue are generally sterile, or contain relatively few organisms which generally do not grow at low temperatures. The spoilage flora, therefore, is restricted to the surfaces and hide where it is deposited from water, processing, and handling. The surfaces of fresh poultry stored in an environment of high humidity are very susceptible to the growth of aerobic bacteria such as pseudomonads. These organisms grow well on the surfaces where they form minute colonies which later coalesce to produce the sliminess characteristic of spoiled poultry meat. May *et al. (57)* showed that poultry skin supports the growth of the poultry spoilage flora better than even the muscle tissue. In the advanced stages of poultry meat spoilage, the surfaces will very often fluoresce when illuminated with ultraviolet light. The fluorescence is due to the presence of large numbers of fluorescent pseudomonads. Surface spoilage organisms can be recovered directly from the slime for plating, or one can prepare slides for viewing by smearing with portions of slime. Upon gram staining, one may note the uniform appearance of organisms indistinguishable from those listed above. Tetrazolium (2,3,5-triphenyltetrazolium chloride) can be used also to assess microbial activity on poultry surfaces. Upon spraying the eviscerated carcass with this compound, a red pigment develops in areas of high microbial activity. These areas generally consist of cut muscle surfaces and other damaged areas such as feather follicles *(70)*.

As poultry meats undergo spoilage, off-odors are generally noted before sliminess, with the former being first detected when log numbers/cm^2 are about 7.2–8.0. Sliminess generally occurs shortly after the appearance of off-odors with the log counts/cm^2 about 8 *(4)*. Total aerobic plate counts/cm^2 of slimy surface rarely go higher than log 9.5. With the initial growth first confined to poultry surfaces, the tissue below the skin remains essentially free of bacteria for some time. Gradually, however, bacteria begin to enter the deep tissues, bringing about increased hydration of muscle proteins, much as occurs with beef. Whether autolysis plays an important role in the spoilage of inner poultry tissues is not clear at this time.

Recent work on the specific origins of off-odors associated with spoiling poultry meat has shown *Ps. putrefaciens* to be one of the more important organisms. Of 159 isolates of pseudomonads from meat and poultry plants, 48 were found to be *Ps. putrefaciens* capable of good growth at 5 C with the capacity to produce potent off-odors in 7 days when growing on chicken muscle *(60)*. In another study, McMeekin *(59)* found that 80% of 250 isolates from poultry stored at 2 C produced off-odors, and he suggested that there is a selection for types that produce strong off-odors among the varied flora that one finds on fresh poultry. Since Shewan's group II pseudomonads grow faster than the pigment producing group I strains, it appears that the strong odor-producing capacity is a property of these strains. Group II pseudomonads have been shown to be consumers of free amino acids in chicken skin, while group I types effected increases in the quantities of free amino acids and related nitrogenous compounds *(2, 3)*. With respect to the nature of the compounds that constitute the odors of spoiling poultry, Freeman *et al. (24)* identified 15 volatiles consisting of mercaptans, sulfides, alcohols, etc., produced by a normal spoilage flora at refrigerator temperatures. These investigators recovered *Ps. putrefaciens, Moraxella,* and *Acinetobacter* spp. from the spoiling products, but which isolates were responsible for the production of given volatile compounds was not clear. Findings similar to these are given below for fish spoilage. Fewer volatiles are produced under anaerobic storage conditions and this finding is consistent with the aerobic features of the spoilage organisms.

When New York dressed poultry undergoes microbial spoilage, the organisms make their way through the gut walls and invade inner tissues of the intestinal cavity. The characteristic sharpness that is associated with the spoilage of this type of poultry is referred to as "visceral taint."

SPOILAGE OF FISH AND SHELLFISH

1. Fish.   Both salt-water and fresh-water fish contain comparatively high levels of proteins and other nitrogenous constituents (Table 7-7). The carbohydrate content of these fish is nil, while fat content varies from very low to rather high values depending upon species. Of particular importance in fish flesh is the nature of the nitrogenous compounds. The relative percentages of total-N and protein-N are presented in Table 7-8, from which it can be seen that not all nitrogenous compounds in fish are in the form of proteins. Among the nonprotein nitrogen compounds are the free amino acids, volatile nitrogen bases such as ammonia and trimethylamine, creatine, taurine, the betaines, uric acid, anserine, carnosine, and histamine.

TABLE 7-7.   Approximate percentage chemical composition of fish and shellfish (86).

Bony Fish	Water	Carbohyd.	Proteins	Fat	Ash
Bluefish	74.6	0	20.5	4.0	1.2
Cod	82.6	0	16.5	0.4	1.2
Haddock	80.7	0	18.2	0.1	1.4
Halibut	75.4	0	18.6	5.2	1.0
Herring (Atlantic)	67.2	0	18.3	12.5	2.7
Mackerel (Atlantic)	68.1	0	18.7	12.0	1.2
Salmon (Pacific)	63.4	0	17.4	16.5	1.0
Swordfish	75.8	0	19.2	4.0	1.3
Crustaceans					
Crab	80.0	0.6	16.1	1.6	1.7
Lobster	79.2	0.5	16.2	1.9	2.2
Mollusks					
Clams, meat	80.3	3.4	12.8	1.4	2.1
Oysters	80.5	5.6	9.8	2.1	2.0
Scallops	80.3	3.4	14.8	0.1	1.4

TABLE 7-8.   Distribution of nitrogen in fish and shellfish flesh *(35).*

Species	Percentage Total N	Percentage Protein N	Ratio of Protein N/total N
Cod (Atlantic)	2.83	2.47	0.87
Herring (Atlantic)	2.90	2.53	0.87
Sardine	3.46	2.97	0.86
Haddock	2.85	2.48	0.87
Lobster	2.72	2.04	0.75

It is generally recognized that the internal flesh of healthy, live fish is sterile *(31, 83),* although a few reports to the contrary exist. Bacteria that exist on fresh fish are generally found in three places: the outer slime, gills, and the intestines of feeding fish.

The microorganisms known to cause fish spoilage are indicated in Tables 7-3, 7-4, and 7-5. Fresh iced fish are invariably spoiled by bacteria while salted and dried fish are more likely to undergo fungal spoilage. The bacterial flora of spoiling fish is found to consist of asporogenous, gram negative rods of the *Pseudomonas* and *Acinetobacter-Moraxella* types. Many fish spoilage bacteria are capable of good growth between 0–1 C. Shaw and Shewan *(75)* found that a large number of *Pseudomonas* spp. are capable of causing fish spoilage at −3 C although at a slow rate.

The spoilage of salt- and fresh-water fish appears to occur in essentially the same manner, with the chief differences being the requirement of the salt-water flora for a sea-water type of environment and the differences in chemical composition between various fish with respect to nonprotein nitrogenous constituents. The most susceptible part of fish is the gill region, including the gills. The earliest signs of organoleptic spoilage may be noted by examining the gills for the presence of off-odors. If feeding fish are not eviscerated immediately, intestinal bacteria soon make their way through the intestinal walls and into the flesh of the intestinal cavity. This process is believed to be aided by the action of proteolytic enzymes which are from the intestines, and which may be natural enzymes inherent in the intestines of the fish, or enzymes of bacterial origin from the inside of the intestinal canal, or both. Fish spoilage bacteria apparently have little difficulty in growing in the slime and on the outer integument of fish. Slime is composed of mucopolysaccharide components, free amino acids, trimethylamine oxide, piperidine derivatives, and other related compounds. As is the case with poultry spoilage, plate counts are best done on the surface of fish with numbers of organisms expressed per cm² of examined surface.

It appears that the spoilage organisms first utilize the simpler compounds and in the process release various volatile off-odor components. According to Shewan *(80),* trimethylamine oxide, creatine, taurine, anserine, and related compounds along with certain amino acids decrease during fish spoilage with the production of trimethylamine, ammonia, histamine, hydrogen sulfide, indole, and others (Figure 7-3). Fish flesh appears to differ from mammalian flesh in regard to autolysis, where flesh of the former type seems to undergo autolysis at more rapid rates. While the occurrence of this process along with microbial spoilage is presumed by some investigators to either aid the spoilage flora or the spoilage process *(30),* attempts to separate and isolate the events of the two have proved difficult. In a detailed study of fish isolates with respect to the capacity to cause typical fish spoilage by use of sterile fish muscle press juice, Lerke *et al. (55)* found that the spoilers belonged to the genera *Pseudomonas* and *Acinetobacter-Moraxella,* with none of the coryneforms, micrococci, or flavobacteria being spoilers. In characterizing the spoilers with respect to their ability to utilize certain compounds, these workers found that most spoilers were unable to degrade gelatin or digest egg albumin. This suggests that fish spoilage proceeds much as does that of beef—in the general absence of complete proteolysis by the spoilage flora. Pure culture inoculations of cod and haddock muscle blocks have failed to effect tissue softening *(30).* In those fish that contain high levels of lipids (herrings, mackerel, salmon, and others), these compounds undergo rancidity as microbial spoilage occurs. It should be noted that the skin of fish is rich in collagen. The scales of most fish are composed of a scleroprotein belonging to the keratin group, and it is quite probable that these are among the last parts of fish to be decomposed.

The interplay of the bacterial flora of fish undergoing spoilage has been studied by Lee and Harrison *(52)* and Laycock and Regier *(51)* who found that *Pseudomonas* spp. of Shewan's group II became the dominant types of

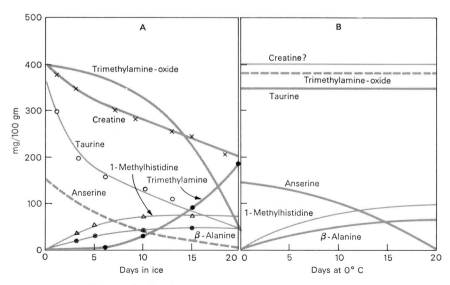

FIGURE 7-3. Changes in the nitrogenous extractives in *(A)* spoiling and *(B)* autolysing cod muscle *(80).*

all bacteria after 14 days at 5 C (Table 7-9). A similar result was found for poultry spoilage (8). Hydrogen sulfide producers also increase late in the spoilage process (30). While Acinetobacter and Moraxella spp. constituted the higher percentage of the initial flora, these organisms could not be isolated after 14 days. Miller and co-workers (61, 62, 63, 64) have attempted to identify the volatile compounds and the respective organisms responsible for the characteristic odors that accompany incipient fish spoilage. Employing a combination of gas-liquid chromatography and mass spectrometry, these investigators have identified a large number of volatiles including mercaptans, di- and trimethyl sulfides, alcohols (ethyl and n-propyl), di- and trimethylamines, aldehydes, ethyl acetate, butanone, etc. Each of the strains employed (Ps. perolens, Ps. fragi, Ps. putrefaciens, Ps. fluorescens, Acinetobacter-Moraxella) produced some of the same volatiles as well as some compounds not produced by the others. Just why this occurred is not clear, but it may indicate the need to have each strain growing in fish in a particular sequence. Attempts thus far to assign specific volatiles and specific spoilage roles to specific strains of fish spoilers have not met with much success. The fruity odors of spoiling fish do appear to be due to the production of ethyl esters of acetate, butyrate, and hexanoate (63), but the organisms responsible for odors variously described as being like those of "evaporated milk", "strong ammonia", "amines", "sulfidy", "rotten vegetables", "sweaty socks", "musty", and "potato-like" have not been clearly indicated. One instance of a compound consistently produced in fish by a specific organism is phenethyl alcohol which Chen et al. (16) and Chen and Levin (15) found to be produced so consistently by an "Achromobacter" as to be of taxonomic significance. This compound along with phenol was recovered from a high-boiling fraction of haddock fillets held at

TABLE 7-9. **Microbial population change in Pacific hake stored at 5 C** (52).

Microorganisms	Microbial Population After Incubation (%)			
	0 Day	5 Days	8 Days	14 Days
Pseudomonas				
Type I	14.0	7.3	2.7	15.1
Type II	14.0	52.4	53.4	77.4
Types III or IV	3.5	12.2	31.5	7.5
Acinetobacter-Moraxella				
Acinetobacter	31.6	17.0	8.2	0
Moraxella	19.3	9.8	2.7	0
Flavobacterium	17.6	0	0	0
Coliforms	0	1.2	1.4	0
Microbial count of sample	1.5×10^4	3.4×10^7	9.3×10^8	2.7×10^9
No. of microorganisms identified[a]	57	82	73	53

[a]All isolated colonies on initial isolation plates were picked and identified.

2 C. None of 10 known *Acinetobacter* and only one of 9 known *Moraxella* spp. produced phenethyl alcohol under similar conditions.

2. Shellfish.   *Crustaceans.* The most widely consumed shellfish within this group are shrimp, lobsters, crabs, and crayfish, and the spoilage of all will be treated together. Unless otherwise specified, spoilage of each is presumed or known to be essentially the same. The chief differences in spoilage of these various foods is referable generally to the way in which they are handled and their specific chemical composition.

An inspection of Table 7-7 reveals that crustaceans differ from fish in having about 0.5% carbohydrate as opposed to none for the fish presented. Shrimp has been reported to have a higher content of free amino acids than fish *(73)* and to contain catheptic-like enzymes which rapidly break down proteins *(23)*.

The bacterial flora of freshly caught crustaceans should be expected to reflect the waters from which these foods are caught, contaminants from the deck, handlers, and washing waters. Many of the organisms reported for fresh fish have been reported on these foods, with pseudomonads, *Acineto-bacter-Moraxella,* and yeast spp. being predominant on microbially spoiled crustacean meats *(13, 47)*.

The spoilage of crustacean meats appears to be quite similar to that of fish flesh. Spoilage would be expected to begin at the outer surfaces of these foods due to the anatomy of the organisms. It has been reported that the crustacean muscle contains over 300 mg of nitrogen/100 g of meat, which is considerably higher than that for fish *(85)*. The presence of higher quantities of free amino acids in particular and of higher quantities of nitrogenous extractives in crustacean meats in general makes them quite susceptible to rapid attack by the spoilage flora. Initial spoilage of crustacean meats is accompanied by the production of large amounts of volatile base nitrogen much as is the case with fish. Some of the volatile base nitrogen arises from the reduction of trimethylamine oxide present in crustacean shellfish (lacking in most mollusks). Creatine is lacking among shellfish, both crustacean and molluscan, while arginine is prevalent. Shrimp microbial spoilage is accompanied by increased hydration capacity in a manner similar to that for meats or poultry *(79)*.

Mollusks. The molluscan shellfish considered in this section are oysters, clams, squid, and scallops. These animals differ in their chemical composition from both teleost fish and crustacean shellfish in having a significant content of carbohydrate material and a lower total quantity of nitrogen in their flesh. The carbohydrate is largely in the form of glycogen, and with levels of the type that exist in molluscan meats, fermentative activities may be expected to occur as a part of the microbial spoilage. Molluscan meats contain high levels of nitrogen bases much as do other shellfish. Of particular interest in molluscan muscle tissue is a higher content of free arginine, aspartic, and glutamic acids than is found in fish. The most important difference in chemical composition between crustacean shellfish and molluscan shellfish is the higher content of carbohydrate in the latter. For example, clam meat and scallops have been reported to contain 3.4% and oysters 5.6% carbohydrate, mostly as glycogen. The higher content of carbohydrate materials in molluscan shellfish is responsible for the different spoilage pattern of these foods over other seafoods.

The microbial flora of molluscan shellfish may be expected to vary considerably, depending upon the quality of the water from which these fish are taken and upon the quality of wash water and other factors. The following genera of bacteria have been recovered from spoiled oysters: *Serratia, Pseudomonas, Proteus, Clostridium, Bacillus, Escherichia, Enterobacter, Streptococcus, Lactobacillus, Flavobacterium,* and *Micrococcus.* As spoilage sets in and progresses, *Pseudomonas* and *Acinetobacter-Moraxella* spp. predominate, with streptococci, lactobacilli, and yeasts dominating the late stages of spoilage.

Due to the relatively high level of glycogen, the spoilage of molluscan shellfish is basically fermentative. Several investigators, including Hunter and Linden *(32)* and Pottinger *(72),* have proposed the following pH scale as a basis for determining microbial quality in oysters:

pH 6.2–5.9   = good

pH 5.8   = "off"

pH 5.7–5.5   = musty

pH 5.2 & below = sour or putrid.

A measure of pH decrease is apparently a better test of spoilage in oysters and other molluscan shellfish than volatile nitrogen bases. A measure of volatile acids was attempted by Beacham *(9)* and found to be unreliable as a test of oyster freshness. While pH is regarded by most investigators as being the best objective technique for examining the microbial quality of oysters, Abbey *et al. (1)* found that organoleptic evaluations and microbial counts were more desirable indices of microbial quality in this product.

Clams and scallops appear to display essentially the same patterns of spoilage as do oysters, but squid meat does not. In squid meat, volatile base nitrogen increases as spoilage occurs much in the same manner as for the crustacean shellfish *(65).*

REFERENCES

1. **Abbey, A., R. A. Kohler, and S. D. Upham.** 1957. Effect of aureomycin chlortetracycline in the processing and storage of freshly shucked oysters. Food Technol. *11:* 265–271.

2. **Adamčič, M. and D. S. Clark.** 1970. Bacteria-induced biochemical changes in chicken skin stored at 5 C. J. Food Sci. *35:* 103–106.

3. **Adamčič, M., D. S. Clark, and M. Yaguchi.** 1970. Effect of psychrotolerant bacteria on the amino acid content of chicken skin. J. Food Sci. *35:* 272–275.

4. **Ayres, J. C., W. S. Ogilvy, and G. F. Stewart.** 1950. Post mortem changes in stored meats. I. Microorganisms associated with development of slime on eviscerated cut-up poultry. Food Technol. *4:* 199–205.

5. **Ayres, J. C.** 1955. Microbiological implications in handling, slaughtering and dressing of meat animals. Adv. Food Res. *6:* 109–161.

6. **Ayres, J. C.** 1960A. The relationship of organisms of the genus *Pseudomonas* to the spoilage of meat, poultry and eggs. J. Appl. Bacteriol. *23:* 471–486.

7. **Ayres, J. C.** 1960B. Temperature relationships and some other characteristics of the microbial flora developing on refrigerated beef. Food Res. *25:* 1–18.

8. **Barnes, E. M. and C. S. Impey.** 1968. Psychrophilic spoilage bacteria of poultry. J. Appl. Bacteriol. *31:* 97–107.

9. **Beacham, L. M.** 1946. A study of decomposition in canned oysters and clams. J. Assoc. Offic. Agr. Chemists *29:* 89–92.

10. **Borton, R. J., L. J. Bratzler, and J. F. Price.** 1970. Effects of four species of bacteria on porcine muscle. 2. Electrophoretic patterns of extracts of salt-soluble protein. J. Food Sci. *35:* 783–786.

11. **Brown, A. D. and J. F. Weidemann.** 1958. The taxonomy of the psychrophilic meat-spoilage bacteria: A reassessment. J. Appl. Bacteriol. *21:* 11–17.

12. **Callow, E. H. and M. Ingram.** 1955. Bone-taint. Food, Feb. issue.

13. **Campbell, L. L., Jr. and O. B. Williams.** 1952. The bacteriology of Gulf Coast shrimp. IV. Bacteriological, chemical, and organoleptic changes with iced storage. Food Technol. *6:* 125–126.

14. **Cavett, J. J.** 1962. The microbiology of vacuum packed sliced bacon. J. Appl. Bacteriol. *25:* 282–289.

15. **Chen, T. C. and R. E. Levin.** 1974. Taxonomic significance of phenethyl alcohol production by *Achromobacter* isolates from fishery sources. Appl. Microbiol. *28:* 681–687.

16. **Chen, T. C., W. W. Nawar, and R. E. Levin.** 1974. Identification of major high-boiling volatile compounds produced during refrigerated storage of haddock fillets. Appl. Microbiol. *28:* 679–680.

17. **Clark, D. S. and C. P. Lentz.** 1971. Use of carbon dioxide for extending shelf-life of prepackaged beef. Can. Inst. Food Technol. J. *5:* 175–178.

18. **Craven, S. E. and A. J. Mercuri.** 1977. Total aerobic and coliform counts in beef-soy and chicken-soy patties during refrigerated storage. J. Food Protec. *40:* 112–115.

19. **Dainty, R. H., B. G. Shaw, K. A. DeBoer, and E. S. J. Scheps.** 1975. Protein changes caused by bacterial growth on beef. J. Appl. Bacteriol. *39:* 73–81.

20. **Davidson, C. M., M. J. Dowdell, and R. G. Board.** 1973. Properties of gram negative aerobes isolated from meats. J. Food Sci. *38:* 303–305.

21. **Dowdell, M. J. and R. G. Board.** 1968. A microbiological survey of British fresh sausage. J. Appl. Bacteriol. *31:* 378–396.

22. **Drake, S. D., J. B. Evans, and C. F. Niven, Jr.** 1958. Microbial flora of packaged frankfurters and their radiation resistance. Food Res. *23:* 291–296.

23. **Fieger, E. A. and A. F. Novak.** 1961. Microbiology of shellfish deterioration. *Fish as Food 1:* 561–611, Edited by G. Borgstrom, Academic Press, N.Y.

24. **Freeman, L. R., G. J. Silverman, P. Angelini, C. Merritt, Jr., and W. B. Esselen.** 1976. Volatiles produced by microorganisms isolated from refrigerated chicken at spoilage. Appl. Environ. Microbiol. *32:* 222–231.

25. **Gardner, G. A.** 1971. Microbiological and chemical changes in lean Wiltshire bacon during aerobic storage. J. Appl. Bacteriol. *34:* 645–654.

26. **Gill, C. O.** 1976. Substrate limitation of bacterial growth at meat surfaces. J. Appl. Bacteriol. *41:* 401–410.

27. **Halleck, F. E., C. O. Ball, and E. P. Steir.** 1958. Factors affecting quality of pre-packaged meat. IV. Microbiological studies. B. Effect of packaging characteristics and atmosphereic pressure in package upon bacterial flora of meat. Food Technol. *12:* 301–306.

28. **Hanna, M. O., C. Vanderzant, Z. L. Carpenter, and G. C. Smith.** 1977a. Characteristics of psychrotrophic, gram-positive, catalase-positive, pleomorphic coccoid rods from vacuum-packaged wholesale cuts of beef. J. Food Protec. *40:* 94–97.

29. **Hanna, M. O., C. Vanderzant, Z. L. Carpenter, and G. C. Smith.** 1977b. Microbial flora of vacuum-packaged lamb with special reference to psychrotrophic, gram-positive, catalase-positive pleomorphic rods. J. Food Protec. *40:* 98–100.

30. **Herbert, R. A., M. S. Hendrie, D. M. Gibson, and J. M. Shewan.** 1971. Bacteria active in the spoilage of certain sea foods. J. Appl. Bacteriol. *34:* 41–50.

31. **Hess, E.** 1950. Bacterial fish spoilage and its control. Food Technol. *4:* 477–480.

32. **Hunter, A. C. and B. A. Linden.** 1923. An investigation of oyster spoilage. Am. Food J. *18:* 538–540.

33. **Ingram, M.** 1960. Bacterial multiplication in packed Wiltshire bacon. J. Appl. Bacteriol. *23:* 206–215.

34. **Ingram, M. and R. H. Dainty.** 1971. Changes caused by microbes in spoilage of meats. J. Appl. Bacteriol. *34:* 21–39.

35. **Jacquot, R.** 1961. Organic constituents of fish and other aquatic animal foods. *Fish as Food 1:* 145–209, Edited by G. Borgstrom, Academic Press, N.Y.

36. **Jay, J. M.** 1964A. Release of aqueous extracts by beef homogenates, and factors affecting release volume. Food Technol. *18:* 1633–1636.

37. **Jay, J. M.** 1964B. Beef microbial quality determined by extract-release volume (ERV). Food Technol. *18:* 1637–1641.

38. **Jay, J. M.** 1966A. Influence of postmortem conditions on muscle microbiology. In: *The Physiology and Biochemistry of Muscle as a Food.* Edited by E. J. Briskey *et al.,* University of Wisconsin Press, Madison, Chapter 26.

39. **Jay, J. M.** 1966B. Relationship between the phenomena of extract-release volume and water-holding capacity of meats as simple and rapid methods for determining microbial quality of beef. Hlth. Lab. Sci. *3:* 101–110.

40. **Jay, J. M.** 1966C. Response of the extract-release volume and water-holding capacity phenomena to microbiologically spoiled beef and aged beef. Appl. Microbiol. *14:* 492–496.

41. **Jay, J. M. and K. S. Kontou.** 1967. Fate of free amino acids and nucleotides in spoiling beef. Appl. Microbiol. *15:* 957–964.

42. **Jay, J. M.** 1967. Nature, characteristics, and proteolytic properties of beef spoilage bacteria at low and high temperatures. Appl. Microbiol. *15:* 943–944.

43. **Jay, J. M. and L. A. Shelef.** 1976. Effect of microorganisms on meat proteins at low temperatures. J. Agric. Food Chem. *24:* 1113–1116.

44. **Jay, J. M.** 1977. Meats, poultry, and seafoods, Chapter 5, IN: *Food and Beverage Mycology,* edited by L. R. Beuchat (Avi Publishing Co.: Westport, Conn.).

45. **Jaye, M., R. S. Kittake, and Z. J. Ordal.** 1962. The effect of temperature and packaging material on the storage life and bacterial flora of ground beef. Food Technol. *16:* 95–98.

46. **Kirsch, R. H., F. E. Berry, C. L. Baldwin, and E. M. Foster.** 1952. The bacteriology of refrigerated ground meat. Food Res. *17:* 495–503.

47. **Koburger, J. A., A. R. Norden, and G. M. Kampler.** 1975. The microbial flora of rock shrimp—*Sicyonia brevirostris.* J. Milk Food Technol. *38:* 747–749.

48. **Kraft, A. A. and J. C. Ayres.** 1952. Postmortem changes in stored meats. IV. Effect of packaging materials on keeping quality of self-service meats. Food Technol. *6:* 8–12.

49. **Lahellec, C., C. Meurier, and G. Bennejean.** 1975. A study of 5920 strains of psychrotrophic bacteria isolated from chickens. J. Appl. Bacteriol. *38:* 89–97.

50. **Lawrie, R. A.** 1966. *Meat Science.* Pergamon Press, N.Y., Chapters 4 and 10.

51. **Laycock, R. A. and L. W. Regier.** 1970. Pseudomonads and achromobacters in the spoilage of irradiated haddock of different pre-irradiation quality. Appl. Microbiol. *20:* 333–341.

52. **Lee, J. S. and J. M. Harrison.** 1968. Microbial flora of Pacific hake *(Merluccius productus).* Appl. Microbiol. *16:* 1937–1938.

53. **Lepovetsky, B. C., H. H. Weiser, and F. E. Deatherage.** 1953. A microbiological study of lymph nodes, bone marrow and muscle tissue obtained from slaughtered cattle. Appl. Microbiol. *1:* 57–59.

54. **Lerke, P., R. Adams, and L. Farber.** 1963. Bacteriology of spoilage of fish muscle. I. Sterile press juice as a suitable experimental medium. Appl. Microbiol. *11:* 458–462.

55. **Lerke, P., R. Adams, and L. Farber.** 1965. Bacteriology of spoilage of fish muscle. III. Characteristics of spoilers. Appl. Microbiol. *13:* 625–630.

56. **Margitic, S. and J. M. Jay.** 1970. Antigenicity of salt-soluble beef muscle proteins held from freshness to spoilage at low temperatures. J. Food Sci. *35:* 252–255.

57. **May, K. N., J. D. Irby, and J. L. Carmon.** 1961. Shelf life and bacterial counts of excised poultry tissue. Food Technol. *16:* 66–68.

58. **McLean, R. A. and W. L. Sulzbacher.** 1953. *Microbacterium thermosphactum* spec. nov., a non-heat resistant bacterium from fresh pork sausage. J. Bacteriol. *65:* 428–432.

59. **McMeekin, T. A.** 1975. Spoilage association of chicken breast muscle. Appl. Microbiol. *29:* 44–47.

60. **McMeekin, T. A. and J. T. Patterson.** 1975. Characterization of hydrogen sulfide-producing bacteria isolated from meat and poultry plants. Appl. Microbiol. *29:* 165–169.

61. **Miller, A., III, R. A. Scanlan, J. S. Lee, and L. M. Libbey.** 1972. Volatile compounds produced in ground muscle tissue of Canary rockfish *(Sebastes pinniger)* stored on ice. J. Fish. Res. Bd. Canada *29:* 1125–1129.

62. **Miller, A., III, R. A. Scanlan, J. S. Lee, L. M. Libbey, and M. E. Morgan.** 1973a. Volatile compounds produced in sterile fish muscle *(Sebastes malanops)* by *Pseudomonas perolens.* Appl. Microbiol. *25:* 257–261.

63. **Miller, A., III, R. A. Scanlan, J. S. Lee, and L. M. Libbey.** 1973b. Identification of the volatile compounds produced in sterile fish muscle *(Sebastes melanops)* by *Pseudomonas fragi.* Appl. Microbiol. *25:* 952–955.

64. **Miller, A., III, R. A. Scanlan, J. S. Lee, and L. M. Libbey.** 1973c. Volatile compounds produced in sterile fish muscle *(Sebastes melanops)* by *Pseudomonas putrefaciens, Pseudomonas fluorescens,* and an *Achromobacter* species. Appl. Microbiol. *26:* 18–21.

65. **Motohiro, T. and E. Tanikawa.** 1952. Studies on food poisoning of mollusk especially of squid and octopus meat. I. Chemical change and freshness tests of squid and octopus meat during deterioration of freshness. Bull. Fac. Fisheries Hokkaido Univ. *3* (2): 142–153.

66. **Nagel, C. W., K. L. Simpson, H. Ng, R. H. Vaughn, and G. F. Stewart.** 1960. Microorganisms associated with spoilage of refrigerated poultry. Food Technol. *14:* 21–23.

67. **Niven, C. F., Jr., A. G. Castellani, and V. Allanson.** 1949. A study of the lactic acid bacteria that cause surface discolorations of sausages. J. Bacteriol. *58:* 633–641.

68. **Njoku-Obi, A. N., J. V. Spencer, E. A. Sauter, and M. W. Eklund.** 1957. A study of the fungal flora of spoiled chlortetracycline treated chicken meat. Appl. Microbiol. *5:* 319–321.

69. **Pearson, D.** 1968. Assessment of meat freshness in quality control employing chemical techniques: A review. J. Sci. Food & Agr. *19:* 357–363.

70. **Peel, J. L. and J. M. Gee.** 1976. The role of micro-organisms in poultry taints, pp. 151–160, IN: *Microbiology in Agriculture, Fisheries and Food,* edited by F. A. Skinner and J. G. Carr (Academic Press: N.Y.).

71. **Pierson, M. D., D. L. Collins-Thompson, and Z. J. Ordal.** 1970. Microbiological, sensory and pigment changes in aerobically and anaerobically packaged beef. Food Technol. *24:* 1171–1175.

72. **Pottinger, S. R.** 1948. Some data on pH and the freshness of shucked eastern oysters. Comm. Fisheries Rev. *10* (9): 1–3.

73. **Ranke, B.** 1955. Über papier-chromatographische Untersuchungen des freien und eiweissgebundenen Aminosäuren-bestandes bei Krebsen und Fischen. Arch. Fischereiwiss. *6:* 109–113.

74. **Sharpe, M. E.** 1962. Lactobacilli in meat products. Food Manuf. *37:* 582–589.

75. **Shaw, B. G. and J. M. Shewan.** 1968. Psychrophilic spoilage bacteria of fish. J. Appl. Bacteriol. *31:* 89–96.

76. **Shelef, L. A. and J. M. Jay.** 1969. Relationship between meat-swelling, viscosity, extract-release volume, and water-holding capacity in evaluating beef microbial quality. J. Food Sci. *34:* 532–535.

77. **Shelef, L. A. and J. M. Jay.** 1969. Relationship between amino sugars and meat microbial quality. Appl. Microbiol. *17:* 931–932.

78. **Shelef, L. A. and J. M. Jay.** 1970. Use of a titrimetric method to assess the bacterial spoilage of fresh beef. Appl. Microbiol. *19:* 902–905.

79. **Shelef, L. A. and J. M. Jay.** 1971. Hydration capacity as an index of shrimp microbial quality. J. Food Sci. *36:* 994–997.

80. **Shewan, J. M.** 1961. The microbiology of sea-water fish. *Fish as Food 1:* 487–560. Edited by G. Borgstrom, Academic Press, N.Y.

81. **Sulzbacher, W. L. and R. A. McLean.** 1951. The bacterial flora of fresh pork sausage. Food Technol. *5:* 7–8.

82. **Sutherland, J. P., J. T. Patterson, and J. G. Murray.** 1975. Changes in the microbiology of vacuum-packaged beef. J. Appl. Bacteriol. *39:* 227–237.

83. **Tarr, H. L. A.** 1954. Microbiological deterioration of fish post mortem, its detection and control. Bact. Revs. *18:* 1–15.

84. **Tonge, R. J., A. C. Baird-Parker, and J. J. Cavett.** 1964. Chemical and microbiological changes during storage of vacuum packed sliced bacon. J. Appl. Bacteriol. *27:* 252–264.

85. **Velankar, N. K. and T. K. Govindan.** 1958. A preliminary study of the distribution of nonprotein nitrogen in some marine fishes and invertebrates. Proc. Indian Acad. Sci. *B47:* 202–209.

86. **Watt, B. K. and A. L. Merrill.** 1950. Composition of foods—raw, processed, prepared. Agric. Handbook No. 8, U.S. Dept. of Agric., Washington, D.C.

8

Spoilage of Miscellaneous Foods

This chapter covers the microbiological spoilage of the following groups of foods: Eggs, cereals and flour, bakery products, dairy products, sugar and spices, nutmeats, beverages and fermented foods, salad dressings, and canned foods.

SPOILAGE OF EGGS

The hen's egg is an excellent example of a product that normally is well protected by its intrinsic parameters. Externally, a fresh egg has three structures each of which is effective to some degree in retarding the entry of microorganisms: the outer, waxy shell membrane; the shell; and the inner shell membrane (see Fig. 8-1). Internally, lysozyme is present in egg white. This enzyme has been shown to be quite effective against gram positive bacteria. Egg white also contains avidin, which forms a complex with biotin, thereby making this vitamin unavailable to microorganisms. In addition, egg white has a high pH (about 9.3) and contains conalbumin, which forms a complex with iron, thus rendering it unavailable to microorganisms. On the other hand, the nutrient content of the yolk material and its pH in fresh eggs (about 6.8) make it an excellent source of growth for most microorganisms.

Freshly laid eggs are generally sterile. However, in a relatively short period of time after laying, numerous microorganisms may be found on the outside and may enter eggs under the proper conditions where they grow and cause spoilage. Among the bacteria found are members of the following genera: *Pseudomonas, Acinetobacter, Proteus, Aeromonas, Alcaligenes, Escherichia, Micrococcus, Salmonella, Serratia,* and *Enterobacter.* Among the molds generally found are members of the genera *Mucor, Penicillium, Hormodendron, Cladosporium,* and others, while *Torula* is the only yeast found with any degree of consistency.

The most common form of bacterial spoilage of eggs is a condition known as **rotting. Green rots** are caused by *Pseudomonas* spp., especially *Pseu-*

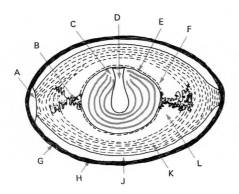

A Air cell
B Chalazae
C Yolk
D Germinal disc and white yolk
E Vitelline membrane

F Film of mucin
G Shell
H Shell membranes
J Outer thin white
K Thick white
L Inner thin white

FIGURE 8-1. Structure of the hen's egg as shown by a section through the long axis (*3,* reproduced with permission of Elsevier Publishing Co.)

domonas fluorescens; **colorless rots** by *Pseudomonas, Acinetobacter,* and other species; **black rots** by *Proteus, Pseudomonas,* and *Aeromonas;* **pink rots** by *Pseudomonas;* **red rots** by *Serratia* spp., and **"custard" rots** by *P. vulgaris* and *P. intermedium (7).* Mold spoilage of eggs is generally referred to as **pinspots** from the appearance of mycelial growth on the inside upon candling. *Penicillium* and *Cladosporium* spp. are among the most common causes of pinspots and fungal rotting in eggs. Bacteria also cause a condition in eggs known as **mustiness.** *Ps. graveolens* and *Proteus* spp. have been implicated in this condition, with *Ps. graveolens* producing the most characteristic spoilage pattern.

The entry of microorganisms into whole eggs is favored by high humidity. Under such conditions, growth of the microorganisms on the surface of eggs is favored, followed by penetration through the shell and inner membrane. The latter structure is the most important barrier to the penetration of bacteria into eggs, followed by the shell and the outer membrane *(15).* More bacteria are found in egg yolk than in egg white, and the reason for a general lack of microorganisms in egg white is quite possibly its content of antimicrobial substances. In addition, upon the storage of eggs, the thick white loses water to the yolk, resulting in a thinning of yolk and a shrinking of the thick white. This phenomenon makes it possible for the yolk to come into direct contact with the inner membrane where it may be infected directly by microorganisms. Once inside of the yolk, bacteria apparently grow in this nutritious medium where they produce by-products of protein and amino acid metabolism such as H_2S and other foul-smelling compounds. The effect of significant growth is to cause the yolk to become "runny" and discolored.

Molds generally show growth first in the region of the air sac where oxygen favors the growth of these forms. Under conditions of high humidity, molds may be seen growing over the outer surface of eggs. Under conditions of low humidity and low temperatures, surface growth is not favored, but eggs lose water at a faster rate and thereby become undesirable as products of commerce.

SPOILAGE OF CEREALS, FLOUR, AND DOUGH PRODUCTS

The microbial flora of wheat, rye, corn, and related products may be expected to be that of soil, storage environments, and those picked up during the processing of these commodities. While these products are high in proteins and carbohydrates, their low a_w is such as to restrict the growth of all microorganisms if stored properly. The microbial flora of flour is relatively low, since some of the bleaching agents reduce the load. When conditions of a_w favor growth, bacteria of the genus *Bacillus* and molds of several genera are usually the only ones that develop. Many aerobic sporeformers are capable of producing amylase which enables them to utilize flour and related products as sources of energy, providing, of course, that sufficient moisture is present to allow growth to occur. With less moisture, mold growth occurs and may be seen as typical mycelial growth and spore formation. Members of the genus *Rhizopus* are common and may be recognized by their black spores.

The spoilage of fresh refrigerated dough products including buttermilk biscuits, dinner and sweet rolls, and pizza dough is caused mainly by lactic acid bacteria. In a study by Hesseltine *et al. (10)*, 92% of isolates were *Lactobacillaceae* with more than one-half belonging to the genus *Lactobacillus*, 36% to the genus *Leuconostoc,* and 3% to *Streptococcus*. Molds were found generally in low numbers in spoiled products. The fresh products showed lactic acid bacterial numbers as high as log 8.38/g.

SPOILAGE OF BAKERY PRODUCTS

Commercially produced and properly handled bread generally lacks sufficient amounts of moisture to allow for the growth of any organisms except molds. One of the most common is *Rhizopus stolonifer,* often referred to as the "bread mold." The "red bread mold," *Neurospora sitophila,* may also be seen from time to time. Storage of bread under conditions of low humidity retards mold growth, and this type of spoilage is generally seen only where bread is stored at high humidities or where wrapped while still warm. Home-made breads may undergo a type of spoilage known as **ropiness,** which is caused by the growth of certain strains of *Bacillus subtilis* (*B. mesentericus*). The ropiness may be seen as stringiness by carefully breaking a batch of dough into two parts. The source of the organisms is flour and their growth is favored by holding the dough for sufficient periods of time at suitable temperatures.

Cakes of all types rarely undergo bacterial spoilage due to their unusually high concentrations of sugars which restrict the availability of water. The most common form of spoilage displayed by these products is moldiness. Common sources of spoilage molds are any and all cake ingredients, especially sugar, nuts, and spices. While the baking process is generally sufficient to destroy these organisms, many are added in icings, meringues, toppings, and so forth. Also, molds may enter baked cakes from handling and from the air. Growth of molds on the surface of cakes is favored by conditions of high humidity. On some fruit cakes, growth often originates underneath nuts and fruits if they are placed on the surface of such products after baking. Continued growth of molds on breads and cakes results in a hardening of the products.

SPOILAGE OF DAIRY PRODUCTS

Dairy products such as milk, butter, cream, and cheese are all susceptible to microbial spoilage because of their chemical composition. Milk is an excellent growth medium for all of the common spoilage organisms including molds and yeasts. Fresh, nonpasteurized milk generally contains varying numbers of microorganisms, depending upon the care employed in milking, cleaning, and handling of milk utensils. Raw milk held at refrigerator temperatures for several days invariably shows the presence of several or all bacteria of the following genera: *Streptococcus, Leuconostoc, Lactobacillus, Microbacterium, Propionibacterium, Micrococcus,* coliforms, *Proteus, Pseudomonas, Bacillus,* and others. Those unable to grow at the usual low temperature of holding tend to be present in very low numbers. The pasteurization process eliminates all but thermoduric strains, primarily streptococci and lactobacilli, and sporeformers of the genus *Bacillus* (and clostridia if present in raw milk). The spoilage of pasteurized milk is caused by the growth of heat-resistant streptococci utilizing lactose to produce lactic acid, which depresses the pH to a point (about pH 4.5) where curdling takes place. If present, lactobacilli are able to grow at pH values below that required by *S. lactis*. These organisms continue the fermentative activities and may bring the pH to 4.0 or below. If mold spores are present, these organisms begin to grow at the surface of the sour milk and raise the pH towards neutrality, thus allowing the more proteolytic bacteria such as *Pseudomonas* spp. to grow and bring about the liquefaction of the milk curd.

The same general pattern outlined above may be expected to occur in raw milk, especially if held at refrigerator temperatures. Another condition sometimes seen in raw milk is referred to as **ropiness.** This condition is caused by the growth of *Alcaligenes viscolactis* and is favored by low-temperature holding of raw milk for several days. The rope consists of slime-layer material produced by the bacterial cells and it gives the product a stringy consistency. This condition is not as common today as it was in years past.

Butter contains around 15% water, 81% fat, and generally less than 0.5% carbohydrate and protein (Table 8-1). Although it is not a highly perishable product, it does undergo spoilage by bacteria and molds. The main source of microorganisms to butter is cream, whether sweet or sour, pasteurized or nonpasteurized. The flora of whole milk may be expected to be found in cream since as the fat droplets rise to the surface of milk, they carry up microorganisms. The processing of both raw and pasteurized creams to yield butter brings about a reduction in the numbers of all microorganisms with values for finished cream ranging from several hundred to over 100,000/g having been reported for the finished salted butter *(18)*.

Bacteria cause two principle types of spoilage in butter. The first is a condition known as **"surface taint"** or putridity. This condition is caused by *Ps. putrefaciens* as a result of its growth on the surface of finished butter. It develops at temperatures within the range 4–7 C and may become apparent within 7–10 days. The odor of this condition is apparently due to certain organic acids, especially isovaleric acid *(5)*. The second most common bacterial spoilage condition of butter is **rancidity.** This condition is caused by the hydrolysis of butterfat with the liberation of free fatty acids. It should be recognized, of course, that lipase from sources other than microorganisms can cause the effect. The causative organism is *Ps. fragi* although *Ps. fluorescens* is sometimes found. Bacteria may cause 3 other less common spoilage conditions in butter. **Malty flavor** is reported to be due to the growth of *S. lactis* var. *maltigenes*. **Skunk-like** odor is reported to be caused by *Ps. mephitica,* while black discolorations of butter have been reported to be caused by *Ps. nigrifaciens (8).*

Butter undergoes fungal spoilage rather commonly by species of the following genera: *Cladosporium, Alternaria, Aspergillus, Mucor, Rhizopus, Penicillium,* and *Geotrichum,* especially *G. candidum (Oospora lactis).* These organisms can be seen growing on the surface of butter where they produce colorations referable to their particular spore colors. Black yeasts

TABLE 8-1.   Percentage composition of 9 miscellaneous foods *(30)*.

Foods	Water	Carbohy.	Proteins	Fat	Ash
Beer (4% alcohol)	90.2	4.4	0.6	0.0	0.2
Bread, enriched white	34.5	52.3	8.2	3.3	1.7
Butter	15.5	0.4	0.6	81.0	2.5
Cake (pound)	19.3	49.3	7.1	23.5	0.8
Figbars	13.8	75.8	4.2	4.8	1.4
Jellies	34.5	65.0	0.2	0.0	0.3
Margarine	15.5	0.4	0.6	81.0	2.5
Mayonnaise	1.7	21.0	26.1	47.8	3.4
Peanut butter	16.0	3.0	1.5	78.0	1.5

of the genus *Torula* also have been reported to cause discolorations on butter. The microscopic examination of moldly butter reveals the presence of mold mycelia some distances from the visible growth. The generally high lipid content and low water content make butter more susceptible to spoilage by molds than by bacteria.

Cottage cheese undergoes spoilage by bacteria, yeasts, and molds. The most common spoilage pattern displayed by bacteria is a condition known as **slimy curd.** *Alcaligenes* spp. have been reported to be among the most frequent causative organisms although *Pseudomonas, Proteus, Enterobacter,* and *Acinetobacter* spp. have been implicated. *Penicillium, Mucor, Alternaria,* and *Geotrichum* all grow well on cottage cheese, to which they impart stale, musty, moldy, and yeasty flavors *(8)*. The shelf life of commercially produced cottage cheese in Alberta, Canada was found to be limited by yeasts and molds *(27)*. While 48% of fresh samples contained coliforms, these organisms did not increase upon storage in cottage cheese at 40 F for 16 days.

The low moisture content of ripened cheeses makes them insusceptible to spoilage by most organisms, although molds can and do grow on these products as would be expected. Some ripened cheeses have sufficiently low O/R potentials to support the growth of anaerobes. It is, therefore, not surprising to find that anaerobic bacteria sometimes cause the spoilage of these products when a_w permits growth to occur. *Clostridium* spp., especially *C. pasteurianum, C. butyricum,* and *C. sporogenes,* have been reported to cause **gassiness** of cheeses. One aerobic sporeformer, *B. polymyxa,* has been reported to cause gassiness. All of these organisms utilize lactic acid with the production of CO_2, which is responsible for the gassy condition of these products.

SPOILAGE OF SUGARS, CANDIES, AND SPICES

These products rarely undergo microbial spoilage if properly prepared, processed, and stored due primarily to the lack of sufficient moisture for growth. Both cane and beet sugars may be expected to contain microorganisms. The important bacterial contaminants are members of the genera *Bacillus* and *Clostridium* which sometimes cause trouble in the canning industry (see Chapter 12). If sugars are stored under conditions of extremely high humidity, growth of some of these organisms is possible, usually at the exposed surfaces. The successful growth of these organisms depends, of course, upon their getting an adequate supply of moisture and essential nutrients other than carbohydrates. *Torula* and osmophilic strains of *Saccharomyces* (formerly *Zygosaccharomyces* spp.) have been reported to cause trouble in high-moisture sugars. These organisms have been reported to cause inversion of sugar. One of the most troublesome organisms in sugar refineries is *Leuconostoc mesenteroides*. This organism hydrolyzes sucrose and synthesizes a glucose polymer referred to as

dextran. This gummy and slimy polymer sometimes clogs the lines and pipes through which sucrose solutions pass.

Among candies that have been reported to undergo microbial spoilage are chocolate creams, which sometime undergo explosions. The causative organisms have been reported to be *Clostridium* spp., especially *C. sporogenes,* which finds its way into these products through sugars, starch, and possibly other ingredients.

Although spices do not undergo microbial spoilage in the usual sense of the word, molds and a few bacteria do grow on those that do not contain antimicrobial principals, providing sufficient moisture is available. Prepared mustard has been reported to undergo spoilage by yeasts and by *Proteus* and *Bacillus* spp., usually with a gassy fermentation. The usual treatment of spices with propylene oxide reduces their content of microorganisms, and those that remain are essentially sporeformers and molds. No trouble should be encountered from microorganisms as long as the moisture level is kept low.

SPOILAGE OF NUTMEATS

Due to the extremely high-fat and low-water content of products such as pecans and walnuts (Table 8-2), these products are quite refractory to spoilage by bacteria. Molds can and do grow upon them if they are stored under conditions that permit sufficient moisture to be picked up. Molds of many genera may be found on examining nutmeats which are picked up by the products during collecting, cracking, sorting, packaging, etc. See Chapter 19 for a discussion of aflatoxins as related to nutmeats.

SPOILAGE OF BEERS, WINES, AND FERMENTED FOODS

The products covered in this section are beers and ales, table wines, sauerkraut, pickles, and olives. All of these products are themselves the products of microbial actions.

The industrial spoilage of beers and ales is commonly referred to as beer infections. This condition is caused by yeasts and bacteria. The spoilage pat-

TABLE 8-2.   Percentage composition of various nuts (30).

Nuts	Water	Carbohy.	Protein	Fat	Ash
Almonds (dried)	4.7	19.6	18.6	34.1	3.0
Brazil nuts	5.3	11.0	14.4	65.9	3.4
Cashews	3.6	27.0	18.5	48.2	2.7
Peanuts	2.6	23.6	26.9	44.2	2.7
Pecans	3.0	13.0	9.4	73.0	1.6
Mean	3.8	18.8	17.6	57.1	2.7

terns of beers and ales may be classified into 4 groups: ropiness, sarcinae sickness, sourness, and turbidity. **Ropiness** in beer is a condition in which the liquid becomes characteristically viscous and pours as an "oily" stream. It is caused by *Acetobacter, Lactobacillus,* and/or *Pediococcus cerevisiae (26, 31)*. **Sarcinae sickness** is caused by *P. cerevisiae* which produces a honey-like odor. This characteristic odor is the result of diacetyl production by the spoilage organism in combination with the normal odor of beef. **Sourness** in beers is caused by *Acetobacter* spp. These organisms are capable of oxidizing ethanol to acetic acid, and the sourness that results is referable to increased levels of acetic acid. **Turbidity** and off-odors in beers are caused by *Zymomonas anaerobia* (formerly *Achromobacter anaerobium*) and several yeasts such as *Saccharomyces* spp. Growth of bacteria is possible in beers because of a normal pH range of 4–5 and a good content of utilizable nutrients.

With respect to spoiled packaged beer, one of the major contaminants found is *Saccharomyces diastaticus,* which is able to utilize dextrins which normal brewers yeasts *(S. carlsbergensis* and *S. cerevisiae)* cannot utilize *(11)*. Pediococci, *Flavobacterium proteus* (formerly *Obesumbacterium*), and *Brettanomyces* are sometimes found in spoiled beer.

Table wines undergo spoilage by bacteria and yeasts with *Candida mycoderma* being the most important yeast. Growth of this organism occurs at the surface of wines where a thin film is formed. The organisms attack alcohol and other constituents from this layer and create an appearance that is sometimes referred to as **"wine flowers."** Among the bacteria that cause wine spoilage are members of the genus *Acetobacter,* which oxidize alcohol to acetic acid (produce vinegar). The most serious and the most common disease of table wines is referred to as **tourne disease** *(24)*. Tourne disease is caused by a facultative anaerobe or anaerobe which utilizes sugars and seems to prefer conditions of low alcohol content. This type of spoilage is characterized by an increased volatile acidity, a silky type of cloudiness, and later in the course of spoilage, a "mousey" odor and taste.

The **malo-lactic** fermentation is a spoilage condition in wines of great importance. Malic and tartaric acids are two of the predominant organic acids in grape must and wine, and in the malo-lactic fermentation, contaminating bacteria degrade malic acid to lactic acid and CO_2:

$$L(-)\text{-Malic acid} \xrightarrow{\text{"malo-lactic enzyme"}} L(+)\text{-Lactic acid} + CO_2.$$

L-malic acid may be decarboxylated also to yield pyruvic acid *(12)*. The effect of these conversions is to reduce the acid content and affect flavor. The malo-lactic fermentation (which may occur also in cider) can be carried out by many lactic acid bacteria including leuconostocs, pediococci, and lactobacilli *(16, 25)*. While the function of the malo-lactic fermentation to the fermenting organism is not well understood, it has been shown that *L. oenos* is actually stimulated by the process *(22)*. The decomposition in wines of tartaric acid is undesirable also, and this process can be achieved by some strains of *L. plantarum* in the following general manner:

$$\text{Tartaric acid} \rightarrow \text{Lactic acid} + \text{acetic acid} + CO_2$$

The effect of the above is to reduce the acidity of wine. Unlike the malolactic fermentation, few lactic acid bacteria break down tartaric acid *(25)*.

Root beer undergoes bacterial spoilage on occasions. Lehmann and Byrd *(14)* investigated spoiled root beer which was characterized by a musty odor and taste. The causative organism was found to be *"Achromobacter"* sp. By inoculating this organism into the normal product, these authors found that the characteristic spoilage appeared in 2 weeks.

Sauerkraut is the product of lactic acid fermentation of fresh cabbage, and while the finished product has a pH in the range of 3.1–3.7, it is still subject to spoilage by bacteria, yeasts, and molds. The microbial spoilage of sauerkraut generally falls into the following categories: soft kraut, slimy kraut, rotted kraut, and pink kraut. **Soft kraut** results when bacteria that normally do not initiate growth until the late stages of kraut production actually grow earlier. **Slimy kraut** is caused by the rapid growth of *Lactobacillus cucumeris* and *L. plantarum,* especially at elevated temperatures *(24)*. **Rotted** sauerkraut may be caused by bacteria, molds, and/or yeasts, while **pink kraut** is caused by the surface growth of *Torula* spp., especially *T. glutinis*. Due to the high acidity, finished kraut is generally spoiled by molds growing on the surface. The growth of these organisms effects an increase in pH to levels where a large number of bacteria can grow that were previously inhibited by conditions of high acidity.

Pickles result from lactic acid fermentation of cucumbers. The finished product has a pH of around 4.0. These products undergo spoilage by bacteria and molds. **Pickle blackening** may be caused by *Bacillus nigrificans* where the dark color is due to the production of a water-soluble pigment. *Enterobacter* spp., lactobacilli, and pediococci have all been implicated in the cause of **bloaters.** This condition is caused by gas formation within the individual pickles. **Pickle softening** is caused by pectolytic organisms of the genus *Bacillus, Fusarium, Penicillium, Phomaceae, Cladosporium, Alternaria, Mucor, Aspergillus,* and others. The actual softening of pickles may be caused by any one or several of these or related organisms. Pickle softening results from the production of pectinases which break down the cement-type substance in the wall of the product.

Among the types of microbial spoilage that olives undergo, one of the most characteristic is **zapatera spoilage.** This condition sometimes occurs in brined olives and is characterized by an undesirable malodorous fermentation. The odor is due apparently to propionic acid which is produced by certain species of *Propionibacterium (23)*.

A **softening** condition of Spanish-type green olives has been found to be caused by the yeasts *Rhodotorula glutinis* var. *glutinis, R. minuta* var. *minuta,* and *R. rubra (29).* All of these organisms produced polygalacturonases which effect olive tissue softening. Under appropriate cultural conditions, the organisms were shown to produce pectin methyl esterase as well as polygalacturonase. A **sloughing** type spoilage of California ripe olives was shown by Patel and Vaughn *(19)* to be caused by *Cellulomonas flavigena*. This organism showed high cellulolytic activity and its activities were enhanced by the growth of other organisms such as *Xanthomonas, Enterobacter,* and *Escherichia* spp.

SPOILAGE OF SALAD DRESSINGS

Mayonnaise may be defined as a semisolid emulsion of edible vegetable oil, egg yolk or whole egg, vinegar, and/or lemon juice, with one or more of the following: salt, other seasoning used in preparation of the mayonnaise, and/or glucose, such that the finished product contains not less than 50% edible oil *(17).* The pH of mayonnaise has been found to range from 3.0 to 4.1, while for tartar sauce, french dressing, and salad dressing, from 2.9–4.4. While the nutrient content of these products is suitable as food sources for many spoilage organisms, the low pH values restrict the spoilers to yeasts, a few bacteria, and molds. The sugar content of these products is generally about 4–5%. It is, therefore, not surprising to note that yeasts of the genus *Saccharomyces* have been implicated in the spoilage of mayonnaise, salad dressing, and French dressing. One of the few bacteria reported to cause spoilage of products of this type is *Lactobacillus brevis* which was reported to produce gas in salad dressing. Appleman *et al. (1)* investigated spoiled mayonnaise and recovered a strain of *B. subtilis* and a yeast which they believed to be the etiologic agents. *Bacillus vulgatus* has been recovered from spoiled thousand islands dressing where it caused darkening and separation of the emulsion. In one particular study of the spoilage of thousand island dressing, pepper and paprika were shown to be the sources of *B. vulgatus (20).* Mold spoilage of products of this type occur only at the surfaces when sufficient oxygen is available. Separation of the emulsion is generally one of the first signs of spoilage of these products although bubbles of gas and the rancid odor of butyric acid may precede emulsion separation. The spoilage organisms apparently attack the sugars fermentatively. It appears that the pH remains low, thereby preventing the activities of proteolytic and lipolytic organisms. It is not surprising to find yeasts and lactic acid bacteria under these conditions. In a study of 17 samples of spoiled mayonnaise, mayonnaise-like, and blue cheese dressings, Kurtzman *et al. (13)* found high yeast counts in most samples and high lactobacilli counts in 2. The pH of samples ranged from 3.6 to 4.1. Two-thirds of the spoiled samples yielded *Saccharomyces bailii*. Common in some samples was *L. fructivorans,* with aerobic sporeformers being found in only two samples. Of 10 unspoiled samples tested, microorganisms were in low numbers or not detectable at all.

SPOILAGE OF CANNED FOODS

Although the objective in the canning of foods is the destruction of microorganisms, these products nevertheless undergo microbial spoilage under certain conditions. The main reasons for this are the following: under-processing, inadequate cooling, infection of can resulting from leakage through seams, and pre-process spoilage *(2)*. Since some canned foods receive low heat treatments, it is to be expected that a rather large number of different types of microorganisms may be found upon examining such foods.

As a guide to the type of spoilage that canned foods undergo, the following classification of canned foods based upon acidity is helpful.

Low-acid: pH > 4.6. Meat and marine products, milk, some vegetables (corn, lima beans), meat and vegetable mixtures, and so on. Spoiled by thermophilic flat-sour group *(B. stearothermophilus, B. coagulans)*, sulfide spoilers *(C. nigrificans, C. bifermentans)*, and/or gaseous spoilers *(C. thermosaccharolyticum)*. Mesophilic spoilers include putrefactive anaerobes (especially P. A. 3679 types). Spoilage and toxin production by *C. botulinum* types A and B may occur if these organisms are present. Medium acid foods are those with pH range of 5.3–4.6 while low acid foods are those with pH $\geq$ 5.4.

Acid: pH 3.7–4.0 to 4.6. Within this category are many fruits such as tomatoes, pears, figs, etc. Thermophilic spoilers include *B. coagulans* types. Mesophiles include *B. polymyxa, B. macerans (B. betanigrificans), C. pasteurianium, C. butyricum,* lactobacilli, and others.

High-Acid: pH < 4.0–3.7. This category includes some fruits and fruit juices such as grapefruit; rhubarb, sauerkraut, pickles, and so forth. Generally spoiled by nonsporeforming mesophiles—yeasts, molds, and/or lactic acid bacteria.

Canned food spoilage organisms may be further characterized as follows:

1. *Mesophilic organisms*
   a. Putrefactive anaerobes
   b. Butyric anaerobes
   c. Aciduric flat sours
   d. Lactobacilli
   e. Yeasts
   f. Molds
2. *Thermophilic organisms*
   a. Flat-sour spores
   b. Thermophilic anaerobes producing sulfide
   c. Thermophilic anaerobes not producing sulfide

The canned food spoilage manifestations of these organisms are presented in Table 8-3.

With respect to the spoilage of high acid and other canned foods by yeasts, molds, and bacteria, several of these organisms have been repeatedly associated with certain specific foods. The yeasts *Torula lactis-condensi* and *T. globosa* cause blowing or gaseous spoilage of sweetened condensed milk, which is not heat processed. The mold *Aspergillus repens* is associated with the formation of "buttons" on the surface of sweetened condensed milk. *Lactobacillus brevis (L. lycopersici)* causes a vigorous fermentation in tomato ketchup, Worcestershire sauce, and similar products. *Leuconostoc mesenteroides* has been reported to cause gaseous spoilage of canned

TABLE 8-3.   Spoilage manifestations in acid and low-acid canned foods *(28)*.

Type of organism	Appearance & manifestations of can	Appearance of product
	Acid Products	
1. *B. thermoacidurans* (flat sour: tomato juice).	Can flat. Little change in vacuum.	Slight pH change. Off-odor and flavor.
2. Butyric anaerobes (tomatoes & tomato juice).	Can swells. May burst.	Fermented; butyric odor.
3. Nonspore formers (mostly lactics).	Can swells, usually bursts, but swelling may be arrested.	Acid odor.
	Low-Acid Products	
1. Flat sour	Can flat. Possible loss of vacuum on storage.	Appearance not usually altered. pH markedly lowered—sour. May have slightly abnormal odor. Sometimes cloudy liquor.
2. Thermophilic anaerobe.	Can swells. May burst.	Fermented, sour, cheesey, or butyric odor.
3. Sulfide spoilage.	Can flat. H_2S gas absorbed by product.	Usually blackened. "Rotten egg" odor.
4. Putrefactive anaerobe.	Can swells. May burst.	May be partially digested. pH slightly above normal. Typical putrid odor.
5. Aerobic sporeformers (odd types).	Can flat. Usually no swelling, except in cured meats when NO_3 and sugar are present.	Coagulated evaporated milk, black beets.

pineapples and ropiness in peaches. The mold *Byssochlamys fulva* causes spoilage of bottled and canned fruits. Its actions cause disintegration of fruits as a result of pectin breakdown *(2)*. *Torula stellata* has been reported to cause the spoilage of canned bitter lemon, and to grow at a pH of 2.5 *(21)*. *et al.*, 1964).

Frozen concentrated orange juice sometimes undergoes spoilage by yeasts and bacteria. Hays and Riester *(9)* investigated samples of this product spoiled by bacteria. The orange juice was characterized as having a vinegary to buttermilk off-odor with an accompanying off-flavor. From the spoiled product were isolated *L. plantarum* var. *mobilis*, *L. brevis*, *Leuconostoc mesenteroides*, and *Leuconostoc dextranicum*. The spoilage characteristics could be reproduced by inoculating the above isolates into fresh orange juice.

Minimum growth temperatures of spoilage thermophiles are of some importance in diagnosing the cause of spoiled canned foods. *B. coagulans* (*B. thermoacidurans*) has been reported to grow only slowly at 25 C but grows well between 30 and 55 C. *B. stearothermophilus* does not grow at 37, while it has its optimum temperature around 65 C with smooth varients having a shorter generation time at this temperature than rough varients *(6)*. *C. thermosaccharolyticum* does not grow at 30 but has been reported to grow at 37 C.

Also of importance in diagnosing the cause of canned food spoilage is the appearance of the unopened can or container. The ends of a can of food are normally flat or slightly concave. When microorganisms grow and produce gases, the can goes through a series of changes visible from the outside. The change is designated a **flipper** when one end can be made to become convex by striking or heating the can. A **springer** is a can with both ends bulged when one or both remain concave if pushed in, or when one end is pushed in the other pops out. A **soft swell** refers to a can with both ends bulged but may be dented by pressing with the fingers. A **hard swell** has both ends bulged so that neither end can be dented by hand. The above events tend to develop successively and become of value in predicting the type of spoilage that might be in effect. Flippers and springers may be incubated under wraps at a temperature appropriate to the pH and type of food in order to allow for further growth of any organisms that might be present. These types of cans do not always represent microbial spoilage. Soft swells often represent microbial spoilage as do hard swells. In high acid foods, hard swells are very often **hydrogen swells,** which result from the release of hydrogen gas by the action of food acids on the iron of the can. The other two most common gases in cans of spoiled foods are CO_2 and H_2S, both of which are the result of the metabolic activities of microorganisms. Hydrogen sulfide may be noted by its characteristic odor, while CO_2 and hydrogen may be determined by the following test. Construct an apparatus of glass or plastic tubing attached to a hollow punch fitted with a large rubber stopper. To a test tube filled with dilute KOH, insert the free end of this apparatus inside the tube and invert in a beaker filled with dilute KOH. When an opening is made in one end of the can with the hollow punch, the gases will displace the dilute KOH inside the tube. Before removing the open end from the beaker, close by placing thumb over tube. To test for CO_2, shake the tube and note for a vacuum as evidenced by suction against the finger. To test for hydrogen, repeat above and apply a match near the top of the tube and then quickly remove thumb. A "pop" indicates the presence of hydrogen. Both gases may be found in some cans of spoiled foods.

"**Leakage-type**" spoilage of canned foods is characterized by a flora of nonspore-forming organisms that would not survive the heat treatment normally given heat-processed foods. These organisms enter cans at the start of cooling through faulty seams which generally result from can abuse. The organisms that cause "leakage-type" spoilage can be found either on the cans or in the cooling water. This type of can spoilage may be further differentiated from that caused by understerilization as follows *(28)*:

	Understerilization	*Leakage*
Can	Flat or swelled. Seams generally normal.	Swelled; may show defects.
Product Appearance	Sloppy or fermented.	Frothy fermentation; viscous.
Odor	Normal, sour or putrid, but generally consistent.	Sour, fecal, generally varying from can to can.
pH	Usually fairly constant.	Wide variation.
Microscopic and Cultural	Pure cultures, spore formers. Growth at 98°F and/or 131°F. May be characteristic on special media, e.g., acid agar for tomato juice.	Mixed cultures, generally rods and cocci. Growth only at usual temperatures.
History	Spoilage usually confined to certain portions of pack. In acid products diagnosis may be less clearly defined. Similar organisms may be involved in understerilization and leakage.	Spoilage scattered.

Thermophilic strains of *Actinomycetes* have been reported to cause spoilage of hay and grain when overheating occurs. These organisms have been found in various milk powders where they may cause spoilage. They are presumed to be of little or no importance in canned food spoilage since spores of the majority of species are killed after a few minutes exposure to temperatures between 65–70 C *(4)*.

REFERENCES

1. **Appleman, M. D., E. P. Hess, and S. C. Rittenberg.** 1949. An investigation of a mayonnaise spoilage. Food Technol. *3:* 201–203.

2. **Baumgartner, J. G. and A. C. Hersom.** 1957. *Canned Foods.* D. Van Nostrand Company, Inc., Princeton, N.J.

3. **Brooks, J. and H. P. Hale.** 1959. The mechanical properties of the thick white of the hen's egg. Biochem. Biophys. Acta *32:* 237–250.

4. **Cross, T.** 1968. Thermophilic *Actinomycetes.* J. Appl. Bacteriol. *31:* 36–53.

5. **Dunkley, W. L., G. Hunter, H. R. Thornton, and E. G. Hood.** 1942. Studies on surface taint butter. II. An odorous compound in skimmilk cultures of *Pseudomonas putrefaciens.* Scientific Agr. *22:* 347–355.

6. **Fields, M. L.** 1970. The flat sour bacteria. Adv. Food Res. *18:* 163–217.

7. **Florian, M. L. E. and P. C. Trussell.** 1957. Bacterial spoilage of shell eggs. IV. Identification of spoilage organisms. Food Technol. *11:* 56–60.

8. **Foster, E. M., F. E. Nelson, M. L. Speck, R. N. Doetsch, and J. C. Olson, Jr.** 1957. *Dairy Microbiology.* Prentice-Hall, Inc., Englewood Cliffs, N.J., Chapters 13 and 14.

9. **Hays, G. L. and D. W. Riester.** 1952. The control of "off-odor" spoilage in frozen concentrated orange juice. Food Technol. *6:* 386–389.

10. **Hesseltine, C. W., R. R. Graves, R. Rogers, and H. R. Burmeister.** 1969. Aerobic and facultative microflora of fresh and spoiled refrigerated dough products. Appl. Microbiol. *18:* 848–853.

11. **Kleyn, J. and J. Hough.** 1971. The microbiology of brewing. Ann. Rev. Microbiol. *25:* 583–608.

12. **Kunkee, R. E.** 1975. A second enzymatic activity for decomposition of malic acid by malo-lactic bacteria, pp. 29–42, IN: *Lactic Acid Bacteria in Beverages and Food,* edited by J. G. Carr *et al.* (Academic Press: N.Y.).

13. **Kurtzman, C. P., R. Rogers, and C. W. Hesseltine.** 1971. Microbiological spoilage of mayonnaise and salad dressings. Appl. Microbiol. *21:* 870–874.

14. **Lehmann, D. L. and B. E. Byrd.** 1953. A bacterium responsible for a musty odor and taste in root beer. Food Res. *18:* 76–78.

15. **Lifshitz, A., R. G. Baker, and H. B. Naylor.** 1964. The relative importance of chicken egg exterior structures in resisting bacterial penetration. J. Food Sci. *29:* 94–99.

16. **London, J.** 1976. The ecology and taxonomic status of the lactobacilli. Ann. Rev. Microbiol. *30:* 279–301.

17. **Longrée, K.** 1967. *Quantity Food Sanitation.* Interscience Publishers, N.Y.

18. **Macy, H., S. T. Coulter, and W. B. Combs.** 1932. Observations on the quantitative changes in the microflora during the manufacture and storage of butter. Minn. Agr. Exp. Sta. Techn. Bull. *82.*

19. **Patel, I. B. and R. H. Vaughn.** 1973. Cellulolytic bacteria associated with sloughing spoilage of California ripe olives. Appl. Microbiol. *25:* 62–69.

20. **Pederson, C. S.** 1930. Bacterial spoilage of a thousand island dressing. J. Bacteriol. *20:* 99–106.

21. **Perigo, J. A., B. L. Gimbert, and T. E. Bashford.** 1964. The effect of carbonation, benzoic acid, and pH on the growth rate of a soft drink spoilage yeast as determined by a turbidostatic continuous culture apparatus. J. Appl. Bacteriol. *27:* 315–332.

22. **Pilone, G. J. and R. E. Kunkee.** 1976. Stimulatory effect of malo-lactic fermentation on the growth rate of *Leuconostoc oenos*. Appl. Environ. Microbiol. *32:* 405–408.

23. **Plastourgos, S. and R. H. Vaughn.** 1957. Species of *Propionibacterium* associated with zapatera spoilage of olives. Appl. Microbiol. *5:* 267–271.

24. **Prescott, S. C. and C. G. Dunn.** 1959. *Industrial Microbiology.* McGraw-Hill, N.Y.

25. **Radler, F.** 1975. The metabolism of organic acids by lactic acid bacteria, pp. 17–27, IN: *Lactic Acid Bacteria in Beverages and Food,* edited by J. G. Carr *et al.* (Academic Press: N.Y.).

26. **Rainbow, C.** 1975. Beer spoilage lactic acid bacteria, pp. 149–158, IN: *Lactic Acid Bacteria in Beverages and Food,* edited by J. G. Carr *et al.* (Academic Press: N.Y.).

27. **Roth, L. A., L. F. L. Clegg, and M. E. Stiles.** 1971. Coliforms and shelf life of commercially produced cottage cheese. Can. Inst. Food Technol. J. *4:* 107–111.

28. **Schmitt, H. P.** 1966. Commercial sterility in canned foods, its meaning and determination. Assoc. Food & Drug Officials of U.S., Quart. Bull. *30:* 141–151.

29. **Vaughn, R. H., T. Jakubczyk, J. D. MacMillan, T. E. Higgins, B. A. Dave, and V. M. Crampton.** 1969. Some pink yeasts associated with softening of olives. Appl. Microbiol. *18:* 771–775.

30. **Watt, B. K. and A. L. Merrill.** 1950. Composition of foods—raw, processed, prepared. Agric. Handbook No. 8, U.S.D.A., Washington, D.C.

31. **Williamson, D. H.** 1959. Studies on lactobacilli causing ropiness in beer. J. Appl. Bacteriol. *22:* 392–402.

9

Food Preservation
By the Use
of Chemicals

The use of chemicals to prevent or delay the spoilage of foods derives in part from the fact that such compounds have been used with great success in the treatment of diseases of man, animals, and plants. This is not to imply that any and all chemotherapeutic compounds can or should be used as food preservatives. On the other hand, there are some chemicals of value as food preservatives that would be ineffective or too toxic as chemotherapeutic compounds. With the exception of certain antibiotics, none of the presently used food preservatives find any real use as chemotherapeutic compounds in man and animals. While a large number of chemicals have been described that show potential as food preservatives, only a relatively small number is allowed in food products. The reason for this is due in large part to the strict rules of safety adhered to by the Food and Drug Administration (FDA), and to a lesser extent to the fact that not all compounds that show antimicrobial activity *in vitro* do so when added to certain foods. Below are described those compounds most widely used, their modes of action where known, and the types of foods in which they are used. These as well as others are summarized in Table 9-1.

BENZOIC ACID AND RELATED COMPOUNDS

Benzoic acid (C_6H_5COOH) and its sodium salt ($C_7H_5NaO_2$) along with methyl- and propylparaben are considered together in this section. Sodium benzoate was the first chemical preservative permitted in foods by the U.S. Food and Drug Administration, and this compound is still in wide use today in a large number of foods. Benzoic acid occurs naturally in certain foods

TABLE 9-1.   Summary of chemical food preservatives.

Preservatives	Maximum tolerance	Organisms affected	Foods
Propionic acid and propionates[a]	0.32%	Molds	Bread, cakes, some cheeses, rope inhibitor in bread dough.
Sorbic acid and sorbates[a]	0.2%	Molds	Hard cheeses, figs, syrups, salad dressings, jellies, cakes, etc.
Benzoic acid and benzoates[a]	0.1%	Yeasts and Molds	Margarine, pickle relishes, apple cider, soft drinks, tomato catsup, salad dressings, etc.
Sulfur dioxide, sulfites, bisulfites, metabisulfites[a]	200–300 ppm	Insects, microorganisms	Molasses, dried fruits, wine making, lemon juice, etc. (Not to be used in meats or other foods recognized as sources of thiamine.)
Ethylene and propylene oxides[b]	700 ppm	Fungi & vermin	Fumigant for spices
Sodium diacetate[a]	0.32%	Molds	Bread
Dehydroacetic acid[a]	65 ppm	Insects	Pecticide on strawberries, squash.
Chlor- and oxytetracycline[c]	7 ppm	Bacteria	Fresh poultry and turkey meats
Polymyxin B	15 ppm	Bacteria	Yeast cultures for beer
Sodium nitrite[b]	200 ppm	Bacteria	Fish fillets, meat-curing preparations, smoked salmon.
Caprylic acid[a]	—	Molds	Cheese wraps
Diethyl pyrocarbonate[c]	200 ppm[d]	Yeasts	Bottled wines, soft drinks
NaCl, Sugars	None	Microorganisms	NaCl: meats. Sugars: preserves, jellies, etc.
Wood smoke	None	Microorganisms	Meats, fish.
Ethyl formate[a]	15–200 ppm[e]	Fungi	Dried fruits, nuts.

[a]On the GRAS (Generally Recognized As Safe) lists per Section 201 (321)(s) of Federal Food, Drug, and Cosmetic Act as amended. Many additives are currently under review.
[b]May be involved in mutagenesis and/or carcinogenesis.
[c]Currently not permissible in U.S. foods.
[d]Must not be present 5 days after bottling.
[e]As formic acid.

such as cranberries. Its approved derivatives have structural formulas as indicated:

Methylparaben
Methyl p-Hydroxybenzoate

Propylparaben
Propyl p-Hydroxybenzoate

$$HO-\langle \rangle -COOCH_3 \qquad HO-\langle \rangle -COO(CH_2)_2CH_3$$

The antimicrobial activity of benzoate is related to pH with the greatest antimicrobial activity being at low pH values. The antimicrobial activity resides in the undissociated molecule. These compounds are most active at the lowest pH values of foods and essentially ineffective at neutral values. The pK of benzoate is 4.20 and at a pH of 4.00, 60% of the compound is undissociated while at a pH of 6.0 only 1.5% is undissociated. This results in the restriction of benzoic acid and its sodium salts to high acid foods such as apple cider, soft drinks, tomato catsup, salad dressings, margarine, syrups, and others. High acidity alone is generally sufficient to prevent growth of bacteria in these foods, but not that of certain molds and yeasts. As used in acidic foods, sodium benzoate acts essentially as a mold and yeast inhibitor. Bosund (6) has shown the similarity in antimicrobial activity between benzoic and salicylic acids. Both compounds were taken up by respiring microbial cells and both were found to block the oxidation of glucose and pyruvate at the acetate level in *Proteus vulgaris*. Bosund also reported that benzoic acid causes an increase in the rate of O_2 consumption during the first part of glucose oxidation in *P. vulgaris*. More recently, it has been shown that the benzoates, like propionate and sorbate, act against microorganisms by inhibiting the cellular uptake of substrate molecules (27).

In certain foods, such as fruit juices, at the maximum level of 0.1%, benzoates may impart disagreeable tastes. The taste has been described as being "peppery" or burning.

Methyl and proplyparaben (esters of parahydroxybenzoic acid) are said to be effective against bacteria, yeasts, and molds. The p-hydroxybenzoic esters have been reported to be superior to benzoic acid and its salts for low acid or neutral foods (79). The esters of p-hydroxybenzoic acid have also been reported to be more specific for molds than yeasts. Like benzoic acid and sodium benzoate, their use is restricted also to up to 0.1% in foods. The parabens have a pK of 8.47 and their antimicrobial activity is not increased to the same degree as for benzoate as pH is lowered.

SORBIC ACID

This compound (CH_3CH=$CHCH$=$CHCOOH$) is employed as a food preservative usually as the calcium, sodium, or potassium salt. These compounds are permissible in foods at levels not to exceed 0.2%. Like sodium benzoate, they are more effective in acid foods than in neutral foods and tend to be on par with the benzoates as fungal inhibitors. Sorbic acid works best below pH 6.0 and is generally ineffective about pH 7.0. These compounds are more effective than sodium benzoate between pH 4.0–6.0 (72). At pH values of 3.0 and below, the sorbates are slightly more effective than the propionates but about the same as sodium benzoate. The pK of sorbate is 4.80 and at a pH of 4.0, 86% of the compound is undissociated while at a pH of 6.0 only 6% is undissociated. Sorbic acid can be employed in cakes at higher levels than propionates without imparting flavor to the product (56).

The undissociated molecule is essential to antimicrobial activity by these compounds, and as noted above, they prevent microbial growth by inhibiting cellular uptake of substrate molecules such as amino acids, phosphate, organic acids, and the like (27). They are primarily effective against molds and yeasts although some activity is displayed against certain bacteria. The inhibition of molds by sorbic acid has been reported to be due to inhibition of the dehydrogenase enzyme systems in molds (20). The sorbates have been reported to inhibit *Salmonella* spp., fecal streptococci, and staphylococci, but not clostridia. The lactic acid bacteria are not affected by sorbates at pH of 4.5 and above. These compounds have been used as mold inhibitors on hard cheeses and in cucumber fermentations where undesirable yeasts are kept down without affecting the desirable lactic acid bacteria. In the latter application, 0.01 percent sorbic acid was shown by Costilow *et al.* (16) to effectively inhibit yeast activity in a medium containing 8% NaCl at pH 4.6. More sorbic acid was required as the pH was increased and/or salt concentration decreased. Bell *et al.* (4) studied the effect of sorbic acid on 66 molds, 32 yeasts, and 6 species of lactic acid bacteria and found that all grew in the presence of 0.1% sorbic acid at pH 7.0. The molds and yeasts were inhibited when pH was lowered to 4.5, but all bacteria were not inhibited until the pH was lowered to 3.5.

Sorbic acid is preferred in apple cider to benzoates, since the latter compounds impart a burning taste that many people find objectionable. The sorbates are effective in chocolate syrups, jellies, cakes, dried fruits, salad dressings, figs, syrup, cheeses, Melba dessert, macaroni salad, and other products of this nature.

With respect to its toxicity, Deuel *et al.* (22) showed that sorbic acid is metabolized by the body to CO_2 and H_2O in the same manner as fatty acids normally found in foods.

SULFUR DIOXIDE

Sulfur dioxide (SO_2) and the sodium and potassium salts of sulfite ($=SO_3$), bisulfite ($—HSO_3$), and metabisulfite ($=S_2O_5$) all appear to act similarly and are here treated together. Sulfur dioxide is used in its gaseous or liquid form, or in the form of one or more of its neutral or acid salts on dried fruits, in lemon juice, molasses, wines, fruit juices, and others. The parent compound has been used as a food preservative since ancient times. Its use as a meat preservative in the United States dates back to at least 1813; however, it is not permitted in meats or other foods recognizable as sources of thiamine. While SO_2 possesses antimicrobial activity, it is also used in certain foods as an antioxidant. Concentrations up to 2,000 ppm may be employed in the preservation of fruit concentrates.

Although the actual mechanism of action of SO_2 is not known, 3 possibilities have been suggested with some experimental evidence in support of each. It has been suggested by various investigators that the undissociated sulfurous acid or molecular SO_2 is responsible for the antimicrobial activity. Its greater effectiveness at low pH values tends to support this. Vas and Ingram (78) suggested the lowering of pH of certain foods by addition of acid as a means of obtaining greater preservation with SO_2. On the other hand, it has been suggested that the antimicrobial action is due to the strong reducing power which allows these compounds to reduce oxygen tension to a point below that at which aerobic organisms can grow, or by direct action upon some enzyme system. Sulfur dioxide is also thought to be an enzyme poison, inhibiting growth of microorganisms by inhibiting essential enzymes. Its use in the drying of foods to inhibit enzymatic browning is based upon this assumption. Acetic and lactic acid bacteria and many molds are more sensitive to SO_2 than yeasts. Among yeasts, the more strongly aerobic species are generally more sensitive to SO_2 than the more fermentative species (48). Sulfite does not inhibit cellular transport; it has two pK values, 1.8 and 7.0.

THE PROPIONATES

Propionic acid is a three-carbon organic acid with the following structure: CH_3CH_2COOH. This acid and its calcium and sodium salts are permitted in breads, cakes, certain cheeses, and other foods primarily as a mold inhibitor. Propionic acid is employed also as a ''rope'' inhibitor in bread dough. The tendency towards dissociation is low with this compound and its salts, and these compounds are consequently active in low acid foods. They tend to be highly specific against molds, with the inhibitory action being primarily fungistatic rather than fungicidal.

With respect to the antimicrobial mode of action of propionates, they act in a manner similar to that of benzoate and sorbate. The pK of propionate is 4.87 and at a pH of 4.00, 88% of the compound is undissociated while at a pH of 6.0, only 6.7% remains undissociated. The undissociated molecule of this lipophilic acid is necessary for its inhibition of uptake of substrate molecules by microbial cells.

In a study of the antimicrobial action of lipids in general, Hoffman et al. (39) found the straight chained acids to be generally more effective than branched acids. These authors found that organic acids with less than 7 carbons were more effective at low pH values, while those with from 8–12 carbons were most effective as fungistats at neutrality and above. Some fatty acids are known to affect cell permeability by attacking the cell membrane of microorganisms.

ETHYLENE AND PROPYLENE OXIDES

Ethylene and propylene oxides along with ethyl and methyl formate ($HCOOC_2H_5$ and $HCOOCH_3$, respectively) are treated together in this section because of their similar actions. The structures of the oxide compounds are as follows:

$$\begin{array}{cc}
\begin{array}{c}
H_2C \\
\quad\diagdown \\
\quad\quad O \\
\quad\diagup \\
H_2C
\end{array}
&
\begin{array}{c}
\quad\quad CH_2 \\
\quad\diagup\; | \\
O \\
\quad\diagdown\; | \\
\quad\quad CH\cdot CH_3
\end{array}
\\
\text{ethylene oxide} & \text{propylene oxide}
\end{array}$$

The oxides exist as gases and are employed as fumigants in the food industry. The oxides are applied to dried fruits, nuts, spices, and so forth, primarily as antifungal compounds.

Ethylene oxide is an alkylating agent *(62)* and its antimicrobial activity is presumed to be related to this action in the following manner. In the presence of labile H-atoms, the unstable 3-membered ring of ethylene oxide splits. The H-atom attaches itself to the oxygen forming a hydroxyl ethyl radical, CH_2CH_2OH, which in turn attaches itself to the position in the organic molecule left vacant by the H-atom. As a result, the hydroxy ethyl group blocks reactive groups within microbial proteins, thus resulting in inhibition. Among the groups capable of supplying a labile H-atom are the following: COOH, NH_2, —SH, and OH. Ethylene oxide appears to affect endospores of *C. botulinum* by alkylation of guanine and adenine components of spore DNA *(57, 82)*.

Ethylene oxide is used as a gaseous sterilant for flexible and semi-rigid containers for packaging aseptically processed foods. All of the gas dissipates from the containers following their removal from treatment chambers. With respect to its action on microorganisms, it is not much more effective against vegetative cells than it is against endospores as can be seen from the D values given in Table 9-2.

ANTIBIOTICS

The antibiotics that have been employed and found effective as food preservatives are the following: chlor- and oxytetracycline, nisin, tylosin, polymyxin B, and subtilin. Chlortetracycline (Aureomycin) and Oxytetracycline (Terramycin) were approved by FDA in 1955 and 1956, respectively, for use in uncooked poultry, both at a concentration of 7 ppm (approval subsequently rescinded). Unlike the preservatives discussed

TABLE 9-2.   D Values for 4 chemical sterilants of some food-borne microorganisms.

Organisms	D^1	Conc.	Temp.[2]	Condition	Ref.
		Hydrogen Peroxide			
C. botulinum 169B	0.03	35%	88		See 77
B. coagulans	1.8	26%	25		76
B. stearothermophilus	1.5	26%	25		76
B. subtilis ATCC 95244	1.5	20%	25		74
B. subtilis A	7.3	26%	25		77
		Ethylene Oxide			
C. botulinum 62A	11.5	700 mg/L	40	47% R.H.	68
C. botulinum 62A	7.4	700 mg/L	40	23% R.H.	82
C. sporogenes ATCC 7955	3.25	500 mg/L	54.4	40% R.H.	49
B. coagulans	7.0	700 mg/L	40	33% R.H.	5
B. coagulans	3.07	700 mg/L	60	33% R.H.	5
B. stearothermophilus ATCC 7953	2.63	500 mg/L	54.4	40% R.H.	49
L. brevis	5.88	700 mg/L	30	33% R.H.	5
M. radiodurans	3.00	500 mg/L	54.4	40% R.H.	49
		Sodium Hypochlorite			
A. niger conidiospores	0.61	20 ppm[3]	20	pH 3.0	12
A. niger conidiospores	1.04	20 ppm[3]	20	pH 5.0	12
A. niger conidiospores	1.31	20 ppm[3]	20	pH 7.0	12
		Iodine ($\frac{1}{2}I_2$)			
A. niger conidiospores	0.86	20 ppm[3]	20	pH 3.0	12
A. niger conidiospores	1.15	20 ppm[3]	20	pH 5.0	12
A. niger conidiospores	2.04	20 ppm[3]	20	pH 7.0	12

[1] In minutes; [2] °C; [3] as Cl.

above, these compounds are bacteriostats and are not as dependent upon pH as are the fungistats.

R^I	R^{II}	R^{III}	
H	CH_3	H	Tetracycline
OH	CH_3	H	Oxytetracycline
H	CH_3	Cl	Chlortetracycline

In 1952, Tarr *et al.* *(75)* established the efficacy of the tetracyclines in delaying fish and flesh food spoilage. These antibiotics were later shown to be effective in fish and shellfish as well as in certain other foods *(25,*

86). Chlortetracycline has been found slightly more effective than oxytetracycline. The efficacy of the tetracycline antibiotics to delay spoilage in flesh foods other than poultry has been confirmed *(18, 29).* Although these antibiotics are effective in delaying refrigerator spoilage of meats for several days, ultimate spoilage is caused by yeasts and molds. In 1956, the combined effect of the tetracyclines and low doses of gamma irradiation was suggested as a means of prolonging shelf life. This treatment resulted in spoilage by some yeasts and molds and could be prevented by the addition of 0.1% sorbic acid *(58).*

The effectiveness of the tetracycline antibiotics as meat preservatives is due in large part to the capacity of these agents to enter into complexes with metal ions in meats and thereby compete with bacteria for these essential growth requirements. This also explains why the tetracycline antibiotics are effective in controlling meat spoilage by organisms which *in vitro* are quite resistant to these antibiotics, namely, *Pseudomonas* and *Proteus* spp. It has been shown that in meats such as beef, the normally tetracycline-resistant bacteria are rather sensitive to these antibiotics and that this sensitivity can be reversed by the addition to meats of certain metal ions such as the divalent cations Ca^{++}, Mg^{++}, Sr^{++}, and others *(42, 43, 44, 45).* The ultimate inhibition of bacteria by these antibiotics appears to be due to an interference with protein synthesis at the ribosomal level of the cells.

The antibiotic **nisin** was actually the first to be permitted in foods. A patent was issued in 1954 for its use in cheese and bread products to prevent clostridial spoilage. It is produced by certain strains of *Streptococcus lactis* and was allowed to be added to cheeses as early as 1954. When nisin-producing strains of streptococci develop in cheeses during the ripening process, this antibiotic exists naturally in the cheese product. Its addition in the making of processed cheeses such as Gruyère considerably reduces blowing caused by clostridia *(55).* This antibiotic along with **tylosin** $(C_{45}H_{77}O_{17}N)$ and subtilin have all been tested as additives to be placed in canned foods in order to reduce heat processing without consequent spoilage. Tylosin has been shown by several investigators to act against canned food spoilage bacteria and to be more effective than nisin in this regard *(8, 19, 32, 51, 81).* Tylosin has been found to be effective in preventing both spoilage and toxin formation by *C. botulinum* in underprocessed cream style corn *(54).* The cans without tylosin spoiled readily in the above study. Similarly, 20–25 ppm tylosin lactate was found by Greenberg and Silliker *(33)* to prevent botulinal toxicity in canned meat products.

In a study by Segmiller *et al. (70),* tylosin was shown to be more effective than nisin as a heat sterilization adjunct in cream-of-chicken soup. However, in the more acid foods given heat treatment, this antibiotic was more effective in preventing spoilage than tylosin. On the other hand, nisin was shown to be ineffective in acid foods that were not heat processed. The explanation offered to explain the effectiveness of nisin in some heated foods and its ineffectiveness in nonheated foods assumes that bacterial endospores must be "damaged" by heat in order to be sensitive to this antibiotic. This theory was advanced in 1963 by Stumbo (cited by 70).

Subtilin has been investigated also as an antibiotic for use in canned foods given mild heat treatment. The mild heat treatment is employed to destroy naturally occurring enzymes of foods while the antibiotic inhibits potential food poisoning and spoilage types (1). Nisin is related to subtilin in that both are polypeptide antibiotics and have similar structures and antibacterial spectra. Most polypeptide antibiotics damage cell membranes and presumably bring about cell inhibition or death by this mechanism. Tylosin is one of the macrolide antibiotics. This group inhibits cells by binding to cellular ribosomes.

In regard to the activity of nisin, subtilin, and tylosin against sporeforming bacteria, spore germination is not inhibited when minimal inhibitory concentrations of these antibiotics are employed. Nisin and subtilin act at a point during the emergence of vegetative cells from spores before that of tylosin (30).

SALTS AND SUGARS

These compounds are grouped together because of the similarity in their modes of action in preserving foods. Salt has been employed as a food preservative since ancient times (see Chapter 1). The early food uses of salt were for the purpose of preserving meats. This use is based upon the fact that at high concentrations, salt exerts a drying effect upon both food and microorganisms. Nonmarine microorganisms may be thought of as normally possessing a degree of intracellular tonicity equivalent to that produced by about 0.85–0.9% NaCl. When microbial cells are suspended in salt (saline) of this concentration, the suspending menstrum can be said to be **isotonic** with respect to the cells. Since the amounts of NaCl and water are equal on both sides of the cell membrane, water moves across the cell membranes equally in both directions. When microbial cells are suspended in, say, a 5% saline solution, the concentration of water is greater inside the cells than outside (concentration of H_2O is highest where solute concentration is lowest). It should be recalled that in diffusion, water moves from its area of high concentration to its area of low concentration. In this case, water would pass out of the cells at a greater rate than it would enter. The result to the cell is **plasmolysis** which results in growth inhibition and possibly death. This is essentially what is achieved when high concentrations of salt are added to fresh meats for the purpose of preservation. Both the microbial cells and those of the meat undergo plasmolysis (shrinkage), resulting in the drying of the meat as well as inhibition or death of microbial cells. To be effective, one must use enough salt to effect **hypertonic** conditions. The higher the concentration, the greater the preservative and drying effects. In the absence of refrigeration, fish and other meats may be effectively preserved by salting. The inhibitory effects of salt are not dependent upon pH as are some other chemical preservatives. Most nonmarine bacteria can be inhibited by 20% or less of NaCl while some molds generally tolerate higher levels (see Tables 3-5 and 3-6). Organisms that can grow in the presence of and require high concentrations of salt are referred to as **halophiles,** while those that can withstand but not grow in high concentrations are referred to as **halodurics.**

Sugars, such as sucrose, exert their preserving effect in essentially the same manner as salt. One of the main differences is in relative concentrations. It generally requires about 6 times more sucrose than NaCl to effect the same degree of inhibition. The most common uses of sugars as preserving agents consist of their use in the making of fruit preserves, candies, condensed milk, and the like. The shelf-stability of certain pies, cakes, and other such products is due in large part to the preserving effect of high concentrations of sugar which, like salt, make water unavailable to microorganisms.

Microorganisms differ in their response to hypertonic concentrations of sugars with yeasts and molds being less susceptible than bacteria. Some yeasts and molds can grow in the presence of as much as 60% sucrose while most bacteria are inhibited by much lower levels. Organisms that are able to grow in high concentrations of sugars are designated **osmophiles** while **osmoduric** microorganisms are those that are unable to grow but are able to withstand high levels of sugars. Some osmophilic yeasts such as *Saccharomyces rouxii* can grow in the presence of extremely high concentrations of sugars.

NITRATES AND NITRITES

Sodium nitrate ($NaNO_3$) and sodium nitrite ($NaNO_2$) as well as the potassium salts of these compounds are used in the curing formula for cured meats, since they stabilize red meat colors and also inhibit some spoilage and food poisoning microorganisms and contribute to product flavor development. Sodium nitrite is permitted in meat curing preparations at a level not to exceed 200 ppm although many processors do not exceed 100 ppm. This compound has been shown to disappear both on heating and storage. It should be recalled that many bacteria are capable of utilizing nitrate as an electron acceptor and in the process effect the reduction of this compound to nitrite. The nitrite ion is by far the most important of the two in preserved meats. The nitrite ion is highly reactive and is capable of serving both as a reducing and an oxidizing agent. In an acid environment, the nitrite ion ionizes to yield nitrous acid (3HONO). The latter further decomposes to yield nitric oxide (NO), which is the important product from the standpoint of color fixation in cured meats although it apparently has no effect on bacteria *(71)*. Ascorbate or erythorbate also acts to reduce NO_2 to NO. Nitric oxide reacts with myoglobin under reducing conditions to produce the desirable red pigment **nitrosomyoglobin** per the following (see also Table 7-3):

$$\begin{array}{ccc}
\text{N} \quad \text{Globin} \quad \text{N} & & \text{N} \quad \text{Globin} \quad \text{N} \\
\diagdown \quad | \quad \diagup & & \diagdown \quad | \quad \diagup \\
\text{Fe} & +\text{ NO} \rightarrow & \text{Fe} \quad\quad + \text{ H}_2\text{O} \\
\diagup \quad | \quad \diagdown & & \diagup \quad | \quad \diagdown \\
\text{N} \quad \text{H}_2\text{O} \quad \text{N} & & \text{N} \quad \text{NO} \quad \text{N} \\
\text{Myoglobin} & & \text{Nitrosomyoglobin}
\end{array}$$

When the meat pigment exists in the form of **oxymyoglobin**, as would be the case for comminuted meats, this compound is first oxidized to **metmyoglobin** (brown color). Upon the reduction of metmyoglobin, nitric oxide reacts to yield nitrosomyoglobin. Since nitric oxide is known to be capable of reacting with other porphyrin-containing compounds such as catalase, peroxidases, cytochromes, and others, it is conceivable that some of the antibacterial effects of nitrites are due to this action, although substantiation is lacking.

It has been shown that the antibacterial effect of NO_2 increases as pH is lowered within the acid range. This effect is accompanied by an overall increase in the undissociated HNO_2 (9). Microbial inhibition by nitrite has been shown to be reversed by the addition of sulfhydryl compounds such as thioglycolic acid and cysteine. The latter suggests an interference of nitrite or one of its breakdown products with the sulfur nutrition of susceptible microorganisms. Nitrite has been reported to enable some strict aerobes to grow under anaerobic conditions. The compound is believed to enable in some way aerobic growth in anaerobic environments, although good evidence is lacking.

When nitrite reacts with secondary amines, **nitrosamines** are formed and many are known to be carcinogenic. The generalized way in which nitrosamines may form is as follows:

$$R_2NH_2 + HONO \xrightarrow{H^+} R_2N-NO + H_2O$$

The amine dimethylamine reacts with nitrite to form N-nitroso-dimethylamine:

$$\begin{array}{ccc}
H_3C & & H_3C \\
\diagdown & & \diagdown \\
& N-H + NO_2 \rightarrow & \quad N-N=O \\
\diagup & & \diagup \\
H_3C & & H_3C
\end{array}$$

In addition to secondary amines, tertiary amines and quarternary ammonium compounds also yield nitrosamines with nitrite under acidic conditions. Nitrosamines have been found in cured meat and fish products at low levels (for reviews, see 17, 31, 69, 83).

It has been shown that lactobacilli, group D streptococci, clostridia, and other bacteria will nitrosate secondary amines with nitrite at neutral pH values *(37)*. The fact that nitrosation occurred at near neutral pH values was taken to indicate that the process was enzymatic, although no cell-free enzyme was obtained *(38)*. Several species of streptococci including *S. faecalis, S. faecium,* and *S. lactis* have been shown to be capable of forming nitrosamines, but the other lactic acid bacteria and pseudomonads tested did not *(15)*. These investigators found no evidence for an enzymatic reaction. *S. aureus* and halobacteria obtained from Chinese salted marine fish (previously shown to contain nitrosamines) produced nitrosamines when inoculated into salted fish homogenates containing 40 ppm of nitrate and 5 ppm nitrite *(26)*.

Historically, nitrates (mainly) and nitrites were employed in curing formulae for the purpose of developing the proper cured meat color. Nitrates have given way largely to nitrites, since it is the latter that are important in color and flavor development as noted above. The almost total absence of botulism in cured, canned, and vacuum-packed meats and fish products led some investigators in the mid 1960s to seek reasons as to why meat products that contained viable endospores did not become toxic. Employing culture medium, it was shown in 1967 that about 10 times more nitrite was needed to inhibit clostridia if it were added after instead of before the medium was autoclaved. It was concluded that the heating of the medium with nitrite produced a substance or agent c. 10 times more inhibitory than nitrite alone *(60, 61)*. This agent is referred to as the **Perigo factor.** The existence of this Perigo factor or effect has been confirmed by many other workers. The factor does not develop in all culture media, and heating to at least 100 C is necessary for its development, although some activity develops in meats when heated to as low as 70 C. The Perigo factor is dialyzable from some culture media and meat suspensions but not from other media *(47)*. It is not found in filter-sterilized solutions of the same medium with nitrite added *(66)*. It has been shown that if meat is added to a medium containing the Perigo factor, the inhibitory activity is lost *(46)*. For this reason, some Canadian workers call the inhibitor that is formed in meat the "Perigo-type factor" *(10)*.

It is this inhibitory or antibotulinal effect that results from the heat processing or smoking of certain meat and fish products containing nitrite that warrants the continued use of nitrite in such products. The antibotulinal activity of nitrite in cured meats is of greater public health importance than the facts of color and flavor development. For the latter, initial nitrite levels as low as 15 to 50 ppm have been reported to be adequate for various meat products including thuringer *(21)*. Nitrite levels of 100 ppm or more have been found to make for maximum flavor and appearance in fermented sausages *(50)*. The antibotulinal effect seems to require at least 120 to 200 ppm for bacon *(7, 14)*, comminuted cured ham *(13)*, and canned, shelf-stable luncheon meat *(10)*. Many of these canned products are given a low heat process (F_0 of 0.1–0.6).

The formulation of frankfurters without nitrite had no significant effect on staphylococci, salmonellae, or the normal spoilage flora (3, 80). The anti-botulinal activity or the Perigo-type factor is interdependent with pH, NaCl content, temperature of incubation, and size of inoculum (13, 34, 40, 63, 64).

It has been found by several investigators that the curing salts in semipreserved meats are more effective in inhibiting heat-injured spores than noninjured (24, 65). More recently, Chang et al. (10) suggested that the inhibitory effect of salt in shelf-stable canned meats against heat-injured spores may be more important than the Perigo-type factor. If this is simply a manifestation of heat-injured cells to salts, other means of cell injury may be expected to produce salt-sensitive cells. In the radappertization (steriliza-tion) of meats with gamma irradiation, it has been found that cured meats normally have lower minimum radiation doses than those without curing salts (2, 67, Table 10-2). Whether there is developed in nitrite-containing meats a factor or factors that have antibotulinal properties or whether the overall effect is one of heat-injured cells not being able to overcome the stress of curing salts is not yet clear.

A summary of the overall role and effect of nitrites in cured meats is given below with emphasis upon the antibotulinal activities.

1. When added to processed meats such as wieners, bacon, smoked fish, and canned cured meats followed by substerilizing heat treatments, nitrite has definite antibotulinal effects. It also forms desirable product color and enhances flavor in cured meat products. The antibotulinal effect consists of an inhibition of vegetative cell growth and the prevention of germination and growth of spores that survive heat processing or smoking during post-processing storage. Clostridia other than C. botulinum are affected in a similar manner. While low initial levels of nitrite are adequate for color and flavor development, considerably higher levels are necessary for the antimi-crobial effects.

2. When nitrite is heated in certain laboratory media, an antibotulinal fac-tor or inhibitor is formed, the exact identity of which is not yet known. The inhibitory factor is the Perigo effect/factor or Perigo inhibitor, mentioned on page 174. It does not form in filter-sterilized media. It develops in canned meats only when nitrite is present during heating. The initial level of nitrite is more important to antibotulinal activity than the residual level. Once formed, the Perigo factor is not affected greatly by pH changes.

3. Measurable pre-heating levels of nitrite decrease considerably during heating in meats and during post-processing storage, more at higher storage temperatures than at lower.

4. The antibotulinal activity of nitrite is interdependent with pH, salt content, temperature of incubation, and numbers of botulinal spores in as yet unexplained ways. Heat-injured spores are more susceptible to nitrite in-hibition than uninjured spores. Nitrite is more effective under Eh− than under Eh+ conditions.

5. Nitrite does not decrease the heat-resistance of spores. It is not affected

by ascorbate in its antibotulinal actions but does act synergistically with ascorbate in pigment formation.

6. Lactic acid bacteria are relatively resistant to nitrite.

7. Endospores remain viable in the presence of the antibotulinal effect and will germinate when transferred to nitrite-free media.

8. Nitrite has a pK of 3.29, and consequently it exists as undissociated nitrous acid at low pH values. Unlike sorbates, benzoates, and propionates, however, it apparently does not inhibit cellular transport. The maximum undissociated state of nitrous acid is between pH 4.5 and 5.5 with the consequent greatest antibacterial activity.

With respect to the fate of initial nitrite levels in meat products, a recent report indicates that at least some nitrite binds to nonheme proteins (85). By using ^{15}N-sodium nitrite, it was found that bovine serum albumin and myosin bound appreciable amounts of nitrite resulting in modifications of these proteins. Whether or not the nitrite-complexes are capable of inhibiting clostridia is not clear. With respect to the depletion or disappearance of nitrite in ham, Nordin (59) found the rate to be proportional to its concentration and to be exponentially related to both temperature and pH. The depletion rate doubled for every 12.2 C increase in temperature or 0.86 pH unit decrease, and was not affected by heat denaturation of the ham. The above relationships did not apply at room temperature unless the product was first heat treated, suggesting that viable organisms aided in its depletion. It is not known if the binding of nitrite to nonheme proteins is consistent with the temperature/pH relationships relative to its depletion in meats, or vice versa.

ACETIC AND LACTIC ACIDS

These two organic acids are among the most widely employed as preservatives. In most instances, their origin to the subject foods is due to their production within the food by lactic acid bacteria. Products such as pickles, sauerkraut, fermented milks, as well as others are the result of fermentative activities by various lactic acid bacteria with acetic, lactic, and other acids being produced (see Chapter 14 for fermented foods).

The antimicrobial effect of organic acids such as propionic and lactic is due both to the depression of pH below the growth range and metabolic inhibition by the undissociated acid molecules. In determining the quantity of organic acids in foods, **titratable acidity** is of more value than pH alone, since the latter is a measure of hydrogen-ion concentration and organic acids do not ionize completely. In measuring titratable acidity, the amount of acid that is capable of reacting with a known amount of base is determined. The titratable acidity of products such as sauerkraut is a better indicator of the amount of acidity present than pH.

OTHER CHEMICAL PRESERVATIVES

Sodium diacetate ($CH_3COONa \cdot CH_3COOH \cdot xH_2O$), a derivative of acetic acid, is used in bread and cakes to prevent moldiness. Organic acids such as **citric,**

$$
\begin{array}{c}
CH_2COOH \\
| \\
HO-C-COOH \\
| \\
CH_2COOH
\end{array}
$$

exert a preserving effect on foods such as soft drinks. **Hydrogen peroxide** (H_2O_2) has received limited use as a food preservative. In combination with heat, it has been used in milk pasteurization and sugar processing. It has potential as an in-line sterilant for food packaging containers. The D values of some food-borne microorganisms are presented in Table 9-2. **Ethanol** (C_2H_5OH) is present in flavoring extracts and effects preservation by virtue of its desiccant and denaturant properties. **Dehydroacetic acid,**

is used to preserve squash. **Diethylpyrocarbonate** is used in bottled wines and soft drinks as a yeast inhibitor. It decomposes to form ethanol and CO_2 by either hydrolysis or alcoholysis *(28)*:

Hydrolysis (reaction with water).

$$
\begin{array}{c}
C_2H_5O-CO \\
\diagdown \\
O \xrightarrow{H_2O} 2C_2H_5OH + 2CO_2 \\
\diagup \\
C_2H_5O-CO
\end{array}
$$

Alcoholysis (reaction with ethyl alcohol).

$$
\begin{array}{c}
C_2H_5O-CO \\
\diagdown \\
O \xrightarrow{C_2H_5OH}
\end{array}
\begin{array}{c}
C_2H_5O \\
\diagdown \\
C=O + CO_2 + C_2H_5OH \\
\diagup \\
C_2H_5O-CO
\end{array}
\begin{array}{c}
\\
C_2H_5O
\end{array}
$$

Saccharomyces cerevisiae and conidia of *Aspergillus niger* and *Byssochlamys fulva* have been shown to be destroyed by this compound during the first one-half hr of exposure, while the ascospores of *B. fulva* required 4 to 6 hr for maximal destruction *(73)*. *Lactobacillus plantarum* and *Leuconostoc mesenteroides* required 24 hr or longer for destruction. Sporeforming bacteria are quite resistant to this compound. Sometimes urethane is formed when this compound is used. Because urethane is a carcinogen, the use of diethylpyrocarbonate is no longer permissible in the U.S.

Ethylenediaminetetraacetic acid (EDTA) is used as a sequestrant in foods but it also has antimicrobial activity. It has been reported to inhibit the germination and outgrowth of spores of *C. botulinum* Type A in fish homogenates at concentrations above 2.5 mM *(83)*. It was found to be more effective at higher pH values than at lower.

Butylated hydroxyanisole (BHA) is used primarily as an antioxidant in foods, but a recent report indicates that it exerts antimicrobial activity *(11)*. At levels of 150–200 ppm, a 10^6 inoculum of *S. aureus* was inhibited by BHA. Growth and aflatoxin production by *Aspergillus parasiticus* spores was totally prevented by 1,000 ppm while 250 ppm showed inhibitory activity against this organism. The most resistant organism of those studied to BHA was *S. typhimurium* followed by *E. coli* which required 200–400 ppm for growth inhibition. Like the parabens, BHA is more effective against molds and gram positive bacteria than against gram negatives. It was further suggested by the above authors that a greater solubility of BHA in food systems may lead to greater antimicrobial activity.

The combined effect of 10 non-antimicrobial food additives on bacterial counts in ground beef was studied and found to have negligble effect on the flora compared to that of 3% NaCl used alone *(53)*. The 10 compounds employed were glucose, sorbitol, BHA, BHT, propyl gallate, citric acid, 5'-nucleotides, monosodium glutamate, saccharin, and nicotinamide. It is conceivable that other food additives not generally regarded as being antimicrobial may possess antimicrobial activity in at least some foods.

Methyl bromide (MeBr) has found some use as a disinfecting agent in poultry houses. Its effectiveness is restricted to fungi and vegetative bacterial cells. It has been shown to be effective in destroying the spores of *A. parasiticus* and *P. rubrum* on rice in 180 days when used at levels of 30–45 mg/L against 10^4–10^5 spores/g *(52)*. Asci of *Byssochlamys fulva* at a level of *c.* 5×10^4/g were destroyed in 30 days in tapioca starch powder at MeBr levels of 90 mg/kg of starch while 180 asci/g were killed by 60 mg of MeBr/kg starch *(41)*. With respect to indicator organisms, a 98–99% reduction of *E. coli* and enterococci in moist fecal suspensions was effected after a 20-hr exposure to 40 ml of MeBr/L. A similar treatment was found to be less effective against *S. typhimurium,* micrococci, and *A. fumigatus* spores *(36)*. Some D values of MeBr against food-borne microorganisms are presented in Table 9-2.

Wood smoke imparts certain chemicals to smoked products that enable these products to resist microbial spoilage. One of the most important is formaldehyde (CH_2O) which has been known for many years to possess antimicrobial properties. This compound acts as a protein denaturant by virtue of its reaction with amino groups. Also in wood smoke are aliphatic acids, alcohols, ketones, phenols, higher aldehydes, tar, methanol, cresols, and others *(23)*, all of which may contribute to the antibacterial actions of meat smoking. Since a certain amount of heat is necessary to produce smoke, part of the shelf-stability of smoked products is due to heat destruction of surface organisms as well as to the drying that occurs. A study of the antibacterial activity of liquid smoke by Handford and Gibbs *(35)* revealed that little activity occurred at concentrations of smoke that produced acceptable smoked flavor. Employing an agar medium containing 1:1 dilution of smoked water, these investigators found that micrococci and staphylococci were slightly more inhibited than the lactic acid bacteria. The overall combined effect of smoking and vacuum packaging results in a reduction of numbers of catalase positive bacteria on the smoked product, while the catalase-negative lactic acid bacteria are better able to withstand the low Eh conditions of vacuum-packaged products.

REFERENCES

1. **Anderson, A. A., and H. D. Michener.** 1950. Preservation of food with antibiotics. I. The complementary action of subtilin and mild heat. Food Technol. *4:* 188–189.

2. **Anellis, A., D. Berkowitz, W. Swantak, and C. Strojan.** 1972. Radiation sterilization of prototype military foods: Low-temperature irradiation of codfish cake, corned beef, and pork sausage. Appl. Microbiol. *24:* 453–462.

3. **Bayne, H. G. and H. D. Michener.** 1975. Growth of *Staphylococcus* and *Salmonella* on frankfurters with and without sodium nitrite. Appl. Microbiol. *30:* 844–849.

4. **Bell, T. A., J. L. Etchells, and A. F. Borg.** 1959. Influence of sorbic acid on the growth of certain species of bacteria, yeasts, and filamentous fungi. J. Bacteriol. *77:* 573–580.

5. **Blake, D. F. and C. R. Stumbo.** 1970. Ethylene oxide resistance of microorganisms important in spoilage of acid and high-acid foods. J. Food Sci. *35:* 26–29.

6. **Bosund, I.** 1962. The action of benzoic and salicyclic acids on the metabolism of microorganisms. Adv. Food Res. *11:* 331–353.

7. **Bowen, V. G. and R. H. Deibel.** 1974. Effects of nitrite and ascorbate on botulinal toxin formation in wieners and bacon. Proc. Meat Ind. Res. Conf., Amer. Meat Inst. Fdn., pp. 63–68.

8. **Campbell, L. L., E. E. Sniff, and R. T. O'Brien.** 1959. Subtilin and nisin as additives that lower the heat-process requirements of canned foods. Food Technol. *13:* 462–464.

9. **Castellani, A. G. and C. F. Niven, Jr.** 1955. Factors affecting the bacteriostatic action of sodium nitrate. Appl. Microbiol. *3:* 154–159.

10. **Chang, P.-C., S. M. Akhtar, T. Burke, and H. Pivnick.** 1974. Effect of sodium nitrite on *Clostridium botulinum* in canned luncheon meat: Evidence for a Perigo-type factor in the absence of nitrite. Can. Inst. Food Sci. Technol. J. *7:* 209–212.

11. **Chang, H. C. and A. L. Branen.** 1975. Antimicrobial effects of butylated hydroxyanisole (BHA). J. Food Sci. *40:* 349–351.

12. **Cheng, M. K. C. and R. E. Levin.** 1970. Chemical destruction of *Aspergillus niger* conidiospores. J. Food Sci. *35:* 62–66.

13. **Christiansen, L. N., R. W. Johnston, D. A. Kautter, J. W. Howard, and W. J. Aunan.** 1973. Effect of nitrite and nitrate on toxin production by *Clostridium botulinum* and on nitrosamine formation in perishible canned comminuted cured meat. Appl. Microbiol. *25:* 357–362.

14. **Christiansen, L. N., R. B. Tompkin, A. B. Shaparis, T. V. Kueper, R. W. Johnston, D. A. Kautter, and O. J. Kolari.** 1974. Effect of sodium nitrite on toxin production by *Clostridium botulinum* in bacon. Appl. Microbiol. *27:* 733–737.

15. **Collins-Thompson, D. L., N. P. Sen, B. Aris, and L. Schwinghamer.** 1972. Non-enzymic in vitro formation of nitrosamines by bacteria isolated from meat products. Can. J. Microbiol. *18:* 1968–1971.

16. **Costilow, R. N., F. M. Coughlin, E. K. Robbins, and Wen-Tah Hsu.** 1957. Sorbic acid as a selective agent in cucumber fermentations. II. Effect of sorbic acid on the yeast and lactic acid fermentations in brined cucumbers. Appl. Microbiol. *5:* 373–379.

17. **Crosby, N. T. and R. Sawyer.** 1976. N-nitrosamines: A review of chemical and biological properties and their estimation in foodstuffs. Adv. Food Res. *22:* 1–71.

18. **Deatherage, F. E.** 1956. The use of antibiotics in the preservation of foods other than fish, pp. 211–222. Pub. No. 397, Nat'l Acad. Sci., Nat'l Res. Coun., Washington, D.C.

19. **Denny, C. B., L. E. Sharpe, and C. W. Bohrer.** 1961. Effects of tylosin and nisin on canned food spoilage bacteria. Appl. Microbiol. *9:* 108–110.

20. **Desrosier, N. W.** 1963. *The Technology of Food Preservation.* Revised Edition. Avi Publishing Co., Westport, Conn., Chapter 9.

21. **Dethmers, A. E., H. Rock, T. Fazio, and R. W. Johnston.** 1975. Effect of added sodium nitrite and sodium nitrate on sensory quality and nitrosamine formation in thuringer sausage. J. Food Sci. *40:* 491–495.

22. **Deuel, H. J., Jr., C. E. Calbert, L. Anisfeld, H. McKeehan, and H. D. Blunden.** 1954. Sorbic acid as a fungistatic agent for foods. II. Metabolism of α, β-unsaturated fatty acids with emphasis on sorbic acid. Food Res. *19:* 13–19.

23. **Draudt, H. N.** 1963. The meat smoking process: A review. Food Technol. *17:* 1557–1562.

24. **Duncan, C. L. and E. M. Foster.** 1968. Role of curing agents in the preservation of shelf-stable canned meat products. Appl. Microbiol. *16:* 401–405.

25. **Farber, L.** 1959. Antibiotics in food preservation. Ann. Rev. Microbiol. *13:* 125–140.

26. **Fong, Y. Y. and W. C. Chan.** 1973. Bacterial production of di-methyl nitrosamine in salted fish. Nature *243:* 421–422.

27. **Freese, E., C. W. Sheu, and E. Galliers.** 1973. Function of lipophilic acids as antimicrobial food additives. Nature *241:* 321–325.

28. **Genth, H.** 1964. On the action of diethylpyrocarbonate on microorganisms, pp. 77–85, IN: *Microbial Inhibitors in Food,* Proc. 4th Intern. Symp. on Food Microbiol. (Almqvist & Wiksell: Stockholm).

29. **Goldberg, H. S., H. H. Weiser, and F. E. Deatherage.** 1953. Studies on meat. IV. The use of antibiotics in preservation of fresh beef. Food Technol. *7:* 165–166.

30. **Gould, G. W.** 1964. Effect of food preservatives on the growth of bacteria from spores, pp. 17–24, IN: *Microbial Inhibitors in Food,* Proc. 4th Intern. Symp. on Food Microbiol. (Almqvist & Wiksell: Stockholm).

31. **Gray, J. I.** 1976. N-Nitrosamines and their precursors in bacon: A review. J. Milk Food Technol. *39:* 686–692.

32. **Greenberg, R. A. and J. H. Silliker.** 1962. The action of tylosin on sporeforming bacteria. J. Food Sci. *27:* 64–68.

33. **Greenberg, R. A. and J. H. Silliker.** 1964. Spoilage patterns in *Clostridium botulinum*-inoculated canned foods treated with tylosin, pp. 97–103, IN: *Microbial Inhibitors in Food,* Proc. 4th Intern. Symp. on Food Microbiol. (Almqvist & Wiksell: Stockholm).

34. **Greenberg, R. A.** 1972. Nitrite in the control of *Clostridium botulinum.* Proc. Meat Ind. Res. Conf., Amer. Meat Inst. Fdn., pp. 25–34.

35. **Handford, P. M. and B. M. Gibbs.** 1964. Antibacterial effects of smoke constituents on bacteria isolated from bacon, pp. 333–346, IN: *Microbial Inhibitors in Food,* Proc. 4th Intern. Symp. on Food Microbiol. (Almqvist & Wiksell: Stockholm).

36. **Harry, E. G., W. B. Brown, and G. Goodship.** 1972. The disinfecting activity of methyl bromide on various microbes and infected materials under controlled conditions. J. Appl. Bacteriol. *35:* 485–491.

37. **Hawksworth, G. and M. J. Hill.** 1971a. The formation of nitrosamines by human intestinal bacteria. Biochem. J. *122:* 28–29P.

38. **Hawksworth, G. and M. ¯. Hill.** 1971b. Bacteria and the N-nitrosation of secondary amines. Brit. J. Cancer. *25:* 520–526.

39. **Hoffman, C., T. R. Schweitzer, and G. Dalby.** 1939. Fungistatic properties of the fatty acids and possible biochemical significance. Food Res. *4:* 539–545.

40. **Ingram, M.** 1976. The microbiological role of nitrite in meat products, pp. 1–18, IN: *Microbiology in Agriculture, Fisheries and Food,* edited by F. A. Skinner and J. G. Carr (Academic Press: N.Y.).

41. **Ito, K. A., M. L. Seeger, and W. H. Lee.** 1972. The destruction of *Byssochlamys fulva* asci by low concentrations of gaseous methyl bromide and by aqueous solutions of chlorine, an iodophor and peracetic acid. J. Appl. Bacteriol. *35:* 479–483.

42. **Jay, J. M., H. H. Weiser, and F. E. Deatherage.** 1956. The effect of chlortetracycline on the microflora of beef, and studies on the mode of action of this antibiotic in meat preservation. IN: *Antibiotics Annual,* 1956–57: 954–965.

43. **Jay, J. M., H. H. Weiser, and F. E. Deatherage.** 1957A. Studies on the mode of action of chlortetracycline in the preservation of beef. Appl. Microbiol. *6:* 400–405.

44. **Jay, J. M., H. H. Weiser, and F. E. Deatherage.** 1957B. Further studies on the preservation of beef with chlortetracycline. Food Technol. *11:* 563–566.

45. **Jay, J. M., H. H. Weiser, and F. E. Deatherage.** 1958. The inhibition in beef of chlortetracycline-resistant bacteria by sub-bacteriostatic concentrations of the antibiotic. Appl. Microbiol. *6:* 343–346.

46. **Johnston, M. A., H. Pivnick, and J. M. Samson.** 1969. Inhibition of *Clostridium botulinum* by sodium nitrite in a bacteriological medium and in meat. Can. Inst. Food Technol. J. *2:* 52–55.

47. **Johnston, M. A. and R. Loynes.** 1971. Inhibition of *Clostridium botulinum* by sodium nitrite as affected by bacteriological media and meat suspensions. Can. Inst. Food Technol. J. *4:* 179–184.

48. **Joslyn, M. A. and J. B. S. Braverman.** 1954. The chemistry and technology of the pretreatment and preservation of fruit and vegetable products with sulfur dioxide and sulfites. Adv. Food Res. *5:* 97–160.

49. **Kereluk, K., R. A. Gammon, and R. S. Lloyd.** 1970. Microbiological aspects of ethylene oxide sterilization. II. Microbial resistence to ethylene oxide. Appl. Microbiol. *19:* 152–156.

50. **Kueper, T. V. and R. D. Trelease.** 1974. Variables affecting botulinum toxin development and nitrosamine formation in fermented sausages. Proc. Meat Ind. Res. Conf., Amer. Meat Inst. Fdn., pp. 69–74.

51. **LeBlanc, F. R., K. A. Devlin, and C. R. Stumbo.** 1953. Antibiotics in food preservation. I. The influence of subtilin on the thermal resistance of spores of *Clostridium botulinum* and the putrefactive anaerobe 3679. Food Technol. *7:* 181–183.

52. **Lee, W. H. and H. Riemann.** 1970. Destruction of toxic fungi with low concentrations of methyl bromide. Appl. Microbiol. *20:* 845–846.

53. **Lin, T.-S., R. E. Levin, and H. O. Hultin.** 1977. Myoglobin oxidation in ground beef: Microorganisms and food additives. J. Food Sci. *42:* 151–154.

54. **Malin, B. and R. A. Greenberg.** 1964. The effect of tylosin on spoilage patterns of inoculated and noninoculated cream style corn, pp. 87–95, IN: *Microbial Inhibitors in Food,* Proc. 4th Intern. Symp. on Food Microbiol. (Almqvist & Wiksell: Stockholm).

55. **McClintock, M., L. Serres, J. J. Marzolf, A. Hirsch, and G. Macquot.** 1952. Inhibitory action of nisin-producing streptococci on the development of anaerobic sporeformers in processed Gruyere cheese. J. Dairy Res. *19:* 187–193.

56. **Melnick, D., H. W. Vahlteich, and A. Hackett.** 1956. Sorbic acid as a fungistatic agent for foods. XI. Effectiveness of sorbic acid in protecting cakes. Food Res. *21:* 133–146.

57. **Michael, G. T. and C. R. Stumbo.** 1970. Ethylene oxide sterilization of *Salmonella senftenberg* and *Escherichia coli:* Death kinetics and mode of action. J. Food Sci. *35:* 631–634.

58. **Niven, C. F., Jr. and W. R. Chesbro.** 1957. Complementary action of antibiotics and irradiation in the preservation of fresh meats. *Antibiotics Annual,* 1956–57. Medical Encyclopedia, Inc., N.Y., pp. 855–859.

59. **Nordin, H. R.** 1969. The depletion of added sodium nitrite in ham. Can. Inst. Food Technol. J. *2:* 79–85.

60. **Perigo, J. A., E. Whiting, and T. E. Bashford.** 1967. Observations on the inhibition of vegetative cells of *Clostridium sporogenes* by nitrite which has been

autoclaved in a laboratory medium, discussed in the context of sub-lethally processed meats. J. Food Technol. *2:* 377–397.

61. **Perigo, J. A. and T. A. Roberts.** 1968. Inhibition of clostridia by nitrite. J. Food Technol. *3:* 91–94.

62. **Phillips, C. R.** 1952. Relative resistance of bacterial spores and vegetative bacteria to disinfectants. Bacteriol. Revs. *16:* 135–138.

63. **Pivnick, H. and F. S. Thatcher.** 1968. Factors affecting the safety of thermally processed canned cured meats. Proc., 98th Annual Meeting of Am. Pub. Hlth. Assoc., p. 125.

64. **Riha, W. E., Jr. and M. Solberg.** 1975. *Clostridium perfringens* inhibition by sodium nitrite as a function of pH, inoculum size and heat. J. Food Sci. *40:* 439–442.

65. **Roberts, T. A. and M. Ingram.** 1966. The effect of sodium chloride, potassium nitrate and sodium nitrite on the recovery of heated bacterial spores. J. Food Technol. *1:* 147–163.

66. **Roberts, T. A. and J. L. Smart.** 1974. Inhibition of spores of *Clostridium* spp. by sodium nitrite. J. Appl. Bacteriol. *37:* 261–264.

67. **Rowley, D. B., A. Anellis, E. Wierbicki, and A. W. Baker.** 1974. Status of the radappertization of meats. J. Milk Food Technol. *37:* 86–93.

68. **Savage, R. A. and C. R. Stumbo.** 1971. Characteristics of progeny of ethylene oxide treated *Clostridium botulinum* Type 62A spores. J. Food Sci. *36:* 182–184.

69. **Sebranek, J. G. and R. G. Cassens.** 1973. Nitrosamines: A review. J. Milk Food Technol. *36:* 76–91.

70. **Segmiller, J. L., H. Xezones, and I. J. Hutchings.** 1965. The efficacy of nisin and tylosin lactate in selected heat-sterilized food products. J. Food Sci. *30:* 166–171.

71. **Shank, J. L., J. H. Silliker, and R. H. Harper.** 1962. The effect of nitric oxide on bacteria. Appl. Microbiol. *10:* 185–189.

72. **Smith, E. S., J. F. Bowen, and D. R. MacGregor.** 1962. Yeast growth as affected by sodium benzoate, potassium sorbate and vitamin K5. Food Technol. *16:* 93–85.

73. **Splittstoesser, D. F. and M. Wilkison.** 1973. Some factors affecting the activity of diethylpyrocarbonate as a sterilant. Appl. Microbiol. *25:* 853–857.

74. **Swartling, P. and B. Lindgren.** 1968. The sterilizing effect against *Bacillus subtilis* spores of hydrogen peroxide at different temperatures and concentrations. J. Dairy Res. *35:* 423–428.

75. **Tarr, H. L. A., B. A. Southcott, and H. M. Bissett.** 1952. Experimental preservation of flesh foods with antibiotics. Food Technol. *6:* 363–368.

76. **Toledo, R. T., F. E. Escher, and J. C. Ayres.** 1973. Sporicidal properties of hydrogen peroxide against food spoilage organisms. Appl. Microbiol. *26:* 592–597.

77. **Toledo, R. T.** 1975. Chemical sterilants for aseptic packaging. Food Technol. *29:* No. 5, 102–107.

78. **Vas, K. and M. Ingram.** 1949. Preservation of fruit juices with less SO_2. Food Manuf., *24:* 414–416.

79. **Von Schelhorn, M.** 1953. Efficacy and specificity of chemical food preservatives. Food Tech. *7:* 97–101.

80. **Wassermann, A. E. and F. Talley.** 1972. The effect of sodium nitrite on the flavor of frankfurters. J. Food Sci. *37:* 536–538.

81. **Wheaton, E. and G. L. Hays.** 1964. Antibiotics and the control of spoilage in canned foods. Food Technol. *18:* 549–551.

82. **Winarno, F. G. and C. R. Stumbo.** 1971. Mode of action of ethylene oxide on spores of *Clostridium botulinum* 62A. J. Food Sci. *36:* 892–895.

83. **Winarno, F. G., C. R. Stumbo, and K. M. Hayes.** 1971. Effect of EDTA on the germination of and outgrowth from spores of *Clostridium botulinum* 62-A. J. Food Sci. *36:* 781–785.

84. **Wolff, I. A. and A. E. Wasserman.** 1972. Nitrates, nitrites, and nitrosamines. Science *177:* 15–19.

85. **Woolford, G., R. G. Cassens, M. L. Greaser, and J. G. Sebranek.** 1976. The fate of nitrite: Reaction with protein. J. Food Sci. *41:* 585–588.

86. **Wrenshall, C. L.** 1959. Antibiotics in food preservation. In: *Antibiotics, Their Chemistry and Non-Medical Uses.* Edited by H. S. Goldberg, D. Van Nostrand Company, Inc., Princeton, N.J., pp. 449–527.

10

Food Preservation
By the Use
of Radiation

Although a patent was issued to a Frenchman in 1929 on the use of radiation as a means of preserving foods, it was not until shortly after World War II that this method of food preservation received any serious consideration. While the application of radiation as a food preservation method has been somewhat slow in reaching its maximum potential use, the full application of this method presents some interesting challenges to food microbiologists and other food scientists.

Radiation may be defined as the emission and propagation of energy through space or through a material medium. The type of radiation of primary interest in food preservation is electromagnetic. The electromagnetic spectrum is presented in Figure 10-1. The various radiations are separated on the basis of their wavelengths, with the shorter wavelengths being the most damaging to microorganisms. The electromagnetic spectrum may be further divided as follows with respect to those radiations of interest in food preservation: microwaves, ultraviolet rays, X rays, and gamma rays. The radiations of primary interest in food preservation are **ionizing radiations.** Ionizing radiations may be defined as those radiations that have wavelengths of 2,000Å or less, e.g., alpha particles, beta rays, gamma rays, X rays, cosmic rays. Their quanta contain enough energy to actually ionize molecules in their paths. Since they destroy microorganisms without appreciably raising the temperature, the process is termed "cold sterilization."

To better understand the application of radiation to foods, there are several useful concepts that should be considered. A **Roentgen** is a unit of measure used for expressing exposure dose of X-ray or gamma radiation. A **milliroentgen** is equal to 1/1,000 of a Roentgen. A **Curie** is a quantity of radioactive substance in which 3.7×10^{10} radioactive disintegrations occur per second. For practical purposes, 1 g of pure radium possesses the radioactivity of 1 Curie of radium. A **rad** is a unit equivalent to the absorption of 100 ergs/g of matter. A **kilorad** is equal to 1,000 rads. A **Megarad (Mrad)** is equal to 1 million rads. The energy gained by an electron in moving through

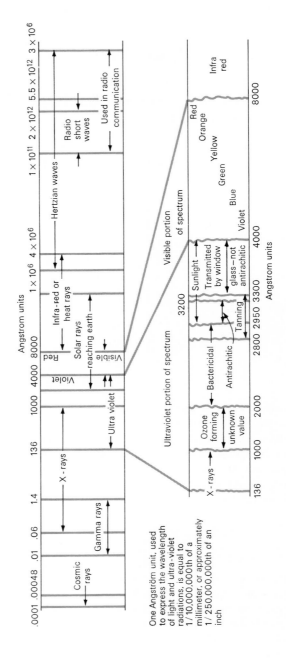

FIGURE 10-1. Spectrum charts (from the Westinghouse Sterilamp and the Rentschler-James Process of Sterilization, courtesy of the Westinghouse Electric & Manufacturing Co., Inc.)

1 volt is designated **ev** (electron volt). A **mev** is equal to 1 million electron volts. Both the rad and ev are measurements of the intensity of irradiation.

CHARACTERISTICS OF RADIATIONS OF INTEREST IN FOOD PRESERVATION

1. Ultraviolet light (UV light).   Ultraviolet light is a powerful bactericidal agent, with the most effective wavelength being about 2,600Å. It is nonioniz- ing and is absorbed by proteins and nucleic acids in which photochemical changes are produced that may lead to cell death. The mechanism of UV death in the bacterial cell is apparently due to the production of lethal muta- tions as a result of action upon cell nucleic acids. UV light has poor penetra- tive capacities which limit its food use to surface applications where it may catalyze oxidative changes that lead to rancidity, discolorations, and other reactions. Small quantities of ozone may also be produced when UV light is used for the surface treatment of certain foods. UV light is sometimes used to treat the surfaces of baked fruit cakes and related products before wrapping.

2. Beta Rays.   Beta rays may be defined as a stream of electrons emitted from radioactive substances. Cathode rays are the same except that they are emitted from the cathode of an evacuated tube. These rays possess poor penetration power. Among the commercial sources of cathode rays are Van de Graaff generators and linear accelerators. The latter seem better suited for food preservation uses. There is some concern over the upper limit of energy level of cathode rays that can be employed without inducing ra- dioactivity in certain constituents of foods.

3. Gamma Rays.   These are electromagnetic radiations emitted from the excited nucleus of elements such as ^{60}Co and ^{137}Cs, which are of importance in food preservation. This is the cheapest form of radiation for food pres- ervation, since the source elements are either by-products of atomic fission or atomic waste products. Gamma rays have excellent penetration power as opposed to beta rays. ^{60}Co has a half-life of about 5 yr, while the half-life for ^{137}Cs is about 30 yr.

4. X rays.   These rays are produced by the bombardment of heavy-metal targets with high-velocity electrons (cathode rays) within an evacuated tube. They are essentially the same as gamma rays in other respects.

5. Microwaves.   Microwave energy may be illustrated in the following way *(8)*. When electrically neutral foods are placed in an electromagnetic field, the charged asymmetric molecules are driven first one way and then another. During this process, each asymmetric molecule attempts to align it- self with the rapidly changing alternating-current field. As the molecules os- cillate about their axes while attempting to go to the proper positive and negative poles, intermolecular friction is created and manifested as a heating effect. This is microwave energy. Most food research has been carried out at two frequencies, 915 and 2450 megacycles. At the microwave frequency of 915 megacycles, the molecules oscillate back and forth 915 million times/sec *(8)*. Microwaves lie between the infrared and radio frequency portion of the electromagnetic spectrum (see Figure 10-1). The destruction of molds in bread, pasteurization of beer, and sterilization of wine have been achieved through the use of microwaves. The most successful application of microwave energy to date has been in the finish frying of potato chips. The general use of microwaves as a means of food preservation is limited by the heating effect that results from their use.

PRINCIPLES UNDERLYING THE DESTRUCTION OF MICROORGANISMS BY RADIATIONS

Several factors come into play when the effects of radiation on microorganisms are considered.

1. The Kinds and Species of Organisms. With respect to kinds of microorganisms, gram positive bacteria are more resistant to radiation than gram negatives. Sporeformers are in general more resistant than nonspore formers with the exception of *Micrococcus radiodurans,* which is one of the most radioresistant bacteria known *(1)*. Among sporeformers, *B. larvae* has been reported to possess a higher degree of resistance than most aerobic sporeformers. Spores of *C. botulinum* type A appear to be the most resistant of all clostridial spores. Apart from *M. radiodurans,* one of the most resistant vegetative bacteria appears to be *Streptococcus faecium* R53. Among the more resistant vegetative forms are *Streptococcus faecalis,* micrococci in general, and the homofermentative lactobacilli. The bacteria most sensitive to radiation belong to the pseudomonad and flavobacteria groups, with other gram negative bacteria being intermediate in radioresistance between these genera and the micrococci (see Figure 10-2). Possible mechanisms of variation in radiosensitivity are discussed in Chapter 22.

With two exceptions, the radioresistance of bacterial endospores in general parallels that of heat resistance. The exceptions are *M. radiodurans,* which is more radioresistant than any endospores, and flat sour and thermophilic anaerobe (T.A.) spores which are more sensitive to radiation than to heat.

With respect to the radiosensitivity of molds and yeasts, the latter have been reported to be more resistant than the former with both groups in general being less sensitive than gram positive bacteria. Some *Candida* strains have been reported to possess resistance comparable to that of some bacterial endospores.

2. The Numbers of Organisms. The numbers of organisms have the same effect upon the efficacy of radiations as on heat, chemical disinfection, and

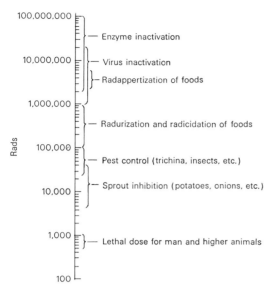

Rads

100,000,000 — Enzyme inactivation

10,000,000 — Virus inactivation
— Radappertization of foods

1,000,000

— Radurization and radicidation of foods

100,000

— Pest control (trichina, insects, etc.)

— Sprout inhibition (potatoes, onions, etc.)

10,000

1,000 — Lethal dose for man and higher animals

100

FIGURE 10-2.   Dose ranges of irradiation for various applications (adapted from Grünewald, *13*).

certain other phenomena—the larger the number of cells, the less effective a given dose of radiation.

3. Composition of Suspending Menstrum (Food).   Microorganisms are in general more sensitive to radiation when suspended in buffer solutions than in protein-containing media. For example, Midura *et al.* *(24)* found radiation D values (see Chapter 12) for a strain of *C. perfringens* to be 0.23 in phosphate buffer while in cooked-meat broth the D value was 0.30 Mrad. Proteins have been shown to exert a protective effect against radiations as well as against certain antimicrobial chemicals and heat. Several investigators have reported that the presence of nitrites tends to make bacterial endospores more sensitive to radiation.

4. Presence or Absence of Oxygen.   The radiation resistance of microorganisms is greater in the absence of oxygen than in its presence. Complete removal of oxygen from the cell suspension of *E. coli* has been reported to increase its radiation resistance up to three-fold *(25)*. The addition of reducing substances such as sulfhydryl compounds generally has the same effect in increasing radiation resistance as an anaerobic environment.

5. Physical State of Food.   The radiation resistance of dried cells is in general considerably higher than that for moist cells. This is most likely a direct consequence of the radiolysis of water by ionizing radiations and is discussed later in this chapter. Radiation resistance of frozen cells has been reported to be greater than that of nonfrozen cells. Grecz *et al. (11)* found that the lethal effects of gamma radiation decreased by 47% when ground beef was irradiated at −196 C as compared to 0 C.

6. *Age of Organisms.* Bacteria tend to be most resistant to radiation in the lag phase just prior to active cell division. The cells become more radiation sensitive as they enter and progress through the log phase and reach their minimum at the end of this phase.

PROCESSING OF FOODS FOR IRRADIATION

Prior to being exposed to ionizing radiations, there are several processing steps that must be carried out in much the same manner as for the freezing or canning of foods.

a. *Selection of Foods.* Foods to be irradiated should be carefully selected for freshness and overall desirable quality. Especially to be avoided are foods that are already in incipient spoilage.

b. *Cleaning of Foods.* All visible debris and dirt should be removed. This will reduce the numbers of microorganisms to be destroyed by the radiation treatment.

c. *Packing.* It is obviously important that foods to be irradiated should be packed in containers that will afford protection against postirradiation contamination. Cans seem to be the best at the present time, although a large amount of research has been carried out on the feasibility of using plastic containers. Clear glass containers undergo color changes when exposed to doses of radiation of around 1 Mrad, and the subsequent color may be undesirable.

d. *Blanching or Heat Treatment.* Sterilizing doses of radiation are insufficient to destroy the natural enzymes of foods (see Figure 10-2). In order to avoid undesirable postirradiation changes, it is necessary to destroy these enzymes. The best method appears to be heat treatment, i.e., the blanching of vegetables and mild heat treatment of meats prior to irradiation.

THE APPLICATION OF RADIATION

As previously stated, the two most widely used techniques of irradiating foods are gamma radiation from either ^{60}Co and ^{137}Cs, and the use of electron beams from linear accelerators.

1. *Gamma Radiation.* The advantage of gamma radiation is that ^{60}Co and ^{137}Cs are relatively inexpensive by-products of atomic fission. In a common experimental radiation chamber employing these elements, the radioactive material is placed on the top of an elevator which can be moved up for use and down under water when not in use. Materials to be irradiated are placed around the radioactive material (the source) at a suitable distance for the desired dosage. Once the chamber has been vacated by all personnel, the source is raised into position, and the gamma rays irradiate the food. Irradiation at desired temperatures may be achieved either by placing the samples in temperature-controlled containers, or by controlling the temperature of the entire concrete- and lead-walled chamber. Among the drawbacks to the use of radioactive material is that the isotope source emits rays in all directions and cannot be turned "on" or "off" as may be desirable (see Figure 10-3). Also, the half-life of ^{60}Co (5.27 yr) requires that the source be changed periodically in order to maintain a given level of radioactive potential. This drawback is overcome by the use of ^{137}Cs which has a half-life of around 30 yr.

2. Electron Beams.   The use of electron accelerators offers certain advantages over radioactive elements that make this form of radiation somewhat more attractive to potential commercial users. Koch and Eisenhower (18) have listed the following:

a.   High efficiency for the direct deposition of energy of the primary electron beams means high plant-product capacity.

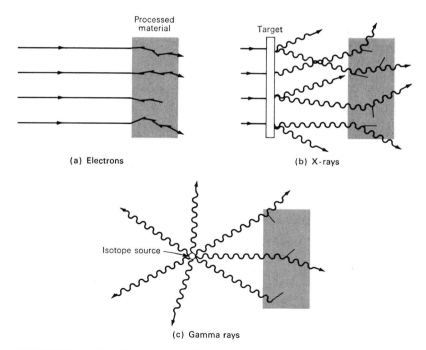

(a) Electrons   (b) X-rays

(c) Gamma rays

FIGURE 10-3.   The three basic techniques for radiation processing—interactions of electrons, X rays, and gamma rays in the medium (18). (Eisenhower, 1965, *Radiation Preservation of Foods,* Publication 1273, Advisory Board on Military Personnel Supplies, National Academy of Sciences, National Research Council).

b. The efficient convertibility of electron power to X-ray power means the capability of handling very thick products that cannot be processed by electron or gamma-ray beams.
c. The easy variability of electron-beam current and energy means a flexibility in the choice of surface and depth treatments for a variety of food items, conditions, and seasons.
d. The monodirectional characteristic of the primary and secondary electrons and X rays at the higher energies permits a great flexibility in the food-package design.
e. The ability to program and to automatically regulate from one instant to the next, with simple electronic detectors and circuits, various beam parameters means the capability of efficiently processing small, intricate, or nonuniform shapes.
f. The ease with which an electron accelerator can be turned off or on means the ability to shut down during off-shifts or off-seasons without a maintenance problem and the ability to transport the radiation source without a massive radiation shield.

There appears to be a definite preference for electron accelerators in countries that lack atomic energy resources. The use of radioactive elements will probably remain popular in the United States for sometime for reasons cited above.

THE RADAPPERTIZATION, RADICIDATION, AND RADURIZATION OF FOODS

Until recently, the destruction of microorganisms in foods by ionizing radiations was referred to by terminology brought over from heat and chemical destruction of microorganisms. While microorganisms can indeed be destroyed by chemicals, heat, and radiations, there is, nevertheless, a lack of precision in the use of this terminology for radiation-treated foods. Consequently, in 1964 an international group of microbiologists suggested the following terminology for radiation treatment of foods *(9)*:

Radappertization. This is equivalent to radiation sterilization, or "commercial sterility" as it is understood in the canning industry.

Radicidation. This is equivalent to pasteurization, e.g., of milk. Specifically, it refers to the reduction of the number of viable specific nonsporeforming *pathogens,* other than viruses, so that none is detectable by any standard method.

Radurization. This process may be considered equivalent to pasteurization. It refers to the enhancement of the keeping quality of a food by causing substantial reduction in the numbers of viable specific *spoilage* microbes by radiation.

1. Radappertization.   Radappertization of any foods may be achieved by application of the proper dose of radiation under the proper conditions. The effect of this treatment on endospores and exotoxins of *C. botulinum* is of obvious interest. Type E spores have been reported to possess radiation D values on the order of 0.12–0.17 Mrad *(10, 27)*. Types A and B spores were found by Kempe *(17)* to have D values of 0.279 and 0.238 Mrad, respectively. Type E spores are the most radiation sensitive of these three types and a value of 0.132 has been reported.

The effect of temperature of irradiation on D values of *C. botulinum* spores is presented in Table 10-1. It can be seen that radiation resistance increases at the colder temperatures and decreases at higher temperatures. The different inoculum levels had no significant effect on D values whose calculations were based on a linear destruction rate. D Values of four *C. botulinum* strains in 3 food products are presented in Table 10-2, from which it can be seen that each strain displayed different degrees of radiation resistance in each product. Also, irradiating in cured meat products

TABLE 10-1.   The effect of irradiation temperature on D values of two load levels of *C. botulinum* 33A in precooked ground beef. Data are based on linear spore destruction *(12)*.

	D (Mrad)	
Temperature (C)	c. 5 × 10⁶ spores/can	c. 2 × 10⁸/can
−196	0.577	0.595
−150	0.532	0.543
−100	0.483	0.486
−50	0.434	0.430
0	0.385	0.373
25	0.360	0.345
65	0.321	0.299

Reproduced by permission of Nat'l. Res. Coun. of Canada from Can. J. Microbiol. *17:* 135–142, 1971.

TABLE 10-2.   Radiation D values of 4 strains of *C. botulinum* at −30 C in three meat products as computed by the Schmidt equation *(2)*.

	D (Mrad)		
Strain Number	Codfish Cake	Corned Beef	Pork Sausage
33A	0.203	0.129	0.109
77A	0.238	0.262	0.098
41B	0.245	0.192	0.184
53B	0.331	0.183	0.076

produced the lowest D values. The possible significance of this fact is discussed in the previous chapter under nitrates and nitrites. The minimum radiation doses (MRD) in Mrad for the radappertization of 9 meat and fish products are indicated below *(2, 4, 16)*. With the exception of bacon (irradiated at ambient temperatures), each was treated at −30 C ± 10:

Bacon	2.3
Beef	4.7
Chicken	4.5
Ham	3.7
Pork	5.1
Shrimp	3.7
Codfish cakes	3.2
Corned beef	2.5
Pork sausage	2.4–2.7

The radiation resistance of *C. botulinum* spores in aqueous media was studied by Roberts and Ingram *(28),* and these values are considerably lower than those obtained in meat products. On 3 Type A strains, D ranged from 0.10 to 0.14; on two strains of Type B, 0.10 to 0.11; on two strains of Type E, 0.08 to 0.16; and the one Type F strain examined by these authors showed a D value of 0.25. All strains were irradiated at 18–23 C and an exponential death rate was assumed in the D calculations.

With respect to the effect of radiation on *C. perfringens,* one each of 5 different strains (Types A, B, C, E, and F) was found to have D values between 0.15 and 0.25 in an aqueous environment *(28)*. More recently, 12D values for 8 strains of this organism were found to range between 3.04 to 4.14 Mrad depending upon the strain and method of computing 12D doses *(5)*.

As indicated in Figure 10-2, viruses are considerably more resistant to radiations than bacteria. Radiation D values of 30 viruses were found by Sullivan *et al. (30)* to range between 0.39 and 0.53 Mrad in Eagle's minimal essential medium supplemented with 2 percent serum. The 30 viruses included coxsackie-, echo-, poliovirus, and others. Of 5 selected viruses subjected to ^{60}Co rays in distilled water, the D values ranged from 0.10 to 0.14 Mrad. D Values of coxsackievirus B-2 in various menstra at −30 and −90 C are presented in Table 10-3. The use of a radiation 12D process for *C. botulinum* in meat products would result in the survival of virus particles unless previously destroyed by other methods such as heating.

Enzymes are also highly resistant to radiation, and a dose from 2 to 6 Mrad has been found to destroy only up to 75% of the proteolytic activity of ground beef *(21)*. When blanching at 65 or 70 C was combined with radiation doses of 4.5 to 5.2, however, at least 95% of the beef proteolytic activity was destroyed.

The main drawbacks to the application of radiations to all foods are color changes and/or the production of off-odors. Consequently, those food

TABLE 10-3.   D values of coxsackievirus B-2 at −30 and −90 C in various menstra (31).

Suspending Menstrum	D (Mrad)	
	−30	−90
Eagle's minimal essential medium + 2 percent serum	0.69	0.64
Distilled water	—	0.53
Cooked ground beef	0.68	0.81
Raw ground beef	0.75	0.68

A linear model was assumed in D calculations.

products that undergo relatively minor changes in color and odor development have received the greatest amount of attention for commercial radappertization. In 1963, the FDA approved the use of ^{60}Co and electron acceleration for the radappertization of bacon. A year later, the same agency approved the use of ^{137}Cs for this use. In 1968, however, the FDA withdrew its previous approval for the radappertization of bacon pending additional information on the safety of this treatment. Bacon is one product that undergoes only slight changes in color and odor development as a result of radappertization. Mean preference scores on radappertized versus control bacon were found by Wierbicki et al. (37) to be rather close, with control bacon scoring just slightly higher. More recent acceptance scores on a larger variety of radiated products are in the favorable range (16).

The world-wide approved uses of ionizing radiations for sprout inhibition, insect disinfection, radurization, and radicidation of a variety of foods and raw materials are listed in Table 10-4. Approvals by the U.S. FDA for packaging materials are listed in Table 10-5.

2. Radurization.   Radurization as a means of extending the shelf-life of seafoods, vegetables, and fruits has been verified in many laboratories. Typical of what can be achieved by this treatment are data reported for various seafoods and meats (Table 10-6). The shelf-life of shrimp, crab, haddock, and clams may be extended from 2- to 6-fold by radurization with doses of radiation from 100,000 to 400,000 rads. Similar results can be achieved for fish and shellfish under various conditions of packaging (26). As was pointed out earlier in this chapter, the gram negative, nonspore-forming rods are among the most radiosensitive of all bacteria to radiation, and they are the main spoilage organisms for these foods. Among the organisms that survive radurization treatments are gram positive bacteria of the following genera: *Micrococcus* (especially *M. roseus* and *M. radiodurans*), *Corynebacterium*, *Streptococcus*, *Microbacterium* (especially *M. thermosphactum*), *Lactobacillus*, and some yeasts. The gram negative coccobacillary rods belonging to the genera *Moraxella* and *Acinetobacter* have

TABLE 10-4. General Survey of Irradiated Food Products Cleared for Human Consumption in Different Countries (16). (Grouped according to product.) December, 1973. Compiled by Dr. K. Vas, International Atomic Energy Agency, Vienna.

Product	Country	Purpose of Irradiation	Source of Radiation	Dose (Krad)	Date of Approval
		Fruits and Vegetables			
Potatoes	USSR	sprout inhibition	^{60}Co	10	14 March, 1958
	Canada	sprout inhibition	^{60}Co	10 max.	9 November, 1960
				15 max.	14 June, 1963
	USA (white potatoes)	sprout inhibition	^{60}Co	5–10	30 June, 1964
			^{137}Cs	5–10	2 October, 1964
			^{60}Co + ^{137}Cs	5–10	1 November, 1965
	Israel	sprout inhibition	^{60}Co	15 max.	5 July, 1967
	Japan	sprout inhibition	^{60}Co	15 max.	30 August, 1972
	WHO**	sprout inhibition	^{60}Co or ^{137}Cs	15 max.	12 April, 1969
	Spain	sprout inhibition	^{60}Co	5–15	4 November, 1969
	Hungary*	sprout inhibition	^{60}Co	10	23 December, 1969
			^{60}Co	15 max.	10 January, 1972
	Denmark	sprout inhibition	10 MeV electrons	15 max.	27 January, 1970
	Netherlands	sprout inhibition	^{60}Co	15 max.	23 March, 1970
			4 MeV electrons	15 max.	23 March, 1970
	Bulgaria*	sprout inhibition	^{60}Co		1971
	Uruguay*	sprout inhibition	^{60}Co		1971
	Philippines**	sprout inhibition	^{60}Co	15 max.	13 September, 1972
	France**	sprout inhibition	^{60}Co	7.5–15	8 November, 1972
	Italy	sprout inhibition	^{60}Co or ^{137}Cs	7.5–15	30 August, 1973
	Canada	sprout inhibition	^{60}Co	15 max.	25 March, 1965
	USSR*	sprout inhibition	^{60}Co	6	25 February, 1967
	Israel	sprout inhibition	^{60}Co	10 max.	25 July, 1968
Onions	Netherlands*	sprout inhibition	^{60}Co	15 max.	5 February, 1971

Garlic	Thailand	sprout inhibition	4 MeV electrons	15 max.	5 February, 1971
	Italy	sprout inhibition	^{60}Co	10 max.	20 March, 1973
Dried fruits	Italy	sprout inhibition	^{60}Co or ^{137}Cs	7.5–15	30 August, 1973
Fresh fruits & vegetables	USSR	insect disinfestation	^{60}Co or ^{137}Cs	7.5–15	30 August, 1973
	USSR*	radurization	^{60}Co	100	15 February, 1966
Mushrooms	Netherlands	growth inhibition	^{60}Co	200–400	11 July, 1964
	Netherlands*	radurization	^{60}Co	250 max.	23 October, 1969
			4 MeV electrons	250 max.	23 October, 1969
Asparagus	Netherlands*	radurization	^{60}Co	200 max.	7 May, 1969
Strawberries	Netherlands*	radurization	^{60}Co	250 max.	7 May, 1969
			4 MeV electrons	250 max.	7 May, 1969
Cocoabeans	Netherlands*	insect disinfestation	^{60}Co	70 max.	7 May, 1969
			4 MeV electrons	70 max.	7 May, 1969
Spices & condiments	Netherlands*	radicidation	^{60}Co	800–1000	13 September, 1971
			4 MeV electrons	800–1000	13 September, 1971
Grain and Grain Products					
Grain	USSR	insect disinfestation	^{60}Co	30	1969
Wheat & wheat flour (changed on 4 March, 1966 from wheat products)	USA	insect disinfestation	^{60}Co	20–50	21 August, 1963
			^{137}Cs	20–50	2 October, 1964
Wheat, flour, whole wheat flour			5 MeV electrons	20–50	26 February, 1966
Wheat flour	Canada	insect disinfestation	^{60}Co	75 max.	25 February, 1969
Wheat & ground wheat products	WHO**	insect disinfestation		75 max.	12 April, 1969
Meat and Fish					
Semi-prepared raw beef, pork & rabbit products (in plastic bags)	USSR*	radurization	^{60}Co	600–800	11 July, 1964

TABLE 10-4. *(Continued)*

Product	Country	Purpose of Irradiation	Source of Radiation	Dose (Krad)	Date of Approval
Poultry, eviscerated (in plastic bags)	USSR* Netherlands*	radurization radurization	^{60}Co ^{60}Co	600 300 max.	4 July, 1966 31 December, 1971
Culinary prepared meat products (fried meat, entrecote) (in plastic bags)	USSR*	radurization	^{60}Co	800	1 February, 1967
Shrimps	Netherlands*	radurization	^{60}Co 4 MeV electrons	50–100 50–100	13 November, 1970 13 November, 1970
Other Products					
Dry food concentrates	USSR	insect disinfestation	^{60}Co	70	6 June, 1966
Any food for consumption by patients who require a sterile diet as an essential factor in their treatment	UK	radappertization		2500 min.	1 December, 1969
Deep-frozen meals	Netherlands***	radappertization	^{60}Co		27 November, 1969
Fresh, tinned & liquid foodstuffs	Netherlands***	radappertization	^{60}Co	2500 min.	8 March, 1972

*experimental batches
**Temporary acceptance for 5 years.
***for hospital patients in reversed barrier isolation
1975 © by Academic Press, N.Y. Used with permission of publisher and authors.

TABLE 10-5. Approvals by the US Food & Drug Administration for Radiation Preservation of Food Packaging Materials, July 1971 (16).*

Packaging Materials

Packaging Material	Petitioner	Source	Dose	Food Additive Petition No.	Filing			Regulation		
					Date	Fed. Reg Vol.	Fed. Reg Page	Date	Fed. Reg Vol.	Page
Nitrocellulose Coated Cellophane	AEC	gamma	1 megarad	1297	2-8-64	29	2318	8-14-64	29	11651
Glassine Paper	AEC	gamma	1 megarad	1297	2-8-64	29	2318	8-14-64	29	11651
Wax Coated Paperboard	AEC	gamma	1 megarad	1297	2-8-64	29	2318	8-14-64	29	11651
Polypropylene Film with or without Adjuvants	AEC	gamma	1 megarad	1297	2-8-64	29	2318	8-14-64	29	11651
Ethylene-Alkene-1 copolymer	AEC	gamma	1 megarad	1297	2-8-64	29	2318	8-14-64	29	11651
Polyethylene Film	AEC	gamma	1 megarad	1297	2-8-64	29	2318	8-14-64	29	11651
Polystyrene Film with or without Adjuvant Substances	AEC	gamma	1 megarad	1297	2-8-64	29	2318	8-14-64	29	11651
Rubber Hydrochloride with or without Adjuvant Substances	AEC	gamma	1 megarad	1297	2-8-64	29	2318	8-14-64	29	11651
Vinylidene Chloride-Vinyl-Chloride Copolymer Film (Saran Wrap)	AEC	gamma	1 megarad	1297	2-8-64	29	2318	8-14-64	29	11651
Polyolefin Film with or without Adjuvant Substances or Vinylidene Chloride coatings (Saran)	AEC	gamma	1 megarad	1297	2-8-64	29	2318	8-14-64	29	11651
Polyethylene Terephthalate	AEC	gamma	1 megarad	5M1674	7-30-65	30	9551	3-19-68	33	4659

TABLE 10-5. *(Continued)*

Packaging Material	Petitioner	Source	Dose	Food Additive Petition No.	Filing			Regulation		
					Date	Fed. Reg Vol.	Page	Fed. Reg Date	Vol.	Page
with or without Adjuvant Substances or Vinylidene Chloride coatings (Saran) or polyethylene coatings	AEC	gamma	1 megarad	5M1674	7-30-65	30	9551	3-19-68	33	4659
Nylon 11	AEC	gamma	1 megarad	6M1820	9-8-65	30	11400	3-19-68	33	4659
Vinylidene Chloride Copolymer (Saran) coated cellophane	AEC	gamma	1 megarad	5B1670	2-18-65	30	9116	6-11-65	30	7599
Vegetable parchment	US Army	gamma	6 megarad	5M1622	1-15-65	30	547	3-12-65	30	3354
Polyethylene Film with or without adjuvants	US Army	gamma	6 megarad	5M1645	7-21-65	30	9116	6-10-67	32	8360
Polyethylene Terephthalate with or without Adjuvants	US Army	gamma	6 megarad	5M1645	7-21-65	30	9116	6-10-67	32	8360
Nylon 6 films with or without Adjuvants	US Army	gamma	6 megarad	5M1645	7-21-65	30	9116	6-10-67	32	8360
Vinyl Chloride-Vinyl Acetate Copolymer film with or without Adjuvant	US Army	gamma	6 megarad	5M1645	7-21-65	30	9116	6-10-67	32	8360
Kraft Paper	US Army	gamma	50,000 rads	7M2172	5-23-67	32	7877	7-19-67	32	10567

*Compiled by Mr. Frank Leone, US Atomic Energy Commission.
1975 © by Academic Press, N.Y. Used by permission of Publisher and authors.

TABLE 10-6. Tentative levels of irradiation for fishery products and meat and poultry*

Fishery Products

Product	Packaging	Dose Krads	Post Irradiation Storage Conditions	Shelf-life Control Days	Shelf-life Irradiated Days	Shelf-life Extension Days
Atlantic haddock fillets	Air packing commercial fillet tins	100–250	33–35°F	10–14	30–37	16–27
Atlantic cod fillets	Same	100–250	33–35°F	12–14	30–37	16–27
Fresh shrimp	Mylar bags	100–200	33–35°F	14–16	21–28	7–12
Fresh shrimp	Various	100–200	33–35°F	13	30	17
Oysters	Pint cans	100–200	33°F	14–16	20–21	6–7
Clam meats	Air packing cans	200	33°F	5	30	25
Petrale sole fillets	Can (vacuum packed)	100–150	33°F	8–11	20–25	12–14
	Can (air) or polyester pouch	200–300	33°F	4–10	21–42	17–32
English sole fillets	Can (vacuum packed)	100	33°F	4–6	14–21	10–15
Pacific cod fillets	Polyester pouch (vacuum packed)	100	33°F	4–7	16–18	10–12
Pacific oysters	Can (vacuum packed) or glass jar	100	33°F	20	31	11
		200	33°F	20	34	11
Dungeness crab meat	Can or heat-sealable polyester pouch	100	33°F	6–14	14–35	8–21
		200	33°F	6–14	14–56	8–42
King crab meat	Can	100	33°F	5–14	21	7–16
		200	33°F	5–14	28–42	14–28

Meats and Poultry

Product	Packing	Dose Krads	Post Storage Condition	Shelf-life Control Days	Shelf-life Irradiated Days	Shelf-life Extension Days
Retail cuts of fresh meat and poultry	Oxygen permeable fresh meat films (i.e., polyvinyl chloride, polyethylene, fresh meat cellophane) (Tentative) Oxygen impermeable films or laminates, or impermeable rigid containers	50–100	33°F	2–3	Up to 21	Up to 19**

*From Eight Annual AEC Food Irradiation Contractors' Meeting, October 16–17, 1968. Conference 681006—U.S.A.E.C., Isotopes-Industrial Technology. Courtesy of Dr. Kevin G. Shea.

**Shelf-life extension reported requires the combination treatment of drip and color control in addition to the radiation dose reported.

been found to possess degrees of radiation resistance higher than for all other gram negatives. In studies on ground beef subjected to doses of 272 krad, Tiwari and Maxcy *(32)* found that 73–75% of the surviving flora consisted of these related genera. In unirradiated meat, they constituted only around 8 percent of the flora. Of the two genera, the *Moraxella* spp. appear to be more resistant than *Acinetobacter* spp. with D_{10} values of 273 to 2,039 krad having been found *(34)*. If this finding is correct, some of these organisms are among the most radiation-resistant of all bacteria. Among specific species, *M. nonliquefaciens* strains showed D_{10} values of 539 and 583 krad while the D_{10} for *M. osloensis* strains was 477 up to 1,000 krad.

In comparing the radio-sensitivity of some nonsporeforming bacteria in phosphate buffer at -80 C, Anellis *et al. (3)* found that *M. radiodurans* survived 1.8 Mrad, *Streptococcus faecium* strains survived 0.9–1.5, *S. faecalis* survived 0.6–0.9, and *S. lactis* did not survive 0.6 Mrad. *S. aureus, L. casei,* and *L. arabinosus* did not survive 0.3 Mrad exposures. It was shown that radiation sensitivity decreased as temperature of irradiation was lowered, as is the case for endospores.

The ultimate spoilage of radurized, low-temperature stored foods is invariably caused by one or more of the *Acinetobacter-Moraxella* or lactic acid types noted above. Radurization of fruits with doses on the order of 200,000 to 300,000 rads also brings about an extension of shelf-life (Table 10-7). It should be noted that the extension of shelf-life is not as great for radurized fruits as it is for meats and seafoods. This is because many of the molds that cause spoilage or market diseases of fruits are more radiation-resistant than the gram negative bacteria that cause spoilage of seafoods.

TABLE 10-7.   Tentative levels of radiation for fruit along with expected storage life (Adapted from Shea, *29,* on data provided to him by Dr. E. C. Maxie, Univ. of Calif., Davis).

Product	Dose, rads	Shelf-life, Control	Shelf-life, Post-irradiation
Cherries	250,000	10–14 days	14–20 days
Figs	200,000	10 days	14 days
Oranges	200,000	2 months	3 months
Papaya	200,000	10 days	14 days
Peaches, nectarines	200,000	10 days	14 days
Pineapples	300,000	10 days	14+ days
Strawberries	200,000	10–14 days	14–18 days

EFFECT OF IRRADIATION ON FOOD CONSTITUENTS

The undesirable changes that occur in certain irradiated foods may be caused directly by irradiation or indirectly as a result of post-irradiation reactions. Water undergoes radiolysis when irradiated in the following manner:

$$3 \text{ H}_2\text{O} \xrightarrow{\text{radiolysis}} \text{H} + \text{OH} + \text{H}_2\text{O}_2 + \text{H}_2$$

In addition, free radicals are formed along the path of the primary electron and react with each other as diffusion occurs (6). Some of the products formed along the track escape and can then react with solute molecules. By irradiating under anaerobic conditions, off-flavors and odors are somewhat minimized due to the lack of oxygen to form peroxides. One of the best ways to minimize off-flavors is to irradiate at subfreezing temperatures (33). The effect of subfreezing temperatures is to reduce or halt radiolysis and its consequent reactants. Other ways to reduce side effects in foodstuffs are presented in Table 10-8.

Other than water, proteins and other nitrogenous compounds appear to be the most sensitive to irradiation effects in foods. The products of irradiation of amino acids, peptides, and proteins depend upon the radiation dose, temperature, amount of oxygen, amount of moisture present, etc. The following are among the products reported: NH_3, hydrogen, CO_2, H_2S, amides, and carbonyls. With respect to amino acids, the aromatics tend to be more sensitive than the others, and undergo changes in ring structure. Among the most sensitive to irradiation are methionine, cysteine, histidine, arginine, and tyrosine. The amino acid most susceptible to electron beam irradiation is cystine; Johnson and Moser (15) reported that about 50% of this amino acid was lost when ground beef was irradiated. Tryptophan suffered a 10 percent loss while little or no destruction of the other amino acids occurred. Amino acids have been reported to be more stable to gamma irradiation than to electron beam irradiation.

Several investigators have reported that the irradiation of lipids and fats results in the production of carbonyls and other oxidation products such as peroxides, especially if irradiation and/or subsequent storage takes place in

TABLE 10-8.   Methods for reducing side effects in foodstuffs exposed to ionizing radiations (7).

Method	Reasoning
Reducing temperature	Immobilization of free radicals
Reducing oxygen tension	Reduction of numbers of oxidative free radicals to activated molecules
Addition of free radical scavengers	Competition for free radicals by scavengers
Concurrent radiation distillation	Removal of volatile off-flavor, off-odor precursors
Reduction of dose	Obvious

the presence of oxygen. The most noticeable organoleptic effect of lipid ir-radiation in air is the development of rancidity.

It has been observed that high levels of irradiation lead to the production of "irradiation odors" in certain foods, especially meats. Wick *et al. (35, 36)* investigated the volatile components of raw ground beef irradiated with from 2–6 Mrads at room temperature and reported finding a large number of odorous compounds (Table 10-9). Of the 45 or more constituents identified by these investigators, there were 17 sulfur-containing, 14 hydrocarbons, 9 carbonyls, and 5 or more were basic and alcoholic in nature. The higher the level of irradiation, the greater the quantity of volatile constituents produced. It should be noted that many of these constituents have been identified in various extracts of nonirradiated, cooked ground beef.

With regard to B-vitamins, Liuzzo *et al. (20)* found that levels of ^{60}Co irradiation between 0.2–0.6 Mrads effected partial destruction of the follow-ing B-vitamins in oysters: thiamine, niacin, pyridoxine, biotin, and B_{12}. Riboflavin, pantothenic acid, and folic acid were reported to be increased by irradiation, probably owing to release of bound vitamins.

In addition to flavor and odor changes produced in certain foods by ir-

TABLE 10-9.   Components of irradiated raw ground beef *(36)* *

Basic and Alcoholic Constituents		
Ammonia[a]	Ethylamine	Methanol[a]
Methylamine	4 unknown amines	Ethanol[a]

Sulfur-Containing Components		
Hydrogen sulfide[a]	3-(Methylthio)-propionaldehyde	Dimethyl disulfide[a]
Methyl mercaptan[a]	Dimethyl sulfide[a]	Diethyl disulfide
Ethyl mercaptan[a]	Carbon disulfide	Ethyl isopropyl disulfide
n-Propyl mercaptan	Methyl ethyl sulfide	Diisopropyl disulfide
Isobutyl mercaptan	Methyl isopropyl sulfide	Carbonyl sulfide
A C_5 mercaptan	Diisopropyl sulfide	

Carbonyl Components		
Acetaldehyde[a]	Pentanal	Propanone[a]
2-Propenal	Hexanal	2-Butanone[a]
2-Butenal	Heptanal	3-Buten-2-one

Hydrocarbons		
Ethylene	1-Heptene	*n*-Hexane
Propene	1-Octene	*n*-Heptane
1-Butene	Propane	*n*-Octane
1-Pentene	*n*-Butane	Benzene
1-Hexene	*n*-Pentane	

[a]Also present in nonirradiated beef isolates, but in smaller quantity.
Radiation Preservation of Foods, Advances in Chemistry Series No. 65.

radiation, certain detrimental effects have been reported for irradiated fruits and vegetables. One of the most serious is the softening of these products caused by the irradiation-degradation of pectin and cellulose, the structural polysaccharides of plants. This effect has been shown by Massey and Bourke *(22)* to be caused by radappertization doses of irradiation. Ethylene synthesis in apples is affected by irradiation such that this product fails to mature as rapidly as nonirradiated controls *(22)*. In green lemons, ethylene synthesis is stimulated upon irradiation, resulting in a faster ripening than in controls *(23)*.

THE STORAGE STABILITY OF IRRADIATED FOODS

Foods subjected to radappertizing doses of ionizing radiation may be expected to be as shelf-stable as commercially heat-sterilized foods. There are, however, two differences between foods processed by these two methods that affect storage stability: radappertization does not destroy inherent enzymes which may continue to act, and some post-irradiation changes may be expected to occur. Employing 4.5 Mrads and enzyme-inactivated chicken, bacon, fresh and barbecued pork, Heiligman *(14)* found these products to be acceptable after storage for up to 24 mon. Those stored at 70 C were more acceptable than those stored at 100 F. The effect of irradiation on beefsteak, ground beef, and pork sausage held at refrigerator temperatures for 12 yr was reported by Licciardello *et al.* *(19)*. These foods were packed with flavor preservatives and treated with 1.08 Mrads. The authors described the appearance of these meats as excellent after 12 yr of storage. A slight irradiation odor was perceptible but was not considered objectionable. The meats were reported to have a sharp, bitter taste which was presumed to be caused by the crystallization of the amino acid tyrosine. The free amino nitrogen content of the beefsteak was 75 and 175 mg %, respectively, before and after irradiation storage, and 67 and 160 mg percent before and after storage for hamburger.

Foods subjected to radurization ultimately undergo spoilage from the surviving flora if stored at temperatures suitable for growth of the organisms in question. The normal spoilage flora of seafoods is so sensitive to ionizing radiations that 99% of the total flora of these products is generally destroyed by doses on the order of 0.25 Mrads. Ultimate spoilage of radurized products is the property of the few microorganisms that survive the radiation treatment.

REFERENCES

1. **Anderson, A. W., H. C. Nordan, R. F. Cain, G. Parrish, and D. Duggan.** 1956. Studies on a radioresistant *Micrococcus.* I. Isolation, morphology, cultural characteristics, and resistance to gamma radiation. Food Technol. *10:* 575–578.

2. **Anellis, A., D. Berkowitz, W. Swantak, and C. Strojan.** 1972. Radiation sterilization of prototype military foods: Low-temperature irradiation of codfish cake, corned beef, and pork sausage. Appl. Microbiol. *24:* 453–462.

3. **Anellis, A., D. Berkowitz, and D. Kemper.** 1973. Comparative resistance of nonsporogenic bacteria to low-temperature gamma irradiation. Appl. Microbiol. *25:* 517–523.

4. **Anellis, A., E. Shattuck, D. B. Rowley, E. W. Ross, Jr., D. N. Whaley, and V. R. Dowell, Jr.** 1975. Low-temperature irradiation of beef and methods for evaluation of a radappertization process. Appl. Microbiol. *30:* 811–820.

5. **Clifford, W. J. and A. Anellis.** 1975. Radiation resistance of spores of some *Clostridium perfringens* strains. Appl. Microbiol. *29:* 861–863.

6. **Doty, D. M.** 1965. Chemical changes in irradiated mats. In: *Radiation Preservation of Foods.* Pub. No. 1273, Nat'l Acad. Sci., Nat'l Res. Coun., Washington, D.C., pp. 121–125.

7. **Goldblith, S. A.** 1963. Radiation preservation of foods—two decades of research and development. In: *Radiation Research,* U.S. Dept. of Commerce, Office of Techn. Services, Washington, D.C., pp. 155–167.

8. **Goldblith, S. A.** 1966. Basic principles of microwaves and recent developments. Advances in Food Research *15:* 277–301.

9. **Goresline, H. E., M. Ingram, P. Macuch, G. Mocquot, D. A. A. Mossel, C. F. Niven, and F. S. Thatcher.** 1964. Tentative classification of food irradiation processes with microbiological objectives. Nature *204:* 237–238.

10. **Graikoski, J. T. and L. L. Kempe.** 1962. Progress Report, U.S. Atomic Energy Comm., Contract No. AT (11-1)-1095.

11. **Grecz, N., O. P. Snyder, A. A. Walker, and A. Anellis.** 1965. Effect of temperature of liquid nitrogen on radiation resistance of spores of *Clostridium botulinum.* Appl. Microbiol. *13:* 527–536.

12. **Grecz, N., A. A. Walker, A. Anellis, and D. Berkowitz.** 1971. Effect of irradiation temperature in the range −196 to 95 C on the resistance of spores of *Clostridium botulinum* 33A in cooked beef. Can. J. Microbiol. *17:* 135–142.

13. **Grünewald, T.** 1961. Behandlung von Lebensmitteln mit energiereichen Strahlen. Ernährungs-Umschau *8:* 239–244.

14. **Heiligman, F.** 1965. Storage stability of irradiated meats. Food Technol. *19:* 114–116.

15. **Johnson, B. and K. Moser.** 1967. Amino acid destruction in beef by high energy electron beam irradiation. In: *Radiation Preservation of Foods.* American Chemical Society, Washington, D.C., pp. 171–179.

16. **Josephson, E. S., A. Brynjolfsson, and E. Wierbicki.** 1975. The use of ionizing radiation for preservation of food and feed products, pp. 96–117. In: *Radiation Research-Biomedical, Chemical, and Physical Perspectives,* edited by O. F. Nygaard, H. I. Adler, and W. K. Sinclair (Academic Press: N.Y.).

17. **Kempe, L. L.** 1965. The potential problems of type E botulism in radiation-preserved seafoods. In: *Radiation Preservation of Foods,* Pub. No. 1273, Nat'l Acad. Sci., Nat'l Res. Council Washington, D.C., pp. 211–215.

18. **Koch, H. W. and E. H. Eisenhower.** 1965. Electron accelerators for food

processing. In: *Radiation Preservation of Foods*. Pub. No. 1273, Nat'l Acad. Sci., Nat'l Res. Council, Washington, D.C., pp. 149–180.

19. **Licciardello, J. J., J. T. R. Nickerson, and S. A. Goldblith.** 1966. Observations on radio-pasteurized meats after 12 years of storage at refrigerator temperatures above freezing. Food Technol. *20:* 1232.

20. **Liuzzo, J. S., W. B. Barone, and A. F. Novak.** 1966. Stability of B-vitamins in Gulf oysters preserved by gamma radiation. Federation Proc. *25:* 722.

21. **Losty, T., J. S. Roth, and G. Shults.** 1973. Effect of irradiation and heating on proteolytic activity of meat samples. J. Agr. Food Chem. *21:* 275–277.

22. **Massey, L. M., Jr. and J. B. Bourke.** 1967. Some radiation-induced changes in fresh fruits and vegetables. In: *Radiation Preservation of Foods,* American Chemical Society, Washington, D.C., pp. 1–11.

23. **Maxie, E. and N. Sommer.** 1965. Irradiation of fruits and vegetables. In: *Radiation Preservation of Foods*. Publication 1273, Nat'l Acad. Sci., Nat'l Res. Coun., Washington, D.C., pp. 39–52.

24. **Midura, T. F., L. L. Kempe, J. T. Graikoski, and N. A. Milone.** 1965. Resistance of *Clostridium perfringens* type A spores to gamma-radiation. Appl. Microbiol. *13:* 244–247.

25. **Niven, C. F., Jr.** 1958. Microbiological aspects of radiation preservation of food. Annual Rev. of Microb. *12:* 507–524.

26. **Novak, A. F., R. M. Grodner, and M. R. R. Rao.** 1967. Radiation pasteurization of fish and shellfish. In: *Radiation Preservation of Foods*. Adv. in Chem. Series, American Chemical Society, Washington, D.C., pp. 142–151.

27. **Roberts, T. A. and M. Ingram.** 1965. The resistance of spores of *Clastridium botulinum* Type E to heat and radiation. J. Appl. Bacteriol. *28:* 125–141.

28. **Roberts, T. A. and M. Ingram.** 1965. Radiation resistance of spores of *Clostridium* species in aqueous suspension. J. Food Sci. *30:* 879–885.

29. **Shea, K. G.** 1965. The USAEC program on food irradiation. In: *Radiation Preservation of Foods*. Pub. No. 1273, Nat'l Acad. Sci., Nat'l Res. Council, Washington, D.C., pp. 283–291.

30. **Sullivan, R., A. C. Fassolitis, E. P. Larkin, R. B. Read, Jr., and J. T. Peeler.** 1971. Inactivation of thirty viruses by gamma radiation. Appl. Microbiol. *22:* 61–65.

31. **Sullivan, R., P. V. Scarpino, A. C. Fassolitis, E. P. Larkin, and J. T. Peeler.** 1973. Gamma radiation inactivation of coxsackievirus B-2. Appl. Microbiol. *26:* 14–17.

32. **Tiwari, N. P. and R. B. Maxcy.** 1972. *Moraxella-Acinetobacter* as contaminants of beef and occurrence in radurized product. J. Food Sci. *37:* 901–903.

33. **Urbain, W. M.** 1965. Radiation preservation of fresh meat and poultry. In: *Radiation Preservation of Foods*. Pub. No. 1273, Nat'l Acad. Sci., Nat'l Res. Council, Washington, D.C., pp. 87–98.

34. **Welch, A. B. and R. B. Maxcy.** 1975. Characterization of radiation-resistant vegetative bacteria in beef. Appl. Microbiol. *30:* 242–250.

35. **Wick, E., E. Murray, J. Mizutani, and M. Koshika.** 1967. Irradiation flavor and the volatile components of beef. In: *Radiation Preservation of Foods,* American Chemical Society, Washington, D.C., pp. 12–25.

36. **Wick, E., E. Murray, J. Mizutani, and M. Koshika.** 1967. Irradiation of meats by sterilizing doses of ionizing radiation. In: *Radiation Preservation of Foods.* Pub. No. 1273, Nat'l Acad. Sci., Nat'l Res. Council, Washington, D.C., pp. 383–409.

37. **Wierbicki, E., M. Simon, and E. S. Josephson.** 1965. Preservation of meats by sterilizing doses of ionizing radiation, pp. 383–409, In: *Radiation Preservation of Food,* Pub. #1273, Nat'l. Acad. Sci., Nat'l. Res. Coun. (Washington, D.C.).

11

Food Preservation
By the Use of
Low Temperatures

The use of low temperatures to preserve foods is based upon the fact that the activities of food-borne microorganisms can be slowed down and/or stopped at temperatures above freezing and generally stopped at subfreezing temperatures. The reason for this is that all metabolic reactions of microorganisms are enzyme catalyzed and that the rate of enzyme-catalyzed reactions is dependent on temperature. With a rise in temperature, there is an increase in reaction rate. The **temperature coefficient** (Q_{10}) may be generally defined as follows:

$$Q_{10} = \frac{(\text{Velocity at a given temp.} + 10°)}{\text{Velocity at T}}$$

The Q_{10} for most biological systems is 1.5 to 2.5, so that for each 10 degrees C rise in temperature within the suitable range, there is a two-fold increase in the rate of reaction. For every 10 degrees decrease in temperature, the reverse is true, of course. Since the basic feature of low-temperature food preservation consists of its effect upon spoilage organisms, most of the discussion that follows will be devoted to the effect of low temperatures on food-borne microorganisms. It should be remembered, however, that temperature is related to relative humidity (R.H.) and that subfreezing temperatures affect R.H. as well as pH and possibly other parameters of microbial growth as well.

The organisms that grow well at low temperatures were referred to in Chapter 2 as **psychrophiles.** This term was apparently coined by Schmidt-Nielsen in 1902 for microorganisms that can grow at 0 C (*11*). The existence of such organisms in foods, and especially in soils, was first noted and described by Forster in 1887. Eddy (*4*) suggested that the term psychrophile should be used only when a low optimum growth temperature is implied. In this regard, psychrophiles would be those organisms that display maximum growth temperatures below 35 C. For organisms able to grow at 5 C or less,

Eddy suggested that they be called **psychrotrophs.** While more and more researchers have adopted the latter terminology, others continue to designate as psychrophile any organism capable of growth at or around 5 C.

Many of the bacteria listed in Table 7-3 for meat, poultry, and seafood products are psychrotrophic and are common to a large variety of other food products. Most psychrotrophic bacteria of importance in foods belong to the genus *Pseudomonas,* and to lesser extents to the genera *Acinetobacter, Alcaligenes, Flavobacterium,* and others *(5, 11, 20).* Among the molds and yeasts, species and strains from a large number of genera are capable of growth at refrigerator temperatures. Among the low-temperature growing molds are the genera *Penicillium, Mucor, Cladosporium, Botrytis,* and *Geotrichum.* Among yeast, species and strains of *Debaryomyces, Torulopsis, Candida, Rhodotorula,* and others are known to be psychrotrophic.

There are at least two distinct low temperature ranges in which foods may be stored for preservation. **Chilling temperatures** are those between the usual refrigerator temperatures and room temperature, usually about 10–15 C. These temperatures are suitable for the storage of certain vegetables and fruits, e.g., cucumbers, potatoes, limes, etc. **Refrigerator temperatures** are those between 0–2 and 5–7 C and are suitable for the storage of a large number of perishable and semiperishable foods (see Table 11-1).

Temperatures below 6 C will prevent the growth of all food poisoning organisms except *C. botulinum* type E, and will effectively retard the growth of spoilage organisms. The mechanism by which low temperatures effect retardation of microbial growth is discussed in Chapter 20.

Freezer temperatures are those at or below −18 C. Under normal circumstances, these temperatures are sufficient to prevent the growth of all microorganisms, but as will be seen later, some can and do grow within the freezer range, but at an extremely slow rate.

TABLE 11-1. Recommended storage temperatures, relative humidity, and approximate storage life of various fresh, dried, and processed foods.

Vegetables	°F Storage Temp.	Percent Relative Humidity	Approx. Storage Life
Artichokes, Globe	31–32	90–95	1–2 weeks
Asparagus	32	90–95	3–4 weeks
Beans (green or snap)	45	85–90	8–10 days
Beans (Lima)	32–40	85–90	10–15 days
Beets (bunch)	32	90–95	10–14 days
Brussels sprouts	32	90–95	3–4 weeks
Cabbage, late	32	90–95	3–4 months
Carrots (bunch)	32	90–95	10–14 days
Cauliflower	32	85–90	2–3 weeks
Celery	32	90–95	2–4 months
Corn, sweet	31–32	85–90	4–8 days
Cucumbers	45–50	90–95	10–14 days

TABLE 11-1. Recommended storage temperatures, relative humidity, and approximate storage life of various fresh, dried, and processed foods *(continued)*

Vegetables	°F Storage Temp.	Percent Relative Humidity	Approx. Storage Life
Endive	32	90–95	2–3 weeks
Lettuce	32	90–95	3–4 weeks
Onions	32	70–75	6–8 months
Peas, green	32	85–90	1–2 weeks
Potatoes, early crop	50–55	85–90	—
Potatoes, sweet	55–60	90–95	4–6 months
Radishes (spring, bunched)	32	90–95	10 days
Rhubarb	32	90–95	2–3 weeks
Rutabagas	32	90–95	2–4 months
Spinach	32	90–95	10–14 days
Squash, summer	32–40	85–95	10–14 days
Tomatoes, ripe	32	85–90	7 days

Dairy Products			
Butter	32–36	80–85	2 months
Cheese, process American, process Swiss	40–45	75	—
Eggs, shell	29–31	85–90	8–9 months
Eggs (dried, yolk)	35	minimum	6–12 months
Eggs (spray dried albumin)	35	minimum	6 months

Meat, Poultry, and Fish			
Beef, fresh	32–34	88–92	1–6 weeks
Hams & shoulders, fresh	32–34	85–90	7–12 days
Hams & shoulders, cured	60–65	50–60	0–3 years
Poultry, fresh	32	—	1 week
Fish, fresh	33–40	90–95	5–20 days

Fruits			
Apples	30–32	85–90	2–7 months
Dried fruits	32	50–60	9–12 months
Figs (fresh)	28–32	85–90	5–7 days
Grapefruit	32–50	85–90	4–8 weeks
Lemons	32, 55–58	85–90	1–4 months
Limes	48–50	85–90	6–8 weeks
Melons (honey dew, honey ball)	45–50	85–90	2–4 weeks
Melons, watermelons	36–40	85–90	2–3 weeks
Oranges	32–34	85–90	8–12 weeks
Peaches	31–32	85–90	2–4 weeks
Berries, blue	31–32	85–90	3–6 weeks
Berries, cranberries	36–40	85–90	1–3 months
Grapes (American type)	31–32	85–90	3–8 weeks
Strawberries, fresh	31–32	85–90	7–10 days

TEMPERATURE GROWTH MINIMA OF FOOD-BORNE MICROORGANISMS

In their excellent review of the literature on the temperature growth minima of microorganisms, Michener and Elliott *(15)* summarized the findings of various authors who had reported the growth of microorganisms at and below -10 C. Of the 13 organisms reported, there were 6 bacteria, 4 yeasts, and 3 molds. The yeasts grew at lower temperatures than the others, with one pink yeast reported to grow at -34 C and two others at -18 C. The lowest recorded temperature of growth for a bacterium is -20 C, and several have been reported to grow at about -12 C. The molds grew at about -12 C. Some foods that tend to support microbial growth at subfreezing temperatures are fruit juice concentrates, bacon, ice cream, and certain fruits.

The reported temperature minima for fecal indicator organisms vary between -2 to 10 C *(15)*. In general, the enterococci grow better at refrigerator temperatures than do coliform bacteria. The lowest reported temperatures for growth of staphylococci and salmonellae in foods is 6.7 C *(1)*. As noted above, *C. botulinum* type E has been reported to grow at a temperature lower than all other food poisoning microorganisms; Schmidt *et al. (16)* reported 3.3 C for this organism in beef stew. Minimum temperature of growth and toxin production reported by many investigators for types A and B strains of *C. botulinum* is 10 C. The reader is referred to Chapter 15 for further discussion of the temperature growth minima of fecal indicator and food-poisoning microorganisms.

THE PREPARATION OF FOODS FOR FREEZING

The preparation of vegetables for freezing includes selecting, sorting, washing, blanching, and packaging prior to actual freezing. In selecting foods to be frozen, those in any state of detectable spoilage should be rejected. Meats, poultry, seafoods, eggs, and others should be as fresh as possible.

Blanching is achieved either by brief immersion of foods into hot water or the use of steam. Its primary functions are as follows: (1) Inactivation of enzymes which might cause undesirable changes during freezing storage, (2) enhancement or fixing of the green color of certain vegetables, (3) reduction in the numbers of microorganisms on the foods, (4) facilitating the packing of leafy vegetables by inducing wilting, and (5) the displacement of entrapped air in the plant tissues. The method of blanching employed depends upon the products in question, size of packs, and other related information. When water is used, it is important that bacterial spores not be allowed to build up sufficiently to contaminate foods. With respect to the reduction in numbers of microorganisms by blanching, reductions of initial microbial loads as high as 99% have been claimed. It should be remembered that most vegetative bacterial cells can be destroyed at milk pasteurization temperatures (145 F for 30 min.). This is especially true of most bacteria of importance in the spoilage of vegetables. While it is not the primary function of blanching to destroy microorganisms, the amount of heat necessary to effect destruction of most food enzymes is also sufficient to significantly reduce vegetative cells.

THE FREEZING OF FOODS AND FREEZING EFFECTS

The two basic ways to achieve the freezing of foods are quick and slow freezing. **Quick** or **fast freezing** is the process by which the temperature of foods is lowered to about −20 C within 30 min. This treatment may be achieved by direct immersion or indirect contact of foods with the re-frigerant, and the use of air-blasts of frigid air blown across the foods being frozen.

Slow freezing refers to the process whereby the desired temperature is achieved within 3 to 72 hr. This is essentially the type of freezing utilized in the home freezer.

Quick freezing possesses more advantages than slow freezing from the standpoint of overall product quality. The two methods are compared below *(2)*:

Quick freezing	*Slow freezing*
1. Small ice crystals formed.	1. Large ice crystals formed.
2. Blocks or suppresses metabolism.	2. Breakdown of metabolic rapport.
3. Brief exposure to concentration of adverse constituents.	3. Longer exposure to adverse or injurious factors.
4. No adaptation to low temperatures.	4. Gradual adaptation.
5. Thermal shock (too brutal a transition).	5. No shock effect.
6. No protective effect.	6. Accumulation of concentrated so-lutes with beneficial effects.
7. Microorganisms frozen into crystals?	
8. Avoid internal metabolic imbalance.	

With respect to crystal formation upon freezing, slow freezing favors large extracellular crystals, while rapid or quick freezing favors the formation of small, intracellular ice crystals. Crystal growth is one of the factors that limits the freezer life of certain foods, since ice crystals grow in size and cause cell damage by disrupting membranes, cell walls, and internal struc-tures to the point where the thawed product is quite unlike the original in texture and flavor. Upon thawing, foods frozen by the slow freezing method tend to lose more **drip** (drip for meats but **leakage** for vegetables) than quick frozen foods held for comparable periods of time. The overall advantages of small crystal formation to frozen food quality may be viewed also from the standpoint of what takes place when a food is frozen. During the freezing of foods, water is removed from solution and transformed into ice crystals of a variable but high degree of purity (6). In addition, the freezing of foods is accompanied by changes in properties such as pH, titratable acidity, ionic strength, viscosity, osmotic pressure, vapor pressure, freezing point, surface and interfacial tension, and O/R potential (see below). Some of the many complexities of this process are discussed thoroughly by Fennema *et al. (8)*.

THE STORAGE STABILITY OF FROZEN FOODS

A large number of microorganisms have been reported by many investigators to grow at and below 0 C as previously mentioned. In addition to factors inherent within these organisms, their growth at and below freezing temperatures is dependent upon several factors of foods, namely, nutrient content, pH, and the availability of liquid water. The a_w of foods may be expected to decrease as temperatures fall below the freezing point. The relationship between temperature and a_w of water and ice is presented in Table 11-2. For water at 0 C, a_w is 1.0 but falls to about 0.8 at −20 C and to 0.62 at about −50 C. Organisms that grow at subfreezing temperatures, then, must be able to grow at the reduced a_w levels, unless a_w is favorably affected by food constituents with respect to microbial growth. In fruit juice concentrates which contain comparatively high levels of sugars, these compounds tend to maintain a_w at levels higher than would be expected in pure water, thereby making microbial growth possible even at subfreezing temperatures. The same type of effect can be achieved by the addition of glycerol to culture media. It should be borne in mind also that not all foods

TABLE 11-2.   Vapor pressures of water and ice at various temperatures (17).

°C	Liquid Water mm. Hg	Ice mm. Hg	$a_w = \dfrac{p_{ice}}{p_{water}}$
0	4.579	4.579	1.00
− 5	3.163	3.013	0.953
−10	2.149	1.950	0.907
−15	1.436	1.241	0.864
−20	0.943	0.776	0.823
−25	0.607	0.476	0.784
−30	0.383	0.286	0.75
−40	0.142	0.097	0.68
−50	0.048	0.030	0.62

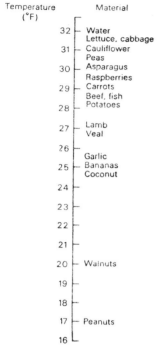

FIGURE 11-1.   Freezing point of selected foods *(3)*.

freeze at the same initial point (see Figure 11-1). The initial freezing point of a given food is due in large part to the nature of its solute constituents and the relative concentration of those that have freezing-point depressing properties.

Although the metabolic activities of all microorganisms can be stopped at freezer temperatures, frozen foods may not be kept indefinitely if the thawed product is to retain the original flavor and texture. Most frozen foods are assigned a freezer life, and suggested maximum holding periods for various foods are presented in Table 11-3. The suggested maximum holding time for frozen foods is not based upon the microbiology of such foods but upon such factors as texture, flavor, tenderness, color, and overall nutritional quality upon thawing and subsequent cooking.

Some foods that are improperly wrapped during freezer storage undergo **freezer burn.** This condition is characterized by a browning of light-colored foods such as the skin of chicken meats. The browning results from the loss of moisture at the surface, leaving the product more porous than the original at the affected site. The condition is irreversible and is known to affect certain fruits, poultry, meats, and fish, both raw and cooked.

TABLE 11-3. Storage life of frozen foods[a]—average life in "good condition" for products processed and packaged in a normal manner (19).

Product	Months at:			
	−10°F	0°F	+5°F	+10°F
Apricots[b]	24	18–24	—	6–8
Asparagus	16–18	8–12	—	4–6
Beans, green	16–18	8–12	—	4–6
Beans, Lima	24	14–16	—	6–8
Broccoli	24	14–16	—	6–8
Brussels sprouts	16–18	8–12	—	4–6
Cauliflower	24	14–16	—	6–8
Corn, on cob	12–14	8–10	—	4–6
Corn, cut	36	24	—	12
Carrots	36	24	—	12
Fish, fatty	10–12	6–8	—	4
Fish, lean	14–16	10–12	—	6
Lobsters	10–12	8–10	—	3–4
Peaches[b]	24	18–24	—	6–8
Peas	24	14–16	—	6–8
Raspberries, sugared	24	18	—	8–10
Spinach	24	14–16	—	6–8
Strawberries, sliced	24	18	—	8–10
Beef				
Roasts, steaks	—	12–14	8–10	—
Cubed, small pieces	—	10–12	6–8	—
Ground	—	8	4	—
Veal				
Roasts, chops	—	10–12	6–8	—
Thin cutlets, cubes	—	8–10	4–6	—
Ground	—	6	3	—
Lamb				
Roasts, chop	—	12–14	8–10	—
Cubed	—	10–12	6	—
Ground	—	8	4	—
Pork				
Roasts, chops	—	6–12	3–5	—
Ground, sausage	—	4	2	—
Pork or ham, smoked	—	5–7	3	—
Bacon	—	3	1	—
Lard				
Unsalted	—	3–6	1–2	—
Salted	—	1–3	1/2	—
Variety meats				
Beef or lamb liver, heart	—	4	2	—
Veal liver, heart	—	3	1	—
Pork liver, heart	—	2	1	—
Tongue	—	4	2	—
Kidneys	—	3	1	—
Sweet breads	—	1	—	—
Brains	—	1	—	—
Oxtails	—	4	2	—
Tripe	—	1	—	—
Spiced sausage or delicatessen meats	—	2–3	1	—

[a]Courtesy of F. W. Williams Publications.
[b]Contains ascorbic acid.

THE EFFECT OF FREEZING UPON MICROORGANISMS

In considering the effect of freezing upon those microorganisms that are unable to grow at freezing temperatures, it is well known that freezing is one means of preserving microbial cultures, with freeze-drying being perhaps the best method known. However, freezing temperatures have been shown to effect the killing of certain microorganisms of importance in foods. Ingram *(12)* has summarized the salient facts of what happens to certain microorganisms upon freezing:

1. There is a sudden mortality immediately on freezing, varying with species.
2. The proportion of cells surviving immediately after freezing is nearly independent of the rate of freezing.
3. The cells which are still viable immediately after freezing die gradually when stored in the frozen state.
4. This decline in numbers is relatively rapid at temperatures just below the freezing point, especially about −2 C, but less so at lower temperatures, and it is usually slow below −20 C.

Bacteria differ in their capacity to survive during freezing, with the cocci being generally more resistant than gram negative rods. Of the food poisoning bacteria, salmonellae are less resistant than *S. aureus* or vegetative cells of clostridia, while endospores and food poisoning toxins are apparently unaffected by low temperatures *(9)*. The effect of freezing several species of *Salmonella* to −25.5 C and holding up to 270 days is presented in Table 11-4. Although a significant reduction in viable numbers occurred over the 270-day storage period with most species, in no instance did all cells die off.

From the strict standpoint of food preservation, freezing should not be regarded as a means of destroying food-borne microorganisms. The type of organisms that lose their viability in this state differ from strain to strain and depend upon the type of freezing employed, the nature and composition of

TABLE 11-4.   The survival of pure cultures of enteric organisms in chicken chow mein at −25.5 C *(10)*

Organism	Bacterial count (10^5/g) after storage for (days)								
	0	2	5	9	14	28	50	92	270
Salm. newington	75.5	56.0	27.0	21.7	11.1	11.1	3.2	5.0	2.2
Salm. typhi-murium	167.0	245.0	134.0	118.0	11.0	95.5	31.0	90.0	34.0
Salm. typhi	128.5	45.5	21.8	17.3	10.6	4.5	2.6	2.3	0.86
Salm. gallinarum	68.5	87.0	45.0	36.5	29.0	17.9	14.9	8.3	4.8
Salm. anatum	100.0	79.0	55.0	52.5	33.5	29.4	22.6	16.2	4.2
Salm. paratyphi B	23.0	205.0	118.0	93.0	92.0	42.8	24.3	38.8	19.0

*1948 Copyright © by Institute of Food Technologists.

the food in question, length of time of freezer storage, and other factors, such as temperature of freezing. Low freezing temperatures of about -20 C are less harmful to microorganisms than the median range of temperatures such as -10 C. For example, more microorganisms are destroyed at -4 C than at -15 C or below. Temperatures below -24 C seem to have no additional effect. Food constituents such as egg white, sucrose, corn syrup, fish, glycerol, and undenatured meat extracts have all been reported to increase freezing viability, especially of food poisoning bacteria, while acid conditions have been reported to decrease cell viability (9).

To consider further the effects of freezing upon microorganisms, it would be of value to consider some of the events that are known to occur when cells freeze:

1. The water that freezes is the so-called "free water." Upon freezing, the free water forms ice crystals. The growth of ice crystals occurs by accretion so that all of the free water of a cell might be represented by a relatively small number of ice crystals. In slow freezing, ice crystals are extracellular but in fast freezing they are intracellular. Bound water remains unfrozen. The freezing of cells depletes them of usable liquid water and thus dehydrates them.

2. Freezing results in an increase in the viscosity of cellular matter, a direct consequence of water being concentrated in the form of ice crystals.

3. Freezing results in a loss of cytoplasmic gases such as O_2 and CO_2. A loss of O_2 to aerobic cells suppresses respiratory reactions. Also, the more diffuse state of O_2 may make for greater oxidative activities within the cell.

4. Freezing causes changes in pH of cellular matter. Various authors have reported changes ranging from 0.3–2.0 pH units. Increases and decreases of pH upon freezing and thawing have been reported.

5. Freezing effects a concentration of cellular electrolytes. This effect is also a consequence of the concentration of water in the form of ice crystals.

6. Freezing causes a general alteration of the colloidal state of cellular protoplasm. It should be recalled that many of the constituents of cellular protoplasm such as proteins exist in a dynamic colloidal state in living cells. A proper amount of water is necessary to the well-being of this state.

7. Freezing causes some denaturation of cellular proteins. Precisely how this effect is achieved is not clear, but it is known that some —SH groups disappear upon freezing. It is also known that such groups as lipoproteins break apart from others upon freezing. The lowered water content along with the concentration of electrolytes no doubt affect this change in state of cellular proteins.

8. Freezing induces temperature shock in some microorganisms. This is true more for thermophiles and mesophiles than psychrophiles. It has

been shown that more cells die when the temperature decline above freezing is sudden than when it is slow.

9. Freezing causes metabolic injury to some microbial cells such as some *Pseudomonas* spp. It has been reported that some bacteria have increased nutritional requirements upon thawing from the frozen state and that as much as 40% of a culture may be affected in this way (See Chapter 4 for other effects of freeze injury.)

The above should serve to illustrate the complexity of the freezing process upon living cells such as bacteria and other microorganisms as well as upon foods. According to Mazur *(14)*, the response of microorganisms to subzero temperatures appears to be largely determined by solute concentration and intracellular freezing, although there are only a few cases of clear demonstration of this conclusion.

Why are some bacteria killed by freezing but yet not all cells? Some small and microscopic organisms are unable to survive freezing as can most bacteria. Examples of these include the foot-and-mouth disease virus and the causative agent of trichinosis *(T. spiralis)*. Protozoa are generally killed when frozen below −5 or −10 C, if protective compounds are not present *(14)*.

Of great importance in the freezing survival of microorganisms is the process of thawing. It is well established that repeated freezing and thawing will destroy bacteria by disrupting cell membranes. It is also known that the faster the thaw, the greater the number of bacterial survivors. Just why this is so is not entirely clear. From the changes listed above that occur during freezing, it can be seen that the thawing process becomes complicated if it is to lead to the restoration of viable activity. Fennema *(7)* has pointed out that thawing is inherently slower than freezing and follows a pattern which is potentially more detrimental. Among the problems attendant to the thawing of specimen and products that transmit heat energy primarily by conduction are those summarized below *(8)*:

1. Thawing is inherently slower than freezing when conducted under comparable temperature differentials,
2. In practice, the maximum temperature differential permissible during thawing is much less than that which is feasible during freezing,
3. The time-temperature pattern characteristic of thawing is potentially more detrimental than that of freezing. During thawing, the temperature rises rapidly to near the melting point and remains there throughout the long course of thawing, thus affording considerable opportunity for chemical reactions, recrystallization, and even microbial growth if thawing is extremely slow.

It has been stated that microorganisms do not die upon freezing *per se,* but rather, during the thawing process. Whether or not this is the case remains to be proven. As to why some organisms are able to survive freezing while others are not, Luyet *(13)* has suggested that it is a question of the ability of an organism to survive dehydration and to undergo dehydration when the medium freezes. Luyet has further stated that the small size of bacterial cells permits them to undergo dehydration upon freezing. With respect to survival after freeze-drying, this author has stated that it might be due to the fact that bacteria do not freeze at all but merely dry. See Chapters 4 and 13 for further discussion of the effect of freeze-drying upon microorganisms.

Most frozen foods processors advise against the refreezing of foods once they have thawed. While the reasons for this are more related to the texture, flavor, and other nutritional qualities of the frozen product, the microbiology of thawed frozen foods is pertinent. Some investigators have pointed out that foods thawed from the frozen state spoil faster than similar fresh products. There are textural changes associated with freezing that would seem to aid the invasion of surface organisms into deeper parts of the product and consequently facilitate the spoilage process. Upon thawing, surface condensation of water is known to occur as well as a general concentration of water-soluble nutrients at the surface such as amino acids, minerals, B-vitamins, and possibly others. Freezing has the effect of destroying many thermophilic and some mesophilic organisms, which makes for less competition among the survivors upon thawing. It is conceivable that a greater relative number of psychrophiles on thawed foods might increase the spoilage rate. Some psychrophilic bacteria have been reported to have Q_{10} values in excess of 4.0 at refrigerator temperatures. For example *Ps. fragi* has been reported to possess a Q_{10} of 4.3 at 0 C. Organisms of this type are capable of doubling their growth rate with only a 4–5 degree rise in temperature. Whether or not frozen thawed foods do in fact spoil faster than fresh foods would depend upon a large number of factors, such as the type of freezing, the relative numbers and types of organisms on the product prior to freezing, and the temperature at which the product is held to thaw. Although there are no known toxic effects associated with the refreezing of frozen and thawed foods, this act should be minimized in the interest of the overall nutritional quality of the products. One effect of freezing and thawing animal tissues is the release of lysosomal enzymes consisting of cathepsins, nucleases, phosphatases, glycosidases, and others *(18)*. Once released, these enzymes may act to degrade macromolecules and thus make available simpler compounds which are more readily utilized by the spoilage flora.

REFERENCES

1. **Angelotti, R., M. J. Foter, and K. H. Lewis.** 1961. Time-temperature effects on salmonellae and staphylococci in foods. Am. J. Public Health *51:* 76–88.
2. **Borgstrom, G.** 1961. Unsolved problems in frozen food microbiology. In: *Proceedings, Low Temperature Microbiology Symposium,* Campbell Soup Company, Camden, N.J., pp. 197–251.

3. **Desrosier, N. W.** 1963. *The Technology of Food Preservation*. The Avi Publishing Company, Westport, Conn., Chapter 4.

4. **Eddy, B. P.** 1960. The use and meaning of the term 'psychrophilic'. J. Appl. Bacteriol. *23:* 189–190.

5. **Farrell, J. and A. Rose.** 1967. Temperature effects on microorganisms. Annual Rev. of Microb. *21:* 101–120.

6. **Fennema, O. and W. Powrie.** 1964. Fundamentals of low-temperature food preservation. Advances in Food Research *13:* 219–347.

7. **Fennema, O.** 1966. An over-all view of low temperature food preservation. Cryobiol. *3:* 197–213.

8. **Fennema, O. R., W. D. Powrie, and E. H. Marth.** 1973. *Low-Temperature Preservation of Foods and Living Matter* (Marcel Dekker: N.Y.).

9. **Georgala, D. L. and A. Hurst.** 1963. The survival of food poisoning bacteria in frozen foods. J. Appl. Bact. *26:* 346–358.

10. **Gunderson, M. F. and K. D. Rose.** 1948. Survival of bacteria in a precooked fresh-frozen food. Food Res. *13:* 254–263.

11. **Ingraham, J. L. and J. L. Stokes.** 1959. Psychrophilic bacteria. Bacteriological Reviews *23:* 97–108.

12. **Ingram, M.** 1951. The effect of cold on microorganisms in relation to food. Proc. Soc. Appl. Bact. *14:* 243.

13. **Luyet, B.** 1962. Recent developments in cryobiology and their significance in the study of freezing and freeze-drying of bacteria. In: *Proceedings, Low Temperature Microbiology Symposium,* Campbell Soup Company, Camden, N.J., pp. 63–87.

14. **Mazur, P.** 1966. Physical and chemical basis of injury in single-celled microorganisms subjected to freezing and thawing. In: *Cryobiology*. Edited by H. T. Meryman, Academic Press, N.Y., Chapter 6.

15. **Michener, H. and R. Elliott.** 1964. Minimum growth temperatures for food-poisoning, fecal-indicator, and psychrophilic microorganisms. Advances in Food Research *13:* 349–396.

16. **Schmidt, C. F., R. V. Lechowich, and J. F. Folinazzo.** 1961. Growth and toxin production by Type E *Clostridium botulinum* below 40°F. J. Food Sci. *26:* 626–630.

17. **Scott, W. J.** 1962. Available water and microbial growth. In: *Proceedings, Low Temperature Microbiology Symposium,* Campbell Soup Co., Camden, N.J., pp. 89–105.

18. **Tappel, A. L.** 1966. Effects of low temperature and freezing on enzymes and enzyme systems. In: *Cryobiology*. Edited by H. T. Meryman, Academic Press, N.Y., Chapter 4.

19. **Tressler, D. K.** 1960. Storage life of selected frozen foods. Recommended maximum storage periods for meat. Frozen Food Almanac, Quick Frozen Foods *23:* 148.

20. **Witter, L. D.** 1961. Psychrophilic bacteria—A review. J. Dairy Sci. *44:* 983–1015.

12

Food Preservation
By Use of
High Temperatures

The use of high temperatures to preserve food is based on their destructive effects on microorganisms. By high temperatures are meant any and all temperatures above ambient. With respect to food preservation, there are two temperature categories in common use: pasteurization and sterilization. **Pasteurization** by use of heat implies either the destruction of all disease-producing organisms (e.g., pasteurization of milk) or the destruction or reduction in number of spoilage organisms in certain foods, e.g., the pasteurization of vinegar. The pasteurization of milk is achieved by heating at 145 F for 30 min, or at 161 F for 15 sec (high temperature short time—HTST method). These treatments are sufficient to destroy the most heat-resistant of the nonsporeforming pathogenic organisms—*Mycobacterium tuberculosis* and *Coxiella burnetti*. Milk pasteurization temperatures are sufficient to also destroy all yeasts, molds, gram negative bacteria, and many gram positives. The two groups of organisms that survive milk pasteurization are placed into one of two groups: thermodurics and thermophiles. **Thermoduric** organisms are those that can survive heat treatment at relatively high temperatures but do not necessarily grow at these temperatures. The nonsporeforming organisms that survive milk pasteurization generally belong to the genera *Streptococcus* and *Lactobacillus,* and sometimes to other genera. **Thermophilic** organisms are those that not only survive relatively high temperatures but *require* high temperatures for their growth and metabolic activities. The genera *Bacillus* and *Clostridium* contain the thermophiles of greatest importance in foods.

Sterilization means the destruction of all viable organisms as may be measured by an appropriate plating or enumerating technique. Canned foods are sometimes called "commercially sterile" to signify that no viable organisms can be detected by the usual cultural methods employed, or that the number of survivors is so low as to be of no significance under the conditions of canning and storage. Also, microorganisms may be present in canned foods that cannot grow in the product by reason of undesirable pH, Eh, or temperature of storage.

A more recent development in food processing is the use of **ultra-high temperature** (UHT). The UHT range begins at around 190–210 F and may extend to 300 F or above. Exposure times may be anywhere from 1 sec to many sec. UHT is employed mainly in milk processing and milk flavor is said to be improved by use of this method over conventional processing temperatures. Sterility can be achieved by heating the product to 285 F for 15 sec and then to 300 F for 0.5 sec. Pressures up to 100 lb/in.^{2} are achieved under these conditions.

FACTORS THAT AFFECT HEAT RESISTANCE IN MICROORGANISMS

It is well known that equal numbers of bacteria placed in physiologic saline and nutrient broth at the same pH are not destroyed with the same ease by heat. Some 11 factors or parameters of microorganisms and their environment have been studied for their effects on heat destruction and are presented below *(12)*.

1. Effect of Water.   The heat resistance of microbial cells increases with decreasing humidity or moisture. Dried microbial cells placed into test tubes and then heated in a water bath are considerably more heat resistant than moist cells of the same type. Since it is well established that protein denaturation occurs at a faster rate when heated in water than in air, it is suggested that protein denaturation is either the mechanism of death by heat or is closely associated with it (see Chapter 21 for further discussion of the mechanism of heat death). The precise manner in which water facilitates heat denaturation of proteins is not entirely clear, but it has been pointed out that the heating of wet proteins causes the formation of free —SH groups with a consequent increase in the water-binding capacity of proteins. The presence of water allows for thermal breaking of peptide bonds, a process which requires more energy in the absence of water and consequently confers a greater refractivity to heat.

2. Effect of Fat.   In the presence of fats, there is a general increase in the heat resistance of some microorganisms (see Table 12-1). This is sometimes referred to as fat protection and is presumed to increase heat resistance by directly affecting cell moisture. Sugiyama *(23)* demonstrated the heat-protective effect of long-chain fatty acids on *C. botulinum*. It appears that the long-chain fatty acids are better protectors than short-chain acids.

3. Effect of Salts.   The effect of salt on the heat resistance of microorganisms is variable and dependent upon the kind of salt, concentration employed, and other factors. It has been observed that some salts have a protective effect upon microorganisms while others tend to make cells more

TABLE 12-1. The effect of the medium upon the thermal death point of *Escherichia coli* (4)***

Medium	Thermal death point (°C)
Cream	73
Whole milk	69
Skim milk	65
Whey	63
Bouillon (broth)	61

*Heating time: 10 min.
**Courtesy of W. B. Saunders Co., Philadelphia.

heat sensitive. It has been suggested that some salts may decrease water activity and thereby increase heat resistance by a mechanism similar to that of drying, while others may increase water activity (e.g., Ca^{++} and Mg^{++}) and consequently increase sensitivity to heat. It has been shown that supplementation of the growth medium of *B. megaterium* spores with $CaCl_2$ yields spores with increased heat resistance, while the addition of L-glutamate, L-proline, or increased phosphate content decreases heat resistance *(17)*.

4. Effect of Carbohydrates.   The presence of sugars in the suspending menstrum causes an increase in the heat resistance of microorganisms suspended therein. This effect is at least in part due to the decrease in water activity that is caused by high concentrations of sugars. There is great variation, however, among sugars and alcohols relative to their effect on heat resistance as may be seen in Table 12-2 for D values of *Salmonella senftenberg* 775W. At identical a_w values obtained by use of glycerol and sucrose, wide differences in heat se isitivity occurs *(3, 11)*. Corry *(5)* found that sucrose increased the heat resistance of *S. senftenberg* more than any of four other carbohydrates tested. The following decreasing order was found for the five tested substances: sucrose > glucose > sorbitol > fructose > glycerol.

5. Effect of pH.   Microorganisms are most resistant to heat at their optimum pH of growth, which is generally about 7.0. As pH is lowered or raised from this optimum value, there is a consequent increase in heat sensitivity (Figure 12-1). Advantage is taken of this fact in the heat processing of high acid foods, where considerably less heat is applied to achieve sterilization compared to foods at or near neutrality. The heat pasteurization of egg white provides an example of an alkaline food product that is neutralized prior to heat treatment, a practice which is not done with other foods. The pH of egg white is about 9.0. When this product is subjected to pasteurization conditions of 60–62 C for 3.5–4 min, coagulation of proteins occurs along with a marked increase in viscosity. These changes affect the volume and texture of cakes made from such pasteurized egg white. Cun-

TABLE 12-2. Reported D values of *Salmonella senftenberg* 775W as functions of various parameters of growth and other conditions.

Temp. (C)	D Values	Conditions	Ref.
61	1.1 min	Liquid whole egg	(20)
61	1.19 min	Tryptose broth	(20)
60	9.5 min[1]	Liquid whole egg, pH c. 5.5	(2)
60	9.0 min[1]	Liquid whole egg, pH c. 6.6	(2)
60	4.6 min[1]	Liquid whole egg, pH c. 7.4	(2)
60	0.36 min[1]	Liquid whole egg, pH c. 8.5	(2)
65.6	34–35.3 sec	Milk	(19)
71.7	1.2 sec	Milk	(19)
70	360–480 min	Milk chocolate	(10)
55	4.8 min	TSB[2], log phase, grown 35 C	(18)
55	12.5 min	TSB[2], log phase, grown 44 C	(18)
55	14.6 min	TSB[2], stationary, grown 35 C	(18)
55	42.0 min	TSB[2], stationary, grown 44 C	(18)
57.2	13.5 min[1]	a_w 0.99 (4.9% glyc.), pH 6.9	(11)
57.2	31.5 min[1]	a_w 0.90 (33.9% glyc.), pH 6.9	(11)
57.2	14.5 min[1]	a_w 0.99 (15.4% sucro.), pH 6.9	(11)
57.2	62.0 min[1]	a_w 0.90 (58.6% sucro.), pH 6.9	(11)
60	0.2–6.5 min[3]	HIB[4], pH 7.4	(3)
60	2.5 min	a_w 0.90, HIB, glycerol	(3)
60	75.2 min	a_w 0.90, HIB, sucrose	(3)
65	0.29 min	0.1 M phosphate buf., pH 6.5	(5)
65	0.8 min	30% sucrose	(5)
65	43.0 min	70% sucrose	(5)
65	2.0 min	30% glucose	(5)
65	17.0 min	70% glucose	(5)
65	0.95 min	30% glycerol	(5)
65	0.70 min	70% glycerol	(5)
55	35 min	a_w 0.997, tryptone soya agar, pH 7.2	(13)

[1]Mean/average values.
[2]Trypticase soy broth
[3]Total of 76 cultures
[4]Heart infusion broth

ningham and Lineweaver *(6)* have reported that egg white may be pasteurized the same as whole egg if the pH is reduced to about 7.0. This reduction of pH makes both microorganisms and egg-white proteins more heat-stable. The addition of salts of iron or aluminum increases the stability of the highly heat-labile egg protein conalbumin sufficiently to permit pasteurization at 60–62 C. Unlike their resistance to heat in other materials, bacteria are more resistant to heat in liquid whole egg at pH values of 5.4–5.6 than at values of 8.0–8.5 (Table 12-2). This is true when pH is lowered with an acid such as HCl. When organic acids such as acetic or lactic acid are used to lower pH, a decrease in heat resistance occurs.

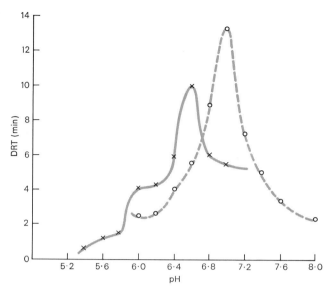

FIGURE 12-1.   The effect of pH on the DRT of *Strep. faecalis* (C & G) exposed to 60°
in citrate-phosphate buffer (crosses) and phosphate buffer (circles) solutions at
various pH levels *(24)*. DRT = decimal reduction time.

6. Effect of Proteins and Other Substances.   It is well known that proteins
in the heating menstrum have a protective effect upon microorganisms. Consequently, high protein-content foods must be heat processed to a greater
degree than low protein-content foods in order to achieve the same end
results. For identical numbers of organisms, the presence of colloidal-size
particles in the heating menstrum also offers protection against heat. For
example, under identical conditions of pH, numbers of organisms, etc., it
would take longer to sterilize pea purée than nutrient broth.

7. Effect of Numbers of Organisms.   The larger the number of organisms,
the higher is the degree of heat resistance. This is well illustrated in Table
12-3. It has been assumed that the mechanism of heat protection by large
microbial populations was due to the production of protective substances
excreted by the cells, and some authors claim to have demonstrated the
existence of such substances. Since proteins are known to offer some protection against heat, many of the extracellular compounds in a culture would
be expected to be protein in nature and consequently capable of affording
some protection. Of perhaps equal importance in the higher heat resistance
of large cell populations over smaller ones is the greater chance for the
presence of organisms with differing degrees of natural heat resistance.

8. Effect of Age of Organisms.   Bacterial cells tend to be most resistant to
heat while in the stationary phase of growth (old cells) and less resistant dur-

TABLE 12-3. Effect of the number of spores of
Clostridium botulinum on the thermal death time at
100 C *(4).**

Number of spores	Thermal death time (minutes)
72,000,000,000	240
1,640,000,000	125
32,000,000	110
650,000	85
16,400	50
328	40

*Courtesy of W. B. Saunders Co., Philadelphia.

ing the logarithmic phase. This is true for *S. senftenberg* (see Table 12-2) where stationary phase cells may be several times more resistant than log phase cells *(18)*. Heat resistance has been reported to be high also at the beginning of the lag phase but decreases to a minimum as the cells enter the log phase. Old bacterial spores have been reported to be more heat resistant than young spores. The mechanism of increased heat resistance of less active microbial cells is undoubtedly complex and at this time is not well understood.

9. Effect of Temperature of Growth. The heat resistance of microorganisms tends to increase as the temperature of incubation increases. Lechowich and Ordal *(15)* showed that as sporulation temperature was increased for *B. subtilis* and *B. coagulans,* the thermal resistance of spores of both organisms also increased. Although the precise mechanism of this effect is unclear, it is conceivable that genetic selection favors the growth of the more heat-resistant strains at succeedingly high temperatures. *S. senftenberg* grown at 44 C has been found to be approximately 3 times more resistant than when grown at 35 C (Table 12-2).

10. Effect of Inhibitory Compounds. As might be expected, a decrease in heat resistance of most microorganisms occurs when heating takes place in the presence of microbial inhibitors such as heat-resistant antibiotics, SO_2, and others. The use of heat + antibiotics and heat + nitrite together has been found to be more effective in controlling the spoilage of certain foods than either alone. The practical effect of adding inhibitors to foods prior to heat treatment is to reduce the amount of heat that would be necessary if used alone (see Chapter 9).

11. Effect of Time and Temperature. One would expect that the longer the time of heating, the greater the killing effect of heat. All too often, though, there are exceptions to this basic rule. A more dependable rule is that the higher the temperature, the greater is the killing effect of heat. This is illustrated in Table 12-4 for bacterial spores. As temperature increases, time necessary to achieve the same effect decreases.

These rules assume that heating effects are immediate and not mechanically obstructed or hindered. Also important is the size of the heating vessel or container and its composition (glass, metal, plastic, etc.). It should be obvious that it would take longer to effect pasteurization or sterilization in large containers than in smaller ones. The same would be true of containers with walls that do not conduct heat as readily as others.

RELATIVE HEAT RESISTANCE OF MICROORGANISMS

In general, the heat resistance of microorganisms is related to their optimum growth temperatures. Psychrophilic microorganisms are the most heat sensitive of the three temperature groups, followed by mesophiles and thermophiles. Sporeforming bacteria are more heat resistant than nonsporeformers, while thermophilic sporeformers are in general more heat resistant than mesophilic sporeformers. With respect to gram reaction, gram positive bacteria tend to be more heat resistant than gram negative, with cocci in general being more resistant than nonsporeforming rods. Yeasts and molds tend to be fairly sensitive to heat with yeast ascospores being only slightly more resistant than vegetative yeasts. The asexual spores of molds tend to be slightly more heat resistant than mold mycelia. Sclerotia are the most heat resistant of these types and sometimes survive and cause trouble in canned fruits.

The heat resistance of bacterial endospores is of special interest in the thermal processing of foods. These structures are produced by *Bacillus* and *Clostridium* spp. usually upon the exhaustion of nutrients essential for continued vegetative growth, although other factors appear to be involved. Only one spore is produced per cell, and it may occur in various parts of the vegetative cell and possess various shapes and sizes, all of which are of taxonomic value. The endospore is not only resistant to heat but to drying,

TABLE 12-4.   Effect of temperature upon the thermal death times of spores *(4)**

Temperature	Clostridium botulinum (60,000,000,000 spores suspended in buffer at pH 7)	A thermophile (150,000 spores per ml. of corn juice at pH 6.1)
	Minutes	
100°C	360	1140
105°C	120	
110°C	36	180
115°C	12	60
120°C	5	17

*Courtesy of W. B. Saunders Co., Philadelphia.

cold, chemicals, and other adverse environmental factors. It is a highly refractive body that resists staining by ordinary methods. The refractivity is due in part to the spore coats which consist of at least two layers—outer (exine) and inner (intine). Heat resistance is due also in part to the dehydrated nature of the cortex and spore core. Endospores are known to contain DNA, RNA, water, various enzymes, metal ions, and other compounds, especially dipicolinic acid (DPA). Although the precise mechanism of heat resistance of endospores is not yet fully understood, numerous authors have related this resistance to spore DPA and calcium contents. DPA may constitute from 5–15% of the dry weight of endospores, while these bodies may contain from 2–10 times more calcium than the corresponding vegetative cell. There is a general increase in heat resistance as the ratio of cations to DPA increases *(14, 16)*. The addition of chelating agents with high affinities towards calcium and manganese has been shown to decrease the heat resistance of endospores *(1)*, while the addition of calcium and manganese generally restores thermal resistance. The thermal death of endospores is accompanied by a release of DPA, divalent cations, and ninhydrin-positive material into the medium *(8)*.

THE THERMAL DESTRUCTION OF MICROORGANISMS

In order to better understand the thermal destruction of microorganisms relative to food preservation and canning, it is necessary to understand certain basic concepts associated with this technology. Below are listed some of the more important concepts, but for a more extensive treatment of thermobacteriology, the excellent monograph by Stumbo *(22)* should be consulted.

1. Thermal Death Time (TDT).   This is the time necessary to kill a given number of organisms at a specified temperature. By this method, the temperature is kept constant and the time necessary to kill all cells is determined. Of less importance is the **thermal death point,** which is the temperature necessary to kill a given number of microorganisms in a fixed time, usually 10 min. The following 7 methods have all been proposed for determining TDT: tube, can, "tank," flask, thermo-resistometer, unsealed tube, and capillary tube methods. The general procedure for determining TDT by either of these methods is to place a known number of cells or spores in a sufficient number of sealed containers in order to get the desired number of survivors for each test period. The organisms are then placed in an oil bath and heated for the required time period. At the end of the heating period, containers are removed and cooled quickly in cold water. The organisms are then placed on a suitable growth medium, or the entire heated containers are incubated if the organisms are suspended in a suitable growth substrate. The suspensions or containers are incubated at a temperature suitable for growth of the specific organisms. Death is defined as the inability of the organisms to form a visible colony.

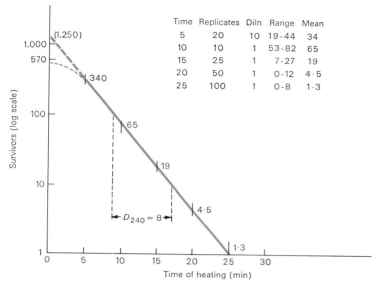

Time	Replicates	Diln	Range	Mean
5	20	10	19-44	34
10	10	1	53-82	65
15	25	1	7-27	19
20	50	1	0-12	4·5
25	100	1	0-8	1·3

FIGURE 12-2.   Rate of destruction curve. Spores of strain F.S.7 heated at 240°F in canned pea brine pH 6.2 (*9*, courtesy of Butterworth Publishers, London).

2. D Value.   This is the decimal reduction time, or the time required to destroy 90% of the organisms. This value is numerically equal to the number of min required for the survivor curve to traverse one log cycle (see Figure 12-2). Mathematically, it is equal to the reciprocal of the slope of the survivor curve and is a measure of the death rate of an organism. When D is determined at 250 F, it is often expressed as D_r. The effect of pH on the D value of *C. botulinum* in various foods is presented in Table 12-5, and D

TABLE 12-5.   Effect of pH on D values for spores of *C. botulinum* 62A suspended in three food products at 240°F (*25*)*

| | D Value (in Min.) | | |
pH	Spaghetti, tomato sauce, and cheese	Macaroni creole	Spanish rice
4.0	0.128	0.127	0.117
4.2	0.143	0.148	0.124
4.4	0.163	0.170	0.149
4.6	0.223	0.223	0.210
4.8	0.226	0.261	0.256
5.0	0.260	0.306	0.266
6.0	0.491	0.535	0.469
7.0	0.515	0.568	0.550

*1965 Copyright © by Institute of Food Technologists.

values for *S. senftenberg* 775W under various conditions are presented in Table 12-2. D values of 0.20 to 2.20 min at 150 F have been reported for *S. aureus* strains and D_{150} of 0.50 to 0.60 min for *Coxiella burnetti*.

3. z Value.   The z value refers to the degrees Fahrenheit required for the thermal destruction curve to traverse one log cycle. Mathematically, this value is equal to the reciprocal of the slope of the TDT curve (see Fig. 12-3). While D reflects the resistance of an organism to a specific temperature, z provides information on the relative resistance of an organism to different destructive temperatures; it allows for the calculation of equivalent thermal processes at different temperatures. If, for example, 3.5 min at 140 F is considered to be an adequate process and z = 8.0, either 0.35 min at 148 F or 35 min at 132 F would be considered equivalent processes.

4. F Value.   This value is the equivalent time, in minutes at 250 F, of all heat considered, with respect to its capacity to destroy spores or vegetative cells of a particular organism. The integrated lethal value of heat received by all points in a container during processing is designated F_s or F_0. This represents a measure of the capacity of a heat process to reduce the number of spores or vegetative cells of a given organism per container. When we assume instant heating and cooling throughout the container of spores, vegetative cells, or food, F_0 may be derived as follows:

$$F_o = D_r(\log a - \log b)$$

where a = number of cells in the initial population, and b = the number of cells in the final population.

5. Thermal Death Time Curve.   For the purpose of illustrating a thermal destruction curve and D value, data are employed from Gillespy *(9)* on the killing of flat sour spores at 240 F in canned pea brine at pH 6.2. Counts were determined at intervals of 5 min with the mean viable numbers indicated below:

Time (Min.)	Mean viable count
5	340
10	65
15	19
20	4.5
25	1.3

The time of heating in minutes is plotted on semi-log paper along the linear axis, and the number of survivors is plotted along the log scale to produce the TDT curve presented in Figure 12-2. The curve is essentially linear, indicating that the destruction of bacteria by heat is logarithmic and obeys a first order reaction. Although difficulty is encountered at times at either end of

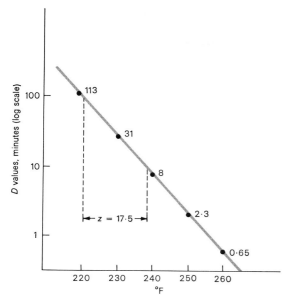

FIGURE 12-3.   Thermal death time curve. Spores of strain F.S.7 heated in canned pea brine pH 6.2 (*9*, courtesy of Butterworths Publishers, London).

the TDT curve, process calculations in the canning industry are based upon a logarithmic order of death. From the data presented in Figure 12-2, the D value is calculated to be 8 min, or $D_{240} = 8.0$.

D Values may be used to reflect the relative resistance of spores or vegetative cells to heat. The most heat-resistant strains of *C. botulinum* types A and B spores have a D_r value of 0.21, while the most heat-resistant thermophilic spores have D_r values of around 4.0–5.0. Putrefactive anaerobe (P.A.) 3679 was found by Stumbo *et al.* (*21*) to have a D_r value of 2.47 in cream-style corn, while flat sour (F.S.) spores 617 were found to have a D_r of 0.84 in whole milk.

The approximate heat resistance of spores of thermophilic and mesophilic spoilage organisms may be compared by use of D_r values as below.

B. *stearothermophilus* .... 4.0 –5.0
C. *thermosaccharolyticum* .... 3.0 –4.0
C. *nigrificans* .... 2.0 –3.0
C. *botulinum* (types A & B) 0.10–0.2
C. *sporogenes* (including P.A. 3679) .. 0.10–1.5
B. *coagulans* 0.01–0.07

The effect of pH and suspending menstrum on D values of *C. botulinum* spores are presented in Table 12-5. As noted above, microorganisms are more resistant at and around neutrality and show different degrees of heat resistance in different foods.

In order to determine the z value, D values are plotted on the log scale and degrees F are plotted along the linear axis. From the data presented in Figure 11-3, the z value is seen to be 17.5. Values of z for *C. botulinum* range from 14.7 to 16.3, while for P.A. 3679 the range of 16.6–20.5 has been reported. Some spores have been reported to have z values as high as 22. Peroxidase has been reported to have a z value of 47 while 50 has been reported for riboflavin and 56 for thiamine.

6. The 12-D Concept.   The 12-D concept refers to the process lethality requirement long in effect in the canning industry and implies that the minimum heat process should reduce the probability of survival of the most resistant *C. botulinum* spores to 10^{-12}. Since *C. botulinum* spores do not germinate and produce toxin below pH 4.6, this concept is observed only for foods above this pH value. An example from Stumbo *(22)* illustrates this concept from the standpoint of canning technology. If it is assumed that each container of food contains only 1 spore of *C. botulinum*, F_o may be calculated by use of the general survivor curve equation with the other assumptions noted above in mind:

$$F_o = D_r (\log a - \log b)$$
$$F_o = 0.21 (\log 1 - \log 10^{-12})$$
$$F_o = 0.21 \times 12 = 2.52$$

Processing for 2.52 min at 250 F, then, should reduce the *C. botulinum* spores to one spore in one of one million million containers (10^{12}). When it is considered that some flat-sour spores have D_r values of about 4.0 and some canned foods receive F_o treatments of 6.0–8.0, the potential number of *C. botulinum* spores is reduced even more.

REFERENCES

1.   **Amaha, M. and Z. J. Ordal.** 1957. Effect of divalent cations in the sporulation medium on the thermal death rate of *Bacillus coagulans* var. *thermoacidurans.* J. Bacteriol. *74:* 596–604.

2.   **Anellis, A., J. Lubas, and M. M. Rayman.** 1954. Heat resistance in liquid eggs of some strains of the genus *Salmonella.* Food Res. *19:* 377–395.

3.   **Baird-Parker, A. C., M. Boothroyd, and E. Jones.** 1970. The effect of water activity on the heat resistance of heat sensitive and heat resistant strains of salmonellae. J. Appl. Bacteriol. *33:* 515–522.

4.   **Carpenter, P. L.** 1967. *Microbiology,* 2nd edition. W. B. Saunders & Co., Philadelphia.

5.   **Corry, J. E. L.** 1974. The effect of sugars and polyols on the heat resistance of salmonellae. J. Appl. Bacteriol. *37:* 31–43.

6. **Cunningham, F. E. and H. Lineweaver.** 1965. Stabilization of egg-white proteins to pasteurizing temperatures above 60°C. Food Technol. *19:* 1442–1447.

7. **El-Bisi, H. M. and Z. J. Ordal.** 1956. The effect of certain sporulation conditions on the thermal death rate of *Bacillus coagulans* var. *thermoacidurans.* J. Bacteriol. *71:* 1–9.

8. **El-Bisi, H. M., R. V. Lechowich, M. Amaha, and Z. J. Ordal.** 1962. Chemical events during death of bacterial endospores by moist heat. J. Food Sci. *27:* 219–231.

9. **Gillespy, T. G.** 1962. The principles of heat sterilization. Recent Advances in Food Science *2:* 93–105.

10. **Goepfert, J. M. and R. A. Biggie.** 1968. Heat resistance of *Salmonella typhimurium* and *Salmonella senftenberg* 775W in milk chocolate. Appl. Microbiol. *16:* 1939–1940.

11. **Goepfert, J. M., I. K. Iskander, and C. H. Amundson.** 1970. Relation of the heat resistance of salmonellae to the water activity of the environment. Appl. Microbiol. *19:* 429–433.

12. **Hansen, N. H. and H. Riemann.** 1963. Factors affecting the heat resistance of nonsporing organisms. J. Appl. Bacteriol. *26:* 314–333.

13. **Horner, K. J. and G. D. Anagnostopoulos.** 1975. Effect of water activity on heat survival of *Staphylococcus aureus, Salmonella typhimurium* and *Salm. senftenberg.* J. Appl. Bacteriol. *38:* 9–17.

14. **Lechowich, R. V. and Z. J. Ordal.** 1960. The influence of sporulation temperature on the thermal resistance and chemical composition of endospores. Bacteriol. Proc., p. 44.

15. **Lechowich, R. V. and Z. J. Ordal.** 1962. The influence of the sporulation temperature on the heat resistance and chemical composition of bacterial spores. Can. J. Microbiol. *8:* 287–295.

16. **Levinson, H. S., M. T. Hyatt, and F. E. Moore.** 1961. Dependence of the heat resistance of bacterial spores on the calcium: dipicolinic acid ratio. Biochem. Biophys. Res. Commun. *5:* 417–419.

17. **Levinson, H. S. and M. T. Hyatt.** 1964. Effect of sporulation medium on heat resistance, chemical composition, and germination of *Bacillus megaterium* spores. J. Bacteriol. *87:* 876–886.

18. **Ng, H., H. G. Bayne, and J. A. Garibaldi.** 1969. Heat resistance of *Salmonella:* The uniqueness of *Salmonella senftenberg* 775W. Appl. Microbiol. *17:* 78–82.

19. **Read, R. B., Jr., J. G. Bradshaw, R. W. Dickerson, Jr., and J. T. Peeler.** 1968. Thermal resistance of salmonellae isolated from dry milk. Appl. Microbiol. *16:* 998–1001.

20. **Solowey, M., R. R. Sutton, and E. J. Calesnick.** 1948. Heat resistance of *Salmonella* organisms isolated from spray-dried whole-egg powder. Food Technol. *2:* 9–14.

21. **Stumbo, C. R., J. R. Murphy, and J. Cochran.** 1950. Nature of thermal death time curves for P.A. 3679 and *Clostridium botulinum.* Food Technol. *4:* 321–326.

22. **Stumbo, C. R.** 1973. *Thermobacteriology in Food Processing,* 2nd Edition. Academic Press, N.Y.

23. **Sugiyama, H.** 1951. Studies on factors affecting the heat resistance of spores of *Clostridium botulinum*. J. Bacteriol. *62:* 81–96.

24. **White, H. R.** 1963. The effect of variation in pH on the heat resistance of cultures of *Streptococcus faecalis*. J. Appl. Bacteriol. *26:* 91–99.

25. **Xezones, H. and I. J. Hutchings.** 1965. Thermal resistance of *Clostridium botulinum* (62A) spores as affected by fundamental food constituents. Food Technol. *19:* 1003–1005.

13

Preservation of Foods By Drying

The preservation of foods by drying is based upon the fact that microorganisms and enzymes need water in order to be active. In preserving foods by this method, one seeks to lower the moisture content of foods to a point where the activities of food spoilage and food poisoning microorganisms are inhibited. Dried, desiccated, or low moisture (LM) foods are those that generally do not contain more than 25% moisture and have an a_w between 0.00 and 0.60. These are the traditional dried foods. Freeze-dried foods are also in this category. Another category of shelf stable foods are those that contain between 15 and 50% moisture and an a_w between 0.60 and 0.85. These are the intermediate moisture (IM) foods. Some of the microbiological aspects of IM and LM foods are dealt with in this chapter.

THE PREPARATION AND DRYING OF LOW MOISTURE FOODS

The earliest uses of food desiccation consisted simply of exposing fresh foods to sunlight until drying had been achieved. This type of drying is referred to as sun drying and certain foods may be successfully preserved by this method if the temperature and relative humidity (R.H.) allow. Fruits such as raisins, prunes, figs, apricots, and others may be dried by this method, and require a large amount of space for large quantities of the product. The drying methods of greatest commercial importance consist of spray, drum, evaporation, and freeze-drying.

Preparatory to drying, foods are handled in much the same manner as for freezing, with a few exceptions. In the drying of fruits such as prunes, alkali dipping is employed by immersing the fruits into hot lye solutions of between 0.1–1.5%. This is especially true when sun drying is employed. Light-colored fruits and certain vegetables are treated with SO_2 in order that levels of between 1,000–3,000 ppm may be absorbed. The latter treatment helps to maintain color, conserve certain vitamins, prevent storage changes, and reduce the microbial load. After drying, fruits are usually heat pasteurized at 150–185 F for 30–70 min.

Similar to the freezing preparation of vegetable foods, blanching or scalding is a vital step prior to dehydration. This may be achieved by immersion from 1–8 min, depending upon the particular type of product. The primary function of this step in drying is to destroy enzymes that may become active and bring about undesirable changes in the finished products. Leafy vegetables generally require less time than peas, beans, or carrots. For drying, temperatures of 140–145 F have been found to be safe for many vegetables. The moisture content of vegetables should be reduced below 4% in order to have satisfactory storage life and quality. Many vegetables may be made more stable if given a treatment with SO_2 or a sulfite. The drying of vegetables is usually achieved by use of tunnel, belt, or cabinet-type driers.

Meat is usually cooked before being dehydrated. The final moisture content after drying should be approximately 4% for beef and pork.

Milk is dried as either whole milk or nonfat skim milk. The dehydration may be accomplished by either the drum or spray methods. The removal of about 60% water from whole milk results in the production of **evaporated** milk, which has about 11.5% lactose in solution. **Sweetened condensed** milk is produced by the addition of sucrose or glucose before evaporation so that the total average content of all sugar is about 54%, or over 64% in solution. The stability of sweetened condensed milk is due in part to the fact that the sugars tie up some of the water and make it unavailable for microbial growth.

Eggs may be dried as whole egg powder, yolks, or egg white. Dehydration stability is increased by reducing the glucose content prior to drying. Spray drying is the method most commonly employed.

In **freeze-drying** (lyophilization, cryophilization), actual freezing is preceded by the blanching of vegetables and the precooking of meats. The rate at which a food material freezes or thaws is influenced by the following factors *(11):* (1) The temperature differential between the product and the cooling or heating medium, (2) the means of transferring heat energy to, from, and within the product (conduction, convection, radiation), (3) the type, size, and shape of the package, and (4) the size, shape, and thermal properties of the product. Rapid freezing has been shown to produce products that are more acceptable than slow freezing. As discussed in Chapter 10, rapid freezing allows for the formation of small ice crystals and consequently less mechanical damage to food structure. Upon thawing, fast-frozen foods take up more water and in general display characteristics more like the fresh product than slow-frozen foods. After freezing, the water in the form of ice is removed by sublimation. This process is achieved by various means of heating plus vacuum. The water content of protein foods can be placed into two groups: freezable and unfreezable. Unfreezable (bound) water has been defined as that which remains unfrozen below

−30 C. The removal of freezable water takes place during the first phases of drying and this phase of drying may account for the removal of anywhere from 40–95% of the total moisture. The last water to be removed is generally bound water, some of which may be removed throughout the drying process. Unless heat treatment is given prior to freeze-drying, freeze-dried foods retain their enzymes. In studies on freeze-dried meats, it has been shown that 40–80% of the enzyme activity is not destroyed and may be retained after 16 mon storage at −20 C *(24)*.

Freeze-drying is generally preferred to high-temperature vacuum drying. Among the disadvantages of the latter compared to the former are the following *(17)*.

1. Pronounced shrinkage of solids.
2. Migration of dissolved constituents to the surface when drying solids.
3. Extensive denaturation of proteins.
4. Case-hardening. The formation of a relatively hard, impervious layer at the surface of a solid is caused by one or more of the first three changes. This impervious layer slows rates of both dehydration and reconstitution.
5. Formation of hard, impervious solids when drying liquid solution.
6. Undesirable chemical reactions in heat-sensitive materials.
7. Excessive loss of desirable volatile constituents.
8. Difficulty of rehydration as a result of one or more of the above changes.

EFFECT OF DRYING UPON MICROORGANISMS

Although some microorganisms are destroyed in the process of drying, this process is not *per se* lethal to microorganisms, and indeed, many types may be recovered from dried foods, especially if poor quality foods are used for drying and if proper practices are not followed in the drying steps.

As indicated in Chapter 3, bacteria require relatively high levels of moisture for their growth, with yeasts requiring less and molds still less. Since most bacteria require a_w values above 0.90 for growth, these organisms play no role in the spoilage of dried foods. With respect to the stability of dried foods, Scott *(32)* has related a_w levels to the probability of spoilage in the following manner. At a_w values of between 0.80 and 0.85, spoilage occurs readily by a variety of fungi in from 1–2 weeks. At a_w values of 0.75, spoilage is delayed, with fewer types of organisms in those products that spoil. At a_w 0.70, spoilage is greatly delayed and may not occur during prolonged holding. At a_w of 0.65, very few organisms are known to grow, and spoilage is most unlikely to occur for even up to 2 yr. Some authors have suggested that dried foods to be held for several yr should be processed so that the final a_w should be between 0.65–0.75, with 0.70 being suggested by most.

At a_w levels of about 0.90, the organisms most likely to grow are yeasts and molds, with this value being near the minimum for most normal yeasts.

Osmophilic yeasts such as *Saccharomyces rouxii* strains have been reported to grow at an a_w of 0.65 under certain conditions. The most troublesome group of microorganisms in dried foods are the molds, with the *Aspergillus glaucus* group being the most notorious at low a_w values. The minimum a_w values reported for the germination and growth of molds and yeasts are presented in Table 13-1. Pitt and Christian *(28)* found the predominant spoilage molds of dried and high-moisture prunes to be members of the *A. glaucus* group and *Xeromyces bisporus*. Aleuriospores of *X. bisporus* were able to germinate in 120 days at an a_w of 0.605. Generally higher moisture levels were required for both asexual and sexual sporulation.

As a guide to the storage stability of dried foods, the **"alarm water"** content has been suggested. The "alarm water" content is the water content which should not be exceeded if mold growth is to be avoided. While these values may be used to advantage, they should be followed with caution, as a rise of only 1% may be disastrous in some instances *(32)*. The "alarm water" content for some miscellaneous foods is presented in Table 13-2. In freeze-dried foods, the rule of thumb has been to reduce the moisture level to 2%. Burke and Decareau *(7)* have pointed out that this low level is probably too severe for some foods that might keep well at higher levels of moisture without the extra expense of removing the last low levels of water.

Although drying destroys some microorganisms, bacterial endospores survive as do yeasts, molds, and many gram negative and positive bacteria. In their study of bacteria from chicken meat after freeze-drying and rehydration at room temperature, May and Kelly *(25)* were able to recover about 32% of the original flora. These workers showed that *S. aureus* added prior to freeze-drying could survive under certain conditions. Some or all food-

TABLE 13-1.   Minimum a_w reported for the germination and growth of various food spoilage yeasts and molds.

Organisms	Minimum a_w
Candida utilis	0.94
Botrytis cinerea	0.93
Rhizopus stolonifer (nigricans)	0.93
Mucor spinosus	0.93
Candida scottii	0.92
Trichosporon pullulans	0.91
Candida zeylanoides	0.90
Saccharomyces rouxii	0.90
Endomyces vernalis	0.89
Alternaria citri	0.84
Aspergillus glaucus	0.70
Aspergillus echinulatus	0.64
Saccharomyces rouxii (Z. barkeri)	0.62

TABLE 13-2. The "alarm water" content for miscellaneous foods, assuming R.H. of 70 percent and a temperature of 20 C *(27)*.

Foods	Percent water
Whole milk powder	ca. 8
Dehydrated whole eggs	10–11
Wheat flour	13–15
Rice	13–15
Milk powder (separated)	15
Fat-free dehydrated meat	15
Pulses	15
Dehydrated vegetables	14–20
Starch	18
Dehydrated fruit	18–25

borne parasites, such as *T. spiralis,* have been reported to survive the drying process *(10)*. The present goal is to produce dried foods with a total count of not more than 100,000/g. It can be seen from Table 5-1 (pp. 79–81) that the total counts of dehydrated soups and vegetables are often considerably above this level. It is generally agreed that the coliform count of dried foods should be zero or nearly so, and no food poisoning organisms should be allowed, with the possible exception of low numbers of *C. perfringens* and *S. faecalis.* With the exception of those that may be destroyed by blanching or precooking, relatively fewer organisms are destroyed during the freeze-drying process. More are destroyed during freezing than during dehydration. During freezing, between 5–10% of water remains "bound" to other constituents of the medium. This water is removed by drying. Death or injury from drying may result from: (1) denaturation in the still frozen, undried portions due to concentration resulting from freezing, (2) the act of removing the "bound" water, and/or (3) recrystallization of salts or hydrates formed from eutectic solutions *(26)*. When death occurs during dehydration, the rate is highest during the early stages of drying. Young cultures have been reported to be more sensitive to drying than old cultures *(12)*. The freeze-drying method is, of course, one of the best known ways of preserving microorganisms. Once the process has been completed, the cells may remain viable indefinitely. Upon examining the viability of 277 cultures of bacteria, yeasts, and molds that had been lyophilized for 21 yr, Davis *(9)* found that only 3 failed to survive.

STORAGE STABILITY OF DRIED FOODS

In the absence of fungal growth, desiccated foods are subject to certain chemical changes which may result in the food becoming undesirable upon holding. In dried foods that contain fats and oxygen, oxidative rancidity is a common form of chemical spoilage. Foods that contain reducing sugars undergo a color change known as the **Maillard** reaction or nonenzymic browning. This process is brought about when the carbonyl groups of reducing sugars react with amino groups of proteins and amino acids, followed by a series of other more complicated reactions. Maillard-type browning is quite undesirable in fruits and vegetables not only because of the unnatural color but also because of the bitter taste imparted to susceptible foods. Freeze-dried foods also undergo browning if the moisture content is above 2%. Thus, the moisture content should be held below 2% *(13)*.

Other chemical changes that take place in dried foods include a loss of vitamin C in vegetables, general discolorations, structural changes leading to the inability of the dried product to fully rehydrate, and toughness in the rehydrated, cooked product.

Conditions that favor one or more of the above changes in dried foods generally tend to favor all, so that preventative measures against one are also effective against others to varying degrees. At least 4 methods of minimizing chemical changes in dried foods have been offered: (1) Keep the moisture content as low as possible. Gooding *(14)* has pointed out that lowering the moisture content of cabbage from 5 to 3 percent doubles its storage life at 37 C. (2) Reduce the level of reducing sugars as low as possible. These compounds, of course, are directly involved in nonenzymic browning and their reduction has been shown to increase storage stability. (3) When blanching, use water in which the level of leached soluble solids is kept low. Gooding *(14)* has shown that the serial blanching of vegetables in the same water increases the chances of browning. The explanation given is that the various extracted solutes (presumably reducing sugars and amino acids) are impregnated on the surface of the treated products at relatively high levels. (4) Use of sulfur dioxide. The treatment of vegetables prior to dehydration with this gas protects vitamin C along with retarding the browning reaction. The precise mechanism of this gas in retarding the browning reaction is not well understood, but it apparently does not block reducing groups of hexoses. It has been suggested that it may act as a free radical acceptor.

One of the most important considerations in preventing fungal spoilage of dried foods is the R.H. of the storage environment. If improperly packed and stored under conditions of high R.H., dried foods will pick up moisture from the atmosphere until some degree of equilibrium has been established. Since the first part of the dried product to gain moisture is the surface, spoilage would be inevitable, as surface growth tends to be characteristic of molds due to their oxygen requirements.

INTERMEDIATE MOISTURE FOODS (IMF)

As noted earlier in this chapter, intermediate moisture foods (IMF) are characterized by a moisture content around 15 to 50% and an a_w between 0.60 and 0.85. These foods are shelf stable at ambient temperatures for varying periods of time. While impetus was given to this class of foods during the early 1960s with the development and marketing of intermediate moisture dog food, foods for human consumption that meet the basic criteria of this class have been produced for many years. These are referred to as traditional IMF's to distinguish them from the newer IMF's. In Table 13-3 are listed some traditional IMF's along with their a_w values. All of these foods have, of course, lowered a_w values which are achieved by withdrawal of water by desorption, adsorption, and/or the addition of permissible additives such as salts and sugars. The newly developed IMF's are characterized not only by a_w values of 0.60–0.85 but by the use of additives such as glycerol, glycols, sorbitol, sucrose, and so forth, as humectants, and by their content of fungistats such as sorbate and benzoate. The remainder of this chapter is devoted to the newly developed IMF's.

The Preparation of IMF

Since *S. aureus* is the only bacterium of public health importance that can grow at a_w values near 0.86, an IMF can be prepared by: (1) formulating the product so that its moisture content is between 15–50 percent, (2) adjusting the a_w to a value below 0.86 by use of humectants, and (3) adding an antifungal agent to inhibit the rather large number of yeasts and molds that are known to be capable of growth at a_w values above 0.70. Additional storage stability is achieved by reducing pH. While this is essentially all that one needs to produce an IMF, the actual process and the storage stability of the product is considerably more complicated.

TABLE 13-3.   Some traditional intermediate moisture foods (a_w values taken from the literature).

Food Products	a_w Range
Dried fruits	0.60–0.75
Cake and pastry	0.60–0.90
Frozen foods	0.60–0.90
Sugars, syrups	0.60–0.75
Some candies	0.60–0.65
Commercial pastry fillings	0.65–0.71
Cereals (some)	0.65–0.75
Fruit cake	0.73–0.83
Honey	0.75
Fruit juice concentrates	0.79–0.84
Jams	0.80–0.91
Sweetened condensed milk	0.83
Fermented sausages (some)	0.83–0.87
Maple syrup	0.90
Ripened cheeses (some)	0.96
Liverwurst	0.96

The determination of the a_w of a food system is discussed in Chapter 3. One can use also Raoult's law of mole fractions where the number of moles of water in a solution is divided by the total number of moles in the solution (4):

$$a_w = \frac{\text{Moles of } H_2O}{\text{Moles of } H_2O + \text{Moles of solute}}$$

For example, a liter of water contains 55.5 moles. Assuming that the water is pure,

$$a_w = \frac{55.5}{55.5 + 0} = 1.00$$

If, however, one mole of sucrose is added,

$$a_w = \frac{55.5}{55.5 + 1} = 0.98$$

This equation can be rearranged to solve for the number of moles of solute required to give a specified a_w value. While the foregoing is not incorrect, it is highly oversimplified, since food systems are complex by virtue of their content of ingredients that interact with water and with each other in ways that are difficult to predict. Sucrose, for example, decreases a_w more than expected by the above so that calculations based upon Raoult's law may be meaningless (3). The development of techniques and methods to more accurately predict a_w in IMF has been the concern of several investigators (4, 16, 31), while an extensive evaluation of available a_w-measuring instruments and techniques has been carried out by Labuza et al. (23).

In preparing IMF, water may be removed either by adsorption or desorption. By adsorption, food is first dried (often freeze dried), and then subjected to controlled re-humidification until the desired composition is achieved. By desorption, the food is placed in a solution of higher osmotic pressure so that at equilibrium, the desired a_w is reached (30). While identical a_w values may be achieved by these two methods, IMF produced by adsorption is more inhibitory to microorganisms than that produced by desorption (see below). When sorption isotherms of food materials are determined, adsorption isotherms sometimes reveal that less water is held than for desorption isotherms at the same a_w. The sorption isotherm of a food material is a plot of the amount of water adsorbed as a function of the relative humidity or activity of the vapor space surrounding the material. It is the amount of water that is held after equilibrium has been reached at a constant temperature (21). Sorption isotherms may be either adsorption or desorption, and when the former procedure results in the holding of more water than the latter, the difference is ascribed to an hysteresis effect. This as well as other physical properties associated with the preparation of IMF has been discussed by Labuza *(21)*, Sloan *et al. (34)*, and others, and will not be dealt with further here. The sorption properties of an IMF recipe, the interaction of each ingredient with water, with each other, and the order of mixing of ingredients all add to the complications of the overall IMF preparation procedures and both direct and indirect effects on the microbiology of these products may result.

The general techniques employed to change the water activity in producing an IMF are summarized below *(20)*:

1. Moist-infusion. Solid food pieces are soaked and/or cooked in an appropriate solution to result in a final product having the desired water level (desorption).
2. Dry-infusion. Solid food pieces are first dehydrated, following which they are infused by soaking in a solution containing the desired osmotic agents (adsorption).
3. Component blending. All IMF components are weighed, blended, cooked, and extruded or otherwise combined to result in a finished product of desired a_w.
4. Osmotic drying. Foods are dehydrated by immersion in liquids with a water activity lower than that of the food. When salts and sugars are used, two simultaneous counter-current flows develop: solute diffuses from solution into food while water diffuses out of food into solution.

The foods in Table 13-4 were prepared by moist-infusion for military use. The 1-cm-thick slices equilibrated following cooking at 95–100 C in water and holding overnight in a refrigerator. Equilibration is possible without cooking over prolonged periods under refrigeration *(6)*. IMF deep-fried catfish, with raw samples *ca.* 2 g each, has been prepared by the moist infusion method *(8)*. Pet foods are more often prepared by component blending. The general composition of one such product is given in Table 13-5. The general way in which a product of this type is made is as follows. The meat and meat products are ground and mixed with liquid ingredients. The resulting slurry is cooked or heat treated and later mixed with the dry ingredient mix (salts, sugars, dry solids, etc.). Once the latter are mixed into the slurry, an additional cook or heat process may be applied prior to extrusion and packaging. The extruded material may be further shaped in the form of patties or packaged in loose form. The composition of a model IMF product called "Hennican" is given in Table 13-6. According to Acott and Labuza *(1)*, this is an adaptation of Pemmican, an Indian trail and winter storage food made of buffalo meat and berries. Hennican is the name given to the chicken-based IMF. Both moisture content and a_w of this system can be altered by adjustment of ingredient mix.

The humectants commonly used in pet food formulations are propylene glycol, polyhydric alcohols (sorbitol, e.g.), polyethylene glycols, glycerol,

TABLE 13-4. Preparation of representative intermediate moisture foods by equilibration (6). The solutions in which the products were equilibrated are given in the six columns on the right side of the table.

Initial material	% H₂O	Processing	Equilibrated product H₂O%	a_w	Ratio: Solution Wt./Initial Wt.	% Components of solution Glycerol	Water	NaCl	Sucrose	K sorbate	Na benzoate
Tuna, canned water pack pieces 1 cm thick	60.0	cold soak	38.8	0.81	0.59	53.6	38.6	7.1	—	0.7	—
Carrots diced 0.9 cm cooked	88.2	95–98 C cook refrig	51.5	0.81	0.48	59.2	34.7	5.5	—	0.6	—
Macaroni, elbow cook, drain	63.0	95–98 C cook refrig	46.1	0.83	0.43	42.7	48.8	8.0	—	0.5	—
Pork loin, raw 1 cm thick	70.0	95–98 C cook refrig	42.5	0.81	0.73	45.6	43.2	10.5	—	0.7	—
Pineapple canned, chunks	73.0	cold soak	43.0	0.85	0.46	55.0	21.5	—	23.0	0.5	—
Celery 0.6 cm cross cut blanch	94.7	cold soak	39.6	0.83	0.52	68.4	25.2	5.9	—	0.5	—
Beef, ribeye 1 cm thick	70.8	95–98 C cook refrig		0.86	2.35	87.9	—	10.1	—	—	2.0

Copyright © 1970, Institute of Food Technologists.

TABLE 13-5. Typical composition of soft moist or intermediate moisture dog food (19).

Ingredient	Percent
Meat By-products	32.0
Soy Flakes	33.0
Sugar	22.0
Skimmed Milk, Dry	2.5
Calcium and Phosphorus	3.3
Propylene Glycol	2.0
Sorbitol	2.0
Animal Fat	1.0
Emulsifier	1.0
Salt	0.6
Potassium Sorbate	0.3
Minerals, Vitamins & Color	0.3
	100.0%

Copyright © 1970, Institute of Food Technologists.

sugars (sucrose, fructose, lactose, glucose, and corn syrup), and salts (NaCl, KCl, etc.). The commonly used mycostats are propylene glycol, K-sorbate, Na-benzoate, and others. The pH of these products may be as low as 5.4 and as high as 7.0.

Microbial Aspects of IMF

The general a_w range of IMF products makes it unlikely that gram negative bacteria will proliferate. This is true also for most gram positive bacteria with the exception of cocci, some sporeformers, and lactobacilli. In addition to the inhibitory effect of lowered a_w, antimicrobial activity results from an interaction of pH, Eh, added preservatives (including some of the humectants), the competative microflora, generally low storage tempera-

TABLE 13-6. Composition of IM food: Hennican* (1).

Components	Amount (wt basis)
Raisins	30%
Water	23%
Peanuts	15%
Chicken (freeze dried)	15%
Non-fat dry milk	11%
Peanut butter	4%
Honey	2%

*Moisture content = 41 g water/100 g solids, a_w = 0.85.
Copyright © 1975, Institute of Food Technologists.

tures, and the pasteurization or other heat processes applied during processing.

The fate of *S. aureus* S-6 in IM pork cubes with glycerol at 25 C is illustrated in Figure 13-1. In this desorption IM pork at a_w 0.88, the numbers remained stationary for about 15 days and then increased slightly, while in the adsorption IM system at the same a_w the cells died off slowly during the first three weeks and thereafter more rapidly. At all a_w values below 0.88, the organisms died off with the death rate considerably higher at 0.73 than at higher values *(29)*. Findings similar to these have been reported by Haas *et al. (15)* who found that an inoculum of 10^5 staphylococci in a meat-sugar system at a_w 0.80 decreased to 3×10^3 after 6 days and to 3×10^2 after one mon. Although growth of *S. aureus* has been reported to occur at an a_w of 0.83, enterotoxin is not produced below a_w 0.86 *(35)*. It appears that enterotoxin A is produced at lower values of a_w than enterotoxin B *(36)*.

Using the model IM Hennican at pH 5.6 and a_w 0.91, Boylan *et al. (5)* showed that the effectiveness of the IM system against *S. aureus* F265 was a function of both pH and a_w. As noted above, adsorption systems are more destructive to microorganisms than desorption systems. Labuza *et al. (22)* found that the reported minimum a_w's apply in IMF systems when desorption systems are involved but that growth minima are much higher if the food is prepared by an adsorption method. *S. aureus* was inhibited at a_w 0.9 in adsorption while values between 0.75–0.84 were required for desorption systems. A similar effect was noted for molds, yeasts, and pseudomonads.

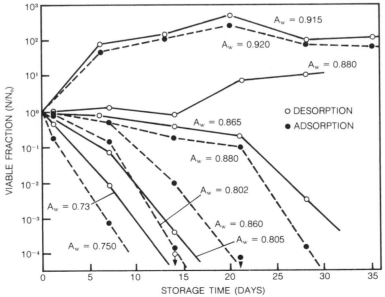

FIGURE 13-1.   Viability of *Staphylococcus aureus* in IMF systems: pork cubes and glycerol at 25°C *(29)*. Copyright © 1973, Institute of Food Technologists.

In regard to the effect of IMF systems on the heat destruction of bacteria, it was noted in Chapter 12 that heat resistance increases as a_w is lowered and that the degree of resistance is dependent upon the compounds employed to control a_w (see Table 12-2). In a study of the death rate of salmonellae and staphylococci in the IM range of *ca.* 0.8 at pasteurization temperatures (50–65 C), it has been found that cell death occurs under first order kinetics *(18)*. These investigators confirmed the findings of many others that the heat destruction of vegetative cells is at a minimum in the IM range, especially when a solid menstrum is employed. Some D values for the thermal destruction of *S. senftenberg* 775W at various a_w values are given in Table 12-2.

With respect to molds in IMF systems, these products would be made quite stable if a_w were reduced to around 0.70, but a dry-type product would then result. A large number of molds are capable of growth in the 0.80 range and the shelf life of IM pet foods is generally limited by the growth of these organisms. The interaction of various IM parameters on the inhibition of molds was shown by Acott *et al. (2)*. In their evaluation of 7 chemical inhibitors used alone and in combination to inhibit *Aspergillus niger* and *A. glaucus* inocula, propylene glycol was the only approved agent that was effective alone. None of the agents tested could inhibit alone at a_w 0.88, but in combination the product was made shelf stable. All inhibitors were found to be more effective at pH 5.4 and a_w 0.85 than at pH 6.3. Growth of the two fungi occurred in 2 weeks in the a_w 0.85 formulation without inhibitors but did not occur until 25 weeks when K-sorbate and Ca-propionate were added (Table 13-7). Growth of *S. epidermidis* was inhibited by both fungistats with the inhibition being greater at a_w 0.85 than at 0.88. This is probably an example of the combined effects of pH, a_w, and other growth parameters on the growth inhibition of microorganisms in IMF systems as previously noted.

TABLE 13-7.   Time for growth of microbes[1] in inoculated dog food with inhibitors, pH 5.4 *(2)*.

	Storage conditions	
Inhibitor	$a_w = 0.85$ 9 mo storage	$a_w = 0.88$ 6 mo storage
No Inhibitor added	*A. niger*-2 wk *A. glaucus*-1 wk *S. epider.*-2 wk	*A. niger*-1 wk *A. glaucus*-1 wk *S. epider.*-½ wk
K-sorbate (0.3%)	No mold *S. epider.*-25 wk	*A. niger*-5 wk *S. epider.*-3½ wk
Ca-propionate (0.3%)	*A. niger*-25 wk *A. glaucus*-25 wk *S. epider.*-3½ wk	*A. glaucus*-2 wk *S. epider.*-1½ wk

[1]Mold − first visible sign; Bacteria − 2 log cycle increase.

Storage Stability of IMF

The undesirable chemical changes that occur in dried foods occur also in IM foods. Lipid oxidation and Maillard browning are at their optima in the general IMF ranges of a_w and percent moisture. A recent report, however, indicates that the maximum rate for Maillard browning occurs in the 0.4–0.5 a_w range, especially when glycerol is used as the humectant *(37)*.

The storage of IMF's under the proper conditions of humidity is imperative in preventing moldiness and for overall shelf stability. The measurement of equilibrium relative humidity (ERH) is of importance in this regard. ERH is an expression of the desorbable water present in a food product, and is further defined by the following equation:

$$ERH = (P_{equ}/P_{sat})T, P = 1 \text{ atm}$$

where P_{equ} is partial pressure of water vapor in equilibrium with the sample in air at 1 atmosphere total pressure and temperature T; P_{sat} is the saturation partial vapor pressure of water in air at a total pressure of 1 atmosphere and temperature T *(16)*. A food in moist air exchanges water until the equilibrium partial pressure at that temperature is equal to the partial pressure of water in the moist air, so that the ERH value is a direct measure of whether moisture will be sorbed or desorbed. In the case of foods packaged or wrapped in moisture-impermeable materials, the relative humidity of the food-enclosed atmosphere is determined by the ERH of the product, which in turn is controlled by the nature of the dissolved solids present, ratio of solids to moisture, and the like *(33)*. Both traditional and newer IMF products have longer shelf stability under conditions of lower ERH.

In addition to the direct effect of packaging on ERH, gas impermeable packaging affects the Eh of packaged products with consequent inhibitory effects upon the growth of aerobic microorganisms.

REFERENCES

1. **Acott, K. M. and T. P. Labuza.** 1975. Inhibition of *Aspergillus niger* in an intermediate moisture food system. J. Food Sci. *40:* 137–139.

2. **Acott, K., A. E. Sloan, and T. P. Labuza.** 1976. Evaluation of antimicrobial agents in a microbial challenge study for an intermediate moisture dog food. J. Food Sci. *41:* 541–546.

3. **Bone, D. P.** 1969. Water activity—its chemistry and applications. Food Prod. Dev. *3:* No. 5, 81–94.

4. **Bone, D.** 1973. Water activity in intermediate moisture foods. Food Technol. *27:* No. 4, 71–76.

5. **Boylan, S. L., K. A. Acott, and T. P. Labuza.** 1976. *Staphylococcus aureus* challenge study in an intermediate moisture food. J. Food Sci. *41:* 918–821.

6. **Brockmann, M. C.** 1970. Development of intermediate moisture foods for military use. Food Technol. *24:* 896–900.

7. **Burke, R. F. and R. V. Decareau.** 1964. Recent advances in the freeze-drying of food products. Adv. Food Res. *13:* 1–88.

8. **Collins, J. L. and A. K. Yu.** 1975. Stability and acceptance of intermediate moisture, deep-fried catfish. J. Food Sci. *40:* 858–863.

9. **Davis, R. J.** 1963. Viability and behavior of lyophilized cultures after storage for twenty-one years. J. Bacteriol. *85:* 486–487.

10. **Desrosier, N. W.** 1963. *The Technology of Food Preservation.* Chapter 5. Avi Publishing Co., Westport, Conn.

11. **Fennema, O. and W. D. Powrie.** 1964. Fundamentals of low-temperature food preservation. Adv. Food Res. *13:* 219–347.

12. **Fry, R. M. and R. I. N. Greaves.** 1951. The survival of bacteria during and after drying. J. Hyg. *49:* 220–246.

13. **Goldblith, S. A.** 1964. In: *Lyophilisation-Freeze-Drying.* Edited by L. Rey, Hermann, Paris, France, p. 555.

14. **Gooding, E. G. B.** 1962. The storage behaviour of dehydrated foods. In: *Recent Advances in Food Science,* Vol. *2:* 22–38.

15. **Haas, G. J., D. Bennett, E. B. Herman, and D. Collette.** 1975. Microbial stability of intermediate moisture foods. Food Prod. Dev. *9:* No. 4, 86–94.

16. **Hardman, T. M.** 1976. Measurement of water activity. Critical appraisal of methods, pp. 75–88 In: *Intermediate Moisture Foods,* edited by R. Davies, G. G. Birch, and K. J. Parker (Applied Sci. Publ: London).

17. **Harper, J. C. and A. L. Tappel.** 1957. Freeze-drying of food products. Adv. Food Res. *7:* 171–234.

18. **Hsieh, F.-H., K. Acott, and T. P. Labuza.** 1976. Death kinetics of pathogens in a pasta product. J. Food Sci. *41:* 516–519.

19. **Kaplow, M.** 1970. Commercial development of intermediate moisture foods. Food Technol. *24:* 889–893.

20. **Karel, M.** 1976. Technology and application of new intermediate moisture foods, pp. 4–31 In: *Intermediate Moisture Foods,* edited by R. Davies, G. G. Birch, and K. J. Parker (Applied Sci. Publ: London).

21. **Labuza, T. P.** 1968. Sorption phenomena in foods. Food Technol. *22:* 263–272.

22. **Labuza, T. P., S. Cassil, and A. J. Sinskey.** 1972. Stability of intermediate moisture foods. 2. Microbiology. J. Food Sci. *37:* 160–162.

23. **Labuza, T. P., K. Acott, S. R. Tatini, R. Y. Lee, J. Flink, and W. McCall.** 1976. Water activity determination: A collaborative study of different methods. J. Food Sci. *41:* 910–917.

24. **Matheson, N. A.** 1962. Enzymes in dehydrated meat. *Recent Advances in Food Science,* Vol. *2:* 57–64.

25. **May, K. N. and L. E. Kelly.** 1965. Fate of bacteria in chicken meat during freeze-dehydration, rehydration, and storage. Appl. Microbiol. *13:* 340–344.

26. **Meryman, H. T.** 1966. Freeze-drying. In: *Cryobiology.* Edited by H. T. Meryman, Academic Press, N.Y., Chapter 13.

27. **Mossel, D. A. A. and M. Ingram.** 1955. The physiology of the microbial spoilage of foods. J. Appl. Bacteriol. *18:* 232–268.

28. **Pitt, J. I. and J. H. B. Christian.** 1968. Water relations of xerophilic fungi isolated from prunes. Appl. Microbiol. *16:* 1853–1858.

29. **Plitman, M., Y. Park, R. Gomez, and A. J. Sinskey.** 1973. Viability of *Staphylococcus aureus* in intermediate moisture meats. J. Food Sci. *38:* 1004–1008.

30. **Robson, J. N.** 1976. Some introductory thoughts on intermediate moisture foods, pp. 32–42 In: *Intermediate Moisture Foods,* edited by R. Davies, G. G. Birch, and K. J. Parker (Applied Sci. Publ: London).

31. **Ross, K. D.** 1975. Estimation of water activity in intermediate moisture foods. Food Technol. *29:* No. 3, 26–34.

32. **Scott, W. J.** 1957. Water relations of food spoilage microorganisms. Adv. Food Res. *1:* 83–127.

33. **Seiler, D. A. L.** 1976. The stability of intermediate moisture foods with respect to mould growth, pp. 166–181 In: *Intermediate Moisture Foods,* edited by R. Davies, G. G. Birch, K. J. Parker (Applied Sci. Publ: London).

34. **Sloan, A. E., P. T. Waletzko, and T. P. Labuza.** 1976. Effect of order-of-mixing on a_w lowering ability of food humectants. J. Food Sci. *41:* 536–540.

35. **Tatini, S. R.** 1973. Influence of food environments on growth of *Staphylococcus aureus* and production of various enterotoxins. J. Milk Food Technol. *36:* 559–563.

36. **Troller, J. A.** 1972. Effect of water activity on enterotoxin A production and growth of *Staphylococcus aureus.* Appl. Microbiol. *24:* 440–443.

37. **Warmbier, H. C., R. A. Schnickels, and T. P. Labuza.** 1976. Effect of glycerol on nonenzymatic browning in a solid intermediate moisture model food system. J. Food Sci. *41:* 528–531.

14

Fermented Foods and Related Products of Fermentation

There are numerous food products which owe their production and characteristics to the activities of microorganisms. Many of these foods such as ripened cheeses, pickles, sauerkraut, and fermented sausages, are preserved products in that their shelf life is extended considerably over that of the raw materials from which they are made. In addition to being made more shelf stable, all fermented foods have aroma and flavor characteristics that result directly or indirectly from the fermenting organisms. In some instances, the vitamin content of the fermented food is increased along with an increased digestibility of the raw materials. The fermentation process reduces the toxicity of some foods (e.g., gari and peujeum), while others may become extremely toxic during fermentation (e.g., bongkrek). From all indications, no other single group or category of foods or food products is as important as these are and have been relative to man's nutritional well-being throughout the world. Included in this chapter along with the classical fermented foods are such products as coffee beans, wines, and distilled spirits, for these and similar products either result from or are improved by microbial fermentation activities.

The microbial ecology of food and related fermentations has been studied for many years in the case of ripened cheeses, sauerkraut, wines, and so on, and the activities of the fermenting organisms are dependent upon the intrinsic and extrinsic parameters of growth discussed in Chapter 2. For example, when the natural raw materials are acidic and contain free sugars, yeasts develop readily and the alcohol which they produce restricts the activities of most other naturally contaminating organisms (e.g., the fermentation of fruits to produce wines). If, on the other hand, the acidity of a plant product permits good bacterial growth and at the same time the product is high in simple sugars, lactic acid bacteria may be expected to develop and the addition of low levels of NaCl will insure their growth preferential to yeasts (e.g., the sauerkraut fermentation).

Products that contain polysaccharides but no significant levels of simple sugars are normally stable to the activities of yeasts and lactic acid bacteria due to the lack of amylase by these organisms. In order to effect their fermentation, an exogenous source of saccharifying enzymes must be supplied. The use of barley malt in the brewing and distilling industries is an example of this. The fermentation of sugars to ethanol that results from malting is then carried out by yeasts. The use of **koji** in the fermentation of soybean products is another example of the way in which alcoholic and lactic acid fermentations may be carried out on products that have low levels of sugars but high levels of starches and proteins. While the saccharifying enzymes of barley malt arise from germinating barley, the enzymes of koji are produced by *Aspergillus oryzae* growing on soaked or steamed rice or other cereals (the commercial product **takadiastase** is prepared by growing *A. oryzae* on wheat bran). The koji hydrolysates may be fermented by lactic acid bacteria and yeasts as is the case for soy sauce, or the koji enzymes may act directly upon soybeans in the production of products such as Japanese miso.

FERMENTATION—DEFINED AND CHARACTERIZED

Fermentation is the metabolic process in which carbohydrates and related compounds are oxidized with the release of energy in the absence of any external electron acceptors. The final electron acceptors are organic compounds produced directly from the breakdown of the carbohydrates. Consequently, only partial oxidation of the parent compound occurs and only a small amount of energy is released during the process. As fermenting organisms, the lactic acid bacteria lack functional heme-linked electron transport systems or cytochromes, and they obtain their energy by substrate level phosphorylation while oxidizing carbohydrates: they do not have a functional Krebs cycle. The products of fermentation consist of some that are more reduced than others.

While the above is a molecular characterization, the word "fermentation" has had many shades of meaning in the past. According to one dictionary definition, it is ". . . . a process of chemical change with effervescence . . . a state of agitation or unrest . . . any of various transformations of organic substances. . . ." The word came into use before Pasteur's studies on wines. Prescott and Dunn *(65)* and Doelle *(20)* have discussed the history of the concept of fermentation and the former authors note that in the broad sense in which the term is commonly used, it is ". . . . a process in which chemical changes are brought about in an organic substrate through the action of enzymes elaborated by microorganisms. . . ." It is in this broad context that the term is used in this chapter. In the brewing industry, a **top fermentation** refers to the use of a yeast strain which carries out its activity at the upper parts of a large vat such as in the production of ale, while a **bottom fermentation** requires the use of a yeast strain which will act in lower parts of the vat—such as in the production of lager beer.

THE LACTIC ACID BACTERIA

The lactic acid bacteria as presently constituted consist of the following four genera: *Lactobacillus, Leuconostoc, Pediococcus,* and *Streptococcus* (Groups D and N). These organisms are rather widespread in nature and the natural habitat of some is unclear. The common occurrence of leuconostocs on plants has been documented *(54).* The lactobacilli appear to be more widespread than the streptococci, and the pediococci tend to be restricted to plants, but exceptions exist for both of these genera *(48).* The common occurrence of some species of streptococci and lactobacilli in the oral cavity of man and animals is well known.

The history of man's knowledge of the lactic streptococci and their ecology have been reviewed by Sandine *et al. (70).* These authors believe that plant matter is the natural habitat of this group, but they note the lack of proof of a plant origin for *S. cremoris.* It has been suggested that plant streptococci may be the ancestral pool from which other species and strains developed *(54).*

Related to the above 4 genera of lactic acid bacteria in some respects but generally not considered to fit the group are the following: some *Aerococcus* spp; some *Erysipelothrix* spp; *Eubacterium, Microbacterium, Peptostreptococcus,* and *Propionibacterium (40, 49).*

While the lactic acid group is loosely defined with no precise boundaries, all members share the property of producing lactic acid from hexoses. Kluyver divided the lactic acid bacteria into 2 groups based upon end products of glucose metabolism. Those that produce lactic acid as the major or sole product of glucose fermentation are designated **homofermentative.** This pattern is summarized in Figure 14-1A. The homofermentative pattern is observed when glucose is metabolized but not necessarily when pentoses are metabolized, for some homolactics produce acetic and lactic acids when utilizing pentoses. Also, the homofermentative character of homolactics may be shifted for some strains by altering cultural conditions such as glucose concentration, pH, and nutrient limitation *(11, 49).* The homolactics are able to extract about twice as much energy from a given quantity of glucose as are the heterolactics. Those lactics that produce equal molar amounts of lactate, carbon dioxide, and ethanol from hexoses are designated **heterofermentative** (Figure 14-1B). All members of the genera *Pediococcus* and *Streptococcus* are homofermenters along with some of the lactobacilli, while all *Leuconostoc* spp. are heterofermenters as well as some lactobacilli (Table 14-1). The heterolactics are more important than the homolactics in producing flavor and aroma components in foods such as acetylaldehyde and diacetyl (Figure 14-2).

While the genus *Microbacterium* is not recognized in the 8th edition of *Bergey's Manual,* both homo- and heterolactics are represented by this genus, e.g., *M. thermosphactum* is homofermentative; *M. lacticum* is heterofermentative.

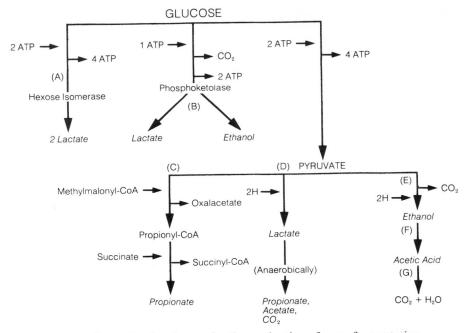

FIGURE 14-1. Generalized pathways for the production of some fermentation products from glucose by various organisms. A = homofermentative lactics; B = heterofermentative lactics; C & D = *Propionibacterium* (see Fig. 14-3); E = *Saccharomyces* spp.; F = *Acetobacter* spp.; G = *Acetobacter* "overoxidizers"

The end product differences between homo- and heterofermenters when glucose is attacked are a result of basic genetic and physiological differences (Figure 14-1). The homolactics possess the enzymes aldolase and hexose isomerase but lack phosphoketolase (Figure 14-1A). They use the Embden-Meyerhof-Parnas (EMP) pathway toward their production of 2 lactates/glucose molecule. The heterolactics, on the other hand, have phosphoketolase but do not possess aldolase and hexose isomerase, and instead of the EMP pathway for glucose degradation, these organisms use the hexose monophosphate or pentose pathway (See Figure 14-1B).

The genus *Lactobacillus* has been sub-divided classically into three subgenera: *Betabacterium, Streptobacterium,* and *Thermobacterium.* All of the heterolactic lactobacilli in Table 1 are betabacteria. The streptobacteria (e.g., *L. casei, L. plantarum*) produce up to 1.5% of lactic acid with an optimal growth temperature of 30 C, while the thermobacteria (e.g., *L. acidophilus, L. bulgaricus*) can produce up to 3% of lactic acid and have an optimal temperature of 40 C *(50).*

In terms of their growth requirements, the lactic acid bacteria require preformed amino acids, B-vitamins, and purine and pyrimidine bases,

TABLE 14-1. Homo- and Heterofermentative Lactic Acid Bacteria with Lactate Configuration and %G + C (taken from the literature)

Homofermentative			Heterofermentative		
Organisms	Lactate Config.	%G + C	Organisms	Lactate Config.	%G + C
Lactobacillus			*Lactobacillus*		
L. acidophilus	DL	36.7	L. brevis	DL	42.7–46.4
L. bulgaricus	D(−)	50.3	L. buchneri	DL	44.8
L. casei	L(+)	46.4	L. cellobiosus	DL	53
L. coryniformis	DL	45	L. confusus	DL	44.5–45.0
L. curvatus	DL	43.9	L. coprophilus	DL	41.0
L. delbrueckii	D(−)	50	L. fermentum	DL	53.4
L. helveticus	DL	39.3	L. hilgardii	DL	40.3
L. jugurti	DL	36.5–39.0	L. sanfrancisco	DL	38.1–39.7
L. jensenii	D(−)	36.1	L. trichodes	DL	42.7
L. lactis	D(−)	50.3	L. viridescens	DL	35.7–42.7
L. leichmannii	D(−)	50.8	*Leuconostoc*		
L. plantarum	DL	45	L. cremoris	D(−)	39–42
L. salivarius	L(+)	34.7	L. dextranicum	D(−)	38–39
L. xylosus	L(+)	39.4	L. lactis	D(−)	43–44
Pediococcus			L. mesenteroides	D(−)	39–42
P. acidilactici	DL	44.0	L. oenos	D(−)	39–40
P. cerevisiae[1]	DL		L. paramesenteroides	D(−)	38–39
P. pentosaceus	DL	38			
Streptococcus					
S. bovis	D(−)	38–42			
S. cremoris	D(−)	38–40			
S. diacetilactis	D(−)	35.1			
S. lactis	D(−)	38.4–38.6			
S. thermophilus	D(−)	40			

[1]*P. damnosus*

hence, their use in microbiological assays for these compounds. Although they are mesophilic, some can grow at 5 C and some as high as 45 C. With respect to growth pH, some can grow as low as 3.2, some as high as 9.6, and most can grow in the pH range 4.0–4.5. The lactic acid bacteria are only weakly proteolytic and lipolytic *(79)*.

In the past, the taxonomy of the lactic acid bacteria has been based largely upon the gram reaction, general lack of ability to produce catalase, the production of lactic acid of a given configuration, along with the ability to ferment various carbohydrates. More recently, the use of DNA base composition, DNA homology, cell wall peptidoglycan type, and immunologic specificity of enzymes have come into use. While these organisms may be thought of as having derived from a common ancestor, high degrees of diversity as well as high degrees of relatedness are found within the group.

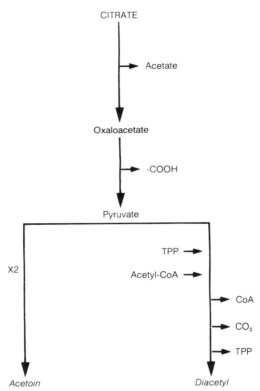

FIGURE 14-2. The general pathway by which acetoin and diacetyl are produced from citrate by Group N streptococci and *Leuconostoc* spp. Pyruvate may be produced from lactate, and acetylCoA from acetate. For further details, see ref. *14, 77, 78*.

The DNA base composition (expressed as moles % G + C) of those lactics studied varies from a low of around 35 for *L. salivarius* to a high of 53 for *L. fermentum (95)*. There is considerable overlap between the homo- and heterolactics (see Table 14-1). On the basis of DNA base composition, Miller *et al. (51, 52)* placed 15 species of lactobacilli into 3 groups. Group I contains those within 32.4 to 38.3 moles % G + C and includes *L. jugurti, L. helveticus, L. salivarius, L. bulgaricus,* and *L. (jugurti) bulgaricus*. Group II contains 42.7 to 48.0 moles % G + C and includes *L. buchneri, L. brevis, L. casei, L. viridescens,* and *L. plantarum,* while Group III contains 49.0 to 51.9 moles % G + C and includes *L. lactis, L. leichmannii, L. delbrueckii, L. fermentum,* and *L. cellobiosus*. While the Group I species are all homofermenters, Groups II and III contain both homo- and heterolactic types. The diversity of relatedness of this group has been shown also by both DNA hybridization and immunologic studies. With respect to DNA hybridization, strains of *L. jugurti* and *L. helveticus* have been shown to share between 85 and 100% DNA homology; *L. leichmannii, L. delbrueckii,*

L. lactis, and *L. bulgaricus* to share between 78 and 100%; and *L. fermentum* and *L. cellobiosus* to share from 77 to 100% DNA homology (See London, *49*). Simonds *et al. (74)* have shown that *L. bulgaricus* shares 86% DNA homology with *L. lactis,* 4.8% with *L. helveticus,* and none with *L. jugurti.*

DNA-DNA hybridization studies in the genus *Leuconostoc* were reported recently *(38)*. Of 45 strains representing 6 species, 6 homology groups were distinguished using 4 reference DNA preparations. Of 19 strains of *L. mesenteroides,* 3 different hybridization groups were determined.

In regard to immunologic methods, Gasser and Gasser *(29)* showed that antisera prepared against NAD-dependent D-lactic dehydrogenases of 3 lactobacilli reacted against crude extracts of almost all other species of lactobacilli containing the enzyme. Extracts of *Leuconostoc* spp. cross reacted with anti-D-lactic dehydrogenases. In a somewhat similar manner, it has been shown that antisera prepared against purified *S. faecalis* fructose diphosphate aldolase reacted to varying degrees with the aldolases of all homofermentative lactics including some other streptococci, pediococci, and lactobacilli *(48)*.

The cell wall mucopeptides of lactics and other bacteria have been reviewed by Schleifer and Kandler *(72)* and Williams *(95)*. While there appears to be wide variation within most of the lactic acid genera, the homofermentative lactobacilli of the subgenus *Thermobacterium* appear to be the most homogeneous in this regard in having L-lysine in the peptidoglycan peptide chain and D-aspartic acid as the interbridge peptide. The Group N streptococci have similar wall mucopeptides.

The measurement of molar growth yields provides information on fermenting organisms relative to their fermentation substrates and pathways. By this concept, the μg dry weight of cells produced per μmole substrate fermented is determined as the **molar yield constant,** indicated by Y. It is tacitly assumed that essentially none of the substrate carbon is used for cell biosynthesis, that oxygen does not serve as an electron or hydrogen acceptor, and that all of the energy derived from the metabolism of the substrate is coupled to cell biosynthesis *(34)*. When the substrate is glucose, e.g., the molar yield constant for glucose, Y_G, is determined by:

$$Y_G = \frac{\text{g dry weight of cells}}{\text{moles glucose fermented}}$$

If the ATP yield or moles ATP produced per mole substrate used is known for a given substrate, the amount of dry weight of cells produced/mole of ATP formed can be determined by:

$$Y_{ATP} = \frac{\text{g dry weight of cells/moles ATP formed}}{\text{moles substrate fermented}}$$

A large number of fermenting organisms have been examined during growth and found to have $Y_{ATP} = 10.5$ or close thereto. This value is assumed to be a constant so that an organism which ferments glucose by the EMP pathway to produce 2 ATP/mole of glucose fermented should have $Y_G = 21$ (that is, it should produce 21 g of cells dry weight/mole of glucose). This has been verified for *Streptococcus faecalis, Saccharomyces cerevisiae, Saccharomyces rosei,* and *L. plantarum* on glucose (all $Y_G = 21$, $Y_{ATP} = 10.5$, within experimental error). A recent study by Brown and Collins *(11)* indicates that Y_G and Y_{ATP} values for *S. diacetilactis* and *S. cremoris* differ when cells are grown aerobically on a partially defined medium with low and higher levels of glucose, and further when grown on a complex medium. On a partially defined medium with low glucose levels (1 to 7 μmol/ml) values for *S. diacetilactis* were $Y_G = 35.3$, $Y_{ATP} = 15.6$, while for S. cremoris $Y_G = 31.4$ and $Y_{ATP} = 13.9$. On the same medium with higher glucose levels (t to 15 μmol/ml), Y_G for *S. diacetilactis* was 21. Y_{ATP} values for these two organisms on the complex medium with 2 μmol glucose/ml were 21.5 and 18.9 for *S. diacetilactis* and *S. cremoris*, respectively. Anaerobic molar growth yields for streptococcal species on low levels of glucose have been studied by Johnson and Collins *(41)*. *Zymomonas mobilis* utilizes the Entner-Doudoroff pathway to produce only 1 ATP/mole of glucose fermented ($Y_G = 8.3$, $Y_{ATP} = 8.3$). If and when the produced lactate is metabolized further, the molar growth yield would be higher. *Bifidobacterium bifidum* produces 2.5–3 ATP/mole of glucose fermented with $Y_G = 37$, $Y_{ATP} = 13$ *(81)*.

In addition to the use of molar growth yields to compare organisms on the same energy substrate, this concept can be applied to assess the metabolic routes used by various organisms in attacking a variety of carbohydrates (for further information, see *25, 61, 71, 81*).

PRODUCTS OF FERMENTATION

Dairy Products

Some of the many fermented foods and products produced and utilized world-wide are listed in Table 14-2. The commercial and sometimes the home production of many of these is begun by use of appropriate **starter** cultures. A lactic starter is a basic starter culture with widespread use in the dairy industry. For cheese making of all kinds, lactic acid production is essential and the lactic starter is employed for this purpose. Lactic starters are also used for butter, cultured buttermilk, cottage cheese, and cultured sour cream and are often referred to by product such as butter starter, buttermilk starter, etc. Lactic starters always include bacteria which convert lactose to lactic acid, usually *S. lactis, S. cremoris,* or *S. diacetilactis.* Where flavor and aroma compounds such as diacetyl are desired, the lactic starter will include a heterolactic such as *Leuconostoc citrovorum,*

S. diacetilactis, or *L. dextranicum* (for biosynthetic pathways, see Figure 14-2 and Reference *14*). Starter cultures may consist of single or mixed strains. They may be produced in quantity and preserved by freezing in liquid nitrogen *(30),* or by freeze drying. The streptococci generally make up around 90% of a mixed dairy starter population, and a good starter culture can convert most of the lactose to lactic acid. The titratable acidity may increase to 0.8 to 1.0%, calculated as lactic acid, and the pH usually drops to 4.3–4.5 *(26).* Some of the many products that result from the use of lactic and other starters are discussed below.

Butter, buttermilk, and **sour cream** are produced generally by inoculating pasteurized cream or milk with a lactic starter culture and holding until the desired amount of acidity is attained. In the case of butter where cream is inoculated, the acidified cream is then churned to yield butter which is washed, salted, and packaged *(63).* Buttermilk, as the name suggests, is the milk that remains after cream is churned for the production of butter. The commercial product, however, is usually prepared by inoculating skim milk with a lactic or buttermilk starter culture and holding until souring occurs. The resulting curd is broken up into fine particles by agitation and this product is termed **cultured buttermilk.** Cultured sour cream is produced generally by fermenting pasteurized and homogenized light cream with a lactic starter. These products owe their tart flavor to lactic acid, and their buttery aroma and taste to diacetyl.

Yogurt (yoghurt) is produced with a yogurt starter which is a mixed culture of *Streptococcus thermophilus* and *L. bulgaricus* in a 1:1 ratio. The coccus grows faster than the rod and is primarily responsible for acid production, while the rod adds flavor and aroma. The product is prepared by first reducing the water content of either whole or skim milk by at least one fourth. This may be done in a vacuum pan following sterilization of the milk. Approximately 5% by weight of milk solids or condensed milk is usually added. The concentrated milk is then heated between 82–93 C for 30–60 min and cooled to around 45 C *(63).* The yogurt starter is now added at a level of around 2% by volume and incubated at 45 C for 3–5 hr followed by cooling to 5 C. The titratable acidity of a good finished product is around 0.85–0.90%, and to get this amount of acidity the fermenting product should be removed from 45 C when the titratable acidity is around 0.65–0.70% *(15).* Good yogurt keeps well at 5 C for 1 to 2 weeks. As noted above, the coccus grows first during the fermentation followed by the rod so that after around 3 hr the numbers of the 2 organisms should be approximately equal. Higher amounts of acidity such as 4% can be achieved by allowing the product to ferment longer with the effect that the rods will exceed the cocci in number. The streptococci tend to be inhibited at pH values of 4.2–4.4, while the lactobacilli can tolerate pHs in the 3.5–3.8 range. The lactic acid of yogurt is produced more from the glucose moiety of lactose than the galactose moiety. Goodenough and Kleyn *(33)* found only a trace of glucose throughout yogurt fermentation, while galactose increased from an initial

TABLE 14-2. Some of the many known fermented foods and related products (compiled from various sources).

Foods & Products	Raw ingredients	Fermenting organisms	Commonly produced
DAIRY PRODUCTS			
Acidophilus milk	Milk	*Lactobacillus acidophilus*	Many countries
Bulgarian buttermilk	Milk	*Lactobacillus bulgaricus*	The Balkans, others
Cheeses (ripened)	Milk curd	Lactic starters; others	World-wide
Kefir	Milk	*Streptococcus lactis, L. bulgaricus, Torula* spp.	Southwestern Asia
Kumiss	Raw mare's milk	*L. bulgaricus, Lactobacillus leichmannii, Torula* spp.	Russia
Taette	Milk	*S. lactis* var. *taette*	Scandinavian Peninsula
Tarhana[1]	Wheat meal & yogurt	Lactics	Turkey
Yogurt[2]	Milk, milk solids	*Streptococcus thermophilus, L. bulgaricus*	World-wide
MEAT & FISHERY PRODUCTS			
Country cured hams	Pork hams	*Aspergillus, Penicillium* spp.	Southern U.S.A.
Dry sausages[3]	Pork, beef	*Pediococcus cerevisiae*	Europe, U.S.A.
Lebanon bologna	Beef	*P. cerevisiae*	U.S.A.
Burong dalag	Dalag fish & rice	*Leuconostoc mesenteroides, P. cerevisiae, L. plantarum*	Philippines
Fish sauces[4]	Small fish	halophilic *Bacillus* spp.	Southeast Asia
Izushi	Fresh fish, rice, vegetables	*Lactobacillus* spp.	Japan
Katsuobushi	Skipjack tuna	*Aspergillus glaucus*	Japan
NONBEVERAGE PLANT PRODUCTS			
Bongkrek	Cocoanut presscake	*Rhizopus oligosporus*	Indonesia
Cocoa beans	Cacao fruits (pods)	*Candida krusei, Geotrichum* spp.	Africa, South America
Coffee beans	Coffee cheeries	*Erwina dissolvens, Saccharomyces* spp.	Brazil, Congo, Hawaii, India
Gari	Cassava	*Corynebacterium manihot, Geotrichum* spp.	West Africa
Kenkey	Corn	*Aspergillus* spp., *Penicillium* spp., Lactobacilli, yeasts	Ghana, Nigeria
Kimchi	Cabbage & other veg.	Lactic acid bacteria	Korea
Miso	Soy beans	*Aspergillus oryzae, Saccharomyces rouxii*	Japan

262

Product	Substrate	Microorganisms	Location
Ogi	Corn	*L. plantarum, S. lactis, Saccharomyces rouxii*	Nigeria
Olives	Green olives	*L. mesenteroides, L. plantarum*	World-wide
Ontjom[5]	Peanut presscake	*Neurospora sitophila*	Indonesia
Peujeum	Cassava	Molds	Indonesia
Pickles	Cucumbers	*P. cerevisiae, L. plantarum*	World-wide
Poi	Taro roots	Lactics	Hawaii
Sauerkraut	Cabbage	*L. mesenteroides, L. plantarum*	World-wide
Soy sauce (shoyu)	Soybeans	*A. oryzae*; or *A. soyae*; *S. rouxii, L. delbrueckii*	Japan
Sufu	Soybeans	*Mucor* spp.	China & Taiwan
Tao-si	Soybeans	*A. oryzae*	Philippines
Tempeh	Soybeans	*Rhizopus oligosporus; R. oryzae*	Indonesia, New Guinea, Surinam

BEVERAGES AND RELATED PRODUCTS

Product	Substrate	Microorganisms	Location
Arrack	Rice	Yeasts, bacteria	The Far East
Beer and ale	Cereal wort	*Saccharomyces cerevisiae; S. carlsbergensis*	World-wide
Binuburan	Rice	Yeasts	Philippines
Bourbon whiskey	Corn, rye	*S. cerevisiae*	U.S.A.
Bouza beer	Wheat grains	Yeasts	Egypt
Cider	Apples; others	*Saccharomyces* spp.	World-wide
Kaffir beer	Kaffircorn	Yeasts, molds, lactics	Nyasaland
Magon	Corn	*Lactobacillus* spp.	Bantus of South Africa
Mezcal	Century plant	Yeasts	Mexico
Oo	Rice	Yeasts	Thailand
Pulque[6]	Agave juice	Yeasts and lactics	Mexico, U.S. Southwest
Sake	Rice	*Saccharomyces sake*	Japan
Scotch whiskey	Barley	*S. cerevisiae*	Scotland
Teekwass	Tea leaves	*Acetobacter xylinum, Schizosaccharomyces pombe*	
Thumba	Millet	*Endomycopsis fibuliges*	West Bengal
Tibi	Dried figs; raisins	*Betabacterium vermiforme, Saccharomyces intermedium*	
Vinegar	Cider, wine	*Acetobacter* spp.	World-wide
Wines	Grapes, other fruits	*Saccharomyces ellipsoideus* strains	World-wide
Palm wine	Palm sap	*Acetobacter* spp., lactics, yeasts	Nigeria

TABLE 14-2.   Some of the many known fermented foods and related products (compiled from various sources) (Continued).

Foods & Products	Raw ingredients	Fermenting organisms	Commonly produced
BREADS			
Idli	Rice & bean flour	*Leuconostoc mesenteroides*	Southern India
Rolls, cakes, etc.	Wheat flours	*S. cerevisiae*	World-wide
S. F. sour-dough bread	Wheat flour	*S. exiguus, L. sanfrancisco*	Northern California
Sour pumpernickel	Wheat flour	*L. mesenteroides*	Switzerland, others

[1]Similar to Kishk in Syria and Kushuk in Iran.
[2]Also yoghurt (matzoon in Armenia; Leben in Egypt; Naja in Bulgaria; Gioddu in Italy; Dadhi in India).
[3]Such as Genoa, Milano, Siciliano, etc.
[4]See text for specific names.
[5]*N. sitophila* is used to make red ontjom; *R. oligosporus* for white ontjom.
[6]Distilled to produce tequila.

trace to 1.2%. Samples of commercial yogurts showed only traces of glucose while galactose varied from around 1.5 to 2.5%.

The antimicrobial qualities of yogurt, buttermilk, sour cream, and cottage cheese were examined by Goel *et al. (32)* who inoculated *Enterobacter aerogenes* and *Escherichia coli* separately into commercial products and studied the fate of these organisms when the products were stored at 7.2 C. A sharp decline of both coliforms was noted in yogurt and buttermilk after 24 hr. Neither could be found in yogurt generally beyond 3 days. While the numbers of coliforms were reduced also in sour cream, they were not reduced as rapidly as in yogurt. Some cottage cheese samples actually supported an increase in coliform numbers, probably because these products had higher pH values. The initial pH ranges for the products studied by these workers were as follows: 3.65–4.40 for yogurts; 4.1–4.9 for buttermilks; 4.18–4.70 for sour creams, and 4.80–5.10 for cottage cheese samples. In another study, commercially produced yogurts in Ontario were found to contain the desired 1: 1 ratio of coccus to rod in only 15% of 152 products examined *(6)*. Staphylococci were found in 27.6% and coliforms in around 14% of these yogurts. Twenty-six percent of the samples had yeast counts > 1,000/g and almost 12% had psychrotroph counts > 1,000/g. In his study of commercial unflavored yogurt in Great Britain, Davis *(15)* found *S. thermophilus* and *L. bulgaricus* counts to range from a low of around 82, 000,000 to a high of over 1 billion/g, and final pH to range from 3.75 to 4.20.

Kefir is prepared by use of kefir grains which contain *S. lactis, L. bulgaricus,* and a lactose-fermenting yeast held together by layers of coagulated protein. Acid production is controlled by the bacteria while the yeast produces alcohol. The final concentration of lactic acid and alcohol may be as high as 1%. **Kumiss** is similar to kefir except that mare's milk is used, the culture organisms do not form grains, and the alcohol content may reach 2%.

Acidophilus milk is produced by the inoculation into sterile skim milk of an intestinal implantable strain of *L. acidophilus*. The inoculum of from 1–2% is added followed by holding of product at 37 C until a smooth curd develops. **Bulgarian buttermilk** is produced in a similar manner by the use of *L. bulgaricus* as the inoculum or starter, but unlike *L. acidophilus, L. bulgaricus* is not implantable in the human intestines.

All **cheeses** result from a lactic fermentation of milk. In general, the process of manufacture consists of the following important steps. (1) Milk is prepared and inoculated with an appropriate lactic starter. The starter produces lactic acid which along with added rennin gives rise to curd formation. The starter for cheese production may differ depending upon the amount of heat applied to the curds. *Streptococcus thermophilus* is employed for acid production in cooked curds, since it is more heat tolerant than either of the other more commonly used lactic starters, or a combination of *S. thermophilus* and *S. lactis* is employed for curds that receive an intermediate cook. (2) The curd is shrunk and pressed followed by salting and, in the case of ripened cheeses, allowed to ripen under conditions appropriate to the cheese in question. While most ripened cheeses are the product of metabolic activities of the lactic acid bacteria, several well-known cheeses owe their particular character to other related organisms. In the case of **Swiss cheese,** a mixed culture of *L. bulgaricus* and *S. thermophilus* is usually employed along with a culture of *Propionibacterium shermanii,* which is added to function during the ripening process in flavor development and eye formation. (See Figure 14-1C, D for summary of propionibacteria pathways and Figure 14.3 for pathway in detail.) These organisms have been reviewed extensively by Hettinga and Reinbold *(37).* For blue cheeses such as **Roquefort,** the curd is inoculated with spores of *Penicillium roqueforti* which effect ripening and impart the blue-veined appearance characteristic of this type of cheese. In a similar fashion, either the milk or the surface of **Camembert** cheese is inoculated with spores of *Penicillium camemberti.*

There are over 400 varieties of cheeses representing fewer than 20 distinct types, and these are grouped or classified according to texture or moisture content, whether ripened or unripened, and if ripened, whether by bacteria or molds. The 3 textural classes of cheeses are hard, semihard, and soft. Examples of hard cheeses include Cheddar, Provolono, Romano, Edam, and so forth. All hard cheeses are ripened by bacteria over periods ranging from 2 to 16 mon. Semihard cheeses include Muenster, Gouda, and so forth, and are ripened by bacteria over periods of 1 to 8 mon. Blue and Roquefort are two examples of semihard cheeses that are mold ripened for 2 to 12 mon. Limburger is an example of a soft bacteria-ripened cheese, while Brie and Camembert are examples of soft mold-ripened cheeses. Among unripened cheeses are cottage, cream, and Neufchatel.

Several other less widely produced fermented dairy products are listed in Table 14-2.

Meat and Fishery Products

Fermented sausages are produced generally as dry or semidry products although some are intermediate. Dry or Italian-type sausages contain between 30–40% moisture, are generally not smoked or heat processed, and are eaten usually without cooking *(63).* In their preparation, curing and seasoning agents are added to ground meat followed by its stuffing into casings and incubation for varying periods of time at 80–95 F. Incubation times are shorter when starter cultures are employed. The curing mixtures include glucose as substrate for the fermenters, and nitrates and/or nitrites as color stabilizers. When only nitrates are used, it is necessary that the sausage contains bacteria that reduce nitrates to nitrites, usually micrococci present in the sausage flora or added to the mix. Following the incubation, during which fermentation occurs, the products are placed in drying rooms with a relative humidity of 55–65% for periods ranging from 10 to 100 days, or, in the case of Hungarian salami, up to 6 mon *(55).* Genoa and Milano salamis are other examples of dry sausages.

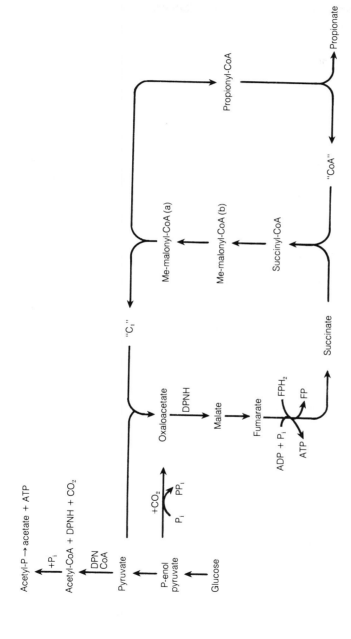

FIGURE 14-3. Reactions of the propionic acid fermentation and the formation of acetate, CO_2, propionate, and ATP. Me-malonyl-CoA is methylmalonyl-CoA and (a) and (b) are the two isomers. FP is flavoprotein and FPH_2 is reduced flavoprotein. (Summary: 1.5 glucose $+ 6\,Pi + 6\,ADP \rightarrow 6\,ATP + 2\,H_2O + CO_2 + $ acetate $+ 2$ propionate). Allen *et al.* *(5)*. Copyright © 1964, American Society for Microbiology.

In a recent study of dry sausages, the pH was found to decrease from an initial of 5.8 to 4.8 during the first 15 days of ripening and remained constant thereafter *(19)*. Nine different brands of commercially produced dry sausages were found by these investigators to have pH values ranging from 4.5 to 5.2 with a mean of 4.87. With respect to the changes that occur in the flora of fermenting dry sausage when starters are not used, Urbaniak and Pezacki *(86)* found the homofermenters to predominate overall with *L. plantarum* being the most commonly isolated species. The heterofermenters such as *L. brevis* and *L. buchneri* increased during the 6-day incubation period as a result of changes in pH and Eh brought about by the homofermenters.

Semidry sausages are prepared essentially as above but are subjected to less drying. They contain ca. 50% moisture and are finished by heating to an internal temperature of 140–154 F during smoking. Thuringer, cervelat, summer, and Lebanon bologna are some examples of semidry sausages. "Summer sausage" refers to those traditionally of Northern European origin, made during colder months, stored, aged, and then eaten during summer months. They may be dry or semidry.

Lebanon bologna is typical of a semidry sausage. This product, originally produced in the Lebanon, Pa. area, is an all-beef, heavily smoked and spiced product which may be prepared by use of a *Pediococcus cerevisiae* starter. The product is made by the addition of approximately 3% NaCl along with sugar, seasoning, and either nitrate, nitrite, or both to raw cubed beef. The salted beef is allowed to age at refrigerator temperatures for about 10 days during which time the growth of naturally occurring lactic acid bacteria or the starter organisms is encouraged and gram negatives are inhibited. A higher level of microbial activity along with some drying occurs during the smoking step at higher temperatures. A controlled production process for this product has been studied *(59)*, and it consists of aging salted beef at 5 C for 10 days and smoking at 35 C with high R.H. for 4 days. Fermentation may be carried out either by the natural flora of the meat or by use of a commercial starter of *P. cerevisiae* or *P. acidilactici*. The amount of acidity produced in Lebanon bologna may reach 0.8–1.2% *(63)*.

The hazard of eating improperly prepared home-made, fermented sausage is pointed up by a recent outbreak of trichinosis. Of the 50 persons who actually consumed the raw summer sausage, 23 became ill with trichinosis *(64)*. The sausage was made on two different days in three batches according to a family recipe which called for smoking at cooler smoking temperatures, believed to produce a better flavored product. All three batches of sausages contained home-raised beef. In addition, two batches eaten by victims contained USDA-inspected pork in one case and home-raised pork in the other, but *Trichinella spiralis* larvae were found only in the USDA-inspected pork. This organism can be destroyed by a heat treatment that results in internal temperatures of at least 137 F.

Fermented sausages produced without use of starters have been found to contain large numbers of lactobacilli such as *L. plantarum (18)*. The use of a

P. cerevisiae starter leads to the production of a more desirable product *(17, 36)*. In their study of commercially produced fermented sausages, Smith and Palumbo *(75)* found total aerobic plate counts to be in the 10^7 to $10^{8/g}$ range with a predominance of lactic acid types. When starter cultures were used, the final pH of the products ranged from 4.0 to 4.5, while those produced without starters ranged between 4.6 and 5.0. pH values of 4.5 to 4.7 for summer-type sausages have been reported for a 72-hr fermentation *(1)*. These investigators found that fermentation at 30 and 37 C led to a lower final pH than at 22 C and that the final pH was directly related to the amount of lactic acid produced. The pH of fermented sausage may actually increase by 0.1–0.2 units during long periods of drying due to uneven buffering produced by increases in amounts of basic compounds *(93)*. The ultimate pH attained following fermentation depends upon the type of sugar added. While glucose is most widely used, sucrose has been found to be an equally effective fermentable sugar for low pH production *(2)*. The effect of a commercial frozen concentrate starter *(P. acidilactici)* in fermenting various sugars added to a sausage preparation is illustrated in Figure 14-4.

Prior to the late 1950s, the production of fermented sausages was facilitated by either back inoculations, or a producer took the chance of the desired organisms being present in the raw materials. The manufacture of these as well as of many other fermented foods has been, until recently, more of an art than a science. With the advent of pure culture starters not only has production time been shortened but more uniform and safer products can be produced *(22)*. While the use of starter cultures has been in effect for many years in the dairy industry, their use in many non-dairy products world-wide is a recent development with great promise for the future. *Micrococcus aurantiacus* has been employed along with starters in the production of some European sausages *(55)*.

Molds are known to contribute to the quality of dry European-type sausages such as Italian salame. In an extensive study of the fungi of ripened meat products, Ayres *et al. (7)* found 9 species of penicillia and 7 of aspergilli on fermented sausages and concluded that these organisms play a role in the preservation of products of this type. Fewer species of other mold genera were also found.

Country cured hams are dry-cured hams produced in the southern United States, and during the curing and ripening period of 6 mon to 2 yr, heavy mold growth occurs on the surfaces. Although Ayres *et al. (7)* noted that the presence of molds is incidental and that a satisfactory cure does not depend upon their presence, it seems quite likely that some aspects of flavor development of these products derive from the heavy growth of such organisms, and to a lesser extent from yeasts. Heavy mold growth obviates the activities of food poisoning and food spoilage bacteria, and in this sense the mold flora aids in preservation. The above investigators found aspergilli and penicillia to be the predominant types of molds on country cured hams.

The processing of country cured hams of the above type takes place during the early winter and consists of rubbing sugar cure into the flesh side and

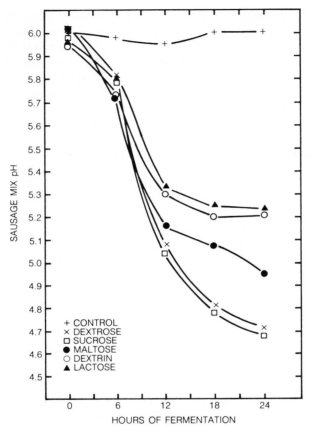

FIGURE 14-4.   Rate of pH reduction in fermenting sausage containing 0% or 1% of various carbohydrates. Acton *et al. (2)*. Copyright © 1977, Institute of Food Technologists.

onto the hock end. This is followed some time later by rubbing NaCl into all parts of the ham not covered by skin. The hams are then wrapped in paper and individually placed in cotton fabric bags and left lying flat for several days between 32 and 40 C. The hams are hung shank end down in ham houses for 6 weeks or longer and may be given a hickory smoke during this time, although smoking is said not to be essential to a desirable product.

Italian-type country cured hams are produced with NaCl as the only cure. Curing is carried out for about a month followed by washing, drying, and ripening for 6–12 mon or longer *(31)*. While halophilic and halotolerant bacteria increase as Italian hams ripen, the microflora in general is thought to play only a minor role *(31)*.

Fish sauces are popular products in Southeast Asian countries where they are known by various names such as ngapi (Burma), nuoc-mam (Cambodia and Vietnam), nam-pla (Laos and Thailand), ketjap-ikan (Indonesia), and so on. The production of some of these sauces begins with the addition of salt to uneviscerated fish at a ratio of approximately 1:3, salt to fish. The salted fish are then transferred either to fermentation tanks generally constructed of concrete and built into the ground, or placed in earthenware pots and buried in the ground. The tanks or pots are filled and sealed off for at least 6 mon to allow the fish to liquefy. The liquid is collected, filtered, and transferred to earthenware containers and ripened in the sun for 1 to 3 mon. The finished product is described as being clear dark-brown in color with a distinct aroma and flavor *(69)*. In a study of fermenting Thai fish sauce by the latter authors, pH from start to finish ranged from 6.2 to 6.6 with NaCl content around 30% over the 12-mon fermentation period *(69)*. These parameters along with the relatively high fermentation temperature result in the growth of halophilic aerobic sporeformers as the predominant microorganisms of these products. Lower numbers of streptococci, micrococci, and staphylococci were found and they along with the *Bacillus* spp. were apparently involved in the development of flavor and aroma. Some part of the liquefaction that occurs is undoubtedly due to the activities of fish proteases. While the temperature and pH of the fermentation are well within the growth range of a large number of undesirable organisms, the safety of products of this type is due to the 30–33% NaCl.

Fish pastes are also common in Southeast Asia, but the role of fermenting microorganisms in these products appears to be minimal. Among the many other fermented fish, fish-paste, and fish-sauce products are the following: Mam-tom of China; mam-ruoc of Cambodia; bladchan of Indonesia; shiokara of Japan; belachan of Malaya; bagoong of the Philippines; kapi, hoi-dong, and pla-mam of Thailand; Fessik of Africa; and nam-pla of Thailand. Some of these as well as other fish products of Southeast Asia have been reviewed and discussed by van Veen and Steinkraus *(91)* and Sundhagul *et al.* *(84)*. See Table 14-2 for other related products.

Nonbeverage Food Products of Plant Origin

Sauerkraut is a fermentation product of fresh cabbage. The starter for sauerkraut production is usually the normal mixed flora of cabbage. The addition of 2.25–2.5% salt restricts the activities of gram negative bacteria while the lactic acid rods and cocci are favored. *Leuconostoc mesenteroides* and *L. plantarum* are the two most desirable lactic acid bacteria in sauerkraut production, with the former having the shortest generation time and the shortest life span. The activities of the coccus usually cease when the acid content increases to 0.7–1.0%. The final stages of kraut production are effected by *L. plantarum* and *L. brevis*. *P. cerevisiae* and *S. faecalis* may also contribute to product development. Final total acidity is generally 1.6–1.8% with lactic acid being 1.0–1.3%.

Pickles are fermentation products of fresh cucumbers, and as is the case for sauerkraut production, the starter culture normally consists of the normal mixed flora of cucumbers. In the natural production of pickles the following lactic acid bacteria are involved in the process in the order of increasing prevalence: *L. mesenteroides, S. faecalis, P. cerevisiae, L. brevis,* and *L. plantarum (21).* Of these the pediococci and *L. plantarum* are the most involved with *L. brevis* being undesirable because of its capacity to produce gas. *L. plantarum* is the most essential species in pickle production as it is for sauerkraut.

In the production of pickles, selected cucumbers are placed in wooden brine tanks with initial brine strengths as low as 5% NaCl (20° salinometer). Brine strength is increased gradually during the course of the 6–9 week fermentation until it reaches around 60° salinometer (15.9% NaCl). In addition to exerting an inhibitory effect on the undesirable gram negative bacteria, the salt extracts from the cucumbers water and water-soluble constituents such as sugars which are converted by the lactic acid bacteria to lactic acid. The product that results is a salt-stock pickle from which pickles such as sour, mixed sour, chowchow, and so forth, may be made.

The general technique of producing brine-cured pickles briefly outlined above has been in use for many years, but it often leads to serious economic loss because of pickle spoilage from such conditions as bloaters, softness, off colors, and so on. More recently, the controlled fermentation of cucumbers brined in bulk has been achieved, and this process not only reduces economic losses of the type noted above but leads to a more uniform product over a shorter period of time *(21).* In brief, the controlled fermentation method employs a chlorinated brine of 25° salinometer, acidification with acetic acid, the addition of sodium acetate, and inoculation with *P. cerevisiae* and *L. plantarum,* or the latter alone. The course of the 10- to 12-day fermentation is represented in Figure 14-5 (for more detailed information, see Etchells *et al., 21).* For more information on different types of pickles, see Prescott and Dunn *(65).*

Olives to be fermented (Spanish, Greek, or Sicilian) are done so by the natural flora of green olives which consists of a variety of bacteria, yeasts, and molds. The olive fermentation is quite similar to that of sauerkraut except that it is slower, involves a lye treatment, and may require the addition of starters. The lactic acid bacteria become prominent during the intermediate stage of fermentation. *L. mesenteroides* and *P. cerevisiae* are the first lactics to become prominent, and these are followed by lactobacilli with *L. plantarum* and *L. brevis* being the most important *(92).*

The olive fermentation is preceded by a treatment of green olives with from 1.6 to 2.0% lye, depending upon type of olive, at 21 to 24 C for 4 to 7 hr for the purpose of removing some of the bitter principal. Following the complete removal of lye by soaking and washing, the green olives are placed in oak barrels and brined so as to maintain a constant 28–30° salinometer level. Inoculation with *L. plantarum* may be necessary because of destruction of

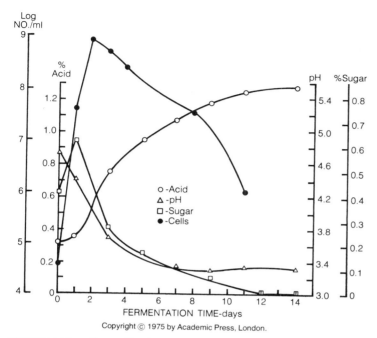

FIGURE 14-5.   Controlled fermentation of cucumbers brined in
bulk. Equilibrated brine strength during fermentation, 6.4% NaCl;
incubation temperature, 27°C *(21)*. Copyright © 1975 by Academic Press.

organisms during the lye treatment. The fermentation may take as long as 6
to 10 mon, and the final product has a pH of 3.8–4.0 following up to 1% lactic
acid production.

Soy sauce or shoyu is produced in a two-stage manner. The first stage, the
koji (analogous to malting in the brewing industry), consists of inoculating
either soybeans or a mixture of beans and wheat flour with *A. oryzae* or
A. soyae and allowing to stand for 3 days. This results in the production of
large amounts of fermentable sugars, peptides, and amino acids. The second
stage, the moromi, consists of adding the fungal-covered product to around
18% NaCl and incubating at room temperatures for at least a year. The liquid
that is obtained at this time is soy sauce. During the incubation of the moromi, lactic acid bacteria, *L. delbrueckii* in particular, and yeasts such as
S. rouxii carry out an anaerobic fermention of the koji hydrolysate. Pure cultures of *A. oryzae* for the koji, and *L. delbrueckii* and *S. rouxii* for the moromi stages have been shown to produce good quality soy sauce *(96)*.

Tempeh is a fermented soybean product. While there are many variations
in its production, the general principle of the Indonesian method for tempeh
consists of soaking soybeans overnight in order to remove the seedcoats or
hulls. Once seedcoats are removed, the beans are cooked in boiling water for
ca. 30 min and then spread on a bamboo tray to cool and surface dry. Small

pieces of tempeh from a previous fermentation are incorporated as starter followed by wrapping with banana leaves. The wrapped packages are kept at room temperature for 1–2 days during which mold growth occurs and binds the beans together as a cake—the tempeh. An excellent product can be made by storing in perforated plastic bags and tubes with fermentations completed in 24 hr at 31 C *(24)*. The desirable organism in the fermentation is *Rhizopus oligosporus,* especially for wheat tempeh. Good soybean tempeh can be made with *R. oryzae* or *R. arrhizus.* During the fermentation, the pH of soybeans rises from around 5.0 to values as high as 7.5.

Miso is a fermented soybean product common in Japan. It is prepared by mixing or grinding steamed or cooked soybeans with koji and salt, and allowing fermentation to take place usually over a 4–12 mon period. White or sweet miso may be fermented for only a week, while the higher quality dark brown product (mame) may ferment for 2 yr *(73)*. In Israel, Ilany-Feigenbaum *et al. (38)* have prepared miso-type products by using defatted soybean flakes instead of whole soybeans and fermenting for around 3 mon. The koji for these products was made by growing *A. oryzae* on corn, wheat, barley, millet or oats, potatoes, sugar beets, or bananas, and the authors found that the miso-type products compared favorably to Japanese-prepared miso.

Ogi is a staple cereal of the Yorubas of Nigeria and is the first native food given to babies at weaning. It is produced generally by soaking corn grains in warm water for 1 to 2 days followed by wet-milling and sieving through a screen mesh. The sieved material is allowed to sediment and ferment, and is marketed as wet cakes wrapped in leaves. Various food dishes are made from the fermented cakes or the ogi *(8)*. During the steeping of corn, *Corynebacterium* spp. become prominent and appear to be responsible for the diastatic action necessary for the growth of yeasts and lactic acid bacteria *(4)*. Along with the corynebacteria, *S. cerevisiae, Enterobacter cloacae,* and *L. plantarum* have been found to be prominent in the traditional ogi fermentation. Most of the acid produced is lactic which depresses the pH of desirable products to around 3.8. The corynebacteria develop early and their activities cease after the first day, while those of the lactobacilli and yeasts continue beyond the first day of fermentation. More recently, a new process for making ogi has been developed, tested, and found to produce an excellent quality product better than the traditional *(9)*. By the new method, corn is dry-milled into whole corn and dehulled corn flour. Upon the addition of water, the mixture is cooked, cooled, and then inoculated with a mixed culture (starter) of the following 3 organisms: *L. plantarum, S. lactis,* and *Saccharomyces rouxii.* The inoculated preparation is incubated at 32 C for 28 hr during which time the pH of corn drops from 6.1 to 3.8. This process eliminates the need for starch hydrolyzing bacteria. In addition to the shorter fermentation time, there is also less chance for faulty fermentations.

Gari is a staple food of West Africa prepared from the root of the cassava plant. Cassava roots contain a cyanogenic glucoside which makes them poisonous if eaten fresh or raw. The roots are rendered safe by a fermentation during which the toxic glucoside decomposes with the liberation of gaseous hydrocyanic acid. In the home preparation of gari, the outer peel and the thick cortex of the cassava roots are removed, followed by grinding or grating the remainder. The pulp is pressed to remove the remaining juice, and placed in bags for 3–4 days to allow fermentation to occur. The fermented product is cooked by frying. The fermentation has been found to occur in a two-stage manner. In the first stage, *Corynebacterium manihot* ferments starch with the production of acids and the consequent lowering of pH. Under acidic conditions, the cyanogenic glucoside then undergoes spontaneous hydrolysis with the liberation of gaseous hydrocyanic acid *(13)*. The acidic conditions favor the growth of *Geotricum* spp., and these organisms appear to be responsible for the characteristic taste and aroma of gari.

Bongkrek is an example of a fermented food product that has in the past led to a large number of deaths. Bongkrek or semaji is a cocoanut presscake product of Central Indonesia and it is the homemade product that may become toxic. The safe products are fermented by *Rhizopus oligosporus* with the finished cakes covered and penetrated by the white fungus. In order to obtain the desirable fungal growth, it appears to be essential that conditions permit good growth within the first 1 to 2 days of incubation. If, however, bacterial growth is favored during this time, and if the bacterium *Pseudomonas cocovenenans* is present, it grows and produces two toxic substances, toxoflavin and bongkrekic acid *(88, 89)*. Both of these compounds show antifungal and antibacterial activity, are toxic for man and animals, and are heat stable. Production for both is favored by growth of the organisms on cocoanut (toxoflavin can be produced in complex culture media). The structural formulae of the two antibiotics are indicated below. Toxoflavin acts as an electron carrier and bongkrekic acid inhibits oxidative phosphorylation in mitochondria.

Toxoflavin

Bongkrekic acid

The latter antibiotic has been shown to be cidal to all 17 molds studied by Subik and Behun *(82)* by preventing spore germination and mycelial outgrowth. The growth of *P. cocovenenans* in the preparation of bongkrek is not favored if the acidity of starting materials is kept at or below pH 5.5 *(90)*.

Ontjom is a somewhat similar but more popular fermented product of Indonesia made from peanut press cake. Peanut press cake is the material that remains after oil has been extracted from peanuts. The press cake is soaked in water for about 24 hr, steamed and pressed into molds. The molds are covered with banana leaves and inoculated with either *Neurospora sitophila* or *Rhizopus oligosporus*. The product is ready for consumption 1 or 2 days later. A more detailed description of the ontjom fermentation and the nutritive value of this product have been provided by Beuchat *(10)*.

Beverage and Distilled Products

Beer and **ale** are examples of malt beverages which are produced by brewing. An essential step in the brewing process is the fermentation of carbohydrates to ethanol. Since most of the carbohydrates in grains used for brewing exist as starches, and since the fermenting yeasts do not produce amylases to degrade the starch, a necessary part of beer brewing includes a step whereby malt or other exogenous sources of amylase are provided for the hydrolysis of starches to sugars. The malt is first prepared by allowing barley grains to germinate and this serves as a source of amylases (fungal amylases may be used also). Both β- and α-amylases are involved, with the latter acting to liquefy starch and the former to increase sugar formation *(35)*. In brief, the brewing process begins with the mixing of malt, malt adjuncts, hops, and water. Malt adjuncts include certain grains, grain products, sugars, and other carbohydrate products to serve as fermentable substances. Hops are added as sources of pyrogallol and catechol tannins, resins, essential oils, and other constituents for the purpose of precipitating unstable proteins during the boiling of wort and to provide for biological stability, bitterness, and aroma. The process by which the malt and malt adjuncts are dissolved, and heated, and the starches digested is called mashing. The soluble part of the mashed materials is called wort (compare with koji). In some breweries, lactobacilli are introduced into the mash to lower the pH of wort through lactic acid production. The species generally used for this purpose is *L. delbrueckii (43)*.

Wort and hops are mixed and boiled for 1.5 to 2.5 hr for the purpose of enzyme inactivation, extraction of soluble hop substances, precipitation of coagulable proteins, concentration, and sterilization. Following the boiling of wort and hops, the wort is separated, cooled, and fermented. The fermentation of the sugar-laden wort is carried out by the inoculation of *S. cerevisiae* (Fig. 14-1E). Ale results from the activities of top-fermenting yeasts which depress the pH to around 3.8 while bottom-fermenting yeasts *(S. carlsbergensis* strains) give rise to lager and other beers with pH values of 4.1–4.2. A top-fermentation is complete in 5 to 7 days while a bottom-fermentation requires 7 to 12 days *(65)*. The freshly fermented product is aged and finished by the addition of CO_2 to a final content of 0.45 to 0.52% before it is ready for commerce. The pasteurization of beer may be carried out at 140 F or higher for the purpose of destroying spoilage organisms. When lactic acid bacteria are found in beers, the lactobacilli are found more commonly in top fermentations while pediococci are found in bottom fermentations *(43)*.

Distilled spirits are alcoholic products that result from the distillation of yeast fermentations of grain, grain products, molasses, or fruit or fruit products. Whiskies, gin, vodka, rum, cordials, and liquers are examples of distilled spirits. While the process for producing most products of these types is quite similar to that for beers, the content of alcohol in the final products is considerably higher than for beers. Rye and bourbon are examples of whiskies. In the former, rye and rye malt or rye and barley malt are used in different ratios but at least 51% rye is required by law. Bourbon is made from corn, barley malt or wheat malt, and usually another grain in different proportions, but at least 51% corn is required by law. A sour wort is maintained in order to keep down undesirable organisms, and the souring of wort may occur naturally or by the addition of acid. The mash is generally soured by inoculating with a homolactic such as *L. delbrueckii,* which is capable of lowering the pH to around 3.8 in 6 to 10 hr *(63)*. The malt enzymes (diastases) convert the starches of the cooked grains to dextrins and sugars, and upon completion of diastatic action and lactic acid production, the mash is heated to destroy all microorganisms. It is then cooled to 75–80 F, and pitched (inoculated) with a suitable strain of *S. cerevisiae* for the production of ethanol. Upon completion of fermentation, the liquid is distilled to recover the alcohol and other volatiles, and these are handled and stored under special conditions relative to the type of product being made. **Scotch whiskey** is made primarily from barley and is produced from barley malt dried in kilns over peat fires. **Rum** is produced from the distillate of fermented sugar cane or molasses. **Brandy** is a product prepared by distilling grape or other fruit wines.

Wines are normal alcoholic fermentations of sound grapes followed by aging. A large number of other fruits such as peaches, pears, and so forth, may be fermented for wines, but in these instances the wine is named by the fruit such as "pear wine", "peach wine", and the like. Since fruits already contain fermentable sugars, the use of exogenous sources of amylases is not necessary as it is when grains are used for beers or whiskies. Wine making begins with the selection of suitable grapes which are crushed and then treated with a sulfite such as potassium metabisulfite to retard the growth of acetic acid bacteria, wild yeasts, and molds. The pressed juice, called must, is inoculated with a suitable wine strain of *S. ellipsoideus*. The fermentation is allowed to go for 3 to 5 days at temperatures between 70 and 90 F, and good yeast strains may produce up to 14–18% ethanol *(63)*. Following fermentation, the wine is racked, i.e., drawn off from the lees or sediment, which among other things contains potassium bitartrate (cream of tartar). The clearing and development of flavor occur during the storage and aging process. Red wines are made by initially fermenting the crushed grape must "on the skins" during which pigment is extracted into the juice, while white wines are prepared generally from the juice of white grapes. Champagne, a sparkling wine made by a secondary fermentation of wine, is produced, briefly, by adding sugar, citric acid, and a champagne yeast starter to bottles of previously prepared, selected table wine. The bottles are corked, clamped, and stored horizontally at suitable temperatures for about 6 mon. They are then removed, agitated, and aged for an additional period of up to 4 yr. The final sedimentation of yeast cells and tartrates is accelerated by reducing the temperature of the wine to around 25 C and holding for 1 to 2 weeks. Clarification of the champagne is brought about by working the sediment down the bottle onto the cork over a period of 2 to 6 weeks by frequent rotation of the bottle. Finally, the sediment is frozen and disgorged upon removal of the cork. There are numerous types of wines, and the reader is referred to other references such as Prescott and Dunn *(65)* for more details of their production, classification, and so on.

Palm wine or Nigerian palm wine is an alcoholic beverage consumed throughout the tropics and is produced by a natural fermentation of palm sap. The sap is sweet, dirty brown in color, and contains 10–12% sugar, mainly sucrose. The fermentation process results in the sap becoming milky-white in appearance due to the presence of large numbers of fermenting bacteria and yeasts. This product is unique in that the microorganisms are alive when the wine is consumed. The fermentation has been reviewed and studied by Faparusi and Bassir *(23)* and Okafor *(57)* who found the following genera of bacteria to be the most predominant in finished products: *Micrococcus, Leuconostoc, Streptococcus, Lactobacillus,* and *Acetobacter.* The predominant yeasts found are *Saccharomyces* and *Candida* spp., with the former being the more common *(56).* The fermentation occurs over a 36–48 hr period during which the pH of sap falls from 7.0–7.2 to <4.5. Fermentation products consist of organic acids in addition to ethanol. During the early phases of fermentation, *Serratia* and *Enterobacter* spp. increase in numbers followed by lactobacilli and leuconostocs. After a 48-hr fermentation, *Acetobacter* spp. begin to appear *(23, 58).*

Saké is an alcoholic beverage commonly produced in Japan. The substrate is the starch from steamed rice and its hydrolysis to sugars is carried out by *A. oryzae* to yield the koji. Fermentation is carried out by *Saccharomyces sake* over periods of 30–40 days resulting in a product containing 12–15% alcohol and around 0.3% lactic acid *(63).* The latter is produced by hetero- and homolactic lactobacilli.

Cider, in the United States, is a product that represents a mild fermentation of apple juice by naturally occurring yeasts. In making apple cider, the fruits are selected, washed, and ground into a pulp followed by pressing the pulp "cheeses" to release the juice. The juice is now strained and placed in a storage tank where sedimentation of particulate matter occurs, usually for 12–36 hr or several days if the temperature is kept at 40 F or below. The clarified juice is cider. If pasteurization is desired, this is accomplished by heating at 170 F for 10 min. The chemical preservative most often used is sodium sorbate at a level of 0.10%. Preservation may be effected also by chilling or freezing. The finished product contains small amounts of ethanol in addition to acetaldehyde. The holding of nonpasteurized or unpreserved cider at suitable temperatures invariably leads to the development of cider vinegar, which indicates the presence of acetic acid bacteria in these products. The pathway employed by acetic acid bacteria is summarized in Figure 14-1F, G.

In their study of the ecology of the acetic acid bacteria in cider manufacture, Passmore and Carr *(60)* found 6 species of *Acetobacter* and noted that those which display a preference for sugars tend to be found early in the cider process, while those which are more acid-tolerant and capable of oxidizing alcohols appear after the yeasts have converted most of the sugars to ethanol. *Zymomonas* spp., gram negative bacteria that ferment glucose to ethanol, have been isolated from ciders, but they are presumed to be present in low numbers.

Coffee beans develop as berries or cherries in their natural state and as such they have an outer pulpy and mucilaginous envelope which must be removed before the beans can be dried and roasted. The wet method of removal of this layer seems to produce the most desirable product, and it consists of depulping and demucilaging followed by drying. While depulping is done mechanically, demucilaging is accomplished by natural fermentation. The mucilage layer is composed largely of pectic substances *(27)* and pectinolytic microorganisms are important in their removal. *Erwinia dissolvens* has been found to be the single most important bacterium during the demucilaging fermentation in Hawaiian *(28)* and Congo coffee cherries *(87)*, although Pederson and Breed *(62)* indicated that the fermentation of coffee berries from Mexico and Colombia was carried out by typical lactic acid bacteria (leuconostocs and lactobacilli). Agate and Bhat *(3)* in their study of coffee cherries from the Mysore State of India found that the following pectinolytic yeasts predominated and played important roles in the loosening and removal of the mucilaginous layers: *Saccharomyces marxianus, S. bayanus, S. ellipsoideus,* and *Schizosaccharomyces* spp. Microorganisms appear not to contribute to the development of flavor and aroma in coffee beans as they do in cocoa beans.

Cocoa beans (actually cacao beans–cocoa is the powder and chocolate is the manufactured product), from which chocolate is derived, are obtained from the fruits or pods of the cacao plant in parts of Africa, Asia, and South America. The beans are extracted from the fruits and fermented in piles, boxes, or tanks for 2–12 days depending upon type and size of beans. During the fermentation, high temperatures (45–50 C) and large quantities of liquid develop. Following sun or air drying, during which the water content is reduced to <7.5%, the beans are roasted to develop the characteristic flavor and aroma of chocolate. The fermentation occurs in two phases. In the first, sugars from the acidic pulp (ca. pH 3.6) are converted to alcohol. The second phase consists of the alcohol being oxidized to acetic acid. In a study of Brazilian cocoa beans by Camargo *et al. (12),* the flora on the first day of fermentation at 21 C consisted of yeasts. On the third day, the temperature had risen to 49 C and the yeast count had decreased to no more than 10% of the total flora. Over the 7-day fermentation, the pH increased from 3.9 to 7.1. The cessation of yeast and bacterial activities around the third day is due in part to the unfavorable temperature, lack of fermentable sugars, and the increase in alcohol. While some decrease in acetic acid bacteria occurs because of high temperature, not all of these organisms are destroyed. Yeasts and acetic acid bacteria are clearly the most important fermenters of cocoa beans. Of the 142 yeasts isolated by Camargo *et al.,* 105 were asporogenous and 37 were ascosporogenous. *Candida krusei* (a thermotolerant yeast) was the most frequently encountered species, and it became predominant after the second day. High numbers of *Geotrichum candidum* and *Candida mycoderma* were found and both were shown to be pectinolytic. The gram negative bacterium, *Zymomonas mobilis,* has been tried without success as a starter in the cocoa bean fermentation. This organism can convert sugars to ethanol and the latter to acetic acid. Heterofermentative lactics and *Acetobacter* spp. are involved in the fermentation *(66).*

While yeasts play important roles in producing alcohol in the cocoa bean fermentation, their presence appears even more essential to the development of the final, desirable chocolate flavor of roasted beans. Levanon and Rossetini *(45)* found that the endoenzymes released by autolyzing yeasts are responsible for the development of chocolate precursor compounds. The acetic acid apparently makes the bean tegument permeable to the yeast enzymes. It has been shown that chocolate aroma occurs only after cocoa beans are roasted and that the roasting of unfermented beans does not produce the characteristic aroma *(67).* Reducing sugars and free amino acids are in some way involved in the final chocolate aroma development *(68).*

Breads

San Francisco sour dough bread is similar to sour dough breads produced in various countries. Historically, the starter for sour dough breads consists of the natural flora of baker's barm (sour ferment or mother sponge, with a portion of each inoculated dough saved as starter for next batch). The barm generally contains a mixture of yeasts and lactic acid bacteria. In the case of San Francisco sour dough bread, the yeast has been identified as *Saccharomyces exiguus (83),* and the responsible bacterium as *Lactobacillus sanfrancisco (44).* The souring is caused by the acids produced by the bacterium, while the yeast is responsible for the leavening action even though some CO_2 is produced by the bacterium. The pH of these sour doughs ranges from 3.8 to 4.5. Both acetic and lactic acids are produced with the former accounting for 20–30% of the total acidity *(44).*

Idli is a fermented bread-type product common in Southern India. It is made from rice and black gram mungo (urd beans). These two ingredients

are soaked in water separately for 3–10 hr and then ground in varying proportions, mixed, and allowed to ferment overnight. The fermented and
raised product is cooked by steaming and served hot. It is said to resemble a
steamed, sour-dough bread (80). During the fermentation, the initial pH of
around 6.0 falls to values of 4.3–5.3. In a particular study, a batter pH of 4.70
after a 20-hr fermentation was associated with 2.5% lactic acid, based on dry
grain weight (53). In their studies of idli, Steinkraus et al. (80) found total
bacterial counts of 10^8–10^9/g after 20–22 hr of fermentation. Most of the
organisms consisted of gram positive cocci or short rods with L. mesenteroides being the single most abundant species followed by S. faecalis. The
leavening action of idli is produced by L. mesenteroides and is the only
known instance of a lactic acid bacterium having this role in a naturally fermented bread (53). The latter authors confirmed the work of others before
them in finding the urd beans to be a more important source of lactic acid
bacteria than rice. L. mesenteroides reaches its peak at around 24 hr with
S. faecalis becoming active only after around 20 hr. Only after idli has fermented for more than 30 hr does P. cerevisiae become active. The product is
not fermented generally beyond 24 hr because maximum leavening action
occurs at this time and decreases with longer incubations. When idli is
allowed to ferment longer, more acidity is produced. It has been found that
total acidity (expressed as g lactic acid/g of dry grains) increased from 2.71%
after 24 hr to 3.70% after 71 hr while pH decreased from 4.55 to 4.10 over the
same period (53).

SINGLE CELL PROTEIN

The cultivation of unicellular microorganisms as a direct source of food
for man was suggested in the early 1900s. The expression "single cell protein" (SCP) was coined at the Massachusetts Institute of Technology around
1966 to depict the idea of microorganisms as food sources (76). Although
SCP is a misnomer in that proteins are not the only food constituent
represented by microbial cells, it obviates the need to refer to each product
generically as in "algal protein", "yeast cell protein", and so on. Although
SCP as a potential and real source of food for man differs from the other
products covered in this chapter, with the exception of that from algal cells it
is produced in a similar manner—by fermentation.

Rationale for SCP Production

It is imperative that new food sources be found in order that future genera-
tions of mankind be adequately fed. A food source that is nutritionally complete, and that requires a minimum of land, time, and cost to produce is
highly desirable. In addition to meeting these criteria, SCP can be produced
on a variety of waste materials. Among the overall advantages of SCP over
plant and animal sources of proteins are the following (42): (1)

Microorganisms have a very short generation time, and can thus provide a rapid mass increase. (2) Microorganisms can be easily modified genetically—to produce cells that bring about desirable results. (3) The protein content is high. (4) The production of SCP can be based upon raw materials readily available in large quantities, and (5) SCP production can be carried out in continuous culture and thus be independent of climatic changes.

The greater speed and efficiency of microbial protein production compared to plant and animal sources may be illustrated as follows: A 1,000-lb steer produces about 1 lb of new protein/day; soybeans (prorated over a growing season) produce about 80 lb; while yeasts produce about 50 tons!

Organisms and Fermentation Substrates

A large number of algae, yeasts, molds, and bacteria have been studied as SCP sources and among the most promising genera and species are the following: **Algae**—*Spirulina maxima, Chlorella* spp., and *Scenedesmus* spp. **Yeasts**—*Candida guilliermondii, C. utilis, C. lipolytica,* and *C. tropicalis; Debaryomyces kloeckeri; Torulopsis candida, T. methanosorbosa; Pichia* spp.; *Kluyveromyces fragilis; Hansenula polymorpha; Rhodotorula* spp.; and *Saccharomyces* spp. **Filamentous fungi**—*Agaricus* spp.; *Aspergillus* spp.; *Fusarium* spp.; *Penicillium* spp.; *Endomycopsis fibuligera;* and *Trichosporon cutaneum.* **Bacteria**—*Bacillus* spp.; *Cellulomonas* spp.; *Acinetobacter calcoaceticus; Nocardia* spp.; *Methylomonas* spp.; *Aeromonas hydrophilia; Alcaligenes eutrophus (Hydrogenomonas eutropha), Mycobacterium* sp.; and *Rhodopseudomonas* sp. Of these groups, yeasts have received by far the most attention.

The choice of a given organism is dictated in large part by the type of substrate or waste material in question. The blue-green alga *Spirulina maxima* grows in shallow waters high in bicarbonate at a temperature of 30 C and between pH 8.5–11.0. It can be harvested from pond waters and dried for food use. This alga has been eaten by the people of the Chad Republic for many years (76). Other algal cells require only sunlight, CO_2, minerals, water, and proper growth temperatures. However, the large scale use of algal cells as SCP sources is said to be practical only in areas below 35° latitude where sunlight is available most of the year (47).

Bacteria, yeasts, and molds can be grown on a wide variety of fermentable materials including food processing wastes (e.g., cheese whey, brewery, potato processing, cannery, and coffee); industrial wastes (e.g., sulfite liquor in the paper industry and combustion gases); and cellulosic wastes (e.g., bagasse, newsprint mill, and barley straw). In the case of cellulosic wastes, it is necessary to use organisms that can utilize cellulose such as *Cellulomonas* sp. or *Trichoderma viride.* A mixed culture of *Cellulomonas* and *Alcaligenes* has been employed. For starchy materials, a combination of *Endomycopsis fibuligera* and a *Candida* sp. such as *C. utilis* has been employed where the former effects hydrolysis of starches and the latter subsists upon the hydrolyzed products to produce biomass. Some other representative substrates and fermenting organisms are listed in Table 14-3.

TABLE 14-3. Some examples of substrate materials that support the growth of microorganisms in the production of SCP.

Substrates	Microorganisms
CO_2 and sunlight	*Chlorella pyrenoidosa*
	Scenedesmus quadricauda
	Spirulina maxima
n-Alkanes, kerosene	*Candida intermedia, C. lipolytica,*
	C. tropicalis
	Nocardia sp.
Methane	*Methylomonas* sp. *(Methanomonas)*
	Methylococcus capsulatus
	Trichoderma sp.
H_2 and CO_2	*Alcaligenes eutrophus (Hydrogenomonas eutropha)*
Gas oil	*Acinetobacter calcoaceticus (Micrococcus cerificans)*
	Candida lipolytica
Methanol	*Methylomonas methanica (Methanomonas methanica)*
Ethanol	*Candida utilis*
	Acinetobacter calcoaceticus
Sulfite liquor wastes	*Candida utilis*
Cellulose	*Cellulomonas* sp.
	Trichoderma viride
Starches	*Endomycopsis fibuligera*
Sugars	*Saccharomyces cerevisiae*
	Candida utilis
	Kluyveromyces fragilis

SCP Products

While the products of fermentation of other foods in this chapter consist of an altered or modified substrate, the product of SCP fermentation consists solely of the microbial cells that grow upon the substrate. The cells may be used directly as a protein source in animal feed formulations thereby freeing animal feed, such as corn, for human consumption, or they may be used as a protein source or food ingredient for human food. In the case of animal feed or feed supplements, the dried cells may be used without further processing. As noted above, whole cells of *Spirulina maxima* are consumed by man in at least one part of Africa.

For human use, the most likely products are SCP concentrates or isolates which may be further processed into textured or functional SCP products.

To produce functional protein fibers, cells are mechanically disrupted, cell walls removed by centrifugation, proteins precipitated from disrupted cells, and the resulting protein extruded from syringe-like orifices into suitable menstra such as acetate buffer, $HClO_4$, acetic acid, and the like. The SCP fibers may now be used to form textured protein products. Baker's yeast protein is the only product of this type presently approved for human food ingredient use in the U.S., but as Litchfield (47) has noted, it is reasonable to assume that similar products from other food-grade yeasts such as *Candida utilis* and *Kluyveromyces fragilis* will ultimately be approved.

The Nutrition and Safety of SCP

Chemical analyses of the microorganisms evaluated for SCP reveal that they are comparable in amino acid content and type to plant and animal sources with the possible exception of methionine, which is lower in some SCP sources. All are relatively high in nitrogen. For example, the approximate percentage composition of N on a dry weight basis is as follows: Bacteria 12–13, yeast 8–9, algae 8–10, and filamentous fungi 5–8 (42). In addition to proteins, microorganisms contain adequate levels of carbohydrates, lipids, and minerals, and are excellent sources of B-vitamins. The fat content varies among these sources with algal cells containing the highest levels and bacteria the lowest. On a dry weight basis, nucleic acids average 3–8% for algae, 6–12 for yeasts, and 8–16 for bacteria (42). B-Vitamins are high in all SCP sources. The digestibility of SCP in experimental animals has been found to be lower than for animal proteins such as casein. A thorough review of the chemical composition of SCP from a large variety of microorganisms has been made by Waslien (94).

Success has been achieved in rat-feeding studies with a variety of SCP products, but human-feeding studies have been less successful with the exception of certain yeast cell products. Gastrointestinal disturbances are common complaints in man following the consumption of algal and bacterial SCP, and these and other problems associated with the consumption of SCP have been reviewed elsewhere (94). When gram negative bacteria are used as SCP sources for human use, the endotoxins must be removed or detoxified.

The high nucleic acid content of SCP leads to kidney stone formation and/or gout in man. As noted above, the nucleic acid content of bacterial SCP may be as high as 16% while the recommended daily intake for man is about 2 g. The problems are caused by an accumulation of uric acid, which is sparingly soluble in plasma. Upon the breakdown of nucleic acids, purine and pyrimidine bases are released. Adenine and guanine (purines) are metabolized to uric acid. Lower animals can degrade uric acid to the soluble compound allantoin (they possess the enzyme uricase) and consequently the consumption of high levels of nucleic acids does not present metabolic problems to these animals as it does to man. While high nucleic acid contents presented problems in the early development and use of SCP, these compounds can be reduced to levels below 2% by techniques such as acid precipitation, acid or alkaline hydrolysis, or by use of endogenous and bovine pancreatic RNases *(47)*.

For further information on SCP, see References *16, 42, 46, 47, 76, 85, 94*.

REFERENCES

1. **Acton, J. C., J. G. Williams, and M. G. Johnson.** 1972. Effect of fermentation temperature on changes in meat properties and flavor of summer sausage. J. Milk Food Technol. *35:* 264–268.

2. **Acton, J. C., R. L. Dick, and E. L. Norris.** 1977. Utilization of various carbohydrates in fermented sausage. J. Food Sci. *42:* 174–178.

3. **Agate, A. D. and J. V. Bhat.** 1966. Role of pectinolytic yeasts in the degradation of mucilage layer of *Coffea robusta* cherries. Appl. Microbiol. *14:* 256–260.

4. **Akinrele, I. A.** 1970. Fermentation studies on maize during the preparation of a traditional African starch-cake food. J. Sci. Fd. Agric. *21:* 619–625.

5. **Allen, S. H. G., R. W. Kellermeyer, R. L. Stjernholm, and H. G. Wood.** 1964. Purification and properties of enzymes involved in the propionic acid fermentation. J. Bacteriol. *87:* 171–187.

6. **Arnott, D. R., C. L. Duitschaever, and D. H. Bullock.** 1974. Microbiological evaluation of yogurt produced commercially in Ontario. J. Milk & Fd. Technol. *37:* 11–13.

7. **Ayres, J. C., D. A. Lillard, and L. Leistner.** 1967. Mold ripened meat products, pp. 156–168 IN: *Proceedings, 20th Annual Reciprocal Meat Conference* (Nat'l. Live Stock & Meat Bd.: Chicago).

8. **Banigo, E. O. I. and H. G. Muller.** 1972. Manufacture of ogi (a Nigerian fermented cereal porridge): Comparative evaluation of corn, sorghum and millet. Can. Inst. Fd. Sci. Technol. J. *5:* 217–221.

9. **Banigo, E. O. I., J. M. deMan, and C. L. Duitschaever.** 1974. Utilization of high-lysine corn for the manufacture of ogi using a new, improved processing system. Cereal Chem. *51:* 559–572.

10. **Beuchat, L. R.** 1976. Fungal fermentation of peanut press cake. Econ. Bot. *30:* 227–234.

11. **Brown, W. V. and E. B. Collins.** 1977. End products and fermentation balances for lactic streptococci grown aerobically on low concentrations of glucose. Appl. Environ. Microbiol. *33:* 38–42.

12. **Camargo, R. de, J. Leme, Jr., and A. M. Filho.** 1963. General observations on the microflora of fermenting cocoa beans *(Theobroma cacao)* in Bahia (Brazil). Food Technol. *17:* 1328–1330.

13. **Collard, P. and S. Levi.** 1959. A two-stage fermentation of cassava. Nature *183:* 620–621.

14. **Collins, E. B.** 1972. Biosynthesis of flavor compounds by microorganisms. J. Dairy Sci. *55:* 1022–1028.

15. **Davis, J. G.** 1975. The microbiology of yoghourt, pp. 245–263 IN: *Lactic Acid Bacteria in Beverages and Food,* edited by J. G. Carr *et al.* (Academic Press: N.Y.).

16. **Davis, P., editor.** 1974. *Single Cell Protein* (Academic Press: N.Y.).

17. **Deibel, R. H. and C. F. Niven, Jr.** 1957. *Pediococcus cerevisiae:* A starter culture for summer sausage. Bacteriol. Proc., pp. 14–15.

18. **Deibel, R. H., C. F. Niven, Jr., and G. D. Wilson.** 1961. Microbiology of meat curing. III. Some microbiological and related technological aspects in the manufacture of fermented sausages. Appl. Microbiol. *9:* 156–161.

19. **DeKetelaere, A., D. Demeyer, P. Vandekerckhove, and I. Vervaeke.** 1974. Stoichiometry of carbohydrate fermentation during dry sausage ripening. J. Food Sci. *39:* 297–300.

20. **Doelle, H. W.** 1975. *Bacterial Metabolism,* Chapter 9 (Academic Press: N.Y.).

21. **Etchells, J. L., H. P. Fleming, and T. A. Bell.** 1975. Factors influencing the growth of lactic acid bacteria during the fermentation of brined cucumbers, pp. 281–305 IN: *Lactic Acid Bacteria in Beverages and Food,* edited by J. G. Carr *et al.* (Academic Press: N.Y.).

22. **Everson, C. W., W. E. Danner, and P. A. Hammes.** 1970. Improved starter culture for semi-dry sausage. Food Technol. *24:* 42–44.

23. **Faparusi, S. I. and O. Bassir.** 1971. Microflora of fermenting palm-wine. J. Fd. Sci. Technol. *8:* 206–000.

24. **Filho, A. M. and C. W. Hesseltine.** 1964. Tempeh fermentation: Package and tray fermentations. Food Technol. *18:* 761–765.

25. **Forrest, W. W. and D. J. Walker.** 1971. The generation and utilization of energy during growth. Adv. Microbial Physiol. *5:* 213–274.

26. **Foster, E. M., F. E. Nelson, M. L. Speck, R. N. Doetsch, and J. C. Olson.** 1957. *Dairy Microbiology* (Prentice-Hall: N.J.).

27. **Frank, H. A. and A. S. Dela Cruz.** 1964. Role of incidental microflora in natural decomposition of mucilage layer in Kona coffee cherries. J. Food Sci. *29:* 850–853.

28. **Frank, H. A., N. A. Lum, and A. S. Dela Cruz.** 1965. Bacteria responsible for mucilage-layer decomposition in Kona coffee cherries. Appl. Microbiol. *13:* 201–207.

29. **Gasser, F. and C. Gasser.** 1971. Immunological relationships among lactic dehydrogenases in the genera *Lactobacillus* and *Leuconostoc.* J. Bacteriol. *106:* 113–125.

30. **Gilliland, S. E. and M. L. Speck.** 1974. Frozen concentrated cultures of lactic starter bacteria. A review. J. Milk & Fd. Technol. *37:* 107–111.

31. **Giolitti, G., C. A. Cantoni, M. A. Bianchi, and P. Renon.** 1971. Microbiology and chemical changes in raw hams of Italian type. J. Appl. Bacteriol. *34:* 51–61.

32. **Goel, M. C., D. C. Kulshrestha, E. H. Marth, D. W. Francis, J. G. Bradshaw, and R. B. Read, Jr.** 1971. Fate of coliforms in yogurt, buttermilk, sour cream, and cottage cheese during refrigerated storage. J. Milk Food Technol. *34:* 54–58.

33. **Goodenough, E. R. and D. H. Kleyn.** 1976. Qualitative and quantitative changes in carbohydrates during the manufacture of yoghurt. J. Dairy Sci. *59:* 45–47.

34. **Gunsalus, I. C. and C. W. Shuster.** 1961. Energy yielding metabolism in bacteria, pp. 1–58 IN: *The Bacteria,* Vol. 2, edited by I. C. Gunsalus and R. Y. Stanier (Academic Press: N.Y.).

35. **Haas, G. J.** 1976. Alcoholic beverages and fermented foods, pp. 165–191 IN: *Industrial Microbiology,* edited by B. M. Miller and W. Litsky (McGraw-Hill: N.Y.).

36. **Harris, D. A., L. Chaiet, R. P. Dudley, and P. Ebert.** 1957. The development of a commercial starter culture for summer sausages. Bacteriol. Proc., p. 15.

37. **Hettinga, D. H. and G. W. Reinbold.** 1972. The propionic-acid bacteria—A review. J. Milk Food Technol. *35:* 295–301, 358–372, 436–447.

38. **Hontebeyrie, M. and F. Gasser.** 1977. Deoxyribonucleic acid homologies in the genus *Leuconostoc.* Intern. J. Syst. Bacteriol. *27:* 9–14.

39. **Ilany-Feigenbaum, J. Diamant, S. Laxer, and A. Pinsky.** 1969. Japanese miso-type products prepared by using defatted soybean flakes and various carbohydrate-containing foods. Food Technol. *23:* 554–556.

40. **Ingram, M.** 1975. The lactic acid bacteria—A broad view, pp. 1–13 IN: *Lactic Acid Bacteria in Beverages and Food,* edited by J. G. Carr *et al.* (Academic Press: N.Y.).

41. **Johnson, M. G. and E. B. Collins.** 1973. Synthesis of lipoic acid by *Streptococcus faecalis* 10Cl and end-products produced anaerobically from low concentrations of glucose. J. Gen. Microbiol. *78:* 47–55.

42. **Kihlberg, R.** 1972. The microbe as a source of food. Ann. Rev. Microbiol. *26:* 427–466.

43. **Kleyn, J. and J. Hough.** 1971. The microbiology of brewing. Ann. Rev. Microbiol. *25:* 583–608.

44. **Kline, L. and T. F. Sugihara.** 1971. Microorganisms of the San Francisco sour dough bread process. II. Isolation and characterization of undescribed bacterial species responsible for the souring activity. Appl. Microbiol. *21:* 459–465.

45. **Levanon, Y. and S. M. O. Rossetini.** 1965. A laboratory study of farm processing of cocoa beans for industrial use. J. Food Sci. *30:* 719–722.

46. **Lipinsky, E. S., J. H. Litchfield, and D. I. C. Wang.** 1970. Algae, bacteria, and yeasts as food or feed. CRC Crit. Rev. Food Technol. *1:* 581–618.

47. **Litchfield, J. H.** 1977. Single-cell proteins. Food Technol. *31:* No. 5, 175–179.

48. **London, J. and K. Kline.** 1973. Aldolase of lactic acid bacteria: A case history in the use of an enzyme as an evolutionary marker. Bacteriol. Rev. *37:* 453–478.

49. **London, J.** 1976. The ecology and taxonomic status of the lactobacilli. Ann. Rev. Microbiol. *30:* 279–301.

50. **Marth, E. H.** 1974. Fermentations, Chapter 13 IN: *Fundamentals of Dairy Chemistry,* edited by B. H. Webb *et al.* (Avi Publishing: Conn.).

51. **Miller, A., III, W. E. Sandine, and P. R. Elliker.** 1970. Deoxyribonucleic acid base composition of lactobacilli determined by thermal denaturation. J. Bacteriol. *102:* 278–280.

52. **Miller, A., III, W. E. Sandine, and P. R. Elliker.** 1971. Deoxyribonucleic acid homology in the genus *Lactobacillus.* Can. J. Microbiol. *17:* 625–634.

53. **Mukherjee, S. K., M. N. Albury, C. S. Pederson, A. G. van Veen, and K. H. Steinkraus.** 1965. Role of *Leuconostoc mesenteroides* in leavening the batter of idli, a fermented food of India. Appl. Microbiol. *13:* 227–231.

54. **Mundt, J. O.** 1975. Unidentified streptococci from plants. Int. J. Syst. Bacteriol. *25:* 281–285.

55. **Niinivaara, F. P., M. S. Pohja, and S. E. Komulainen.** 1964. Some aspects about using bacterial pure cultures in the manufacture of fermented sausages. Food Technol. *18:* 147–153.

56. **Okafor, N.** 1972. Palm-wine yeasts from parts of Nigeria. J. Sci. Fd. Agric. *23:* 1399–1407.

57. **Okafor, N.** 1975a. Microbiology of Nigerian palm wine with particular reference to bacteria. J. Appl. Bacteriol. *38:* 81–88.

58. **Okafor, N.** 1975b. Preliminary microbiological studies on the preservation of palm wine. J. Appl. Bacteriol. *38:* 1–7.

59. **Palumbo, S. A., J. L. Smith, and S. A. Kerman.** 1973. Lebanon bologna. I. Manufacture and processing. J. Milk Food Technol. *36:* 497–503.

60. **Passmore, S. M. and J. G. Carr.** 1975. The ecology of the acetic acid bacteria with particular reference to cider manufacture. J. Appl. Bacteriol. *38:* 151–158.

61. **Payne, W. J.** 1970. Energy yields and growth of heterotrophs. Ann. Rev. Microbiol. *24:* 17–52.

62. **Pederson, C. S. and R. S. Breed.** 1946. Fermentation of coffee. Food Res. *11:* 99–106.

63. **Pederson, C. S.** 1971. *Microbiology of Food Fermentations* (Avi Publishing: Conn.).

64. **Potter, M. E., M. B. Kruse, M. A. Matthews, R. O. Hill, and R. J. Martin.** 1976. A sausage-associated outbreak of trichinosis in Illinois. Amer. J. Pub. Hlth. *66:* 1194–1196.

65. **Prescott, S. C. and C. G. Dunn.** 1957. *Industrial Microbiology.* (McGraw-Hill: N.Y.).

66. **Roelofsen, P. A.** 1958. Fermentation, drying, and storage of cacao beans. Adv. Food Res. *8:* 225–296.

67. **Rohan, T. A.** 1964. The precursors of chocolate aroma: A comparative study of fermented and unfermented cocoa beans. J. Food Sci. *29:* 456–459.

68. **Rohan, T. A. and T. Stewart.** 1966. The precursors of chocolate aroma: Changes in the sugars during the roasting of cocoa beans. J. Food Sci. *31:* 206–209.

69. **Saisithi, P., B.-O. Kasemsarn, J. Liston, and A. M. Dollar.** 1966. Microbiology and chemistry of fermented fish. J. Food Sci. *31:* 105–110.

70. **Sandine, W. E., P. C. Radich, and P. R. Elliker.** 1972. Ecology of the lactic streptococci. A review. J. Milk Food Technol. *35:* 176–185.

71. **Senez, J. C.** 1962. Some considerations on the energetics of bacterial growth. Bacteriol. Rev. *26:* 95–107.

72. **Schleifer, K. H. and O. Kandler.** 1972. Peptidoglycan types of bacterial cell walls and their taxonomic implications. Bacteriol. Rev. *36:* 407–477.

73. **Shibasaki, K. and C. W. Hesseltine.** 1962. Miso fermentation. Econ. Bot. *16:* 180–195.

74. **Simonds, J., P. A. Hansen, and S. Lakshmanan.** 1971. Deoxyribonucleic acid hybridization among strains of lactobacilli. J. Bacteriol. *107:* 382–384.

75. **Smith, J. L. and S. A. Palumbo.** 1973. Microbiology of Lebanon bologna. Appl. Microbiol. *26:* 489–496.

76. **Snyder, H. E.** 1970. Microbial sources of protein. Adv. Food Res. *18:* 85–140.

77. **Speckman, R. A. and E. B. Collins.** 1968. Diacetyl biosynthesis in *Streptococcus diacetilactis* and *Leuconostoc citrovorum.* J. Bacteriol. *95:* 174–180.

78. **Speckman, R. A. and E. B. Collins.** 1973. Incorporation of radioactive acetate into diacetyl by *Streptococcus diacetilactis.* Appl. Microbiol. *26:* 744–746.

79. **Stamer, J. R.** 1976. Lactic acid bacteria, pp. 404–426, IN: *Food Microbiology: Public Health and Spoilage Aspects,* edited by M. P. deFigueiredo and D. F. Splittstoesser (Ave Publishing Co.: Conn.).

80. **Steinhraus, K. H., A. G. van Veen, and D. B. Thiebeau.** 1967. Studies on idli—an Indian fermented black gram-rice food. Food Technol. *21:* 916–919.

81. **Stouthamer, A. H.** 1969. Determination and significance of molar growth yields. Methods in Microbiol. *1:* 629–663.

82. **Subik, J. and M. Behun.** 1974. Effect of bongkrekic acid on growth and metabolism of filamentous fungi. Arch. Microbiol. *97:* 81–88.

83. **Sugihara, T. F., L. Kline, and M. W. Miller.** 1971. Microorganisms of the San Francisco sour dough bread process. I. Yeasts responsible for the leavening action. Appl. Microbiol. *21:* 456–458.

84. **Sundhagul, M., W. Daengsubha, and P. Suyanandana.** 1975. Thailand's traditional fermented food products: A brief description. Thai J. Agr. Sci. *8:* 205–219.

85. **Tannenbaum, S. R. and D. I. C. Wang, editors.** 1975. *Single-Cell Protein II* (M.I.T. Press: Cambridge, Mass.).

86. **Urbaniak, L. and W. Pezacki.** 1975. Die Milchsäure bildende Rohwurst-Mikroflora und ihre technologisch bedingte Veränderung. Fleischwirtschaft *55:* 229–237.

87. **Van Pee, W. and J. M. Castelein.** 1972. Study of the pectinolytic microflora, particularly the *Enterobacteriaceae,* from fermenting coffee in the Congo. J. Food Sci. *37:* 171–174.

88. **van Veen, A. G. and W. K. Mertens.** 1934a. Die Gifstoffe der sogenannten Bongkrek-vergiftungen auf Java. Rec. Trav. Chim. *53:* 257–268.

89. **van Veen, A. G. and W. K. Mertens.** 1934b. Das Toxoflavin, der gelbe Gifstoff der Bongkrek. Rec. Trav. Chim. *53:* 398–404.

90. **van Veen, A. G.** 1967. The bongkrek toxins, pp. 43–50 IN: *Biochemistry of Some Foodborne Microbial Toxins,* edited by R. I. Mateles and G. N. Wogan (MIT Press: Mass.).

91. **van Veen, A. G. and K. H. Steinkraus.** 1970. Nutritive value and wholesomeness of fermented foods. J. Agr. Food Chem. *18:* 576–578.

92. **Vaughn, R. H.** 1975. Lactic acid fermentation of olives with special reference to California conditions, pp. 307–323 IN: *Lactic Acid Bacteria in Beverages and Food,* edited by J. G. Carr *et al.* (Academic Press: N.Y.).

93. **Wardlaw, F. B., G. C. Skelley, M. G. Johnson, and J. C. Acton.** 1973. Changes in meat components during fermentation, heat processing and drying of a summer sausage. J. Food Sci. *38:* 1228–1231.

94. **Waslien, C. I.** 1976. Unusual sources of proteins for man. CRC Crit. Rev. Food Sci. Nutri. *6:* 77–151.

95. **Williams, R. A. D.** 1975. A review of biochemical techniques in the classification of lactobacilli, pp. 351–367 IN: *Lactic Acid Bacteria in Beverages and Food,* edited by J. G. Carr *et al.* (Academic Press: N.Y.).

96. **Yong, F. M. and B. J. B. Wood.** 1974. Microbiology and biochemistry of the soy sauce fermentation. Adv. Appl. Microbiol. *17:* 157–194.

15

Indices of Food
Sanitary Quality, and
Microbiological
Standards and Criteria

Contamination of foods by disease-producing microorganisms has been known and studied since around 1880. Since that time, numerous instances of food-borne diseases have been recorded in addition to those commonly referred to as food poisoning. Prior to the development of the pasteurization process, diseases such as brucellosis, scarlet fever, typhoid fever, diphtheria, and others were commonplace in milk. Diseases of animals transmissible to man, such as tuberculosis, brucellosis, etc., were also commonplace in meats prior to the mandatory federal inspection of slaughter animals. A third source of food-borne diseases is contamination of foods by food handlers. This source is presently one of the most serious in food poisoning outbreaks, especially with convenience foods. Also of great current importance are pathogenic and food poisoning organisms that tend to be normally associated with certain animals, organisms such as salmonellae, enterococci, and *C. perfringens*.

Among the requirements for foods to be of good sanitary quality, they must be shown to be free of hazardous microorganisms, or those present should be at a safe low level. In general, it would not be feasible to examine each food or food product for the presence of hazardous organisms. The practice which has been in effect for many years and continues to be followed is to determine the sanitary quality of foods by their content of certain indicator organisms. The indicators of sanitary quality now employed for foods consist of two groups of bacteria: coliforms and enterococci. In addition, total numbers are also useful in this regard, and are here treated in this manner.

COLIFORM BACTERIA AS INDICATORS OF FOOD SANITARY QUALITY

The use of *Escherichia coli* as an indicator of water-borne pathogens was apparently first suggested in 1892 by Schardinger. This organism was proposed because it is generally found in the intestinal contents of man and animals. A year later, Theobald Smith *(54)* noted that since this organism is so uniformly present in the intestinal tract, its presence outside the intestines may be regarded as due to contamination with fecal discharges of man or animals. This marked the beginning of the use of the coliforms as indicators of pathogens in water, a practice which has continued.

The primary coliform bacteria consist of *E. coli* and *Enterobacter aerogenes* (see below). Both of these organisms are short, gram negative rods that ferment lactose with the production of gas. *E. coli* has as its primary habitat the intestinal tract of man and other animals, hence, *coli* for colon. Of these organisms, *E. coli* is normally found in the gastrointestinal tract of animals, where they normally do not cause disease. From his studies on the flora of the intestinal tract of man, Haenel *(20)* has placed the percentage of coliforms at less than 1. This investigator found 10^8–10^9/g of these organisms to be common in adult feces. Although *E. aerogenes* may sometimes be found in the intestinal tract of man, this organism is normally associated with vegetation. In a study of 6,577 strains of coliforms from various sources, Griffin and Stuart *(19)* concluded that the occurrence of *E. coli* outside the intestinal tract and *E. aerogenes* and intermediates in places other than non-fecal materials were adventitious. The coliform organisms are well established as fecal indicators for water. Their use as indicators of food sanitary quality derives from their successful use for water. The finding of large numbers of these organisms in foods and water is taken to indicate fecal pollution or contamination. Since the water-borne diseases are generally intestinal diseases, the existence of pollution is taken to indicate the *possibility* that the etiologic agents of these diseases may be present. Whether or not intestinal pathogens are present, the presence of fecal matter in foods or water is undesirable. McCoy *(37)* has stated that in the examination of foods, the presence of intestinal inhabitants should be taken to indicate a lack of cleanliness, not safety. He asserts that the safety of foods can be assessed only by examining for the presence of pathogens.

With respect to the overall choice of indicator organisms, Buttiaux and Mossel *(7)* have stated that they should possess the following properties: (1) Specificity; ideally the bacteria selected should occur only in intestinal environments. (2) They should occur in very high numbers in feces, so as still to be encountered in high dilutions. (3) They should possess a high resistance to the extraenteral environment, the pollution of which is to be indicated. (4) They should permit of relatively easy and fully reliable detection even when present in very low numbers.

Since *E. coli* is more indicative of fecal pollution than *E. aerogenes,* it is somtimes desirable to determine the incidence of this organism in a coliform population. For this purpose, the IMViC formula is employed where I = indole production; M = methyl red reaction; V = Voges-Proskauer reaction (production of acetoin); and C = citrate utilization. The two organisms may then be differentiated as follows:

	I	M	V	C
E. coli	+	+	−	−
E. aerogenes	−	−	+	+

While the coliforms are generally regarded as being *E. coli* and *E. aerogenes*, it should be noted that the genera *Citrobacter* and *Klebsiella* come under this functional classification. *Citrobacter* spp. have been referred to as intermediate coliforms, and delayed lactose fermentation by some strains is known. All strains are MR + and VP −. Most are citrate + while indole production varies. *Klebsiella* isolates are highly variable with respect to IMViC reactions. Although *K. pneumoniae*, the most common klebsiellae in the U.S., is generally MR − VP + and C +, variations are known to occur especially in the MR and indole reactions. *E. coli* isolates tend to be rather consistent in their reactions with the MR reaction being the most stable. The IMViC reactions + + − − designate *E. coli* variety I while variety II strains are − + − −. Some recent studies with coliform organisms give indications that this indicator group is not as easily definable as was once believed. It may be that a redefinition of "coliforms" that would have the effect of broadening the group is indicated (see Other Indicators on p. 302).

Fecal coliforms are distinguished most readily from nonfecal coliforms by use of elevated temperature incubations. This is accomplished by inoculating tubes of EC broth from gas-positive lauryl sulfate tryptose (LST) broth cultures and incubating in a water bath at a temperature between 44–46 C, usually 44.5 or 45.5 C. While a higher percentage of *E. coli* isolates are positive at elevated temperatures than other coliforms, some *Enterobacter* and *Klebsiella* isolates are fecal coliforms by this criterion. In general, about 90% of fecal coliforms are *E. coli* strains. The fecal coliforms produce gas in EC broth at the elevated temperatures while the nonfecals generally do not. The superiority of EC broth over other media for this purpose was shown by Fishbein (12). The study of Fishbein and Surkiewicz (13) indicates a preference for 45.5 C incubation over 44.5. These investigators found two- to threefold fewer false positives at 45.5 than at 44.5 C. Fecal coliforms can be determined by use of other media at elevated temperatures, and some of these techniques have been reviewed elsewhere (10). Methods for the determination of coliforms or fecal coliforms may be obtained from one or more of the standard references in Table 4-1, p. 48.

As stated above, the coliform index was originally proposed for and used as a test of water quality, followed by its use in the dairy industry. Peabody (46) has raised the following questions regarding the justification for using this index for foods:

a. Are we justified in transferring this test which has been useful in the water and dairy fields to foods?
b. Does this group of organisms represent unsanitary conditions when found in a food product?

c.   Does it indicate any hazards of pathogenic bacteria?
d.   Do our present methods recover all of the viable organisms in a food sample?
e.   What interference may there be between selective and indicator systems in the medium and the food material which must be added to it—especially in samples with low bacterial populations?
f.   How can the recovery rates be improved?

Although it may be a bit risky to transfer a procedure from one type of product to another and expect the same performance, the coliform index has found wide use in assessing sanitary quality in foods, and indeed, coliform standards have been established for some foods and recommended for a large number of others. The main difference between water and foods is that the former generally does not support growth of these organisms while many foods do (see Microbial Standards and Food Safety on p. 308).

1. The Growth of Coliforms.   Like most nonpathogenic gram negative bacteria, the coliforms grow well on a large number of media and in many foods. These organisms have been reported to grow as low as −2 C and as high as about 50 C. In foods, growth is poor or very slow at about 5 C although several authors have reported the growth of coliforms between 3–6 C. With respect to pH, these organisms have been reported to grow over a rather wide range, with values ranging from 4.4–9.0. *E. coli* can be grown on a medium containing only an organic carbon source such as glucose and a source of nitrogen such as $(NH_4)_2SO_4$, along with other minerals. Consequently, these organisms grow well on nutrient agar and produce visible colonies within 24 hr at 37 C. They may, therefore, be expected to grow in a large number of foods under the proper conditions. They are capable of growth in the presence of bile salts, which inhibit the growth of gram positive bacteria. Advantage is taken of the latter fact in their isolation from various sources. Unlike most bacteria, they have the capacity to ferment lactose with the production of gas, and this characteristic alone is sufficient to make presumptive determinations of coliforms. By incorporating lactose and bile salts into culture media (such as MacConkey agar), it becomes possible to differentiate between these organisms and most others that may be present in foods as well as water. The general ease with which the coliforms can be cultivated and differentiated makes them somewhat ideal as indicators of sanitary quality. Their identification is sometimes complicated by the presence of atypical strains as well as certain other gram negative enteric bacteria such as *Klebsiella*. The aberrant lactose fermenters, however, appear to be of questionable sanitary significance *(19).*

2. Distribution of Coliforms.   As already stated, the primary habitat of *E. coli* is the intestinal tract of many warm-blooded animals although it may at times be absent from the gut of hogs, while that of *E. aerogenes* is vegetation, and occasionally the intestinal tract. It is not difficult, however, to sometimes demonstrate the presence of these organisms in air, dust, on the hands, and in and on many foods. The problem, then, is not simply the presence of coliforms, but their relative numbers. Most market vegetables can be shown to harbor small numbers of lactose-fermenting, gram negative rods of the coliform type, but if these products have been harvested and handled properly, the numbers tend to be quite low and consequently of no real significance from the standpoint of public health. The bacteria formerly classified as paracolons and characterized by the slow fermentation of lactose are often associated with fresh vegetables (see *Citrobacter*, pp. 15 and 293). The author has found large numbers of these types on raw, preroasted squash seeds. While generally regarded as being nonpathogens, they have been incriminated in mild food poisoning outbreaks where large numbers of cells were required.

One of the attractive properties of *E. coli* as a fecal indicator for water is its period of survival in water. It generally dies off at about the same time as the more common intestinal bacterial pathogens, although some reports indicate that some bacterial pathogens are more resistant than this organism in water. It is not, however, as resistant as intestinal viruses which some investigators feel its presence should also indicate. From studies by a number of workers, Buttiaux and Mossel *(7)* concluded that various pathogens may persist after *E. coli* is destroyed in frozen foods, under refrigerated conditions, in radiated foods, and in treated waters. These authors have reported that only in acid food does *E. coli* have particular value as an indicator organism due to its relative resistance to low pH values.

3. Coliform Standards for Foods.   While the presence of large numbers of coliforms in foods is highly undesirable, it would be virtually impractical to eliminate all of these organisms from fresh and frozen foods. The basic questions at this point are these: (1) Under proper conditions of harvesting, handling, storage, and transport of foods, what is the lowest possible and feasible number of coliforms to maintain? (2) At what point does a good product become unsafe with respect to numbers of coliforms? In the case of water and dairy products, there is a long history behind the safety of allowable coliform numbers. Coliform standards for water, dairy products, and other foods covered by some federal, state, and city laws now in effect are as follows: not over 10/ml for Grade A pasteurized milk and milk products, including cultured products; not over 10 for certified raw, and not over 1 for certified pasteurized milk; not over 10 for precooked and partially cooked frozen foods; not over 100 for crab meat; and not over 100 for custard-filled items (see later section in this chapter). It can be seen that low numbers of coliforms are permitted in many instances ranging from 1 to not over 100/g or ml. Implicit in these standards are answers to the two questions raised above, that is, feasibility and safety.

On the basis of present knowledge, it appears that the coliform index as an index of sanitary quality is applicable to at least some foods. The use of this index for foods brings objections from some investigators who feel that the enterococci rather than coliforms better reflect the sanitary quality of some foods.

ENTEROCOCCI AS INDICATORS OF FOOD SANITARY QUALITY

The enterococci are members of the genus *Streptococcus,* which consists of gram positive cocci, producing long or short chains, and differing from most other gram positive cocci in being catalase negative. Although there is some confusion as to which species of streptococci should be included under the term enterococci, it is now generally agreed that the enterococci are all members of Lancefield's serologic Group D streptococci. The Group D streptococci consist of 4 species and 3 subspecies:

1. *S. faecalis*
   subsp. *faecalis*
     "   *liquefaciens*
     "   *zymogenes*
2. *S. faecium*
3. *S. bovis*
4. *S. equinus*

Species #1 and #2 above constitute the enterococcus group. Species #3 and #4 actually belong to the viridans group but possess antigen D in common with the enterococci. **Fecal streptococci** have been characterized as being "all *Streptococcus* spp. consistently present in significant numbers in fresh fecal excreta . . ." *(23).* Culturally, the enterococci grow at 10 C, in the presence of methylene blue, in 6.5% NaCl, and at pH 9.6 while the other two group D streptococci do not.

Ostrolenk *et al. (45)* were apparently the first to show the feasibility of employing the enterococci as indicators of pollution. The first comparative study of coliforms and enterococci as indicators for foods was apparently that of Burton *(5).*

1. Characteristics and Growth of Enterococci. The enterococci are typical of gram positive bacteria in that they are more fastidious in their nutritional requirements than gram negative bacteria. These organisms are unable to grow on simple, sugar-salts media and require varying numbers of B-vitamins and other organic constituents. With respect to growth temperature, they have been reported to grow from 0 to over 50 C, with at least 5 authors having reported growth between 0–6 C. While these organisms are basically mesophilic in nature, many strains may be said to be facultative psychrophiles. The enterococci grow over a wider pH range than coliforms. As may be seen from Table 15-1, *S. faecalis* and *S. faecium* types grow in the presence of 6.5% NaCl and in the presence of 40% bile. Table 15-1 also presents other physiologic characteristics that are useful in differentiating one species or variety from another. With respect to oxygen tension, the enterococci are like most other streptococci in that they are aerobic but tend to

TABLE 15-1.   Group D Streptococci: selected differential physiological characters *(51).* **

	Division 1	Division 2		Division 3	
	S. faecalis and varieties	*S. faecium*	*S. durans*	*S. bovis*	*S. equinus*
β Hemolysis	−/+	−	+/−	−	−
Growth 10°	+	+	+	−	−
45°	+	+	+	+	+
50°	+	+*	−	−	−
pH 9.6	+	+	+/−	−	−
6.5% NaCl	+/−	+/−	+/−	−	−
40% bile	+	+	+	+	+
Resists 60°C for 30 min.	+	+	+/−	−	−*
NH₃ from arginine	+	+	+	−	−
Gelatin liquefied	−/+	−	−	−	−
Tolerates 0.04% Pot. tellurite	+	−	−	−	−
Acid from:					
Glycerol (anaerobic)	+*	−	−	−	−
Mannitol	+	+	−*	−/+	−
Sorbitol	+*	−*	−	−/+	−
L-arabinose	−	+*	−	+/−	−
Lactose	+	+	+	+	−
Sucrose	+*	+/−	−	+	+*
Raffinose	−*	−*	−	+	−*
Melibiose	−	+*	−*	+	−*
Melezitose	+*	−	−	−	−
Starch hydrolyzed	−	−	−	+*	−*
Tetrazolium reduced at pH 6.0	+	−	−	+/−	−

   + = positive result.
     − = negative result.
   +/− = variation between strains, majority positive.
   −/+ = variation between strains, majority negative.
*occasional strains atypical.
Division 1 and Division 2 fulfill the criteria for the "enterococcus group" of Sherman (1938).
**Reproduced by permission from P. M. F. Shattock, "Enterococci, in *Chemical and Biological Hazards in Food.* J. C. Ayres, A. A. Kraft, A. T. Snyder, and H. W. Walker, Editors. © 1963 by the Iowa State University Press, Ames.

grow best under somewhat reduced conditions. Organisms that behave in this manner are generally referred to as *microaerophiles*.

2. Distribution of Enterococci in Nature.   Much like the coliforms, especially *E. coli,* the enterococci are primarily of fecal origin, with *S. faecalis* and its varieties being associated more with the human intestinal canal than that of other animals *(2)*. *S. faecium* and *S. durans* tend to be associated more with the intestinal tract of swine than *S. faecalis,* while *S. bovis* and *S. equinus* tend to be associated more with the intestinal tract of cattle and horses, respectively. All samples of human and pig feces examined by Buttiaux *(6)* contained these organisms while 66 and 88% of the cow and sheep feces, respectively, were found to contain these organisms. This is a somewhat higher carriage rate than for coliforms. *S. faecalis* appears to be more specific for the intestinal tract of man than other enterococcal species, yet this species is not generally accorded the same relationship to the enterococci as is *E. coli.* In addition to their fecal sources, these organisms exist on plants, insects, and in soils *(35, 41, 42, 43)*. Their sources to insects and plants may be through animal fecal matter. They have been regarded as temporary residents of plants and are disseminated among plants by the action of insects and wind and reach the soil by these agencies, rain, and gravity *(42)*. Although *S. faecalis* is often of fecal origin and its presence in foods may be taken to indicate fecal contamination, some *S. faecalis* strains appear to be common microflora of plants with no sanitary significance when found in foods. Mundt *(43)* studied *S. faecalis* from humans, plants, and other sources and found that the nonsanitary indicators could be distinguished from the more fecal types by their reaction in litmus milk and their fermentation reactions in melezitose and melibiose broths. In another study of 2,334 strains from dried and frozen foods, this investigator found that a high percentage of strains bore a close similarity to the plant-resident streptococci and would therefore not be of any sanitary significance *(44)*. When used as indicators of sanitary quality of foods, it is necessary to ascertain whether isolates are of the plant type or whether they represent those of human origin. The enterococci may also be found in dust. In general, they are rather widely distributed in nature, especially in slaughterhouses and curing rooms where pork products such as cured hams are handled.

With respect to the use of enterococci as indicators of water pollution, some investigators who have studied their persistence in water have found that this group dies off at a faster rate than coliforms, while others have found the reverse to be true. Leininger and McCleskey *(31)* have noted that this group does not multiply in water as coliforms sometimes do. Their more exacting growth requirements may be taken to indicate a less competitive role for these organisms in water environments. In sewage, Litsky *et al. (34)* found coliforms and enterococci to exist in high numbers but found approximately 13 times more coliforms than enterococci.

3. Relationship of Enterococci to Sanitary Quality of Foods. A large number of investigators have examined the use of an enterococcus index for food safety and feels that it is a better index of food sanitary quality than the coliform index, especially for frozen foods. In his study of frozen chicken pies obtained from one processor, Hartman *(22)* found that enterococcus counts were more closely related to total counts than to coliform counts, while coliform counts were more closely related to enterococcus counts than to total counts (see Table 15-2). The same author's findings were in agreement with those of Hahn and Appleman *(21),* Larkin *et al. (28, 29, 30),* and Kereluk and Gunderson *(27),* and were supported by Raj *et al. (49)* that in frozen foods, enterococci occur in greater numbers than coliforms. This fact is further illustrated in Table 15-3 for frozen precooked fish sticks.

TABLE 15-2.   **Relationships between coliform, enterococcus, and total counts in 456 chicken pies obtained from one processor** *(22).*

A. Numbers of samples with various coliform: total count ratios			
	Total count per g		
Coliforms per g	*Below 10³*	*10³ to 10⁴*	*10⁴ to 10⁵*	*Over 10⁵*
Below 3	0	94	82	3
3 to 9	0	41	93	10
10¹ to 10²	0	25	60	25
Over 10²	0	0	8	12

B. Numbers of samples with various enterococcus: total count ratios			
	Total count per g		
Enterococci per g	*Below 10³*	*10³ to 10⁴*	*10⁴ to 10⁵*	*Over 10⁵*
Below 10²	0	15	1	0
10² to 10³	0	104	41	2
10³ to 10⁴	0	36	136	15
Over 10⁴	0	0	23	32

C. Numbers of samples with various coliform: enterococcus ratios			
	Enterococci per g		
Coliforms per g	*Below 10²*	*10² to 10³*	*10³ to 10⁴*	*Over 10⁴*
Below 3	11	84	76	4
3 to 9	4	44	86	10
10¹ to 10²	1	19	63	26
Over 10²	0	1	7	12

TABLE 15-3.   Comparison of enterococci and coliform most probable number (MPN) counts in frozen precooked fish sticks (49).

Sample no.	Enterococci MPN count/100 g	Coliforms MPN count/100 g
1	86,000	6
2	18,600	19
3	86,000	0
4	46,000	300
5	48,000	150
6	46,000	28
7	46,000	150
8	18,600	7
9	8,600	0
10	4,600	186
11	4,600	186
12	48,000	1,280
13	8,600	46
14	4,600	480
15	48,000	240
16	10,750	1,075
17	10,750	17,000
18	60,000	23,250
19	10,750	2,275
Avg	32,339	2,457

In a study of 376 samples of commercial frozen vegetables, Burton (5) found that coliforms were more efficient indicators than enterococci prior to freezing and storage while enterococci were superior indicators in frozen foods. In samples stored at −20 C for 1–3 mon, 81% of enterococci and 75% of coliforms survived. After 1 year, 89% of enterococci survived while only 60% of the coliforms survived. Larkin et al. (29) showed that enterococci remained relatively constant for 400 days when stored at freezing temperatures (see Table 15-3). The same authors inoculated E. coli and S. faecalis into orange juice concentrate, followed by storage at −10 F for 51 days. During this period of time, there was no appreciable decrease in the numbers of S. faecalis whereas E. coli fluctuated. The relative occurrence of coliforms, enterococci, and staphylococci in 0.1 g portions of 391 precooked frozen foods is presented in Table 15.4. It should be noted that both enterococci and staphylococci could be recovered from more samples than coliforms. In this particular study, enterococci were found in 86 and staphylococci in 40 of the samples in which coliforms were not detected. Cuthbert et al. (9) studied the viability of E. coli and S. faecalis in soils and found that in limestone soils, these organisms persisted for several weeks though in acid peat soils, they persisted for only a few days. In general, the enterococci have a greater potential for survival with remoteness from the sources of pollution (4).

TABLE 15-3. Comparison of enterococci and coliform most probable number (MPN) counts in frozen precooked fish sticks *(49), (continued)*
The effect of −6°F storage on the longevity of the coliform bacteria and enterococci**

Days in storage at −6°F	Most probable number*	
	Coliform	*Enterococci*
0	5,600,000	15,000,000
7	6,000,000	20,000,000
14	1,400,000	13,000,000
20	760,000	11,300,000
35	440,000	11,200,000
49	600,000	20,000,000
63	88,000	11,000,000
77	395,000	15,000,000
91	125,000	41,000,000
119	50,000	5,400,000
133	136,000	7,400,000
179	130,000	5,600,000
207	55,000	3,500,000
242	14,000	4,000,000
273	21,000	4,000,000
289	42,000	3,200,000
347	20,000	2,300,000
410	8,000	1,600,000
446	260	2,300,000
481	66	5,000,000

*Avg of four determinations.
**Kereluk and Gunderson *(27)*.

TABLE 15-4. Presence of coliforms, enterococci, and staphylococci in 0.1-g portions (precooked frozen foods) *(33)*.

Processing plant		Samples examined	Samples positive for:		
			Coliforms	*Enterococci*	*Staphylococci*
Military	No.	234	10	96	50
	(%)	(100)	(4.3)	(41.0)	(21.3)
Commercial	No.	57	65	131	65
	(%)	(100)	(41.4)	(83.4)	(41.4)

*1961 Copyright © by Institute of Food Technologists.

From 14 groups of dried foods, Mundt *(44)* recovered streptococci from 57% while 87% of 13 different frozen vegetables yielded these organisms. As noted above, many of these isolates were of the plant-resident type (see Chapter 5 for the incidence of streptococci in other foods).

The enterococci and coliforms as sanitary indicators are compared in Table 15-5. See Table 15-6 for suggested enterococcal standards.

TABLE 15-5.   Comparison of coliforms and enterococci as indicators of food sanitary quality.

Characteristic	Coliforms	Enterococci
Morphology	Rods	Cocci
Gram reaction	Negative	Positive
Incidence in intestinal tract	10^7–10^9/g feces	10^5–10^8/g feces
Incidence in fecal matter of various animal species	Absent from some	Present in most
Specificity to intestinal tract	Generally specific	Generally less specific
Occurrence outside of intestinal tract	Common in low nos.	Common in higher nos.
Ease of isolation and identification	Relatively easy	More difficult
Response to adverse environmental conditions	Less resistant	More resistant
Response to freezing	Less resistant	More resistant
Relative survival in frozen foods	Generally low	High
Relative survival in dried foods	Low	High
Incidence in fresh vegetables	Low	Generally high
Incidence in fresh meats	Generally low	Generally low
Incidence in cured meats	Low or absent	Generally high
Relationship to food-borne intestinal pathogens	Generally high	Lower
Relationship to nonintestinal food-borne pathogens	Low	Low

OTHER INDICATORS

The determination of any or all members of the family *Enterobacteriaceae* as indicators of food sanitary quality has received the attention of more and more food scientists. Some advantages of using this group include: (1) since the definition of the *Enterobacteriaceae* is well established, problems due to ill-defined taxonomy of coliforms are resolved, (2) the possibility of obtaining false negative strains or of slow lactose fermenters as the predominant enteric contaminants is eliminated, and (3) the reliability of monitoring can be increased by the addition of direct testing for pathogens *(40)*. In addition to *Escherichia, Citrobacter, Enterobacter,* and *Klebsiella* spp., the *Enterobacteriaceae* also contain the genera *Serratia, Proteus, Edwardsiella,* and other genera that commonly exist in the intestinal tract. Buttiaux *(6)* has called attention to the changing pattern of *E. coli* in the intestinal tract due to the use of antibiotics. This author found that *Klebsiella* rose from 5.2% in 1947 to 48.4% in 1956–7 in the intestinal tract of humans. While most of the *Enterobacteriaceae* exist in the intestinal canal, some (e.g., *Proteus* and *Serratia*) are widespread in other environments.

TABLE 15-6. Summary of suggested microbial limits in chilled and frozen foods/g, except as indicated (11).

Foods	No. of authors or groups	Range of total Viable aerobes	Range of Coliforms	Range of Staph.	Range of salm. and/or shigel.	Range of enterococci
Frozen precooked foods	12	2,000–500,000	0–100	0–1,000	0 to absent in 50 g	1,000
Precooked meats	3	10–10,000	absent in 0.1 g	—	None	absent in 1 g
Raw meats	13	100,000–10,000,000	absent in 0.01–0.1 g	absent in 0.01–0.1 g	absent in 50 g	—
Frozen whole eggs	4	200,000–10,000,000	—	—	—	—
Fish, shellfish, & waters	10	50,000–1,000,000	0.7–1,600	100	—	1,000
Vegetables	10	50,000–500,000	absent in 0.1 g	—	—	—
Ice cream & frozen desserts	25	100–4,000,000	0–200	—	—	0 to absent in 0.001 ml.

Staph. = staphylococci; *Salm.* = *Salmonella; Shigel.* = *Shigella.*

Beside the above groups, the following are among additional groups of bacteria that have been suggested as fecal indicators: *Bifidibacterium* spp., *Ristella* spp., lactobacilli, clostridia, and group D streptococci. The bifidibacteria have been reported to be specific to the intestinal tract and to exist generally in high numbers. *Ristella* or *Bacteroides* spp. have been reported by Haenel *(20)* to be the second largest group of organisms in the human adult's intestinal tract. This worker has consistently found the anaerobic lactobacilli to constitute over 50% of the human intestinal flora— generally 10^9–10^{10}/g of fecal matter. Clostridia may at times be found in the intestinal tract. They are also found in soils. Their common existence in places other than the intestinal tract makes them unattractive as fecal indicators along with the lactobacilli and some of the *Enterobacteriaceae*. Although the group D streptococci are also found outside the intestinal tract, they have, nevertheless, been suggested by many as being better indicators than coliforms for certain foods (see above).

TOTAL COUNTS AS INDICATORS OF FOOD SANITARY QUALITY

As previously discussed in Chapter 4, total counts (more often aerobic plate counts, APC) on food products not only reflect handling history, state of decomposition or degree of freshness; they may in some instances reflect on the sanitary quality of foods. As Silliker has pointed out *(53)*, total counts most effectively evaluate the sanitary quality of foods which do not support microbial growth. In foods of this type (e.g., dried, frozen), total counts may be taken to indicate the type of sanitary control exercised in their production, transport, and storage. The same may be said of fresh foods when it is desired to set a standard to be used as a guide to storage life *(24)*. Total counts as indicators of sanitary quality in fresh foods are made somewhat untenable by the fact that part of the total count flora may represent pathogens. The latter, however, is minimized by food storage at temperatures that do not permit the growth of pathogens, or by other conditions that are generally unfavorable to the growth of pathogens (see the section on the microbial ecology of staphylococcal growth in Chapter 16). It should be noted that low total counts do not always represent safe products. In their study of commercial frozen-egg preparations, Montford and Thatcher *(39)* were able to isolate salmonellae from one preparation with a total count of only 380/g and from several others with total counts below 5,000/g. Even though the total counts of these preparations were quite low, all were shown to contain coliforms. It is also possible to have low-count foods in which toxin-producing organisms have grown and produced toxins that remain stable to conditions which may not favor the continued survival of the cells. The sanitary quality of foods such as sauerkraut, fermented milks, and related products cannot be ascertained by total plate counts, since these products are produced by the activities of microorganisms.

In a critical discussion of coliforms, enterococci, and total plate counts as indices of food sanitary quality, Levine *(33)* pointed out that all too often a sanitary index becomes one of attainment rather than safety. This author further noted that total plate counts would probably give as much information as any other microbiological index suggested to date as an index of sanitary processing or proper storage of food products. A more recent study of a large number of ready-to-eat foods suggests that the APC is the most suitable method for evaluating the microbial quality of foods and that where food safety is of concern, a search for specific pathogens should be made *(38)*.

In adopting microbiological standards for foods, the primary concerns are those of product safety and shelf-life. It might well be that total plate counts, rather than other indicators, applied primarily to plant sanitation and practices rather than merely to the finished products would be the most suitable approach to this problem (see Chapter 5). Also, not all dangerous contamination can be controlled by the use of total counts or indicator organisms *(56)*. In these instances, it may be necessary to process foods in such a way that certain specific pathogens are specifically destroyed as, for example, in the pasteurization of whole egg. Of obvious value in deciding which method is best for assessing food sanitary quality would be more knowledge about the microbial ecology of microorganisms in foods and the use of standardized isolation and plating procedures. This is discussed further in the next section.

MICROBIOLOGICAL STANDARDS/CRITERIA

Microbiological criteria for foods, as defined for Codex Alimentarius purposes, consist briefly of a statement of the microorganisms and/or products of concern, the appropriate examination methods, a plan defining the number and size of samples to be examined, the appropriate microbiological limits for the food in question, and the proportion of sample units that should conform to the stated limits. While microbiological standards, specifications, and guidelines are categories of microbiological criteria with different Codex meanings, these terms have been used with less precision by other authorities. In general, a microbiological standard is a microbiological criterion that has the connotation of law attached to it. Specifications, limits, and guidelines are criteria established sometimes for use in commerce and by regulatory agencies without the connotation of law (see ref. 57 for further details). In the remaining parts of this section, microbiological standards and criteria are used somewhat interchangeably.

Since the safety and keeping quality of fresh foods are related to microbial content, microbiological standards have been proposed for a variety of foods and some of these have been adopted. Microbiological standards for pasteurized and certified milk have been in effect since around the turn of the century while standards for other foods have appeared at a much slower rate. Elliott and Michener *(11)* have presented the principal precautionary arguments of various individuals relative to the adoption of microbiological standards as follows:

1. A single set of microbiological standards should not be applied to foods as a miscellaneous group, such as "frozen foods" or "pre-cooked foods."
2. Microbiological standards should be applied first to the more hazardous types of foods on an individual basis, after sufficient data are accumulated on expected bacterial levels, with consideration of variations in composition, processing procedures, and time of frozen storage.
3. When standards are chosen, there should be a definite relation between the standard and the hazard against which it is meant to protect the public.
4. Methods of sampling and analysis should be carefully studied for reliability and reproducibility among laboratories, and chosen methods should be specified in detail as part of the standard.
5. Tolerances should be included in the standard to account for inaccuracies of sampling and analysis.
6. At first, the standard should be applied on a tentative basis to allow for voluntary compliance before becoming a strictly enforced regulation.
7. Microbiological standards will be expensive to enforce.
8. If standards are unwisely chosen they will not stand in courts of law.

The above authors have also presented views of some of those who argue in favor of adopting microbiological standards for foods:

1. Bacteriological standards are a convenience and a necessity.
2. Bacteriological standards enhance plant sanitation.
3. Low bacterial counts are attainable.
4. Low bacterial counts are associated with safe foods.
5. Bacterial counts reflect sanitation level.
6. Bacterial counts reflect degree of decomposition.
7. Low bacterial counts will enhance shelf-life.

Among the arguments advanced by those who oppose the adoption of bacterial standards are the following *(11)*:

1. There is no need for bacteriological standards; present control is adequate.
2. Bacteriological standards will not free foods of danger from pathogens.
3. A fecal indicator standard has limitations.
4. Total count is unrelated to danger or to spoilage.
5. Methods of sampling and analysis are inadequate.

6. Existing laboratory facilities and personnel are inadequate.
7. Processing and storage influence viable counts.
8. Excessive sanitation will introduce a food poisoning hazard.
9. Foods will be overcooked or preservatives will be introduced to meet a standard.
10. More background information is needed.
11. Bacterial standards will be hard to defend in court.

In their examination of the advantages and objectives of microbiological standards, Shiffman and Kronick *(52)* found that the arguments fell into the following four categories:

1. Scientific or theoretical validity—are the standards based on verified data?
2. Technical—are standards technically feasible?
3. Administrative—can standards be employed in a regulatory food program?
4. Legal—will standards have legal acceptance?

Recently, Greenberg *et al. (18)* called attention to the increased costs that will result if standards are adopted widely. According to these authors, the increased costs will result from: (1) increased sampling, (2) lower action levels, (3) lengthened sampling time, (4) disposing of noncompliance products, (5) increased processing costs, and (6) reduced competition. They suggested the institution of an adequate audit system for testing products direct from the manufacturer and the field in lieu of lot standards. This practice could be aided by more widespread uses of the hazard analysis, critical control point (HACCP) concept at the manufacturing level. As described by Bauman *(3),* HACCP is a preventive system of control, especially with respect to microbiological hazards. It includes a careful analysis of ingredients, products, and processes in an effort to determine those components or areas that must be maintained under very strict control to assure that the end product meets the microbiological specifications that have been developed. **Hazard analysis** is the identification of sensitive ingredients, critical process points, and relevant human factors as they affect product safety; and **critical control points** are those processing determiners whose loss of control would result in an unacceptable food safety risk *(3).* The application of this concept in the production of frozen *(47)* and canned foods *(25)* has been discussed elsewhere.

It should be noted that the basic idea behind the adoption of microbiological standards for foods is protection of the consumer—protection from unnecessary food poisoning hazards as well as from deteriorating products. Foster *(14)* has called attention to some of the more basic thinking behind the adoption of microbiological standards. This author noted that high numbers of bacteria are associated with potential food hazards, especially in raw foods. He also noted that high counts in processed foods must generally be viewed with suspicion. When high counts are found in precooked foods, a process which should effectively reduce the number of organisms to a low level, there is every reason to be suspicious of such products.

Microbiological standards have been suggested by many investigators and public health officials, beginning in 1903 when Marxer first suggested a microbial standard for hamburger meat of 1,000,000/g. Elliott and Michener (11) have summarized the proposed and suggested standards for various foods, and they are further summarized in Table 15-6. While all authors have suggested total viable count standards, some have stressed the presence, absence, or numbers of indicator or pathogenic organisms as being perhaps more important than total numbers. An inspection of Table 15-6 reveals that the suggested viable aerobe counts are about as variable as are individual laboratory findings on given food items. Although it is true that aliquots from the same batch of food tested under similar conditions sometimes yield highly variable results, the variations in suggested standards from one author to another no doubt reflect different plate count techniques as well as attitudes toward the significance of microorganisms in foods. It is important that standard procedures be employed in examining foods, and for this reason the references in Table 4-1 are recommended. This not only makes more meaningful the findings of all investigators, but also clarifies the feasibility and desirability of standards for some of the many foods not now covered by standards.

The International Commission on Microbiological Specifications for Foods (ICMSF) has made specific suggestions regarding standards for certain groups of foods (56). For frozen egg products, this committee has suggested that all salmonellae be killed by some appropriate processing technique. For frozen sea foods (shrimps, prawns, crab meat, lobster meat) and frozen precooked complete meals and closely related products, the committee suggested as maximum numbers/g the following: total count at 35 C of ≯100,000; coliform count of ≯20; and coagulase positive staphylococci count of ≯100. For frozen comminuted meats, the committee questioned the justification for the use of poor quality trimmings that contribute to the generally high total counts of these type products (see last section of this chapter).

Microbial Standards and Food Safety

As noted above, microbial standards should result in foods that have a longer shelf-life and foods that are free of microbial hazards. That the latter is not necessarily true was shown by Miskimin et al. (38) and Solberg et al. (55) who studied over 1,000 foods consisting of 853 ready-to-eat and 180 raw products. These investigators applied arbitrary standards for APC, coliforms, and E. coli and tested the efficacy of the standards to assess safety of the foods with respect to S. aureus, C. perfringens, and salmonellae. An

APC standard of $<10^6$/g for raw foods resulted in 47% of the samples being accepted even though one or more of the 3 pathogens was present, while 5% were rejected from which pathogens were not isolated, for a total of 52% wrong decisions. An APC of $<10^5$/g for ready-to-eat foods resulted in only 5% being accepted that contained pathogens while 10% that did not yield pathogens were rejected. In a somewhat similar fashion, coliform standards of $<10^2$/g resulted in a total of 34% wrong decisions for raw and 15% for ready-to-eat foods. The lowest percentage of wrong decisions for ready-to-eat foods (13%) occurred with an *E. coli* standard of <3/g while 30% of the decisions were wrong when the same standard was applied to raw foods. It was noted that even though the 3 pathogens were found in both types of foods, no foodborne outbreaks were reported over the 4-yr period of the study during which time more than 16 million meals were consumed *(55)*.

While fecal coliforms are not precise indicators of food safety, their validity as indicators for the presence of salmonellae in irrigation water has been found to be high *(17)*. These investigators found that when the fecal coliform density/100 ml was above 1,000, salmonellae occurrence exceeded a 96% frequency, while a frequency of only about 54% was associated with fecal coliform levels of $<1,000$/100 ml.

A study of 2,211 foodborne outbreaks by the U.S. Center for Disease Control over the 9-yr period 1967–1975 revealed that ground beef was involved in 78 outbreaks, frankfurters in 22, and sliced luncheon meat in 11 for a total of about 5% of the 2,211 *(16)*. It was concluded from this survey that with rare exceptions, the proximate causes of the foodborne outbreaks involved mishandling of food by the food service industry and by homemakers. In other words, a high percentage of cases was due to homemaker negligence. It could be argued that microbiological standards would do little if anything to prevent outbreaks at the home level if foods are improperly cooked or stored at improper temperatures. Even though microbial standards were not in effect to cover the foods surveyed above, when the reported outbreaks are considered in light of the high volume of sales it is apparent that these three meat products are not high risk foods *(16)*. Along similar lines, the U.S. Department of Agriculture surveyed beef patties, frankfurters, and sliced luncheon meat for the presence of salmonellae over the period 1972–1975 and found only 3 positive, raw finished beef patties out of 735 examined *(8)*. No salmonellae were isolated from 690 cooked finished frankfurters or from 456 cooked sliced luncheon meats. In view of findings such as the above, opponents of microbiological standards would question the need for such standards in regard to food safety, especially for ground beef, frankfurters, and sliced luncheon meats.

Microbiological Standards Now in Effect for Various Foods

Milk was the first food product for which microbiological standards were adopted in the United States. In addition to New York City and the state of

Massachusetts, which have imposed standards for certain food products, organizations such as the National Canners Association and the American Bottlers of Carbonated Beverages have for many years set standards for certain relevant groups of organisms on product ingredients. The Association of Food & Drug Officials of the U.S. (AFDOUS), the U.S. Department of Agriculture and other agencies (especially the ICMSF of the International Association of Microbiological Societies) have given much consideration to the adoption of standards for foods and continue to be concerned with this problem. Since foods that are produced in one country may be shipped to and sold in other countries, this aspect of food microbiology has also received the attention of the Food and Agricultural Organization (FAO) of the World Health Organization (WHO). The category of foods most likely to come first under microbiological standards nationwide and worldwide are the precooked foods, especially precooked frozen foods.

Presented below are those foods and food ingredients that are now under microbiological standards by various organizations along with federal, state, and city standards now in effect (after Frazier, *15,* through the courtesy of McGraw-Hill Publishing Company).

1. **Standards for Starch and Sugar** (National Canners Association).
    A. *Total thermophilic spore count:* Of the five samples from a lot of sugar or starch none shall contain more than 150 spores per 10 g, and the average for all samples shall not exceed 125 spores per 10 g.
    B. *Flat sour spores:* Of the five samples none shall contain more than 75 spores per 10 g, and the average for all samples shall not exceed 50 spores per 10 g.
    C. *Thermophilic anaerobe spores:* Not more than three (60%) of the five samples shall contain these spores, and in any one sample not more than four (65+ %) of the six tubes shall be positive.
    D. *Sulfide spoilage spores:* Not more than two (40%) of the five samples shall contain these spores, and in any one sample there shall be no more than five colonies per 10 g (equivalent to two colonies in the six tubes).
2. **Standard for "Bottlers" Granulated Sugar,** Effective July 1, 1953 (American Bottlers of Carbonated Beverages).
    A. *Mesophilic bacteria:* Not more than 200 per 10 g.
    B. *Yeasts:* Not more than 10 per 10 g.
    C. *Molds:* Not more than 10 per 10 g.
3. **Standard for "Bottlers" Liquid Sugar,** Effective in 1959 (American Bottlers of Carbonated Beverages). All figures based on dry-sugar equivalent (D.S.E.).
    A. *Mesophilic bacteria: (a)* Last 20 samples average 100 organisms or less per 10 g D.S.E.; *(b)* 95% of last 20 counts show 200 or less per 10 g; *(c)* 1 of 20 samples may run over 200; other counts as in *(a)* or *(b).*

B.   *Yeasts: (a)* Last 20 samples average 10 organisms or less per 10 g D.S.E.; *(b)* 95% of last 20 counts show 18 or less per 10 g; *(c)* 1 of 20 samples may run over 18; other counts as in *(a)* and *(b)*.

C.   *Molds:* Standards like those for yeasts.

4.   Standards for Dairy Products.

   A.   From 1965 recommendations of the U.S. Public Health Service.

   a.   *Grade A raw milk for pasteurization:* Not to exceed 100,000 bacteria per milliliter prior to commingling with other producer milk; and not exceeding 300,000 per milliliter as commingled milk prior to pasteurization.

   b.   *Grade A pasteurized milk and milk products* (except cultured products): Not over 20,000 bacteria per milliliter, and not over 10 coliforms per milliliter.

   c.   *Grade A pasteurized cultured products:* Not over 10 coliforms per milliliter.

NOTE: Enforcement procedures for *(a)*, *(b)*, and *(c)* require three-out-of five compliance by samples. Whenever two of four successive samples do not meet the standard, a fifth sample is tested; and if this exceeds any standard, the permit from the health authority may be suspended. It may be reinstated after compliance by four successive samples has been demonstrated.

   B.   *Certified milk* (American Association of Medical Milk Commissons, Inc.)

   a.   Certified milk (raw): Bacterial plate count not exceeding 10,000 colonies per milliliter; coliform colony count not exceeding 10 per milliliter.

   b.   Certified milk (pasteurized): Bacterial plate count not exceeding 10,000 colonies per milliliter before pasteurization and 500 per milliliter in route samples. Milk not exceeding 10 coliforms per milliliter before pasteurization and 1 coliform per milliliter in route samples.

   C.   *Milk for manufacturing and processing* (U.S. Dep. Agr. 1955)

   a.   Class 1: Direct microscopic clump count (DMC) not over 200, 000 per milliliter.

   b.   Class 2: DMC not over 3 million per milliliter.

   c.   Milk for Grade A dry milk products: must comply with requirements for Grade A raw milk for pasteurization (see above).

   D.   *Dry milk*

   a.   Grade A dry milk products: at no time a standard plate count over 30,000 per gram, or coliform count over 90 per gram (U.S. Public Health Service).

   b.   Standards of Agricultural Marketing Service (U.S. Dep. Agr.):

   (1)   Instant nonfat: U.S. Extra Grade, a standard plate count not over 35,000 per gram, and coliform count not over 90 per gram.

      (2)   Nonfat (roller or spray): U.S. Extra Grade, a standard plate count not over 50,000 per gram; U.S. Standard Grade, not over 100,000 per gram.

      (3)   Nonfat (roller or spray): Direct microscopic clump count not over 200 million per gram; and must meet the requirements of U.S. Standard Guide. U.S. Extra Grade, such as used for school lunches, has an upper limit of 75 million per gram.

   c.   Standards of American Dry Milk Institute, Inc.

Kind of dry milk	Grade	Process	Maximal SPC,* nos/g
Nonfat	Extra	Spray or roller†	50,000
Nonfat	Standard	Spray or roller	100,000
Nonfat, instant	Extra		35,000
Whole milk	Premium	Gas-packed, spray	30,000
Whole milk	Extra	Bulk spray or roller	50,000
Whole milk	Standard	Bulk spray or roller	100,000
Buttermilk	Extra	Spray or roller	50,000
Buttermilk	Standard	Spray or roller	200,000

*Standard plate count.
†Atmospheric roller throughout table.

   E.   *Frozen desserts*

States and cities that have bacterial standards usually specify a maximal count of 50,000 to 100,000 per milliliter or gram. The U.S. Public Health Ordinance and Code sets the limit at 50,000 and recommends bacteriological standards for cream and milk used as ingredients. Few localities have coliform standards.

5. **Standard for Tomato Juice and Tomato Products—Mold-count Tolerances** (Food and Drug Administration).

The percentage of positive fields tolerated is 2% for tomato juice and 40% for other comminuted tomato products, such as catsup, purée, paste, etc. A microscopic field is considered positive when aggregate length of not more than three mold filaments present exceeds one-sixth of the diameter of the field (Howard mold count method). This method also has been applied to raw and frozen fruits of various kinds, especially to berries.

6. In Table 15-7 are listed a variety of foods for which microbiological standards or administrative guidelines have been established. Oregon was the first state to adopt standards for fresh ground meats and several other states currently have under consideration laws that appear to be based upon the Oregon law *(22)*.

TABLE 15-7. Microbiological standards for various food products (compiled from ref. 1, 26, 32, 48).

Food Products	APC/g	Coliforms/g	E. coli/g	S. aureus	Salmonellae	State/Agency
Raw, fresh, frozen meats	5,000,000	—	Not > 50	—	—	Oregon law, 5/73
Raw, fresh, frozen meats	5,000,000	50	—	—	—	Proposed, North Dakota
Raw, fresh, ground meats	1,000,000	250	—	—	—	Administrative guidelines, Rhode Island
Raw meats	100,000	100	—	absent	absent	Guidelines, Massachusetts
Heat-processed, smoked meats	1,000,000	—	Not > 10	—	—	Oregon law, 5/73
Heat-processed, smoked meats	1,000,000	10	—	—	—	Proposed North Dakota
Heat-processed, smoked meats	<50,000	10	—	absent	absent	Massachusetts law, 1959
Heat-processed, smoked meats	100,000	100	—	—	—	Admin. guidelines, Rhode Island
Precooked frozen meat, poultry, and seafood products	100,000	100	absent/g	—	absent/25g	Military & federal specifications
Ice cream and related products	50,000	20	—	—	—	Military & federal specifications
Ice cream and frozen dessert products	500,000	—	—	—	—	Boston, 1906
Ice cream and frozen dessert products	100,000	—	—	—	—	California
Ice cream and frozen dessert products	—	10	—	—	—	20 States
Dry gelatin	3,000[1]	10 (MPN)	—	—	—	Proposed FDA
Frozen cream-type pies	50,000[1]	50 (MPN)	—	—	—	Proposed FDA
Whole dry milk[2]	30,000–50,000	90	—	—	—	Military & federal specifications
Nonfat dry milk	50,000	—	—	—	absent/100 g	Military & federal specifications
Malted milk	30,000	10	—	—	—	Military & federal specifications
Cottage cheese[3]	<100[4]	10	—	—	—	Military & federal specifications
Precooked frozen meals	50,000	10	—	absent	—	Massachusetts
Precooked frozen meals	100,000	100	—	—	—	Rhode Island
Precooked frozen meals	100,000	10	—	—	—	Armed Forces
Dehydrated cooked beef	150,000	40	—	—	—	Military & federal specifications
Dehydrated cooked beef stew, chicken, chili con carne	75,000	—	absent	—	—	Military & federal specifications
Dehydrated cooked tuna and turkey	200,000	40	—	100	—	Military & federal specifications
Crab meat[5]	100,000	100	—	—	—	New York City
Custard-filled items	100,000	100	—	—	—	New York City

[1]Geometric mean
[2]DMC < 40,000,000 or 75,000,000 depending upon grade
[3]Not > 10 yeast and mold/g
[4]Psychrotrophs
[5]Not > 1,000 enterococci.

Current Trends in Establishing Microbiological Criteria or Standards for Foods

The microbiological criteria or standards listed in Table 15-7 are presented in the traditional ways in which a food product is believed to be free of a microbiological hazard as long as the level of a given pathogen or indicator does not exceed a given no./g or unit. It is well known that bacterial populations may vary widely between subsamples. The ICMSF has recommended sampling plans in order to compensate for this variability and to improve in general the overall reliability of microbiological results. A sampling plan is basically a statement of the criteria of acceptance applied to a lot based upon appropriate examinations of a required number of sample units by specified methods. A sampling plan consists of a sampling procedure and decision criteria, and may be a 2-class or a 3-class plan.

A 2-class plan consists of the following specifications: n, c, m; while a 3-class plan requires n, c, m, and M, where-

n = the number of sample units (packages, beef patties, etc.) from a lot that must be examined to satisfy a given sampling plan.

c = the maximum acceptable number, or the maximum allowable no. of sample units that may exceed the microbiological criterion. It is a measure of the number of sample units with levels of organisms between m and M. When this number is exceeded, the lot is rejected.

m = the maximum number or level of relevant bacteria/g; values above this level are either marginally acceptable or unacceptable. It is used to separate acceptable from unacceptable foods in a 2-class plan, or, in a 3-class plan, to separate good quality from marginally acceptable quality foods. The level of the organism in question which is acceptable and attainable in the food product is m. In presence/absence situations for 2-class plans, it is common to assign $m = 0$. For 3-class plans, m is usually some non-zero value.

M = a quantity that is used to separate marginally acceptable quality from unacceptable quality foods. It is used only in 3-class plans. Values at or above M in any sample are unacceptable relative to either health hazard, sanitary indicators, or spoilage potential.

A 2-class plan is the simpler of the two and in its simplest form may be used to accept or reject a large batch (lot) of food in a presence/absence decision by a plan such as $n = 5$, $c = 0$, where $n = 5$ means that 5 individual units of the lot will be examined microbiologically for, say, the presence of salmonellae, and $c = 0$ means that all five units must be free of the organisms by the method of examination in order for the lot to be acceptable. If any unit is positive for salmonellae, the entire lot is rejected. If it is desired that 2 of the 5 samples may contain coliforms, e.g., in a presence/absence test, the sampling plan would be $n = 5$, $c = 2$. By this plan, if 3 or more of the 5 unit samples contained coliforms, the entire lot would be rejected. While presence/absence situations generally obtain for salmonellae, an allowable upper limit for indicator organisms such as coliforms is more often the case. If it is desired to allow up to 100 coliforms/g in 2 of the 5 units, the sampling plan would be $n = 5, c = 2, m = 10^2$. After the 5 units have been examined for coliforms, the lot is acceptable if no more than 2 of the 5 contain as many as 10^2 coliforms/g, but is rejected if 3 or more of the 5 contain 10^2 coliforms/g. This paticular sampling plan may be made more stringent by increasing n (e.g., $n = 10, c = 2, m = 10^2$), or by reducing c (e.g., $n = 5, c = 1, m = 10^2$). On the other hand, it can be made more lenient for a given size n by increasing c.

While a 2-class plan may be used to designate acceptable-unacceptable foods, a 3-class plan is required to designate acceptable-marginally acceptable-unacceptable foods. To illustrate a typical 3-class plan, assume that for a given food product SPC shall not exceed 10^6/g (M) nor be higher than 10^5/g from 3 or more of 5 units examined. The specifications are thus $n = 5$, $c = 2, m = 10^5, M = 10^6$. If any of the 5 units exceeds 10^6/g, the entire lot is rejected (unacceptable). If not more than c sample units give results above m, the lot is acceptable. Unlike 2-class plans, the 3-class plan distinguishes values between m and M (marginally acceptable).

With either 2- or 3-class attributes plans, the numbers n and c may be employed to find the probability of acceptance (P_a) of lots of foods by reference to appropriate tables (see ref. *57*). The decision to employ a 2-class or 3-class plan may be determined by whether or not presence/absence tests are desirable in which case a 2-class plan is required, or whether count or concentration tests are desired in which case a 3-class plan is preferred. The latter offers the advantages of being less affected by nonrandom variations between sample units, and of being able to measure the frequency of values in the m to M range. The ICMSF report and recommendations *(57)* should be consulted for further details on the background, uses, and interpretations of sampling plans. Some recommended and proposed microbiological limits for various foods employing 2- or 3-class plans are presented in Table 15-8. From all indications, future microbiological standards for foods will be based on both sampling plans and microbiological limits similar to the format for the products in Table 15-8 rather than the format of those in Table 15-7.

SUMMARY

The safety of foods from food poisoning and other pathogenic microorganisms is ascertained by examining foods for the presence of fecal or indicator organisms. To be of value in this regard, it is widely accepted that the indicator organism or group of organisms should exist in fecal matter in high numbers, possess properties of survival outside of the intestinal tract close to that of the intestinal pathogens, possess a certain degree of specificity for fecal matter, and be detectable by simple and rapid means. The coliform bacteria, consisting of *Escherichia coli* and *Enterobacter aerogenes,* have found wide acceptance as indicators for water pollution.

TABLE 15-8. Some examples of sampling plans and recommended or proposed microbiological limits for selected foods (taken from sources identified in footnotes 1, 2, and 4).

Product	Tests	Plan			Limit/g	
		Class	n	c	m	M
Fresh and frozen fish[1]	SPC	3	5	3	10^6	10^7
	Fecal coliforms (MPN)	3	5	3	4	400
	S. aureus	3	5	3	10^3	2×10^3
Breaded pre-cooked fish products[1]	SPC	3	5	2	10^6	10^7
	Fecal coliforms (MPN)	3	5	2	4	400
	S. aureus	3	5	2	10^3	2×10^3
Blanched, frozen vegetables[1]	SPC	3	5	3	10^4	10^6
	Coliforms	3	5	3	10	10^3
Dried egg products[1]	SPC	3	5	2	10^4	10^6
	Coliforms or Enterobacteriaceae	3	5	2	10	10^3
	Salmonellae	2	10	0	0	—
Dried milk[1]	SPC	3	5	2	5×10^4	5×10^5
	Coliforms	3	5	2	< 3	10^2
	S. aureus	3	5	1	10	10^2
Frozen raw comminuted meat[1]	SPC	3	5	3	10^6	10^7
	Salmonellae	2	5	1(0)	0	—
Dried and instant foods for infants and children[2,3]	SPC	3	5	2	10^3	10^4
	Coliforms	3	5	1	< 3	20
	Salmonellae	2	60	0	0	—
Nonfrozen ground beef[4]	SPC	3	5	3	10^7	5×10^7
	E. coli	3	5	3	10^2	5×10^2
	S. aureus	3	5	2	10^2	10^3
	Salmonellae	3	5	0	0^5	0

[1] Recommended by ICMSF, 1974(57).
[2] Recommended to Codex Food Hygiene Committee by joint FAO/WHO Expert Consultation on Microbiological Specifications for foods, 1977.
[3] Includes dried infant formulas and supplementary products such as sweetening agents.
[4] Proposed Canadian Standard, 1975 (see ref. 77B, p. 102).
[5] Absent in 25 g in each of 5 subsamples.

While coliform standards have been adopted for certain foods and proposed for others, much evidence suggests that the enterococci are better indicators of sanitary quality of foods—especially frozen foods. Total counts are of value in assessing the sanitary quality of certain foods. The rationale behind the adoption of microbiological standards for foods is based largely upon the idea that foods with low microbial numbers are likely to be safer to the consumer in terms of absence of food-borne pathogens, and to permit longer shelf-life of stored products capable of permitting growth. Such standards usually specify a total count as well as indicator count maxima. While only a relatively small number of foods now come under microbiological standards, they have been proposed for a larger number of foods. Future recommendations of microbiological limits for adoption as microbiological standards are likely to include sampling plans, especially 3-class plans.

REFERENCES

1. **Anon.** 1975. Proposed microbial standards for ground meat in Canada. J. Milk Food Technol. *38:* 639.

2. **Bartley, C. H. and L. W. Slanetz.** 1960. Types and sanitary significance of fecal streptococci isolated from feces, sewage and water. A. J. Pub. Hlth. *50:* 1545–1552.

3. **Bauman, H. E.** 1974. The HAACP concept and microbiological hazard categories. Food Technol. *28:* No. 9, 30–33.

4. **Burman, N. P.** 1961. Some observations on coli-aerogenes bacteria and streptococci in water. J. Appl. Bacteriol. *24:* 368–376.

5. **Burton, M. C.** 1949. Comparison of coliform and enterococcus organisms as indices of pollution in frozen foods. Food Res. *14:* 434–438.

6. **Buttiaux, R.** 1959. The value of the association Escherichieae-Group D streptococci in the diagnosis of contamination in foods. J. Appl. Bacteriol. *22:* 153–158.

7. **Buttiaux, R. and D. A. A. Mossel.** 1961. The significance of various organisms of faecal origin in foods and drinking water. J. Appl. Bacteriol. *24:* 353–364.

8. **Center for Disease Control.** 1975. Microbiologic standards for raw ground beef, cold cuts, and frankfurters. Morb. Mort. Wkly. Rept. *24:* No. 27, 229–230.

9. **Cuthbert, W. A., J. J. Panes, and E. C. Hill.** 1955. Survival of *Bacterium coli* type I and Streptococcus faecalis in soil. J. Appl. Bacteriol. *18:* 408–414.

10. **deFigueiredo, M. P. and J. M. Jay.** 1976. Coliforms, enterococci, and other microbial inhibitors, pp. 271–297, IN: *Food Microbiology: Public Health and Spoilage Aspects,* edited by M. P. deFigueiredo and D. F. Splittstoesser (Avi Publishing: Westport, Conn.).

11. **Elliott, H. P. and H. D. Michener.** 1961. Microbiological standards and handling Codes for chilled and frozen foods: A review. Appl. Microbiol. *9:* 452–468.

12. **Fishbein, M.** 1962. The aerogenic response of *Escherichia coli* and strains of *Aerobacter* in EC broth and selected sugar broths at elevated temperatures. Appl. Microbiol. *10:* 79–85.

13. **Fishbein, M. and B. F. Surkiewicz.** 1964. Comparison of the recovery of *Escherichia coli* from frozen foods and nutmeats by confirmatory incubation in EC medium at 44.5 and 45.5 C. Appl. Microbiol. *12:* 127–131.

14. **Foster, E. M.** 1966. Significance of bacterial counts in food evaluation. Quart. Bull., Assoc. Food & Drug Officials of U.S. *30:* 20–27.

15. **Frazier, W. C.** 1968. *Food Microbiology.* 2nd Edition, McGraw-Hill, N.Y.

16. **Gangarosa, E. J. and J. M. Hughes.** 1977. A public health perspective of microbial standards for meats. Assoc. Food & Drug. Off. Quart. Bull. *41:* No. 1, 23–28.

17. **Geldreich, E. E. and R. H. Bordner.** 1971. Fecal contamination of fruits and vegetables during cultivation and processing for market. A review. J. Milk Food Technol. *34:* 184–195.

18. **Greenberg, R. A., R. B. Tompkin, and R. S. Geister.** 1974. Who will pay for microbiological quality standards? Food Technol. *28:* No. 10, 48–49.

19. **Griffin, A. M. and C. A. Stuart.** 1940. An ecological study of the coliform bacteria. J. Bacteriol. *40:* 83–100.

20. **Haenel, H.** 1961. Some rules in the ecology of the intestinal microflora of man. J. Appl. Bacteriol. *24:* 242–251.

21. **Hahn, S. S. and M. D. Appleman.** 1952. Microbiology of frozen orange concentrate. I. Survival of enteric organisms in frozen orange concentrate. Food Technol. *6:* 156–157.

22. **Hartman, P. A.** 1960. Enterococcus: coliform ratios in frozen chicken pies. Appl. Microbiol. *8:* 114–116.

23. **Hartman, P. A., G. W. Reinbold, and D. S. Saraswat.** 1966. Indicator organisms—a review. I. Taxonomy of the fecal streptococci. Intern. J. System. Bacteriol. *16:* 197–221.

24. **Ingram, M.** 1961. Microbiological standards for foods. Food Technol. *15*(2): 4–16.

25. **Ito, K.** 1974. Microbiological critical control points in canned foods. Food Technol. *28:* No. 9, 46–47.

26. **Johnston, R. W.** 1975. Microbiological criteria, an update. Proc. Meat Ind. Res. Conf., Amer. Meat Inst. Fdn., 115–118.

27. **Kereluk, K. and M. F. Gunderson.** 1959. Studies on the bacteriological quality of frozen meats. IV. Longevity studies on the coliform bacteria and enterococci at low temperatures. Appl. Microbiol. *7:* 327–328.

28. **Larkin, E. P., W. Litsky, and J. E. Fuller.** 1955a. Fecal streptococci in frozen foods. I. A bacteriological survey of some commercially frozen foods. Appl. Microbiol. *3:* 98–101.

29. **Larkin, E. P., W. Litsky, and J. E. Fuller.** 1955b. Fecal streptococci in frozen foods. II. Effect of freezing storage on *Escherichia coli* and some fecal streptococci inoculated onto green beans. Appl. Microbiol. *3:* 102–104.

30. **Larkin, E. P., W. Litsky, and J. E. Fuller.** 1955c. Fecal streptococci in frozen foods. III. Effect of freezing storage on *Escherichia coli, Streptococcus faecalis,* Microbiol. *3:* 104–106.

31. **Leininger, H. V. and C. S. McCleskey.** 1953. Bacterial indicators of pollution in surface waters. Appl. Microbiol. *1:* 119–124.

32. **Leininger, H. V.** 1976. What do bacteria in food really mean. Assoc. Food & Drug. Off. Quart. Bull. *40:* No. 3, 159–165.

33. **Levine, M.** 1961. Facts and fancies of bacterial indices in standards for water and foods. Food Technol. *15:* 29–34.

34. **Litsky, W., M. J. Rosenbaum, and R. L. France.** 1953. A comparison of the most probable numbers of coliform bacteria and enterococci in raw sewage. Appl. Microbiol. *1:* 247–250.

35. **Mallmann, W. L. and W. Litsky.** 1951. Survival of selected enteric organisms in various types of soil. A. J. Pub. Hlth. *41:* 38–44.

36. **Marxer, A.** 1903. Beitrag zur Frage des Bakteriengehaltes und der Haltbarkeit des Fleisches bei gewöhnlicher Aufbewahrung. Fortschr. Vet.-Hyg. *1:* 328.

37. **McCoy, J. H.** 1961. The safety and cleanliness of waters and foods. J. Appl. Bacteriol. *24:* 365–367.

38. **Miskimin, D. K., K. A. Berkowitz, M. Solberg, W. E. Riha, Jr., W. C. Franke, R. L. Buchanan, and V. O'Leary.** 1976. Relationships between indicator organisms and specific pathogens in potentially hazardous foods. J. Food Sci. *41:* 1001–1006.

39. **Montford, J. and F. S. Thatcher.** 1961. Comparison of four methods of isolating salmonellae from foods, and elaboration of a preferred procedure. J. Food Sci. *26:* 510–517.

40. **Mossel, D. A. A., G. A. Harrewijn, and C. F. M. Nesselrooy-van Zadelhoff.** 1974. Standardization of the selective inhibitory effect of surface active compounds used in media for the detection of *Enterobacteriaceae* in foods and water. Hlth. Lab. Sci. *11:* 260–267.

41. **Mundt, J. O., A. H. Johnson, and R. Khatchikian.** 1958. Incidence and nature of enterococci on plant materials. Food Res. *23:* 186–193.

42. **Mundt, J. O.** 1961. Occurrence of enterococci: Bud, blossom, and soil studies. Appl. Microbiol. *9:* 541–544.

43. **Mundt, J. O.** 1973. Litmus milk reaction as a distinguishing feature between *Streptococcus faecalis* of human and nonhuman origins. J. Milk Food Technol. *36:* 364–367.

44. **Mundt, J. O.** 1976. Streptococci in dried and frozen foods. J. Milk Food Technol. *39:* 413–416.

45. **Ostrolenk, M., N. Kramer, and R. C. Cleverdon.** 1947. Comparative studies of enterococci and *Escherichia coli* as indices of pollution. J. Bacteriol. *53:* 197–203.

46. **Peabody, F. R.** 1963. Microbial indexes of food quality: The coliform group. In: *Microbiological Quality of Foods.* L. W. Slanetz *et al.,* Editors, Academic Press, N.Y., pp. 113–118.

47. **Peterson, A. C. and R. E. Gunnerson.** 1974. Microbiological critical control points in frozen foods. Food Technol. *28:* No. 9, 37–44.

48. **Powers, E. M.** 1976. Microbiological criteria for food in military and federal specifications. J. Milk Food Technol. *39:* 55–58.

49. **Raj, H., W. J. Wiebe, and J. Liston.** 1961. Detection and enumeration of fecal indicator organisms in frozen sea foods. Appl. Microbiol. *9:* 295–308.

50. **Schardinger, F.** 1892. Ueber das Vorkommen Gahrung erregender Spaltpilze im Trinkwasser und ihre Bedeutung für die hygienische Beurtheilung desselben. Wien. Klin. Wschr. *5:* 403, 421.

51. **Shattock, P. M. F.** 1962. Enterococci. In: *Chemical and Biological Hazards in Food.* J. C. Ayres *et al.,* Editors, Iowa State U. Press, pp. 303–319.

52. **Shiffman, M. A. and D. Kronick.** 1963. The development of microbiological standards for foods. J. Milk & Food Technol. *26:* 110–114.

53. **Silliker, J. H.** 1963. Total counts as indexes of food quality. In: *Microbiological Quality of Foods.* L. W. Slanetz *et al.,* Editors, Academic Press, N.Y., pp. 102–112.

54. **Smith, T.** 1893. The fermentation-tube with special reference to anaerobiosis and gas production among bacteria. A. J. Pub. Hlth. *23:* 86.

55. **Solberg, M., D. K. Miskimin, B. A. Martin, G. Page, S. Goldner, and M. Libfeld.** 1977. Indicator organisms, foodborne pathogens and food safety. Assoc. Food & Drug Off. Quart. Bull. *41:* No. 1, 9–21.

56. **Thatcher, F. S.** 1963. The microbiology of specific frozen foods in relation to public health: Report of an International Committee. J. Appl. Bacteriol. *26:* 266–285.

57. **ICMSF (Intern. Comm. on Microbiol. Specific. for Foods).** 1974. *microorganisms in foods* 2. (U. of Toronto Press: Canada).

16

Food Poisoning
Caused by
Gram Positive Cocci

There are two genera of gram positive cocci that have been shown to cause food poisoning in man: *Staphylococcus* and *Streptococcus*.

STAPHYLOCOCCAL FOOD POISONING

This syndrome was first studied in 1894 by Denys *(38)* and later in 1914 by Barber who produced in himself the signs and symptoms of the disease by consuming milk which had been contaminated with a culture of *S. aureus*. The capacity of some strains of *S. aureus* to produce food poisoning was proved conclusively in 1930 by Dack *et al. (31)* who showed that the symptoms could be produced by feeding culture filtrates of *S. aureus*. While some authors refer to food-associated illness of this type as food intoxication rather than food poisoning, the latter designation is employed to cover food intoxications and food infections.

Staphylococcal food poisoning is produced by certain strains of *S. aureus* that also produce coagulase. The latter substance has the capacity to clot blood plasma and is elaborated by growing cells. It is one of many extracellular substances produced by staphylococci and has for many years been associated with virulence. Its purification was achieved by Stutzenberger and San Clemente *(97)*. Not all coagulase positive strains are capable of causing food poisoning. Several reports exist in which coagulase negative staphylococci were found to be the etiologic agents of this syndrome *(5, 12, 81)*, but coagulase positive strains are by far the most frequently encountered in this regard. Coagulase negative food poisoning strains may be tested for their capacity to produce lysozyme which appears to indicate pathogenesis among these strains *(59, 62, 81)*. All of the symptoms of staphylococcal food poisoning are caused by an extracellular substance designated as enterotoxin. It is not known to be produced by any other group of microorganisms.

1. Characteristics of Staphylococci.   The genus *Staphylococcus* belongs to the family *Micrococcaecae* and is represented by three species according to *Bergey's Manual*, 8th edition: *S. aureus, S. epidermidis,* and *S. saprophyticum. S. aureus* generally produces a golden pigment, produces coagulase, and ferments glucose and mannitol under anaerobic conditions; in contrast, *S. epidermidis* is white to off-white, does not produce coagulase, and lacks the capacity to ferment glucose and mannitol anaerobically. The latter property is shared by *S. saprophyticum.* Unlike the other two species, *S. saprophyticum* is resistant to novobiocin. Although coagulase production is generally characteristic of *S. aureus,* not all strains produce this substance, and not all strains possess the golden pigment. A significant number of *S. aureus* strains of animal and food origin are nonpigmented. A classification of staphylococci has been proposed by Baird-Parker *(3),* whose scheme establishes 6 subgroups with classical *S. aureus* strains belonging to Group I.

The staphylococci are typical of other gram positive bacteria in having a requirement for certain organic compounds in their nutrition. Amino acids are required as nitrogen sources, and thiamine and nicotinic acid are required among the B-vitamins. When grown anaerobically, these organisms appear also to require uracil. In one minimal medium for aerobic growth and enterotoxin production, monosodium glutamate serves as the C, N, and energy sources. This medium contains only 3 amino acids (arginine, cystine, and phenylalanine) and 4 vitamins (pantothenate and biotin in addition to niacin and thiamine) plus inorganic salts *(74).* Arginine appears to be essential for enterotoxin B production *(111).*

The staphylococci have been shown capable of growth and enterotoxin production in vacuum-packed bacon *(103).* They are capable of growing in the presence of fairly high levels of salt and this property is utilized in the preparation of media selective for these organisms. Most strains grow well in 10% NaCl, with some being able to grow in up to 20%. These organisms also possess a high degree of tolerance to compounds such as tellurite, mercuric chloride, neomycin, polymyxin, and sodium azide, all of which have been suggested as the basis of selective media for this group. On the other hand, coagulase-positive staphylococci have been shown to be sensitive to low levels of borate while coagulase negatives are more resistant *(60).* It should be noted that while staphylococci can grow in the presence of high concentrations of salt and tolerate relatively high levels of certain other toxic compounds, this same property is shared by many strains of the genus *Micrococcus.* The micrococci are widely distributed in nature and occur in foods generally in greater numbers than staphylococci, thus making the recovery of the latter a bit difficult.

The general characteristics and properties of staphylococci relative to their importance in foods have been reviewed and discussed in greater detail by Bryan *(15)* and Minor and Marth *(75).*

2. Distribution of Staphylococci.   In man, the main reservoir of *S. aureus* is the nose. From this source, these organisms find their way to the skin and into wounds either directly or indirectly. While the nasal carriage rate varies, it is generally about 50% for adults and somewhat higher among children. The most common skin sources are the arms, hands, and face where the carriage rate runs between 5–30% *(42)*. In addition to skin and nasal cavities, *S. aureus* may be found in the eyes, throat, and in the intestinal tract. From these sources, these organisms find their way into air and dust, onto clothing, and in other places from which they may contaminate foods. The two most important sources of this organism to foods are nasal carriers and individuals whose hands and arms are inflicted with boils and carbuncles and who are permitted to handle foods.

The incidence of staphylococci in foods is discussed in Chapter 5 (see Table 5-2). They have been found in a large number of commercial foods by many investigators. Of 91 samples of commercial frozen shrimp, Silverman *et al. (95)* recovered coagulase-positive staphylococci from 68% of the samples. Mickelsen *et al. (73)* found numbers ranging from 0 to over 10,000/g in cottage cheese 144 hr after processing. Donnelly *et al. (39)* found 20% of 343 market cheddar cheese samples to contain these organisms ranging from 50–220,000/g. These organisms have also been isolated from other cheeses and have been shown to die off during the normal aging of these products *(109)*. They were found on 80% of retail store chicken (from 2 to 620/ml of rinse) examined by Messer *et al. (71)*; on 10% of green beans (100 to 110,000/g); and on 35% of hamburger (50 to 3,500/g). Seventy-five percent of fresh pork sausage examined revealed *S. aureus* at the level of 100 or fewer/g (see Table 5-2). Coagulase positive staphylococci have been found in market meats, poultry, vegetables, and other foods and apparently originate from animals as well as man *(55, 56, 57, 80, 89, 96)*. With respect to their freezing survival, Raj and Liston *(85)* found a 10-fold decrease in numbers when held in fish homogenates at 0 F for 393 days.

It should be noted also that most domesticated animals harbor *S. aureus* *(77)*. Staphylococcal mastitis is not unknown among dairy herds and if milk from such cows is consumed or used for cheese making, the chances of contracting food poisoning are excellent. There is little doubt that many strains of this organism that cause bovine mastitis are of human origin; however, some are designated as "animal strains" and the ways in which these may be differentiated from human strains are discussed below. Some German workers found that the staphylococcal strains isolated from parts of raw pork products were essentially all of the animal strain type. In the manufacture of pickled pork products, these animal strains were gradually replaced by human strains during the production process to a point where none of the original animal strains could be detected in finished products *(94)*.

It appears that staphylococci may be expected to exist, at least in low numbers, in any or all food products that are of animal origin or in those which are handled directly by man unless heat processing steps are applied to effect their destruction.

3. Enterotoxigenic S. aureus *Strains and Their Enterotoxins.*   Numerous unsuccessful attempts have been made to differentiate between non-enterotoxigenic and enterotoxin-producing strains of staphylococci by some relatively simple laboratory method. Attempts to correlate enterotoxin production with other staphylococcal toxins such as the alpha, beta, and delta have been without success. Many food poisoning strains produce alpha toxin just as do most coagulase-positive strains. The beta toxin of *S. aureus* is somewhat characteristic of staphylococci of animal origin, and while some food poisoning staphylococci produce this toxin, most enterotoxigenic strains do not. Of 11 well-known enterotoxigenic staphylococci recently investigated, only one produced the beta toxin or "hot-cold" lysis on sheep blood *(59)*. Most if not all enterotoxin producers elaborate heat-stable nuclease, but not all nuclease producers are enterotoxigenic (see Chapter 4).

Attempts to relate enterotoxigenesis to specific bacteriophage types have been generally unsuccessful. Of those food poisoning strains subjected to phage typing, most have been found to belong to phage Group III though a few have been shown to belong to Group IV (phage type 42D). Like most staphylococci from clinical sources, those from foods tend to fall into phage Group III *(57, 80)*, thus making phage typing almost valueless as a means of identifying enterotoxigenic strains (see below).

Staphylococcal food poisoning is caused generally by the ingestion of preformed enterotoxin although it has been reported that symptoms may arise as a result of growth of enterotoxigenic strains in the intestinal tract and conceivably elsewhere in the body *(19)*. The enterotoxins are among several groups of a rather large number of extracellular products produced by certain staphylococci, and cultural conditions necessary for their production have been the subject of a large number of investigations. These toxins have been reported to appear in cultures as early as 4–6 hr (Figure 16-1) and increase proportionately through the stationary phase *(66)* and into the transitional phase (Fig. 16-2). Czop and Bergdoll *(30)* have found enterotoxin production to occur during all phases of growth although earlier studies revealed that with strain S-6 of *S. aureus*, 95% of enterotoxin B was released during the latter part of the log phase of growth *(68, 69)*. The latter authors have also demonstrated the production of enterotoxin B in nongrowing cells. On the other hand, Morse *et al. (78)* found that strain S-6 did not produce detectable quantities of enterotoxin B during exponential growth, but did during the post-exponential phase in association with total protein synthesis. These authors found that chloramphenicol inhibited the appearance of enterotoxin, suggesting that the presence of toxin was dependent upon *de novo* protein synthesis. $NaNO_2$ and $NaNO_3$ (Figure 16-3) at the concentra-

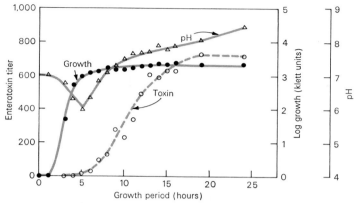

FIGURE 16-1.   Enterotoxin B production, growth, and pH changes in *Staphylococcus aureus* at 37°C *(70)*.

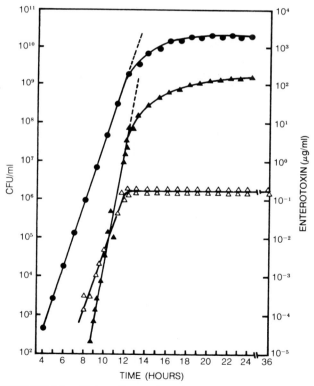

FIGURE 16-2.   Rates of growth and enterotoxins A and B synthesis by *S. aureus* S-6. Symbols: ●, CFU/ml; △, enterotoxin A; ▲, enterotoxin B *(30)*. Copyright © 1974, American Society for Microbiology.

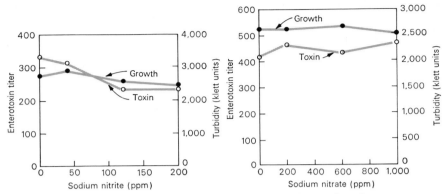

FIGURE 16-3.   The effect of $NaNO_2$ and $NaNO_3$ on total growth and enterotoxin B production by *S. aureus (70).*

tions allowable in cured meats have been reported to have no effect on production of this enterotoxin *(70).*

Though mesophilic in nature, *S. aureus* has been reported to grow at a temperature as low as 6.7 C *(1).* These authors found that 3 food-poisoning strains grew in custard at 114 F but decreased at 116–120 F with time of incubation. The same authors found that these organisms grew in chicken à la king at 112 F but failed to do so in ham salad at this temperature.

With respect to salt tolerance, Genigeorgis *et al. (46)* reported that strain S-6 produced enterotoxin B in cured ham under anaerobic conditions with brine contents of NaCl up to 9.2%. These investigations reported that no anaerobic production of enterotoxin occurred below pH 5.30 at 30 C or below pH 5.58 at 10 C. Although enterotoxin production occurred in cured meats under anaerobic conditions, Genigeorgis *et al. (46)* showed that it appeared earlier in cured meats incubated similarly but under aerobic conditions. These investigators also found that enterotoxin B production decreased as the content of undissociated HNO_2 increased in cured meats. Lechowich *et al. (64)* showed that 5% salt would inhibit staphylococcal growth in broth at pH 4.8. Genigeorgis and Sadler *(45)* reported that growth and enterotoxin B production occurred in 10% salt at pH 6.9, but not with 4% at pH 5.1. These investigators employed strain S-6 of *S. aureus* in broth. The results of these investigators reveal that as salt concentration increases, a higher pH level is needed for growth of staphylococci. More recently, the interrelationship between pH and salt content on growth and enterotoxin C production has been demonstrated *(47).* In a protein hydrolysate culture medium incubated at 37 C for 8 days, enterotoxin was produced over the pH range 4.00–9.83 in the absence of NaCl, but the addition of 4 percent NaCl had the effect of restricting the pH range to 4.4–9.43 (Table 16-1). Toxin production occurred in the presence of 10% NaCl but not below pH 5.45. No toxin was detected when 12% salt was used.

The survival and growth of staphylococci in bacon brines has been ob-

TABLE 16-1.   The effect of pH and NaCl on the production of enterotoxin C by an inoculum of 10^8 cells/ml of *S. aureus* 137 in a protein hydrolysate medium incubated at
37 C for 8 days (47).

pH range	4.00–9.83	4.4–9.43	4.50–8.55	5.45–7.30	4.50–8.55
NaCl content (%)	0	4	8	10^1	12
Enterotoxin production	+	+	+	+	−

[1]Enterotoxin was detected also with an inoculum of 3.6×10^6 at pH 6.38–7.30.
Copyright © 1971, American Society for Microbiology.

served (41). The presence of excess glucose in unbuffered media has been
reported to repress enterotoxin B production (78). Streptomycin,
actinomycin D, acriflavine, Tween 80, and other compounds have been
reported to inhibit enterotoxin B synthesis in broth (44). While actinomycin
D has been shown to inhibit enterotoxin B synthesis in strain S-6, the inhibition occurred about an hr after cellular synthesis ceased. The latter was immediately and completely inhibited. The conclusion from this finding by
Katsuno and Kundo (63) is that the mRNA responsible for enterotoxin synthesis is more stable than that for cellular synthesis. The lowering of incubation temperature has the effect of prolonging the appearance of toxins, with
72–96 hr having been reported for the detection of enterotoxin A in milk
(40).

In general, enterotoxin production tends to be favored by optimum growth
conditions of pH, temperature, Eh, and so forth. These conditions are
generally found to be growth temperatures of 37–40 C, pH 6.0–8.0,
$a_w > 0.95$, and conditions of high Eh. Because of the interrelatedness of
these growth parameters, it is difficult to define precise minima for each just
as it is in the case of intermediate moisture foods discussed in Chapter 13.
The complexity of various growth menstra, the interaction of salts with medium constituents in unknown ways, and the many variations that exist
between staphylococcal strains also add to the difficulty of defining minimal
conditions for growth and enterotoxin production by these organisms. The
interdependence of these parameters has been observed by several investigators (e.g., 50, 101). It seems to be well established, however, that staphylococci can grow under conditions that do not favor enterotoxin production.
Growth of cells has been recorded at an a_w as low as 0.83, at pH as low as
4.0, at NaCl content of 20%, and at temperatures of around 7 C (101), but
not under all of these conditions simultaneously. It is generally agreed that
enterotoxin production occurs over a narrower range of growth parameters
than those above. Enterotoxin A production in pork has been recorded at a_w
0.86 but not 0.83; in beef at 0.88 but not at 0.86 (101). It appears that
enterotoxin A can be produced under a_w conditions that do not favor the
production of enterotoxin B (107). From Table 16-1, it can be seen that
enterotoxin production has been recorded at pH 4.0 in the absence of NaCl.
The buffering of a culture medium at pH 7.0 has been found to lead to more

TABLE 16-2.   Incidence of A, B, and AB enterotoxin producing strains of *S. aureus* (19).*

Source	No. of strains	% Incidence			
		A	B	AB	Total
Clinical specimens. (Wash., D.C.)	396	15.7	7.8	9.1	32.6
Clinical specimens. (Atlanta, Ga.)	53	7.6	1.9	1.9	11.4
Nasal specimens. (Wash., D.C.)	155	17.4	2.6	1.9	21.9
Nasal specimens. (Atlanta, Ga.)	57	22.8	1.8	3.5	28.1
Frozen foods	245	4.5	4.0	0.0	8.5
Raw milk	190	3.2	1.0	0.0	4.2
Mastitic cows	50	0.0	0.0	0.0	0.0

*Copyright, the New York Academy of Sciences, 1965; Reprinted by permission.

enterotoxin B than when the medium was unbuffered or buffered in the acid range (72). A similar result was noted at a controlled pH of 6.5 rather than 7.0 (54). With respect to growth temperature, the production of enterotoxin B in ham at 10 C has been recorded (46); and small amounts of A, B, C, and D in cooked ground beef, ham, and bologna at 10 C (101). Toxin production has been observed at 46 C, but the optimal for enterotoxins B and C has been reported to be 40 C in a protein hydrolysate medium (108), and for enterotoxin E 40 C at pH 6.0 (105). The growth of *S. aureus* on cooked beef at 45.5 C for 24 hr has been demonstrated, but at 46.6 C the initial inoculum decreased by 2 log cycles over the same period (13).

Serologic methods have been used to identify 5 different enterotoxins designated A, B, C, D, and E (4, 7, 16, 20). There is some evidence that enterotoxin C might in fact consist of two different enterotoxins (6). Most food poisoning strains studied to date produce enterotoxin A, and most outbreaks are caused by this serotype. In his study of 1146 strains of *S. aureus* from different sources, Casman (19) found that enterotoxin A was produced by the largest number, followed by enterotoxin B and combinations of A and B (see Table 16-2). Of the 1146 strains studied by Casman, almost 19% were found to be serologically enterotoxigenic. These 214 strains along with 50 known food poisoning strains are compared in Table 16-3 with respect to their incidence of the various enterotoxins. With respect to the phage pattern of 116 strains of *S. aureus* from clinical sources, the Group III pattern was the most common and no meaningful association could be made between phage patterns and enterotoxin types (see Table 16-4). In a study of staphylococci isolated from foods implicated in 75 different food poisoning incidents, Casman (21) found the % incidence of the 4 enterotoxins alone as follows: 49, 4, 4, and 10, respectively, for A, B, C, and D. Enterotoxin A plus others occurred in 28.5% while A plus D combinations occurred in 25% of the strains. In a study by Hall (49) in which 305 cultures were employed from outbreaks and other foods, 149 or 49% were enterotoxigenic with

TABLE 16-3. Relative incidence of A, B, and AB enterotoxigenicity among *S. aureus* strains *(19).**

Source	No. of strains serologically enterotoxigenic	% Incidence		
		A	B	AB
Clinical specimens (Wash., D.C.)	129	48	24.0	27.9
Clinical specimens (Atlanta, Ga.)	6	66.6	16.7	16.7
Nasal specimens (Wash., D.C.)	34	80.0	12.0	9.0
Nasal specimens (Atlanta, Ga.)	16	81.0	6.0	13.0
Frozen goods	21	52.4	47.0	0
Raw milk	8	76.0	24.0	0
Food poisoning	50	90.0	6.0	4

*Copyright, the New York Academy of Sciences, 1965; Reprinted by permission.

62.4% producing enterotoxin A, 17.4% enterotoxin C, 11.4% enterotoxin B, and 8.7% A, B, or C in combination.

Attempts to associate enterotoxigenicity with other biochemical properties of staphylococci such as gelatinase, phosphatase, lysozyme, lecithinase, lipase, and DNase production, or the fermentation of various carbohydrates have been unsuccessful. Food poisoning strains appear to be about the same as most coagulase positive strains in these respects.

Staphylococcal enterotoxins are somewhat unique among bacterial toxins in being proteins that are heat resistant. The biological activities of enterotoxin B were retained after heating for 16 hr at 60 C at pH 7.3 *(91)*. Heating of one preparation of enterotoxin C for 30 min at 60 C resulted in no change in serologic reactions *(10)*. Heating of enterotoxin A at 80 C for 3 min, or at 100 C for 1 min caused it to lose its capacity to react serologically

TABLE 16-4. Phage groups of serologically enterotoxigenic staphylococci from clinical specimens *(19)**

Enterotoxin	A	B	AB
Number of strains	54	27	35
	%	%	%
Phage group I	5.5	14.8	0.0
Phage group II	1.9	18.5	0.0
Phage group III	27.8	26.0	11.4
Unclassified	31.5	22.2	5.7
Phage 187	3.7	0.0	0.0
Phage 83	9.3	3.7	54.3
Not lysed	20.4	14.8	28.6

*Copyright, The New York Academy of Sciences, 1965; Reprinted by Permission.

(6). Heating enterotoxin centrifugates at boiling temperatures for 30–40 min did not completely destroy enterotoxin as measured by ability to induce vomiting in monkeys and kittens (98).

The thermal inactivation of enterotoxin A based on cat emetic response was shown by Denny *et al.* (36) to be 11 min at 250 F (F_{250}^{48} = 11 min). When monkeys were employed, the thermal inactivation was F_{250}^{46} = 8 min. These toxin preparations consisted of a 13.5-fold concentration of casamino acid culture filtrate employing strain 196-E. Using double-gel-diffusion assay, Read and Bradshaw (87) found the heat inactivation of 99+ % pure enterotoxin B in veronal buffer to be F_{250}^{58} = 16.4 min. The end point for enterotoxin inactivation by gel diffusion was identical to that by intravenous injection of cats. More recently, Denny *et al.* (37) found the slope of the thermal inactivation curve for enterotoxin A in beef bouillon at pH 6.2 to be around 27.8 C (50 F) using 3 different toxin concentrations (5, 20, and 60 μg/ml). Some D values for the thermal destruction of enterotoxin B are presented in Table 16-5. Crude toxin preparations have been found to be more resistant than purified toxins (87). It may be noted from Table 16-5 that staphylococcal nuclease displays heat resistance similar to that of enterotoxin B (see Chapter 4 for more information on this enzyme). Satterlee and Kraft (90) showed that enterotoxin B is more heat sensitive at 80 C than at 100 or 110 C. These investigators further showed that thermal loss was more pronounced at 80 C than at either 60 or 100 C when heating was carried out in the presence of meat proteins. In spite of their generally high degree of heat resistance, present thermal process treatments for low-acid foods are adequate to destroy these toxins (36).

The thermal destruction of *S. aureus* cells is considerably easier to accomplish than the destruction of enterotoxins as may be seen by the D values presented in Table 16-6 from various heating menstra. Staphylococcal cells have been shown to be quite heat sensitive in Ringers solution at pH 7.2 ($D_{140\ F}$ = 0.11) and much more resistant in milk at pH 6.9 (D_{140} = 10.0).

Overall, enterotoxin A may be more heat sensitive than B while C is intermediate (9). It should be kept in mind that the heat susceptibility of toxins as

TABLE 16-5.   D values for the heat destruction of staphylococcal enterotoxin B and staphylococcal heat-stable nuclease.

Conditions	$D\,(C)$	References
Veronal buffer	$D_{110} = 29.7^1$	(87)
Veronal buffer	$D_{110} = 23.5^2$	(87)
Veronal buffer	$D_{121} = 11.4^1$	(87)
Veronal buffer	$D_{121} = 9.9^2$	(87)
Veronal buffer, pH 7.4	$D_{110} = 18$	(65)
Beef broth, pH 7.4	$D_{110} = 60$	(65)
Staph. nuclease	$D_{130} = 16.\jmath$	(43)

[1]Crude toxin
[2]99+ percent purified

TABLE 16-6.   D and z values for the thermal destruction of S. *aureus* 196E in various heating menstra at 140 F.

Products	D (F)	z	Reference
Chicken-a-la-king	5.37	10.5	(2)
Custard	7.82	10.5	(2)
Green pea soup	6.7–6.9	8.1	(104)
Skim milk	3.1–3.4	9.2	(104)
0.5 percent NaCl	2.2–2.5	10.3	(104)
Beef bouillon	2.2–2.6	10.5	(104)
Skim milk alone	5.34	—	(61)
Raw skim milk + 10% sugar	4.11	—	(61)
Raw skim milk + 25% sugar	6.71	—	(61)
Raw skim milk + 45% sugar	15.08	—	(61)
Raw skim milk + 6% fat	4.27	—	(61)
Raw skim milk + 10% fat	4.20	—	(61)

well as of enzymes and microorganisms is affected by the overall composition of the suspending menstrum. These toxins would be expected to be more heat resistant in foods than in certain test systems and this can be seen from Tables 16-5 and 16-6.

Other physical properties of enterotoxins are presented in Table 16-7. All

TABLE 16-7.   Some properties of the enterotoxins of S. *aureus* (6).*

	Enterotoxin				
	A	B	C_1	C_2	E
Emetic Dose (ED_{50})					
(Monkey) (μg/animal)	5	5	5	5–10	10–20
Nitrogen Content (%)	16.5	16.1	16.2	16.0	—
Sedimentation Coefficient	3.04	2.89[a]	3.00	2.90	2.60
($s_{20,w}^0$), S		2.78[b]			
Diffusion Coefficient ($D_{20,w}^0$),	7.94	7.72[a]	8.10	8.10	—
$\times 10^{-7}$ cm^2 sec^{-1}		8.22[b]			
Reduced Viscosity (ml/g)	4.07	3.92[a]	3.4	3.7	—
		3.81[b]			
Molecular Weight	27,800[c]	28,366[d]	34,100	34,000	29,600[e]
Partial Specific Volume	0.726	0.743[a]	0.732	0.742	—
		0.726[b]			
Isoelectric Point	6.8	8.6	8.6	7.0	7.0
Maximum absorption (mμ)	277	277	277	277	277
Extinction ($E_{1cm}^{1\%}$)	14.3	14.0[a]	12.1	12.1	12.5
		14.4			

[a](91), [b](4), [c](92), [d](53), [e](11)

*Reproduced in part from *Biochemistry of Some Foodborne Microbial Toxins*, by Richard I. Mateles and Gerald Wogan, by permission of the M.I.T. Press, Cambridge, Massachusetts. Copyright © 1967 by the M.I.T. Press.

are simple proteins which upon hydrolysis yield 18 amino acids with aspartic, glutamic, lysine, and valine being the most abundant. The amino acid sequence for enterotoxin B has been determined *(53)*. The N-terminal acid is glutamic and lysine is the C-terminal. Enterotoxins A, B, C, and E are composed of 239 to 296 amino acid residues. The disulfide bridge in enterotoxin B has been shown to be unessential for biological activity and conformation *(34)*. The biological activity of enterotoxin A was found to be destroyed when the abnormal tyrosyl residues were modified *(24)*. The effect of acetylation, succinylation, guanidination, and carbamylation on the biological activity of enterotoxin B obtained from strains S-6 was recently reported by Chu *et al., (25)*. These investigators found that guanidination of 90% of the lysine residues had no effect on emetic activity or on the combining power of antigen-antibody reactions of this toxin. The latter was reduced, however, when the toxin was acetylated, succinylated, and carbamylated. These investigators concluded that acetylation and succinylation decreased the net positive charge of the toxin which is contributed by amino groups. The normal positive charge of the toxin is thought to play an important role in both its emetic activity and in its combining with specific antibody. In their active state, the enterotoxins have been reported to be resistant to proteolytic enzymes such as trypsin, chymotrypsin, rennin, and papain, but sensitive to pepsin at a pH of about 2 *(6)*. While the various enterotoxins differ in certain physio-chemical properties, each has about the same potency. Although biological activity and serologic reactivity are generally associated, it has been shown that serologically negative enterotoxin may be biologically active.

The maximum amount of enterotoxin A that can be produced in culture media under ideal conditions is *ca.* 5–6 μg/ml, while levels of 350 and 60 μg/ml or more of enterotoxins B and C, respectively, can be produced *(88)*. In protein hydrolysate media, up to 500 μg/ml of enterotoxin B may be produced *(8)*. A recent study by Chesbro *et al.* *(22)* suggests that enterotoxin B is heterogeneous. These investigators found that two electrophoretically distinct toxins can be identified in cultures, one produced in early to mid-log and the other in mid-late log phase growth. Analyses of old cultures should show the presence of both. More details on the production of enterotoxins have been presented elsewhere *(6, 21)*.

4. Detection of Enterotoxins.   The detection of enterotoxins in foods may be achieved by use of animals or *in vitro* procedures. Among the animals that have been employed are dogs, pigs, pigeons, frogs, cats, monkeys, and man. Although man is by far the most suitable for demonstrating the presence of significant levels of enterotoxins, there is usually a problem in obtaining the number of volunteers needed for this purpose at the appropriate time. The use of kittens requires the inactivation of substances other than enterotoxins that may induce emesis. Various investigators have found the kitten to be a suitable test animal. Next to man, young rhesus monkeys appear to be the most susceptible to enterotoxins, although cats have been reported to be more susceptible than monkeys (36). Bergdoll (6) has pointed out that monkeys do not give an emetic response to toxic culture filtrates produced by staphylococci other than the enterotoxins. With the use of monkeys, assays are generally made by administering the enterotoxic preparation in solution by use of catheter. The animals are observed for 5 hr for vomiting. Emesis in at least 2 of 6 animals is considered to indicate a positive reaction (6).

The in vitro methods consist of antigen-antibody reactions in various modifications. The immunologic procedures most often used consist of the gel diffusion, radioimmunoassay (RIA), reverse passive hemagglutination (RPH), and hemagglutionation-inhibition (HI) methods. The minimum times required for results by these methods along with the relative sensitivity of each to enterotoxins are presented in Table 4-3, p. 55. The minimum detectable levels of enterotoxin have been found to be (8):

Single gel diffusion (Oudin) . . . . .ca. 1 μg/ml
Double gel diffusion (Ouchterlony) . . . . .0.05 μg/ml
Microslide method . . . . .0.1 μg/ml
RIA and RPH . . . . .0.0015 μg/ml

The extraction of food for the in vitro assay of enterotoxins by use of the microslide method is presented in Figure 16-4. The production of enterotoxic antisera has been described by Casman and Bennett (17). For the recovery of enterotoxins from foods, one or more of the pertinent references in Table 4-1, p. 48, should be consulted in addition to Casman and Bennett (18) and Bergdoll (8).

5. The Staphylococcal Food Poisoning Syndrome and Its Diagnosis.
Upon the ingestion of contaminated food, the symptoms of staphylococcal food poisoning usually develop within 4 hr, although a range of from 1 to 6 hr has been reported. The symptoms consist of nausea, vomiting, abdominal cramps (which are usually quite severe), diarrhea, sweating, headache, prostration, and sometimes a fall in body temperature. The symptoms generally last from 24–48 hr, and the mortality rate is very low or nil. The usual treatment for healthy persons consists of bed rest and maintenance of fluid balance. Upon cessation of symptoms, the victim possesses no demonstrable immunity to recurring attacks although animals become resistant to enterotoxin after repeated oral doses (8). Since the symptoms are referable to the ingestion of preformed enterotoxin, it is conceivable that stool cultures might be negative for the organisms, although this is rare. Proof of staphylococcal food poisoning is established by recovering coagulase positive, enterotoxigenic staphylococci from left-over food and from the stool cultures of victims. Attempts should be made to extract enterotoxin

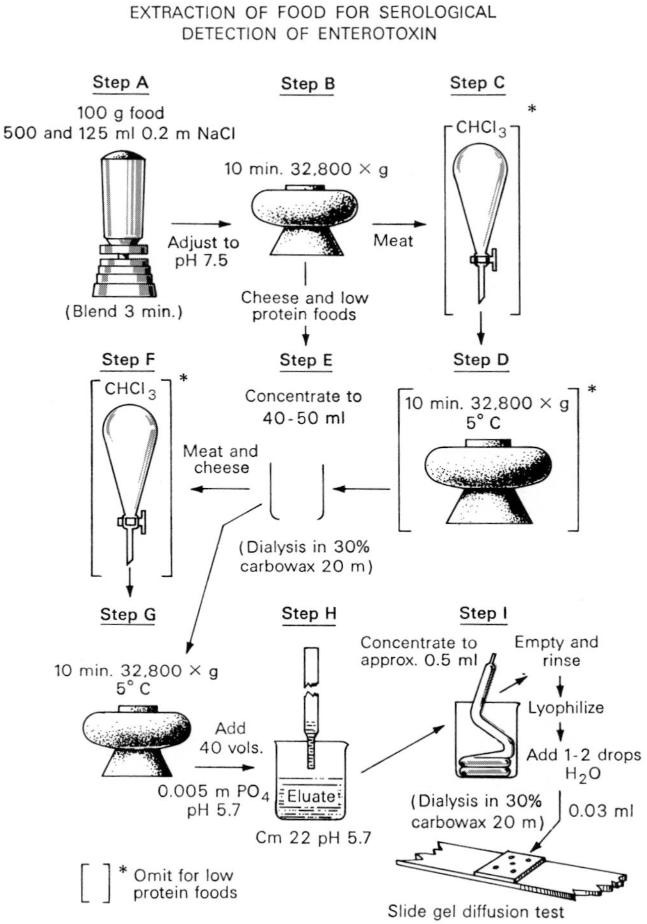

EXTRACTION OF FOOD FOR SEROLOGICAL
DETECTION OF ENTEROTOXIN

FIGURE 16-4.   Extraction of food for serological detection of enterotoxin *(21).*

from suspect foods, especially when the number of recoverable viable cells is low.

The precise minimum quantity of enterotoxin needed to cause illiness in man is not known. Data obtained from the use of 3 human volunteers showed that a dosage of 20–25 μg of pure enterotoxin B can produce a staphylococcal food poisoning syndrome *(84).* From 16 incidents of staphylococcal food poisoning studied by Gilbert and Wieneke *(48),* enterotoxin levels of <0.01–0.25 μg/g of food were found. It has been suggested that <1 μg of these toxins can make a person ill *(8).*

The pathogenesis of enterotoxins in man is not yet clear. The toxins act upon the intestines to induce vomiting and diarrhea and the same effects can be achieved by intravenous injections. The effect of enterotoxins on various animals has been studied and much of this work has been reviewed by Bergdoll *(8)*.

For reasons that are not well understood, enterotoxin A production appears to occur more often than enterotoxin B in strains that produce both toxins. Enterotoxin A producers have been shown to produce this toxin while in the L-phase, but enterotoxin B was not produced by L-forms *(29)*.

6. Common Food Vehicles of Staphylococcal Food Poisoning. A large number of foods have been incriminated in staphylococcal food-poisoning outbreaks and the usual common denominators of such foods consist of a convenience type food made by hand and improperly refrigerated after being prepared. Of selected cases of staphylococcal food poisoning over the period 1961–1973, meat products were involved in over 40% of the cases (Table 16-8). Of the 251 meat products, 76 were baked ham products *(15)*. The precise number of cases of staphylococcal food poisoning in the United States is not known. Various investigators have placed the number in the tens of thousands of cases per year. The problem here is one of reporting, and all too often the small outbreaks that occur in homes are not reported to public health officials. A large percentage of the reported cases of food poisoning of all types are cases that result from banquets generally involving large numbers of persons. For many years, staphylococcal food poisoning

TABLE 16-8.   Foods incriminated in staphylococcal food poisoning outbreaks, 1961–1973 (summarized from Bryan, *15*).

Food Products	No. of Outbreaks
Meat	251
Ham products	137
Beef products	60
Uncured pork products	27
Others/combinations	27
Poultry	102
Turkey products	52
Chicken	50
Custards and cream-filled pastries	55
Fish and shellfish	34
Salads (nonmeat)	31
Eggs and egg products	17
Milk and milk products	14
Vegetables	9
Cereal products	6
Miscellaneous products	59

Condensed from a table in *Food Microbiology: Public Health and Spoilage Aspects*, by de-Figueiredo and Splittstoesser, published by Avi Publishing Co., Westport, Conn. 06880, and used by permission of publisher.

has been the leading cause of all food poisoning in the United States, but *Salmonella* food poisoning, discussed in the next chapter, is currently the most common. Of all reported food poisoning incidents in Great Britain, staphylococcal accounts for only about 2% *(48)*.

With respect to their destruction in foods, Woodburn *et al. (110)* showed that over 2 million/g could be destroyed by boiling plastic pouches containing chicken products for 10 min, or by microwave heating for 2 min.

7. *The Microbial Ecology of* S. aureus *in Foods and Its Recovery.*   In attempting to associate staphylococcal food poisoning with a given food, more than the mere presence of enterotoxigenic strains should be noted. First, the numbers of organisms/g or ml should be determined. Assuming that destruction of cells has not occurred by heat or other means, a rather large number of cells is generally required before the food is rendered hazardous. Although the precise number necessary to produce enough enterotoxin to induce food poisoning symptoms is somewhat debatable, such things as the nature of the food substrate, its pH, presence of enterotoxin-inhibiting compounds, holding temperature, and characteristics of the strain in question all come into play. Hobbs *(51)* has suggested that at least 500,000 to 1 million/g must be present in order to produce food poisoning symptoms in man. Donnelly *et al. (40)* found that 10^8 cells/ml of milk were required before enterotoxin appeared. Genigeorgis *et al. (46)* found maximum cell counts to be in excess of 4×10^6/g in toxigenic hams, and noted that even after as long as 2 mon of incubation at 10 C the hams maintained their normal appearance even in the presence of more than 10^8 staphylococci/g of meat. In inoculated Colby cheese, enterotoxin A was found with cell counts of 15,000,000/g and with 28,000,000/g in Cheddar when starter cultures were used *(100)*. When cheese starters failed because of bacteriophages, enterotoxin A was found with minimal cell populations of 3–5,000,000/g by the above investigators. The foods involved in the 16 staphylococcal food poisoning outbreaks studied by Gilbert and Wieneke *(48)* revealed counts of about 6,000,000/g. In culture media, the exponential growth phase of enterotoxigenic staphylococci ends when the number of cfu/ml reaches *ca.* 10^9 *(102)*, and Czop and Bergdoll *(30)* found that at this cell population 1 to 2 μg/ml of enterotoxins A and B were produced. Other investigators have variously suggested numbers ranging from 1 million to 1 billion/g *(82)* as being necessary to cause staphylococcal food poisoning.

Second, the presence of *S. aureus* in foods should be viewed in relation to the numbers and types of competing microorganisms present. A large number of investigators including Peterson *et al. (82, 83)*, Dack and Lippitz *(32)*, Troller and Frazier *(106)*, and Crisely *et al. (27)* have shown that staphylococci are unable to compete with normal food-borne bacteria in both fresh and frozen foods. That is, at temperatures that favor staphylococcal growth, the normal food saprophytic flora offers protection against staphylococcal growth through antagonism, competition for nutrients, and modification of the environment to conditions less favorable to staphylococcal growth. Bacteria that are known to be antagonistic to staphylococci include *Acinetobacter, Aeromonas, Bacillus, Enterobacteriaceae, Lactobacillaceae, Pseudomonas,* streptococci, and *Staphylococcus epidermidis* (79). Enterotoxin A has been shown to be resistant to a variety of environmental stresses but growth of several lactic acid bacteria did lead to its reduction and to a suggestion by Chordash *et al. (23)* that toxin reduction might have been due to specific enzymes or other metabolites of the lactic acid bacteria. The generally higher incidence of food poisonings of all types over the past few years might simply reflect the general improvement in food plant sanitation and the fact that improving technology enables food producers to produce more low-count foods. More studies on the ecology of staphylococci are necessary before this picture can be made clearer.

The normal saprophytic flora of foods is of great importance in the recovery of staphylococci from foods. Early techniques for recovering staphylococci were designed for clinical specimens such as boils, nasal cavities, skin, and the like. Under such conditions, the relative numbers of staphylococci to other bacteria is generally large so that few problems are encountered in the use of blood agar or nutrient agar fortified with 7.5% NaCl for staphylococcal recovery. However, problems are encountered when one attempts to transfer the use of these media to food products, where the numbers of staphylococci may be expected to be quite low in comparison to the normal saprophytic flora, many members of which are generally capable of growth over a wide range of conditions. It has, therefore, been necessary to devise special media and conditions for the recovery of small numbers of staphylococci from food products. Salt at levels of about 7.5%, tellurite, mercuric chloride, sorbic acid and other compounds may be used as selective agents in staphylococcal media, but there is a problem of differentiation of coagulase-positive staphylococci from micrococci. Members of the family *Micrococcaeae* generally develop on media selective for staphylococci. The micrococci are widely distributed in nature and grow essentially under the same conditions as staphylococci. In recovering staphylococci from foods, differentiation is as important in the choice of a medium as selectivity. The recovery of metabolically injured staphylococci from foods is discussed in Chapter 4, pp. 49–54, and some recovery methods and media are presented in Table 4-2, pp. 52–53. The generally preferred media employ egg yolk as differential agent, and tellurite, NaCl, or antibiotics as selective agents. Comparative studies of various recovery media and methods have been conducted *(28, 26, 58, 86).* The references in Table 4-1 should be consulted for more specific recovery media and methods.

8. The Prevention of Food Poisoning. When susceptible foods are produced with low numbers of staphylococci, they will remain free of enterotoxin and other food poisoning hazards if kept either *below* 40 or *above* 140 F until eaten. For the years 1961–1972, over 700 food-borne disease outbreaks were investigated by Bryan *(14)* relative to the factors that contributed to the outbreaks. Of the 16 factors identified, the five most frequently involved were:

1. Inadequate refrigeration.
2. Preparing foods far in advance of planned service.
3. Infected persons practicing poor personal hygiene.
4. Inadequate cooking or heat processing.
5. Holding food in warming devices at bacterial growth temperatures.

Inadequate refrigeration alone comprised 25.5% of the contributing factors. The 5 listed above contributed to 68 percent of outbreaks. Susceptible foods should not be held within the staphylococcal growth range for more than 3–4 hr.

STREPTOCOCCAL FOOD POISONING

Streptococcal food poisoning was first recorded by Linden *et al.* *(67)* who investigated two outbreaks traced to cheese, Albanian cheese and cheddar cheese. Since that time, a large number of reports on food poisoning outbreaks have incriminated the streptococci, but the precision to be found in the association of staphylococci with food poisoning is lacking here.

The group of streptococci generally associated with food poisoning consists of fecal streptococci, especially *S. faecalis* types (see previous chapter). Food poisoning outbreaks or infections have been reported in which *S. viridans* and *S. pyogenes* were incriminated *(99),* but these may be treated in the same way as the transmission of, say, typhoid fever or scarlet fever by foods. Streptococcal food poisoning is taken here to indicate gastroenteritis caused by *S. faecalis* strains.

The incubation period and symptoms for this syndrome are presented in Table 16-9. These data represent findings taken by Moore *(76)* from five investigators. The incubation periods ranged from 2–22 hr after ingestion of food with a median of 10 hr. Dack *(33)* has reported that the incubation period and symptoms of streptococcal food poisoning are the same as for *Clostridium perfringens* and *Bacillus cereus* food poisoning. Hobbs *et al.* *(52)* found the incubation period for *C. perfringens* food poisoning to be from 8–22 hr with symptoms as follows: acute abdominal pain and diarrhea with nausea and vomiting rare, and a duration of one day or less.

Vehicle foods of streptococcal food poisoning include turkey dressing, cured ham, barbecued beef, Vienna sausage, cheese, evaporated milk, turkey à la king, and others. Table 16-10 summarizes findings on some outbreaks reported between 1926 and 1949.

TABLE 16-9.   Clinical symptoms and incubation periods for streptococcal outbreaks (76).

Authors (see ref. 76)		Symptoms*			Incubation period (hr)
		V	AP	D	
Cary *et al.* (1931)		+	+	+	4–5, occasionally 12
Cary *et al.* (1938)				+	12(range 6–18)
	a	+	+		2–7
Buchbinder *et al.* (1948)	b	+	+	+	5½–12
	c		+	+	10–12 (range 3–15)
	d	+	+	+	3–6
Moore (1948)			+	+	11–13 (range 3–22)
Hobbs *(pers. comm.)*		+	+	+	3, 5, 5, 13, 13
Dack *(pers. comm.)*	a	+		+	4
	b		+	+	4–18

*V, vomiting; AP, abdominal pain; D, diarrhoea.

Attempts to reproduce the symptoms of streptococcal food poisoning by feeding streptococci to human volunteers have been made by several investigators. Deibel and Silliker *(35)* fed a total of 23 strains of enterococci to two and sometimes three human volunteers in an effort to elicit food poisoning symptoms, and in no instance were symptoms produced. When enterococci are suspected as the etiologic agent of food poisoning, attempts should be made to reproduce symptoms in humans by use of suspect strains as soon as possible. It is conceivable that laboratory strains lose this capacity upon aging, or that certain environmental factors are necessary to bring about this syndrome. In view of the lack of success in consistently reproducing the symptoms of this syndrome, some authors have recently questioned whether the pathogenicity of enterococci is sufficient to cause food poisoning. The wide incidence of these organisms in many foods, the relatively high numbers that sometimes exist in edible foods, and the generally low incidence of this syndrome all tend to raise questions about their capacity to cause food poisoning. As was pointed out in the previous chapter, there is need for standardized sampling and plate-count procedures before problems of this type can be effectively solved.

While questions have been raised as to whether or not enterococci are capable of causing food poisoning, the existence of this syndrome in the past can hardly be denied. The causative strains are undoubtedly weak pathogens as is evidenced by the generally large numbers associated with vehicle foods, with numbers/g on the order of several hundred million or several billion being necessary to produce symptoms.

TABLE 16-10. Outbreaks of food poisoning attributed to fecal streptococci (93)*

Reference (see 93)	No. of cases	No. at risk	Food implicated	Author's identification	Comments
Linden, Turner and Thom (1926)	9	?	Albanian cheese	Not identified	Identified as *S. faecalis* by Sherman, Smiley and Niven (1943).
Cary, Dack and Myers (1931)	22 75	? 182	American cheese Vienna sausage	"Green hemolytic strep."	2 biochemical types; one might have been an "enterococcus."
Jordan and Burrows (1934)	?	?	Coconut cream pie	"Str. viridans"	Not possible to name. Cell-free filtrate produced symptoms in susceptible monkeys.
Cary, Dack and Davison (1938)	117	208	Beef croquettes	α hemolytic streptococci; 2 biochemical types	Not possible to name these from the tests recorded. Feeding tests did not show a toxin.
Dack (1943)	294	393	Turkey dressing	α hemolytic streptococci	Identified as *S. faecalis* by Buchbinder *et al.*
Buchbinder, Osler and Steffen (1948)	74	161	Evaporated milk	*S. faecalis*	Physiological reactions were atypical
	3	3	Charlotte russe	*S. faecalis* and *S. liquefaciens*	Both present in large numbers.
	74	87	Barbecued beef	*S. faecalis* and *S. liquefaciens*	17 cultures *S. faecalis* 1 culture *S. liquefaciens*.
	9	9	Ham bologna	*S. faecalis* and *S. liquefaciens*	72 cultures *S. faecalis* 4 cultures *S. liquefaciens*.
Dack, Niven, Kirsner and Marshall (1949)	94	98	Turkey à la king	*S. liquefaciens*	

*Reproduced by permission from P. M. F. Shattock, "Enterococci" in *Chemical and Biological Hazards in Food.* J. C. Ayres, A. A. Kraft, A. T. Snyder, and H. W. Walker, Editors. © 1963 by the Iowa State University Press, Ames.

REFERENCES

1. **Angelotti, R., M. J. Foter, and K. H. Lewis.** 1961. Time-temperature effects on salmonellae and staphylococci in foods. A. J. Pub. Hlth. *51:* 76–88.

2. **Angelotti, R., M. J. Foter, and K. H. Lewis.** 1960. Time-temperature effects on salmonellae and staphylococci in foods. II. Behavior at warm holding temperatures. Thermal-death-time studies. U.S. Dept. Hlth., Educ. & Welfare, Pub. Hlth. Serv. (Cincinnati, O.).

3. **Baird-Parker, A. C.** 1965. The classification of staphylococci and micrococci from world-wide sources. J. Gen. Microbiol. *38:* 363–387.

4. **Bergdoll, M., C. Borja, and R. Avena.** 1965. Identification of a new enterotoxin as enterotoxin C. J. Bacteriol. *90:* 1481–1485.

5. **Bergdoll, M. S., K. F. Weiss, and M. J. Munster.** 1967. The production of staphylococcal enterotoxin by a coagulase-negative microorganism. Bact. Proc., p. 12.

6. **Bergdoll, M. S.** 1967. The staphylococcal enterotoxins. In: *Biochemistry of Some Foodborne Microbial Toxins.* R. I. Mateles and G. N. Wogan, Editors. M.I.T. Press, Cambridge, Mass., pp. 1–25.

7. **Bergdoll, M. S., C. R. Borja, R. N. Robbins, and K. F. Weiss.** 1971. Identification of enterotoxin E. Infect. Immunity *4:* 593–595.

8. **Bergdoll, M. S.** 1972. The enterotoxins, pp. 301–331, IN: *The Staphylococci,* edited by J. O. Cohen (Wiley-Interscience: N.Y.).

9. **Bergdoll, M. S.** 1973. Enterotoxin detection, pp. 287–292, IN: *The Microbiological Safety of Food,* edited by B. C. Hobbs and J. H. B. Christian (Academic Press: N.Y.).

10. **Borja, C. R. and M. S. Bergdoll.** 1967. Purification and partial characterization of enterotoxin C produced by *Staphylococcus aureus* strain 137. J. Biochem. *6:* 1467–1473.

11. **Borja, C. R., E. Fanning, I.-Y. Huang, and M. S. Bergdoll.** 1972. Purification and some physiochemical properties of staphylococcal enterotoxin E. J. Biol. Chem. *247:* 2456–2463.

12. **Breckinridge, J. C. and M. S. Bergdoll.** 1971. Outbreak of foodborne gastroenteritis due to a coagulase-negative enterotoxin-producing staphylococcus. New Engl. J. Med. *284:* 541–543.

13. **Brown, D. F. and R. M. Twedt.** 1972. Assessment of the sanitary effectiveness of holding temperatures on beef cooked at low temperature. Appl. Microbiol. *24:* 599–603.

14. **Bryan, F. L.** 1974. Microbiological food hazards today—based on epidemiological information. Food Technol. *28:* No. 9, 52–59.

15. **Bryan, F. L.** 1976. *Staphylococcus aureus,* pp. 12–128, IN: *Food Microbiology: Public Health and Spoilage Aspects,* edited by M. P. de Figueiredo and D. F. Splittstoesser (Avi Publ.: Conn.).

16. **Casman, E., M. Bergdoll, and J. Robinson.** 1963. Designation of staphylococcal enterotoxins. J. Bacteriol. *85:* 715–716.

17. **Casman, E. P. and R. W. Bennett.** 1964. Production of antiserum for staphylococcal enterotoxins. Appl. Microbiol. *12:* 363–367.

18. **Casman, E. P. and R. W. Bennett.** 1965. Detection of staphylococcal enterotoxin in food. Appl. Microbiol. *13:* 181–189.

19. **Casman, E. P.** 1965. Staphylococcal enterotoxin. In: *The Staphylococci: Ecologic Perspectives.* Ann. N.Y. Acad. Sci. *128:* 124–133.

20. **Casman, E. P., R. W. Bennett, A. E. Dorsey, and J. A. Issa.** 1967. Identification of a fourth staphylococcal enterotoxin, enterotoxin D. J. Bacteriol. *94:* 1875–1882.

21. **Casman, E. P.** 1967. Staphylococcal food poisoning. Hlth. Lab. Sci. *4:* 199–206.

22. **Chesbro, W., D. Carpenter, and G. J. Silverman.** 1976. Heterogeneity of *Staphylococcus aureus* enterotoxin B as a function of growth stage: Implications for surveillance of foods. Appl. Environ. Microbiol. *31:* 581–589.

23. **Chordash, R. A. and N. N. Potter.** 1976. Stability of staphylococcal enterotoxin A to selected conditions encountered in foods. J. Food Sci. *41:* 906–909.

24. **Chu, F. S., K. Thadhani, E. J. Schantz, and M. S. Bergdoll.** 1966. Purification and characterization of staphylococcal enterotoxin A. Biochemistry *5:* 3281–3289.

25. **Chu, F. S., E. Crary, and M. S. Bergdoll.** 1969. Chemical modification of amino groups in staphylococcal enterotoxin B. Biochemistry *8:* 2890–2896.

26. **Collins-Thompson, D. L., A. Hurst, and B. Aris.** 1974. Comparison of selective media for the enumeration of sublethally heated food-poisoning strains of *Staphylococcus aureus.* Can. J. Microbiol. *20:* 1072–1075.

27. **Crisley, F., R. Angelotti, and M. Foter.** 1964. Multiplication of *Staphylococcus aureus* in synthetic cream fillings and pies. Pub. Hlth. Repts. *79:* 369–376.

28. **Crisley, F. D.** 1964. Methods for isolation and enumeration of staphylococci. In: *Examination of Foods for Enteropathogenic and Indicator Bacteria.* K. H. Lewis and R. A. Angelotti, Editors. U.S. Pub. Hlth. Service, Washington, D.C.

29. **Czop, J. K. and M. S. Bergdoll.** 1970. Synthesis of enterotoxins by L-forms of *Staphylococcus aureus.* Infect. Immunity *1:* 169–173.

30. **Czop, J. K. and M. S. Bergdoll.** 1974. Staphylococcal enterotoxin synthesis during the exponential transitional, and stationary growth phases. Infect. Immunity *9:* 229–235.

31. **Dack, G. M., W. E. Cary, O. Woolpert, and H. Wiggers.** 1930. An outbreak of food poisoning proved to be due to a yellow hemolytic staphylococcus. J. Prev. Med. *4:* 167–175.

32. **Dack, G. M. and G. Lippitz.** 1962. Fate of staphylococci and enteric microorganisms introduced into slurry of frozen pot pies. Appl. Microbiol. *10:* 472–479.

33. **Dack, G. M.** 1963. Problems in foodborne diseases. In: *Microbiological Quality of Foods.* L. W. Slanetz, *et al.,* Editors, Academic Press, N.Y., pp. 41–49.

34. **Dalidowicz, J. E., S. J. Silverman, E. J. Schantz, D. Stefanye, and L. Spero.** 1966. Chemical and biological properties of reduced and alkylated staphylococcal enterotoxin B. Biochemistry *5:* 2375–2381.

35. **Deibel, R. H. and J. H. Silliker.** 1963. Food-poisoning potential of the entero-cocci. J. Bacteriol. *85:* 827–832.

36. **Denny, C. B., P. L. Tan, and C. W. Bohrer.** 1966. Heat inactivation of staphylococcal enterotoxin A. J. Food Sci. *31:* 762–767.

37. **Denny, C. B., J. Y. Humber, and C. W. Bohrer.** 1971. Effect of toxin concentration on the heat inactivation of staphylococcal enterotoxin A in beef bouillon and in phosphate buffer. Appl. Microbiol. *21:* 1064–1066.

38. **Denys, J.** 1894. Empoisonnement par de la viande contenant le *Staphylocoque pyogene.* Sem. Med. (Paris) *14:* 441.

39. **Donnelly, C. B., L. A. Black, and K. H. Lewis.** 1964. Occurrence of coagulase-positive staphylococci in cheddar cheese. Appl. Microbiol. *12:* 311–315.

40. **Donnelly, C. B., J. E. Leslie, and L. A. Black.** 1966. Production of enterotoxin A in milk. Bacteriol. Proc., p. 13.

41. **Eddy, B. P. and M. Ingram.** 1962. The occurrence and growth of staphylococci on packed bacon, with special reference to *Staphylococcus aureus.* J. Appl. Bacteriol. *25:* 237–247.

42. **Elek, S. D.** 1959. *Staphylococcus pyogenes and Its Relation to Disease.* E. & S. Livingstone, Ltd. Edinburgh and London.

43. **Erickson, A. and R. H. Deibel.** 1973. Production and heat stability of staphylococcal nuclease. Appl. Microbiol. *25:* 332–336.

44. **Friedman, M. E.** 1966. Inhibition of staphylococcal enterotoxin B formation in broth cultures. J. Bacteriol. *92:* 277–278.

45. **Genigeorgis, C. and W. W. Sadler.** 1966. Effect of sodium chloride and pH on enterotoxin B production. J. Bacteriol. *92:* 1383–1387.

46. **Genigeorgis, C., H. Riemann, and W. W. Sadler.** 1969. Production of enterotoxin B in cured meats. J. Food Sci. *34:* 62–68.

47. **Genigeorgis, C., M. S. Foda, A. Mantis, and W. W. Sadler.** 1971. Effect of sodium chloride and pH on enterotoxin C production. Appl. Microbiol. *21:* 862–866.

48. **Gilbert, R. J. and A. A. Wieneke.** 1973. Staphylococcal food poisoning with special reference to the detection of enterotoxin in food, pp. 273–285, IN: *The Microbiological Safety of Food,* edited by B. C. Hobbs and J. H. B. Christian (Academic Press: N.Y.).

49. **Hall, H. E.** 1968. Enterotoxin and enzyme production by a selected group of *Staphylococcus aureus* cultures. Bact. Proc., pp. 77–78.

50. **Heidelbauer, R. J. and P. R. Middaugh.** 1973. Inhibition of staphylococcal enterotoxin production in convenience foods. J. Food Sci. *38:* 885–888.

51. **Hobbs, B. C.** 1962. Staphylococcal and *Clostridium welchii* food poisoning. In: *Food Poisoning.* Royal Society of Health, London, pp. 49–59.

52. **Hobbs, B. C., M. E. Smith, C. L. Oakley, G. H. Warrack, and J. C. Cruickshank.** 1953. *Clostridium welchii* food poisoning. J. Hyg. *51:* 75–101.

53. **Huang, I.-Y. and M. S. Bergdoll.** 1970. The primary structure of staphylococcal enterotoxin B. III. The cyanogen bromide peptides of reduced and aminoethylated enterotoxin B and the complete amino acid sequence. J. Biol. Chem. *245:* 3518–3525.

54. **Jarvis, A. W., R. C. Lawrence, and G. G. Pritchard.** 1973. Production of staphylococcal enterotoxins A, B, and C under conditions of controlled pH and aeration. Infect. Immunity 7: 847–854.

55. **Jay, J. M.** 1961a. Incidence and properties of coagulase-positive staphylococci in certain market meats as determined on three selective media. Appl. Microbiol. 9: 228–232.

56. **Jay, J. M.** 1961b. Some characteristics of coagulase-positive staphylococci from market meats relative to their origins into the meats. J. Food Sci. 26: 631–634.

57. **Jay, J. M.** 1962. Further studies on staphylococci in meats. III. Occurrence and characteristics of coagulase-positive strains from a variety of non-frozen market cuts. Appl. Microbiol. 10: 247–251.

58. **Jay, J. M.** 1963. The relative efficacy of six selective media in isolating coagulase positive staphylococci from meats. J. Appl. Bacteriol. 26: 69–74.

59. **Jay, J. M.** 1966. Production of lysozyme by staphylococci and its correlation with three other extracellular substances. J. Bacteriol. 91: 1804–1810.

60. **Jay, J. M.** 1970. Effect of borate on the growth of coagulase-positive and coagulase-negative staphylococci. Infect. Immunity 1: 78–79.

61. **Kadan, R. S., W. H. Martin, and R. Mickelsen.** 1963. Effects of ingredients used in condensed and frozen dairy products on thermal resistance of potentially pathogenic staphylococci. Appl. Microbiol. 11: 45–49.

62. **Kashiba, S., K. Nuzu, S. Tanaka, H. Nozu, and T. Amano.** 1959. Lysozyme, an index of pathogenic staphylococci. Biken's J. 2: 50–55.

63. **Katsuno, S. and M. Kondo.** 1973. Regulation of staphylococcal enterotoxin B synthesis and its relation to other extracellular proteins. Jap. J. Med. Sci. Biol. 26: 26–29.

64. **Lechowich, R. V., J. B. Evans, and C. F. Niven, Jr.** 1956. Effect of curing ingredients and procedures on the survival and growth of staphylococci in and on cured meats. Appl. Microbiol. 4: 360–363.

65. **Lee, I. C., K. E. Stevenson, and L. G. Harmon.** 1977. Effect of beef broth protein on the thermal inactivation of staphylococcal enterotoxin B. Appl. Environ. Microbiol. 33: 341–344.

66. **Lilly, H. D., R. A. McLean, and J. A. Alford.** 1967. Effects of curing salts and temperature on production of staphylococcal enterotoxin. Bact. Proc., p. 12.

67. **Linden, B. A., W. B. Turner, and C. Thom.** 1926. Food poisoning from a streptococcus in cheese. Pub. Hlth. Repts. 41: 1647–1652.

68. **Markus, Z. and G. J. Silverman.** 1968. Enterotoxin B production by nongrowing cells of *Staphylococcus aureus*. J. Bacteriol. 96: 1446–1447.

69. **Markus, Z. and G. J. Silverman.** 1969. Enterotoxin B synthesis by replicating and nonreplicating cells of *Staphylococcus aureus*. J. Bacteriol. 97: 506–512.

70. **McLean, R. A., H. D. Lilly, and J. A. Alford.** 1968. Effects of meat-curing salts and temperature on production of staphylococcal enterotoxin B. J. Bacteriol. 95: 1207–1211.

71. **Messer, J. W., J. T. Peeler, R. B. Read, Jr., J. E. Campbell, H. E. Hall, and H. Haverland.** 1970. Microbiological quality survey of some selected market foods in two socioeconomic areas. Bacteriol. Proc., p. 12.

72. **Metzger, J. F., A. D. Johnson, W. S. Collins, II, and V. McGann.** 1973. *Staphylococcus aureus* enterotoxin B release (excretion) under controlled conditions of fermentation. Appl. Microbiol. *25:* 770–773.

73. **Mickelsen, R., V. D. Foltz, W. H. Martin, and C. A. Hunter.** 1963. Staphylococci in cottage cheese. J. Milk Food Technol. *26:* 74–77.

74. **Miller, R. D. and D. Y. C. Fung.** 1973. Amino acid requirements for the production of enterotoxin B by *Staphylococcus aureus* S-6 in a chemically defined medium. Appl. Microbiol. *25:* 800–806.

75. **Minor, T. E. and E. H. Marth.** 1976. *Staphylococci and Their Significance in Foods* (Elsevier Pub. Co.: N.Y.).

76. **Moore, B.** 1955. Streptococci and food poisoning. J. Appl. Bacteriol. *18:* 606–618.

77. **Morrison, S. M., J. F. Fair, and K. K. Kennedy.** 1961. *Staphylococcus aureus* in domestic animals. Pub. Hlth. Repts. *76:* 673–677.

78. **Morse, S. A., R. A. Mah, and W. J. Dobrogosz.** 1969. Regulation of staphylococcal enterotoxin B. J. Bacteriol. *98:* 4–9.

79. **Mossel, D. A. A.** 1975. Occurrence, prevention, and monitoring of microbial quality loss of foods and dairy products. CRC Crit. Rev. Environ. Control *5:* 1–140.

80. **Munch-Peterson, E.** 1963. Staphylococci in food and food intoxication. J. Food Sci. *28:* 692–710.

81. **Omori, G., K. Kato, and S. Lida.** 1960. The lysozyme test in the study of staphylococci associated with food poisoning. Osaka City Med. J. *6:* 33–38.

82. **Peterson, A., J. Black and M. L. Gunderson.** 1962a. Staphylococci in competition. I. Growth of naturally occurring mixed populations in precooked frozen foods during defrost. Appl. Microbiol. *10:* 16–22.

83. **Peterson, A., J. Black, and M. L. Gunderson.** 1962b. Staphylococci in competition. II. Effect of total numbers and proportion of staphylococci in mixed cultures on growth in artificial culture medium. Appl. Microbiol. *10:* 23–30.

84. **Raj, H. D. and M. S. Bergdoll.** 1969. Effect of enterotoxin B on human volunteers. J. Bacteriol. *98:* 833–834.

85. **Raj, H. D. and J. Liston.** 1961. Survival of bacteria of public health significance in frozen sea foods. Food Technol. *15:* 429–434.

86. **Raj, H. D.** 1966. A new procedure for the detection and enumeration of coagulase-positive staphylococci from frozen seafoods. Can. J. Microbiol. *12:* 191–198.

87. **Read, R. B. and J. G. Bradshaw.** 1966. Thermal inactivation of staphylococcal enterotoxin B in veronal buffer. Appl. Microbiol. *14:* 130–132.

88. **Reiser, R. F. and K. F. Weiss.** 1969. Production of staphylococcal enterotoxins A, B, and C in various media. Appl. Microbiol. *18:* 1041–1043.

89. **Richou, R., J. Pantaléon, and M. Cazaillet.** 1959. Recherches sur la détermination de l'origine et du pouvoir pathogène de staphylocoques isolés de produits carnés. Rev. d'Immunol. *23:* 31–38.

90. **Satterlee, L. D. and A. A. Kraft.** 1969. Effect of meat and isolated meat proteins on the thermal inactivation of staphylococcal enterotoxin B. Appl. Microbiol. *17:* 906–909.

91.   Schantz, E. J., W. G. Roessler, J. Wagman, L. Spero, D. A. Dunnery, and M. S. Bergdoll. 1965. Purification of staphylococcal enterotoxin B. J. Biochem. *4:* 1011–1016.

92.   Shantz, E. J., W. G. Roessler, M. J. Woodburn, J. M. Lynth, H. M. Jacoby, S. J. Silverman, J. C. Gorman, and L. Spero. 1972. Purification and some chemical and physical properties of staphylococcal enterotoxin A. Biochem. *11:* 360–366.

93.   Shattock, P. M. F. 1962. Enterococci. In: *Chemical and Biological Hazards in Food.* J. C. Ayres *et al.,* Editors, Iowa State U. Press, pp. 303–319.

94.   Siems, H., D. Kusch, H.-J. Sinell, and F. Untermann. 1971. Vorkommen und Eigenschaften von Staphylokokken in verschiedenen Produktionsstufen bei der Fleischverarbeitung. Fleischwirtschaft *51:* 1529–1533.

95.   Silverman, G. J., J. T. R. Nickerson, D. W. Duncan, N. S. Davis, J. S. Schacter, and M. M. Joselow. 1961. Microbial analysis of frozen raw and cooked shrimp. I. General results. Food Technol. *15:* 455–458.

96.   Splittstoesser, D. F., G. E. R. Hervey, and W. P. Wettergren. 1965. Contamination of frozen vegetables by coagulase-positive staphylococci. J. Milk & Food Technol. *28:* 149–151.

97.   Stutzenberger, F. J. and C. L. San Clemente. 1968. Production and purification of staphylococcic coagulase in a semi-defined medium. Am. J. Vet. Res. *29:* 1109–1116.

98.   Surgalla, J. J. and K. E. Hite. 1945. A study of enterotoxin and alpha and beta hemolysis production by certain *Staphylococcus* cultures. J. Inf. Dis. *76:* 78–82.

99.   Tanner, F. W. 1953. *Food-borne Infections and Intoxications.* 2nd Edition, Garrad Press, III., Chapter 8.

100.   Tatini, S. R., J. J. Jezeski, H. A. Morris, J. C. Olson, Jr., and E. P. Casman. 1971. Production of staphylococcal enterotoxin A in Cheddar and Colby cheeses. J. Dairy Sci. *54:* 815–825.

101.   Tatini, S. R. 1973. Influence of food environments on growth of *Staphylococcus aureus* and production of various enterotoxins. J. Milk Food Technol. *36:* 559–563.

102.   Tatini, S. R., S. A. Stein, and H. M. Soo. 1976. Influence of protein supplements on growth of *Staphylococcus aureus* and production of enterotoxins. J. Food Sci. *41:* 133–135.

103.   Thatcher, F. S., J. Robinson, and I. Erdman. 1962. The 'vacuum pack' method of packaging foods in relation to the formation of the botulinum and staphylococcal toxins. J. Appl. Bacteriol. *25:* 120–124.

104.   Thomas, C. T., J. C. White, and K. Longree. 1966. Thermal resistance of salmonellae and staphylococci in foods. Appl. Microbiol. *14:* 815–820.

105.   Thota, F. H., S. R. Tatini, and R. W. Bennett. 1973. Effects of temperature, pH and NaCl on production of staphylococcal enterotoxins E and F. Bacteriol. Proc., p. 1.

106.   Troller, J. A. and W. C. Frazier. 1963. Repression of *Staphylococcus aureus* by food bacteria. II. Causes of inhibition. Appl. Microbiol. *11:* 163–165.

107. **Troller, J. A.** 1972. Effect of water activity on enterotoxin A production and growth of *Staphylococcus aureus*. Appl. Microbiol. *24:* 440–443.

108. **Vandenbosch, L. L., D. Y. C. Fung, and M. Widomski.** 1973. Optimum temperature for enterotoxin production by *Staphylococcus* aureus S-6 and 137 in liquid medium. Appl. Microbiol. *25:* 498–500.

109. **Walker, G. C., L. G. Harmon, and C. M. Stine.** 1961. Staphylococci in colby cheese. J. Dairy Sci. *44:* 1272–1282.

110. **Woodburn, M., M. Bennion, and G. E. Vail.** 1962. Destruction of salmonellae and staphylococci in precooked poultry products by heat treatment before freezing. Food Technol. *16:* 98–100.

111. **Wu, C.-H. and M. S. Bergdoll.** 1971. Stimulation of enterotoxin B production. Infect. Immunity *3:* 784–792.

17

Food Poisoning Caused By Gram Positive Sporeforming Bacteria

At least three gram positive sporeforming rods are known to cause bacterial food poisoning: *Clostridium perfringens (welchii)*, *C. botulinum*, and *Bacillus cereus*. The incidence of food poisoning caused by each of these organisms is related to certain specific foods as is food poisoning in general.

CLOSTRIDIUM PERFRINGENS FOOD POISONING

The causative organism of this syndrome is a gram positive, anaerobic spore-forming rod widely distributed in nature. Based upon their ability to produce various exotoxins, 6 types are recognized: types A, B, C, D, E, and F. The food poisoning strains of *C. perfringens* primarily belong to type A as do the classical gas gangrene strains, but unlike the latter, the food poisoning strains are heat-resistant and produce only traces of alpha toxin *(58)*. Some type C strains cause a food poisoning syndrome. The food poisoning strains of *C. perfringens* differ from type F strains in not producing beta toxin. The latter strains have been isolated from enteritis necroticans *(135)* and are compared to type A heat-sensitive and heat-resistant strains in Table 17-1.

While *C. perfringens* strains have been associated with gastroenteritis since 1895, the first clear-cut demonstration of their etiological status in food poisoning was made by McClung *(77)* who investigated 4 outbreaks in which chicken was incriminated. The first detailed report of the characteristics of this food poisoning syndrome was that of Hobbs *et al.* *(57)* in Great Britain. While the British workers were more aware of this organism as a cause of food poisoning during the 1940s and 1950s, few incidents of this type of food poisoning were recorded in the United States prior to the publication by Angelotti *et al.* *(6)* of methods of recovery and quantitation of this organism. Recent literature reports indicate that *C. perfringens* food poisoning is widespread in the United States as well as in the British Isles.

TABLE 17-1.   Toxins of Cl. welchii types A and C (58)*.

Cl. welchii	Toxins										
	α	β	γ	δ	ϵ	θ	ι	κ	λ	μ	ν
Heat-sensitive Type A	+++	−	−	−	−	++	−	++	−	+ or −	+
Heat-resistant Type A	± or tr	−	−	−	−	−	−	+ or −	−	+++ or −	−
Heat-resistant Type C	+	+	+	−	−	−	−	−	−	−	+

*Hobbs, B. C., 1962. *Bacterial Food Poisoning,* Royal Society of Health, London.

---

1. Distribution of C. perfringens.   Food poisoning strains of *C. perfringens* exist in soils, water, foods, dust, spices, and the intestinal tract of man and other animals. Various authors have reported the incidence of the heat-resistant, nonhemolytic strains to range from 2–6% in the general population. Between 20–30 percent of healthy hospital personnel and their families have been reported to carry these organisms in their feces while the carrier rate of victims after 2 weeks may be 50% or as high as 88% *(31)*. The heat-sensitive types are common to the gastrointestinal canal of all humans. These organisms may find their way into meats either directly from the slaughter animals or by subsequent contamination of slaughtered meat from containers, handlers, or dust. Since they are sporeformers, they can withstand the adverse environmental conditions of drying, heating, and certain toxic compounds.

2. Characteristics of the Organisms and their Enterotoxins.   Food poisoning as well as most other strains of *C. perfringens* grow well on a variety of media if incubated under anaerobic conditions or if provided with sufficient reducing capacity. Barnes and Ingram *(7)* found that strains of *C. perfringens* isolated from horse muscle grew without increased lag phase at an Eh of −45 or lower, while more positive Eh values had the effect of increasing the lag phase. Although it is not difficult to obtain growth of these organisms on various media, sporulation occurs with difficulty and requires the use of special media such as described by Ellner *(33),* or by use of special techniques such as dialysis sacs *(104).*

With respect to growth temperature, these organisms are mesophilic with an optimum between 37–45 C. The lowest reported temperature for growth is 20 C in beef *(21)*. These authors were unable to obtain growth at 15 C as was Smith *(108)*. The latter author found the temperature growth range of 5 strains to be 20–50 C with no growth at 55 C. With respect to the effect of pH on growth, Smith found that 4 of 5 strains of *C. perfringens* grew from 5.5 to 8.0, with no growth occurring at either pH 5 or 8.5. Optimum growth in thioglycollate medium for 6 strains was found to occur between 30–40 C, while the optimum for sporulation on Ellner's medium was 37–40 C *(99)*. The growth of these organisms at high temperatures has no apparent effect on spore heat resistance. Growth at 45 C under otherwise optimal conditions leads to generation times as short as 10 min. In regard to a_w, the lowest reported values for growth and germination of spores lie between 0.97 and 0.95 with sucrose or NaCl, or about 0.93 with glycerol employing a fluid thioglycollate base *(62)*. Spore production appears to require higher a_w values than the above minima. While growth of type A was demonstrated at pH 5.5 by Labbe and Duncan *(68)*, no sporulation or toxin production occurred. A pH of 8.5 appears to be the highest for growth. At least 13 amino acids are required for the growth of this organism along with biotin, pantothenate, pyridoxal, adenine, and other related compounds. *C. perfringens* is heterofermentative, and a large number of carbohydrates are attacked.

The spores of food-poisoning strains differ widely in their resistance to heat with some being typical of other mesophilic sporeformers and some being highly resistant. A $D_{100\ C}$ value of 0.31 for *C. perfringens* (ATCC 3624) and a value of 17.6 for strain NCTC 8238 have been reported *(129)*. For 8 strains that produced reactions in rabbits, $D_{100\ C}$ values ranged from 0.70 to 38.37; strains that did not produce rabbit reactions were more heat sensitive *(118)*.

While the wide variations in heat resistance recorded for *C. perfringens* spores may be due to many factors, similar variations have not been recorded for *C. botulinum,* especially types A and B. The latter organisms have no history in the gastrointestinal tract of man, while *C. perfringens* strains are common inhabitants of this environment as well as soils. An organism inhabiting environments as diverse as these may be expected to show wide variations among its strains. Another factor that is important in heat resistance of bacterial spores is that of the chemical environment. Alderton and Snell *(3)* have pointed out that spore heat resistance is largely an inducible property, chemically reversible between a sensitive and resistant state. Using this hypothesis, it has been shown that spores can be made more heat resistant by treating them in Ca-acetate solutions such as 0.1 or 0.5 *M* at pH 8.5 for 140 hr at 50 C. The heat resistance of endospores may be increased 5- to 10-fold by this method *(4)*. On the other hand, heat resistance may be decreased by holding spores in 0.1 *N* HCl at 25 C for 16 hr, or as a result of the exposure of endospores to the natural acid conditions of some foods. It is not inconceivable that the high variability of heat resistance of *C. perfringens* spores may be a more or less direct result of immediate environmental history.

The freezing survival of *C. perfringens* in chicken gravy was studied by Strong and Canada *(116)* who found that only around 4% of cells survived when frozen to −17.7 C for 180 days. Dried spores, on the other hand, displayed a survival rate of about 40% after 90 days but only about 11% after 180 days.

The food-poisoning strains are somewhat resistant to polymyxin, cy-closerine, and sulfadiazine, and these inhibitors are employed in various selective media devised for the recovery of the organisms from foods. These organisms have been reported to be inhibited by 5% NaCl.

The causative factor of *C. perfringens* food poisoning is an enterotoxin. It is unusual in that it is a spore-specific protein; its production occurs together with that of sporulation. While most food poisoning cases by this organism in North America are caused by type A strains, type C strains have been found in some European outbreaks. The syndrome produced by type C strains is more severe than that for A. The former has associated with it a mortality rate of 35–40% while the latter is rarely fatal. In spite of these differences, the enterotoxins from A and C strains have been found to be similar serologically and biologically *(106)*.

The enterotoxin of *C. perfringens* was demonstrated by Duncan and Strong *(27)* and also by Hauschild *et al.* *(55)* and Strong *et al.* *(118)*. The purified toxin has a molecular weight of 36,000 and an isoelectric point of 4.3 *(54)*. It is heat sensitive (biological activity destroyed at 60 C for 10 min), pronase sensitive but resistant to trypsin, chymotrypsin, and papain *(31, 113)*. L-forms of *C. perfringens* produce the toxin, and in one study they were shown to produce as much as classical forms *(75)*.

By use of scanning, isoelectric focusing and isotachophoresis in polyacrylamide gels, the enterotoxin was separated into 2 species. The major one had an isoelectric point of 4.5 and possessed antigenic and functional activity, while the minor component possessed no biological activity, displayed an i.p. of 4.6, and represented about 15% of the purified toxin *(134)*.

The toxin is synthesized by sporulating cells in association with late stages of sporulation. The peak for toxin is just before lysis of the cell's sporangium, and the toxin is released along with spores. The toxin has been shown to be a spore structural protein covalently associated with the spore coat. The cells sporulate freely in the gastrointestinal tract. In culture media, the enterotoxin is produced only where endospore formation is permitted. A single gene has been shown to be responsible for the enterotoxin negative trait *(29, 31)*, and enterotoxin and a spore coat protein have been shown to be controlled by a stable mRNA *(69)*.

The toxin may appear in a growth and sporulation medium in about 3 hr after inoculation with vegetative cells *(30)*, and from 1 to 100 μg/ml of toxin production have been shown for 3 strains of *C. perfringens* in Duncan-Strong (DS) medium after 24–36 hr *(40)*. Purified toxin has been shown to contain up to 3,500 mouse MLD/mg N. Its assay may be achieved by a serologic or biologic method. The most sensitive method of all is reverse passive hemagglutination, (RPH) which was shown by Genigeorgis *et al.* *(40)* to be capable of detecting as little as 0.00005 μg of toxin. By the microslide diffusion, single gel diffusion, and electroimmunodiffusion methods, 0.013, 0.30, and 0.01 μg of toxin, respectively, could be detected

(28, 40). The biological tests for this toxin consist of the guinea pig skin test, the rabbit ileal loop test, and a mouse test. These assay methods detected, respectively, 0.06–0.125, 6.25, and 1.8 μg of enterotoxin *(40)*.

Based upon animal studies, the enterotoxin acts upon the small intestine and causes increased capillary permeability, vasodilation, and intestinal motility *(91)*. It induces fluid accumulation in ligated ileal loops of rabbits, lambs, and cattle as well as diarrhea. Diarrhea and vomiting are produced in man and monkeys, and erythema is produced when injected in the skin of guinea pigs and rabbits *(31)*. The toxin is lethal also to mice.

3. Symptoms and Food Vehicles of C. Perfringens Food Poisoning.

Upon the ingestion of contaminated foods, symptoms appear between 6 and 24 hr, especially between 8 and 12 hr. The symptoms are characterized by acute abdominal pain and diarrhea with nausea, fever, with vomiting being rare. Except in elderly, debilitated persons, the illness is of short duration, one day or less. The fatality rate is quite low and no immunity seems to occur.

The true incidence of *C. perfringens* food poisoning is unknown. Because of the relative mildness of this syndrome, it is quite likely that only those outbreaks and cases that affect groups of people are ever reported and recorded. The confirmed outbreaks reported to the Center for Disease Control (CDC) for the years 1973–1975 are listed in Table 17-2. It may be noted from this table that one death occurred for each of the 3 yr.

The foods involved in *C. perfringens* outbreaks are often meat dishes prepared one day and eaten the next. The heat preparation of such foods is presumably inadequate to destroy the heat-resistant spores and upon cooling and rewarming, the spores germinate and grow. Meat dishes are more often the cause of this type of food poisoning than others, although nonmeat dishes may be contaminated by meat gravy. The greater involvement of meat dishes may in part be due to the slower cooling rate of such cooked foods and also to the higher incidence of food-poisoning strains in meats. Strong *et al. (117)* found the overall incidence of *C. perfringens* to be about 6% in 510 American foods. The incidence for various foods was 2.7% for

TABLE 17-2. Confirmed food poisoning outbreaks, cases, and deaths reported to CDC for 1973–1975 *(17)*.

Food Poisoning	Outbreaks—Cases—Deaths								
	1973			1974			1975		
B. cereus	1	2	—	1	11	—	3	45	—
C. botulinum	10	31	4	21	32	7	14	19	2
C. perfringens	9	1424	1	15	863	1	16	419	1
Salmonellae	33	2462	7	35	5499	1	38	1573	2
Staphylococcal	20	1272	—	43	1565	—	45	4067	—
V. parahaemolyticus	1	2	—	0	0	0	2	222	—
Suspect Gp. D. Streptococci	0	0	0	2	38	—	1	50	—

commercially prepared frozen foods, 3.8% for fruits and vegetables, 5% for spices, 1.8% for home-prepared foods, and 16.4% for raw meat, poultry, and fish. Hobbs *et al.* *(57)* found that 14–24% of veal, pork, and beef samples examined contained heat-resistant strains, but all 17 samples of lamb were negative. Recently, an outbreak of food poisoning involving 375 persons where 140 became ill was shown to be caused by both *C. perfringens* and *S. typhimurium (97)*. The organism has been shown to grow well in a large number of foods. A recent study of retail, frozen precooked foods revealed that one half was positive for *C. perfringens* vegetative cells while 15% contained spores *(125)*. These investigators inoculated meat products with this organism and stored them at −29 C for up to 42 days. While spore survival was high, vegetative cells were virtually eliminated during the holding period. The survival of inoculated cells in raw ground beef was studied by Goepfert and Kim *(48)* who found that the organisms actually decreased in numbers upon storage at temperatures between 1 and 12.5 C. The raw beef contained a natural flora, and the above finding suggests that *C. perfringens* is unable to compete under these conditions.

4. *Recovery of* C. perfringens *from Foods.* Foods to be examined for the presence and numbers of food-poisoning *C. perfringens* organisms should be homogenized and plated with a suitable selective medium. The heating of food homogenates to 80–100 C is done at the risk of destruction of cells, although Hobbs *et al.* *(57)* have pointed out that heating of fecal samples may be achieved with success. The first specific medium devised for the recovery of these organisms from foods was that of Mossel *et al.* *(86)* and Mossel *(87)*. These media were modified by Angelotti *et al.* *(6)* and later by Marshall *et al.* *(76)* and Hall *et al.* *(50)*. The medium of Angelotti *et al.* is designated sulfite-polymyxin-sulfadiazine (SPS) agar while that of the latter authors is designated tryptone-sulfite-neomycin (TSN) agar. Both media employ inhibitors of contaminating organisms as well as sulfite, which is reduced by food-poisoning strains of *C. perfringens*. The media are both selective and differential by virtue of the inhibitors and sulfite. Although incubation of SPS agar is best carried out for 24 hr at 37 C, the authors of TSN agar recommend incubating at 46 C for 18 hr. Plates of both must be incubated under anaerobic conditions. Suspect food-poisoning type colonies are characterized by the production of black colonies due to sulfite reduction. Representative suspect colonies should be subcultured and further tested in order to confirm their identification. More recently, tryptose-sulfite-cycloserine (TSC) and Shahidi-Ferguson-perfringens (SFP) media have been developed and found to give good results. Egg-yolk free TSC worked well for some investigators *(56)*, and TSC + egg yolk has been found to be excellent in recovering small numbers of cells from frozen ground beef *(36)*. Harmon *et al.* *(51)* compared 4 media and found SFP to be the best. A chemically defined medium which supports the growth of as few as 10 cells/ml has been developed by Riha and Solberg *(101)*. Other synthetic media for the growth of this organism have been devised but all require large inoculla for growth. (See Table 4-1 for recommended isolation methods and media and ref. *126*).

Environmental stresses are known to injure *C. perfringens,* and recovery procedures are necessary. Heated spores have been shown to recover and germinate upon the addition of lysozyme *(2).* The addition of 1500 units of catalase to the surface of 3-day-old agar plates before incubation was shown to result in 20–90% higher counts of *C. perfringens* than untreated plates *(52).* The latter finding suggests that peroxides are formed in the stored medium and act to inhibit injured cells if not destroyed. Other aspects of the recovery of metabolically injured *C. perfringens* cells are discussed in Chapter 4.

Large numbers of cells/g of food are associated with this syndrome with minimum numbers being around 5×10^8 *(90).* There are similarities between *C. perfringens* and *B. cereus* food poisoning, and these have been reviewed by Gilbert and Taylor *(43).*

5. Prevention of C. perfringens Food Poisoning.   This hazard may be prevented by proper attention to the leading causes of food poisoning of all types given in the previous chapter. Since this syndrome often occurs in institutional cafeterias, some special precautions should be taken. Upon investigating a *C. perfringens* food poisoning outbreak in a school lunchroom in which 80% of students and teachers became ill, Bryan *et al. (14)* constructed a time-temperature chart in an effort to determine when, where, and how the turkey became the vehicle (Fig. 17-1). It was concluded that meat and gravy, but not dressing, were responsible for the illness. As a means of preventing recurrences of such episodes, these authors suggested 9 points for the preparation of turkey and dressing, and they are summarized below.

1. Cook turkeys until internal breast temperature reaches at least 165 F, preferably higher.
2. Thoroughly wash and sanitize all containers and equipment which previously had contact with raw turkeys.
3. Wash hands and use disposable plastic gloves when deboning, deicing, or otherwise handling cooked turkey.
4. Separate turkey meat and stock before chilling.
5. Chill the turkey and stock as rapidly as possible after cooking.
6. Use shallow pans for storing stock and deboned turkey in refrigerators.
7. Bring stock to a rolling boil before making gravy or dressing.
8. Bake dressing until all portions reach 165 F or higher.
9. Just prior to serving, heat turkey pieces submerged in gravy until largest portions of meat reach 165 F.

The importance of cooking temperature and post-cooking storage on *C. perfringens* in turkey stuffing was stressed earlier *(132).*

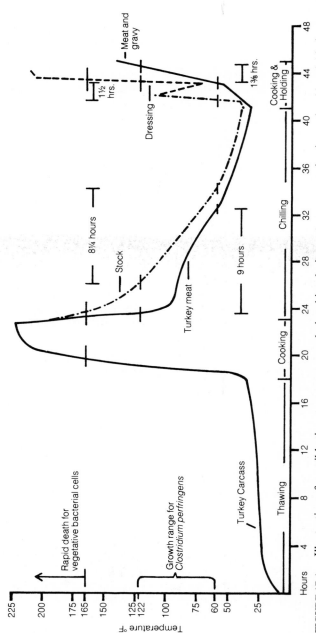

FIGURE 17-1. Illustration of possible time-temperature relationships during turkey preparation in a school lunch kitchen (*14*). Copyright © 1971, Intern. Assoc. Milk, Food & Environ. Sanitarians, Inc.

BOTULISM

Unlike *C. perfringens* food poisoning in which large numbers of viable cells must be ingested, the symptoms of botulism are caused by the ingestion of a highly toxic, soluble exotoxin produced by the organism while growing in foods.

The first recorded case of botulism was in 1793 *(39)*, and the etiologic agent of this disease was first isolated in 1895 by E. Van Ermengen. The outbreak studied by Van Ermengen occurred in Belgium with 34 cases and 3 deaths. The causative organism was named *Bacillus botulinus* from the Latin *botulus* meaning sausage. Botulism is caused by certain strains of *C. botulinum* which is a gram positive, anaerobic sporeforming rod with oval to cylindrical, terminal to subterminal spores. On the basis of the serological specificity of their toxins, 7 types are recognized: A, B, C, D, E, F, and G. Types A, B, E, and F cause disease in man; type C causes botulism of fowls, cattle, mink, and other animals; and type D is associated with forage poisoning of cattle, especially in South Africa. Type G was reported from Argentina in 1970 from a soil culture, and no outbreaks of botulism have been recognized (see *109B*). The types are also differentiated on the basis of their proteolytic capacity. Type A is proteolytic as are some type B and F strains. Type E is nonproteolytic as are some B and F strains.

1. Distribution of C. botulinum.   All of these organisms are basically indigenous to soils from which they find their way into waters. In the United States, type A occurs more frequently in soils in the western states, and type B is found more frequently in the eastern states and in Europe. Soils and manure from various countries have been reported to contain 18% type A and 7% type B spores. Cultivated soil samples examined showed 7% to contain type A and 6% type B spores. Type E spores tend to be confined more to waters, especially marine waters. In a study of mud samples from the harbor of Copenhagen, Pederson *(96)* found 84% to contain type E spores while 26% of soil samples taken from a city park contained these organisms. Type E spores have been known for some time to exist in waters off the shores of Northern Japan. Prior to 1960, the existence of these organisms in Great Lakes and Gulf Coast waters was not known, but their presence in these waters as well as in the Gulf of Maine and the Gulfs of Venezuela and Darien has been established *(15)*. Ten percent of soil samples tested in Russia were found positive for *C. botulinum,* with type E strains being the most predominant. A recent study of sediment samples from the Upper Chesapeake Bay area of the U.S. Atlantic Coast revealed the presence of types A and E spores in 4 of 33 samples *(102)*. The investigators believed these organisms to be randomly distributed in sediment and to be autochthonous.

Nine % of fish caught from Lake Michigan and 57% of those from Green Bay have been found to contain type E spores *(11)*. In a study by Fantasia and Duran *(37)*, 6.2% of 500 commercially dressed fish taken from Lake

Michigan near the Two Rivers, Wisconsin, area yielded type E while only 0.4% of 427 laboratory dressed fish were positive. These authors found the type E organisms to exist in relatively low numbers on freshly caught fish but to increase after evisceration. In a study of the incidence of *C. botulinum* on whitefish chubs in smoking plants, Pace *et al.* *(95)* found the highest incidence (21%) of these organisms at the brine tank processing stage. The build-up of microorganisms on foods through successive processing stages is discussed more fully in Chapter 5.

As to the overall incidence of *C. botulinum* in soils, Riemann *(100)* suggested that the number/g is probably less than one. With respect to their presence in Great Lakes waters, some authors have suggested that this is an event of recent times, but others maintain that these organisms have always been in these waters. The presence of type E strains may have been missed because of the common practice of heating spore-containing preparations in order to destroy vegetative cells, a practice shown to reduce or even completely destroy type E spores. From their soil or water habitats, these organisms find their way into fish and seafoods where they present some problems in terms of their destruction or inhibition.

The first type F strains were isolated by Moller and Scheibel *(83)* from a homemade liver paste incriminated in an outbreak of botulism on the Danish island Langeland with one death. Since that time, Craig and Pilcher *(20)* isolated type F spores from salmon caught in the Columbia River; Eklund and Poysky *(32)* found type F spores in marine sediments taken off the coasts of Oregon and California; Williams-Walls *(131)* isolated two proteolytic strains of type F from crabs collected from the York River in Virginia; and Midura *et al.* *(82)* isolated this organism from venison jerky in California. The isolate was nonproteolytic.

2. Growth Characteristics of **C. botulinum.** *Clostridium botulinum* is mesophilic with an optimum growth temperature of about 37 C for types A and B strains and 30 C for type E strains. The lowest temperature at which types A and B have been consistently reported to grow is 12.5 C, although one report indicates growth as low as 10 C. Ohye and Scott *(94)* found that actively growing cells could initiate growth as low as 12.5 C but not at 10 C, while the lowest temperature at which spore inocula could germinate and grow was 15 C. On the other hand, type E spores have been reported to grow and produce toxin at 3.3 C *(49, 103)*, while type F has been reported to grow and produce toxin at 4 C. In general, type E and nonproteolytic B and F strains have minimum growth temperatures about 10 degrees lower than A and B strains. With respect to maximum tempeatures of growth, types A and B were unable to grow at 50 C when actively growing cells were employed as inocula, or at 45 C using spore inocula *(94)*. Type E strains generally have a maximum growth temperature about 5 degrees lower than types A and B while the optimum is around 30 C. It has been reported that type E strains produce more toxin at 30 than at 37 C. The toxin has been found to be somewhat unstable at 37 C as measured by a decrease in toxicity after a few days at this temperature *(96)*.

With respect to pH, it has been established for some time that growth of *C. botulinum* does not occur at or below 4.5, and it is this fact that determines the degree of heat treatment given to foods with pH values below this level. Numerous authors have shown the relationship between nature of substrate and pH minima for growth of *C. botulinum,* with growth reported on some foods as low as pH 4.8 while on other foods the minimum was about 5.7. At temperatures of 15 C and higher, the outgrowth of Type E botulinal spores was found by Emodi and Lechowich *(35)* to be inhibited at pH 5.0 in the presence of 5% NaCl. The presence of reducing agents such as l-cysteine and sodium thioglycollate have also been shown to affect minimum growth pH of type E organisms *(105)*. This organism has been reported to grow fairly well and produce toxin over the pH range 5.5–8.0. Toxin formation in general seems to be favored by the optimum pH for germination and growth.

The nutritional growth needs for *C. botulinum* are complex, requiring many amino acids, B-vitamins, and minerals. A synthetic medium that supports growth and toxin production of type E has been reported *(127)* while a synthetic medium for type E that supports not only growth and toxin production but up to 60% sporulation of 8 type E strains studied has been reported *(110)*.

C. botulinum types A and B require an a_w of 0.94 and type E an a_w of 0.97 and above for growth as well as anaerobic conditions. Salt at a level of about 10%, or 50 percent sucrose, will inhibit growth of types A and B, and 3–5% has been reported to inhibit toxin production in smoked fish chubs *(18)*. Lower levels of salt are effective in inhibiting these organisms when heating is applied in the presence of $NaNO_2$, and the interdependence of this salt, NaCl, heating, and other factors on *C. botulinum* is discussed more fully in Chapter 9. The low F_o (0.1–0.7) applied to canned shelf-stable cured meats is made possible by the synergistic actions of both chemical and physical agents in preventing the outgrowth of botulinal spores.

Hydroxylamine (which can be produced by reduction of nitrite) has also been reported to be quite inhibitory to *C. botulinum*. Tarr *(121)* reported that levels between 25 and 500 ppm may inhibit type A for more than a month at optimum temperatures over a pH range of 5.9–7.6. Warnecke *et al. (128)* found that type E strains did not produce toxin on bacon held at 3 C but did so at 13 and 23 C. These same authors found that no toxin was produced in bologna and attributed this to the 2.5% salt concentration in this product. Types A and E spores were shown by Thatcher *et al. (122)* to be able to germinate and produce toxin on smoked fish. Fish inoculated with type A spores were offensively spoiled while type E strains caused a less drastic type of spoilage. In a study of type E toxin formation and cell growth in turkey rolls incubated at 30 C, Midura *et al. (80)* found that type E spores germinated and produced toxin within 24 hr. The appearance of toxin coincided with cell growth for 2 weeks after which time toxins outsurvived viable cells. These findings suggest that it is possible to find type E toxin in foods in the absence of type E cells. These investigators were unable to demonstrate toxin after 56 days of incubation.

The radiation resistance of *C. botulinum* spores is presented in Chapter 10. Radiation resistance of types A and E toxins has been reported to be about the same with a D value of 2.1 Mrads *(107)*.

With respect to heat resistance of botulinal spores, estimated $F_{80\ C}$ values for the destruction of approximately 10^6 type E spores/g in fish paste have been reported by Crisley *et al. (22)*. The values for strains Alaska, Beluga, 8E, Iwanai, and Tenno, respectively, were: 34.2, 17.0, 14.0, 12.5, and 13.2 min. Calculated $D_{80\ C}$ values were: 4.3, 2.1, 1.8, 1.6, and 1.6 min, respectively, for strains Alaska, Beluga, 8E, Iwanai, and Tenno. The capacity of type E spores to survive smoking was investigated by Christiansen *et al. (18)*. These investigators inoculated 10^6 type E spores into fish chub, followed by smoking to an internal temperature of 180 F for 30 min. All strains employed survived this treatment, which was followed by incubation at room temperature for 7 days in sealed plastic bags. Fantasia and Duran *(37)* also found that smoking whitefish chubs to an internal temperature of 180 C for 30 min produced a botulism-free product, but Pace *et al. (95)* found that 10 or 1.2% of 858 freshly smoked chubs given the same heat treatment were contaminated, mostly with Type E strains. The heat resistance of botulinal spores is discussed further in Chapter 12.

In regard to the ecology of *C. botulinum* growth, it seems unlikely that this organism can grow and produce its toxins in competition with large numbers of other microorganisms. Toxin-containing foods are generally devoid of other types of organisms because of heat treatments or vacuum packaging. In the presence of yeasts, however, *C. botulinum* has been reported to grow and produce toxin at a pH as low as 4.0 *(79)*. While a synergistic effect between clostridia and lactic acid bacteria has been reported on the one hand *(8)*, it has also been reported that lactobacilli will antagonize growth and toxin production by *C. botulinum*. The latter notion tends to be supported somewhat by the absence of botulinal toxins in milk *(60)*. Yeasts are presumed to produce growth factors needed by the clostridia to grow at low pH, while the lactic acid bacteria presumably aid growth by reducing the O/R potential. In a recent study, C. botulinum type A was inhibited by soil isolates of *C. sporogenes, C. perfringens,* and *B. cereus (109)*. Some *C. perfringens* strains produced an inhibitor that was effective on 11 type A strains; on 7 B proteolytic and 1 nonproteolytic strains; and on 5 E and 7 F strains. Kautter *et al. (64)* found that type E strains are inhibited by other nontoxic organisms whose biochemical properties and morphological characteristics are similar to type E. These organisms were shown to effect inhibition of type E strains by producing a bacteriocin-like substance designated "boticin E". In a more detailed study, proteolytic A, B, and F strains were found to be resistant to boticin E elaborated by a nontoxic type

E, but toxic E cells were susceptible (5). The boticin E was found to be spo-
rostatic for nonproteolytic types B, E, F, and nontoxigenic E. Vacuum-
packed foods such as bacon have been shown to be capable of supporting
growth and toxin production by C. botulinum strains without causing
noticeable off-odors (122). A report on the ecology of C. botulinum type F
made by Wentz et al. (130) showed that the absence of type F in mud samples during certain times of the year was associated with the presence of Bacillus licheniformis in the samples during these periods, thus bringing about
the inhibition of type F strains.

3. The Botulism Syndrome, Its Incidence, and Food Vehicles.   Upon the
ingestion of toxin-containing foods, symptoms of botulism may develop
anywhere between 12 and 72 hr later. The symptoms consist of nausea,
vomiting, fatigue, dizziness, headache, dryness of skin, mouth and throat,
constipation, lack of fever, paralysis of muscles, double vision, and finally
respiratory failure, and death. The duration of the illness is from 1 to 10
days, depending upon host resistance and other factors. The mortality rate
varies between 30–65%, with the rate being generally lower in European
countries than in the United States. All symptoms are caused by the
exotoxin, and treatment consists of administering specific antisera as early
as possible. Although it is assumed that the tasting of toxin-containing foods
allows for absorption from the oral cavity, Lamanna et al. (71) have found
that mice and monkeys are more susceptible to these toxins when
administered by stomach tube than by exposure to the mouth. The botulinal
toxins are neurotoxins and attach irreversibly to nerves. Early treatment by
use of antisera brightens the prognosis.

Prior to 1963, most cases of botulism in the United States in which the vehicle foods were identified were traced to home-canned vegetables and were
caused by types A and B strains. In almost 70% of the 640 cases reported for
the period 1899–1967 the vehicle food was not identified. Of the known vehicle foods, 17.8% were vegetables, 4.1 fruits, 3.6 fish, 2.2 condiments, 1.4
meats and poultry, and 1.1% for all others. Reported cases of botulism in the
U.S. for the period 1950–1974 along with deaths are presented in Figure
17-2. Outbreaks, cases, and deaths from this syndrome for the years 1973–
1975 may be seen from Table 17-2. In 1976, 29 cases were reported, while in
the first four months of 1977, 64 cases were recorded. The latter cases include an outbreak in Pontiac, Michigan in which 45 persons fell victim after
consuming a hot sauce prepared with home-canned jalapeño peppers. The
victims ate at a commercial restaurant where the home-canned peppers were
used against state law. No deaths occurred, and the toxin was identified as
type B (84). This is the largest outbreak of botulism ever recorded in the
U.S., and the number of cases exceeded the yearly totals for any previous
year. The total number of cases per year rarely exceeds 50, with the highest
10-year period being 1930–1939 when 384 cases were reported for noncommercial foods (see Table 17-2). Between 1899 and 1963, 1,561 cases were

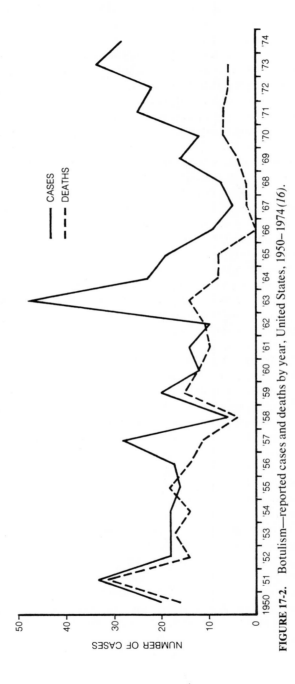

FIGURE 17-2. Botulism—reported cases and deaths by year, United States, 1950–1974 (16).

362

reported from noncommercial foods, while 219 cases were reported from commercial foods between 1906 and 1963, with 24 cases in 1963 alone. The incidence in Germany was constant for several yr with 9–15 outbreaks and 30–40 victims, but seems now to be declining *(100)*. Botulism was seldom reported in France before World War II, but more than 1,000 cases were reported during the war with a leveling off since.

Of 404 verified cases of type E through 1963, 304 or 75% occurred in Japan *(38)*. No outbreak of botulism was recorded in Japan prior to 1951. For the period May, 1951, through January, 1960, 166 cases were recorded with 58 deaths for a mortality rate of 35% *(65)*. Most of the Japanese outbreaks have been traced to a home-prepared food called "izushi", which is a preserved food consisting of raw fish, vegetables, cooked rice, malted rice (koji), and a small amount of salt and vinegar. This preparation is packed tightly in a wooden tub equipped with lid and held for 3 weeks or longer to permit lactic acid fermentation. During this time the O/R potential is lowered, thus allowing for the growth of anaerobes.

In regard to botulism in commercially canned foods, 62 outbreaks were recorded for the period 1899–1973 *(74)*. Of the 62 outbreaks, 41 occurred prior to 1930. Canned mushrooms have been incriminated in several botulism outbreaks in the recent past. A study by the U.S. Food and Drug Administration in 1973 and 1974 turned up 30 cans of mushrooms which contained botulinal toxin (29 were type B). An additional 11 cans contained viable spores of *C. botulinum* without preformed toxin *(74)*. The capacity of the commercial mushroom *(Agaricus bisporus)* to support the growth of inoculated spores of *C. botulinum* was studied by Sugiyama and Yang *(120)*. Following inoculation of various parts of mushrooms, they were sealed with plastic film and incubated. Toxin was detected as early as 3–4 days later when products were incubated at 20 C. The type A strains appeared to be more active than type B, according to these investigators, even though type B strains appear more often in canned products. While the plastic film used to wrap the inoculated mushrooms allowed for gas exchange, the respiration of the fresh mushrooms apparently consumed oxygen at a faster rate than it entered the film. No toxin was detected in products stored in the refrigerator range.

The only recorded outbreak of botulism due to type F is the one cited above involving homemade liver paste. This outbreak involved 5 cases with one death. The only U.S. outbreak occurred in 1966 from home-prepared venison jerky with 3 clinical cases *(82)*.

It is quite clear that the greatest hazards of botulism come from home-prepared and home-canned foods which are improperly handled or given insufficient heat treatment to destroy botulinal spores. Such foods are often consumed without heating. The best preventative measure is the heating of suspect foods to boiling temperatures for a few minutes, which is sufficient to destroy the toxins.

4. Isolation of C. Botulinum from Foods. In examining foods for the presence of type E spores, heating to 80 C is sufficient to destroy all spores as well as vegetative cells. One author has reported killing type E spores at 90 C for 5 min. Kempe *(66)* found that heating type E spores between 70–90 C for 15 min decreased the number to 0.01% of the original population. It appears that different strains of type E spores have different degrees of heat sensitivity, and the nature of the suspending menstrum is of great importance in the heat resistance of these strains.

Increased recoveries of type E spores from foods and soil have been reported to follow a drying process in which vegetative cells die off without destruction of type E spores. Johnston *et al. (61)* reported success by use of 50% ethanol to destroy vegetative cells. For the recovery of types A and B spores, the recommended procedure is heating to about 80 C for 10 min, followed by plating with an appropriate medium, and anaerobic incubation.

The presence of botulinal toxins in foods may be determined by injecting food homogenates into mice via the intraperitoneal route. A sufficient number of mice should be protected with polyvalent antisera prior to being challenged with the food homogenate. In the case of type E and non-proteolytic B and F toxins, treatment with trypsin at a pH of about 6.0 prior to injection is generally necessary (see below). Laboratory mice are quite susceptible to botulinal toxins and generally die within 12 hr after injection, depending upon dosage. Botulinal toxins may be detected also by use of microcapillary agar-gel diffusion, hemagglutination-inhibition, and electroimmunodiffusion methods as discussed in Chapter 4 (see Table 4-3).

In the recovery of botulinal spores and cells, one should be aware of the possibility of metabolically injured organisms and take steps to allow for repair (see Chapter 4 and Table 4-2). Since L-forms of *C. botulinum* (at least types A and E) have been shown to produce toxin *(13)* and since these forms may not develop on ordinary culture media, it is possible to find toxin in the absence of cfu's.

5. Nature of the Botulinal Toxins. The toxins of *C. botulinum* are formed within the organisms and released upon autolysis. They are produced by cells growing under optimal conditions, and resting cells have been reported to form toxin as well *(67)*. Botulinal toxins are the most toxic substances known, with purified A reported to contain about 30,000,000 mouse LD_{50}/mg. When grown in cellophane sacs suspended in culture media, yields as high as 200,000,000 mouse LD_{50} have been reported *(114)*. The first of the botulinal toxins to be purified was type A, which was achieved by Lamanna *et al. (70)* and also Abrams *et al. (1)*. The purification of type E toxin was achieved by Gerwing *et al. (41)*, type B by both Gerwing *et al. (42)* and Boroff *et al. (10)*, and type F by Yang and Sugiyama *(133)*. The latter was achieved by employing a proteolytic strain, and some low-level reciprocal neutralizations between E and F strains were found. Proteolytic A, B, and F strains were shown earlier to possess a common somatic antigen with some cross agglutination with *C. sporogenes, C. tetani,* and *C. histolyticum (111).* Nonproteolytic B, E, and F were shown to share also a common somatic antigen.

The toxin molecule that is produced initially by a toxigenic culture is referred to as a **progenitor** toxin. According to Lamanna and Sakaguchi *(73),* progenitor toxin is the immediate toxic parent of a derivative toxin; also a toxic protein from which material of enhanced potency can be generated. Progenitor toxins consist of a toxic component along with an atoxic component. The potency of progenitor toxins is enhanced by removing the atoxic component. In the case of proteolytic type A, B, and F, this is done by enzymes produced by the toxigenic organisms to yield the derived toxic moiety. In the case of proteolytic type A, the molecular weight (MW) of progenitor toxin is 900,000 with an S (S_{20w}) of around 17.3 *(73),* or 19S *(119).* The derived or neurotoxin is 7.25S with a MW of 150,000. The atoxic portion of the progenitor molecule is a hemagglutinin with MW values varying in size. Progenitor B has a MW of 350,000, S values variously found to range from 12 to 16, and a neurotoxic component with MW >100,000. Progenitor F has MW of 235,000, 10.3S, and upon enhancement two components of equal size (5.9S) result *(93).* One component is toxic while the other is atoxic. In the case of type E, progenitor has a MW of 350,000 and is 11.6S. Its toxicity may be enhanced either by the removal of nontoxic components, or by intramolecular change *(73).* The latter is 11.6S while the former is 7.3S (MW 150,000).

With the above background information, the type E activation phenomenon may be explained in the following manner based upon information presently available. Since the producing organism is nonproteolytic, the progenitor toxin possesses either low or no toxic activity. Activation (enhancement) is brought about by the addition of trypsin which effects what DasGupta and Sugiyama *(24)* have called a "nicked" protein structure. The nicked toxin has MW 145,000 to 167,000 and is composed of H(eavy) and L(ight) chains held together by disulfide bonds. The H chain is approximately twice the MW of the L. On the other hand, the unactivated progenitor toxin is a single-chain unnicked protein which becomes a two-chain protein (nicked) by trypsin. Trypsin has no effect on type A (a proteolytic culture), but it does enhance nonproteolytic B, E, and F cultures.

The toxic component of type A toxin has been shown to be capable of fluorescence under ultraviolet light at 285mμ. The fluorescence is apparently due to the presence of tryptophan residues which confer both toxicity and immunologic specificity to the molecule *(9).* When the tryptophan residues are blocked, fluorescence, toxicity, and antigenicity all disappear. Type A toxin has been reported to be more lethal than B or E. Sterne and Van Heyningen *(115)* have reported that type B toxin has associated with it a much lower case mortality than type A, and that case recoveries from type B have occurred even when appreciable amounts of toxin could be demonstrated in the blood.

The mode of action of botulinal toxins was first described by Dickson and Shevky (25), who showed the similarity between the sites of action of these toxins and those of acetylcholine. Although the toxins act to prevent the release of acetylcholine at the myoneural junction, the release mechanism is not in itself damaged (12, 63). The passage of Ca^{++} into nerve terminals is not blocked (63). The toxins also affect smooth muscles of blood vessels in some as yet unknown way.

A 128,000 MW fraction of A toxin was shown by Lamanna et al. (72) to act more rapidly in vivo than the 900,000 MW toxin. The smaller molecule was about three times more potent than the larger, reflecting, possibly, a faster and more complete passage through cell membranes. The smaller fraction had an activity of around 10^7 LD_{50}/mg.

Symptoms of botulism can be produced by either parenteral or oral administration of botulinal toxins. They may be absorbed into the blood stream through the respiratory mucous membranes as well as through the walls of the stomach and intestines. These toxins are not completely inactivated by the proteolytic enzymes of the stomach and, indeed, the type E and nonproteolytic B and F toxins may be activated. After absorption from the intestinal tract, the toxins can be found in the blood from which they are absorbed by the peripheral nervous system.

Unlike the staphylococcal enterotoxins, botulinal toxins are heat-sensitive and may be destroyed at 80 C (176 F) for 10 min, or at boiling temperatures for a few min. More extensive information on clostridia and C. botulinum may be found elsewhere (78, 123).

BACILLUS CEREUS FOOD POISONING AND RELATED

Bacillus cereus is an aerobic, sporeforming rod normally present in soil, dust, and water. This organism has been associated with food poisoning in Europe since at least 1906, but its incidence in the United States is presumed to be quite low. Among the first to report the existence of this type of food poisoning with precision was Plazikowski (99). His findings were confirmed by several other European workers in the early 1950s, among them Nikodemusz et al. (92). The first documented outbreak in the U.S. occurred in 1969 (81) while the first in Great Britain occurred in 1971 (43).

This organism has a minimum growth temperature around 10–12 C and a maximum of about 48–50 C. Growth has been demonstrated over the pH range 4.9–9.3 (46). Its spores possess a resistance to heat typical of other mesophiles. It grows rapidly in foods held in the 30–40 C range. The food poisoning strains produce at least three well documented extracellular substances that show biological activity: lecithinase, hemolysin, and mouse lethal toxin. The capacity of 3 food products (strained beef and peas; and bananas) to support the production of these substances was investigated recently by Ivers and Potter (59). After 6 hr in beef, cell counts reached about 10^7/g with from 0.5 to 1.0 unit/g of all 3 factors. After 18 hr, 4 units of phospholipase was produced, and 5.4 units of hemolysin and 8 units of lethal toxin after 24 hr. Only lethal toxin was produced in all 3 products indicating, perhaps, that different foods possess varying abilities to induce or permit the production of each factor.

There are 2 syndromes caused by *B. cereus*. The more common food poisoning syndrome is rather mild with symptoms developing within 8–16 hr, more commonly within 12–13, and lasting for 6–12 (53). The symptoms consist of nausea (with vomiting being rare), cramp-like abdominal pains, tenesmus, and watery stools. Fever is generally absent. The similarity between this syndrome and that of *C. perfringens* food poisoning has been noted (43). Vehicle foods consist of cereal dishes that contain corn and corn-starch, mashed potatoes, vegetables, minced meat, liver sausage, Indo-nesian rice dishes, puddings, soups, and others (88). Meat dishes such as meat loaf have been incriminated. The more severe syndrome has been associated with fried or boiled rice, and the incubation range for reported cases is 1–6 hr, with 2–5 being most common (43, 85). This syndrome is, therefore, more acute than the more common one and the strains involved in its etiology may be mutants of the less virulent organisms. Its similarity to the staphylococcal food poisoning syndrome has been noted by Gilbert and Taylor (43).

The pathogenic agent in *B. cereus* food poisoning appears to be an enterotoxin, although the relationship between the three extracellular fac-tors noted above and the enterotoxin is not entirely clear (47, 73). The enterotoxin was prepared by Spira and Goepfert (112) and shown to elicit *in vivo* responses consistent with those of *B. cereus*. The isolated product had activity distinct from the hemolysin and the egg yolk turbididy factor, but its identity to the lethal toxin is unclear. The toxin was found to be most stable over the pH range of 5–10 and rather unstable outside of this range. It is synthesized during the exponential growth phase, and its production is fa-vored at pH 8.0 and 32 C incubation. The toxin was shown to induce fluid accumulation in ligated ileal loops of rabbits, dermal necrotic reactions in guinea pigs, and increased vascular permeability in rabbits (76). The guinea pig factor is unrelated to lecithinolytic or hemolytic activities of *B. cereus* (44).

With respect to numbers of cells required to induce this syndrome, results from foods involved in 11 cases revealed numbers of 0.5–200 × 10^6/g. The meat loaf involved in the U.S. outbreak in 1969 contained 7 × 10^7 cells/g (81). With respect to the more severe form associated with fried rice, in-criminated products involved in 17 cases showed a range of $<10^5->10^9$/g (43).

The existence of the more severe syndrome in the U.S. apparently has not been reported. The more common syndrome, while relatively frequent in parts of Europe, is rarely reported in the U.S. From Table 17-2, it can be seen that only 62 cases from 5 outbreaks were reported to CDC for the 3 years 1973–1975. For the years 1971–1973, 34 outbreaks involving 115 cases were recorded in Great Britain *(43)*.

While *B. cereus* is a sporeformer, the heating of foods prior to plating is not recommended due to the low degree of sporulation by this organism in foods. Although the capacity of this organism to cause a food poisoning syndrome is well established *(43, 46, 89, 90)*, Dack *et al.* *(23)* were unable to produce symptoms in human volunteers fed large numbers of cells. While the strains employed were not previously established as being capable of inducing the syndrome, this finding may also attest to the rather low degree of pathogenicity possessed by this organism and to the high variability among its strains.

The incidence of *B. cereus* in some foods is presented in Chapter 5. Recovery methods and procedures for this organism can be found in some of the references noted in Table 4-1.

Several other aerobic sporeformers have been associated with foodborne illnesses. A food-poisoning outbreak involving 161 cases of 340–350 persons, with one death, was investigated by Tong *et al.* *(124)*. The causative organism was a gram positive, sporeforming rod, and the vehicle food was oven-baked turkey which was found to contain up to 10^7/g of the sporeformer. The mean incubation period was about 11 hr. The causative organism was not specifically identified, but was shown to be lecithinase negative. Elter *(34)* described an outbreak of food poisoning traced to cooked sausage in which the causative organism was identified as *Bacillus sphaericus*. The vehicle food was found to contain 10^6/g of this organism. The incubation period was reported to be only 5 hr. In addition to these organisms, the *B. mesentericus-subtilis* group has also been incriminated in food poisoning outbreaks *(124)*. Some or all of these in fact may have been *B. cereus* cases.

The general involvement of aerobic sporeformers in food-borne gastroenteritis indicates that a very large number of bacterial genera and species may be capable of producing this syndrome under appropriate conditions. This fact makes all the more important the necessity of maintaining low bacterial numbers in foods.

REFERENCES

1. **Abrams, A., G. Keleles, and G. A. Hottle.** 1946. The purification of toxin from *Clostridium botulinum* type A. J. Biol. Chem. *164:* 63–79.

2. **Adams, D. M.** 1974. Requirement for and sensitivity to lysozyme by *Clostridium perfringens* spores heated at ultrahigh temperatures. Appl. Microbiol. *27:* 797–801.

3. **Alderton, G. and N. Snell.** 1969. Bacterial spores: Chemical sensitization to heat. Science *163:* 1212–1213.

4. **Alderton, G., K. A. Ito, and J. K. Chen.** 1976. Chemical manipulation of the heat resistance of *Clostridium botulinum* spores. Appl. Environ. Microbiol. *31:* 492–498.

5. **Anastasio, K. L., J. A. Soucheck, and H. Sugiyama.** 1971. Boticinogeny and actions of the bacteriocin. J. Bacteriol. *107:* 143–149.

6. **Angelotti, R., H. H. Hall, M. Foter, and K. H. Lewis.** 1962. Quantitation of *Clostridium perfringens* in foods. Appl. Microbiol. *10:* 193–199.

7. **Barnes, E. and M. Ingram.** 1956. The effect of redox potential on the growth of *Clostridium welchii* strains isolated from horse muscle. J. Appl. Bacteriol. *19:* 117–128.

8. **Benjamin, M. J. W., D. M. Wheather, and P. A. Shepherd.** 1956. Inhibition and stimulation of growth and gas production by clostridia. J. Appl. Bacteriol. *19:* 159–163.

9. **Boroff, D. A., B. R. DasGupta, and U. S. Fleck.** 1967. Study of the toxin of *Clostridium botulinum*. Japan. J. Microbiol. *11:* 371–379.

10. **Boroff, D. A., B. R. Dasgupta, and U. S. Fleck.** 1968. Homogeneity and molecular weight of toxin of *Clostridium botulinum* type B. J. Bacteriol. *95:* 1738–1744.

11. **Bott, T. L., J. S. Deffner, E. McCoy, and E. M. Foster.** 1966. *Clostridium botulinum* type E in fish from the Great Lakes. J. Bacteriol. *91:* 919–924.

12. **Brooks, V. B.** 1954. Action of botulinum toxin on motor-nerve laments. J. Physiol. (London) *123:* 501–515.

13. **Brown, G. W., Jr., G. King, and H. Sugiyama.** 1970. Penicillin-lysozyme conversion of *Clostridium botulinum* types A and E into protoplasts and their stabilization as L-form cultures. J. Bacteriol. *104:* 1325–1331.

14. **Bryan, F. L., T. W. McKinely, and B. Mixon.** 1971. Use of time-temperature evaluations in detecting the responsible vehicle and contributing factors of foodborne disease outbreaks. J. Milk Food Technol. *34:* 576–582.

15. **Carroll, B. J., E. S. Garrett, G. B. Reese, and B. Ward.** 1966. Presence of *Clostridium botulinum* in the Gulf of Venezuela and the Gulf of Darien. Appl. Microbiol. *14:* 837–838.

16. **Center for Disease Control.** 1974. Reported morbidity and mortality in the United States 1974. *23:* No. 53.

17. **Center for Disease Control.** 1976. Foodborne and waterborne disease outbreaks, Annual Summary 1975. H.E.W. Publ. No. (CDC) 76–8185.

18. **Christiansen, L. N., J. Deffner, E. M. Foster, and H. Sugiyama.** 1968. Survival and outgrowth of *Clostridium botulinum* type E spores in smoked fish. Appl. Microbiol. *16:* 133–137.

19. **Collee, J., J. Knowlden, and B. Hobbs.** 1961. Studies on the growth, sporulation and carriage of *Clostridium welchii* with special reference to food poisoning strains. J. Appl. Bacteriol. *24:* 326–339.

20. **Craig, J. and K. Pilcher.** 1966. *Clostridium botulinum* type F: Isolation from salmon from the Columbia River. Science *153:* 311–312.

21. **Crisley, F. D. and G. E. Helz.** 1961. Some observations of the effect of filtrates of several representative concomitant bacteria on *Clostridium botulinum* type A. Can. J. Microbiol. *7:* 633–639.

22. **Crisley, F. D., J. T. Peeler, R. Angelotti, and H. E. Hall.** 1968. Thermal resistance of spores of five strains of *Clostridium botulinum* type E in ground whitefish chubs. J. Food Sci. *33:* 411–416.

23. **Dack, G. M., H. Sugiyama, F. J. Owens, and J. B. Kirsner.** 1954. Failure to produce illness in human volunteers fed *Bacillus cereus* and *Clostridium perfringens.* J. Inf. Dis. *94:* 34–36.

24. **DasGupta, B. R. and H. Sugiyama.** 1976. Molecular forms of neurotoxins in proteolytic *Clostridium botulinum* type B cultures. Infect. Immunity *14:* 680–686.

25. **Dickson, E. C. and E. Shevky.** 1923. Botulism studies on the manner in which the toxin of *Clostridium botulinum* acts on the body. I. The effect on the autonomic nervous system. J. Exp. Med. *37:* 711–731.

26. **Duncan, C. L. and D. H. Strong.** 1968. Improved medium for sporulation of *Clostridium perfringens.* Appl. Microbiol. *16:* 82–89.

27. **Duncan, C. L. and D. H. Strong.** 1969. Ileal loop fluid accumulation and production of diarrhea in rabbits by cell-free products of *Clostridium perfringens.* J. Bacteriol. *100:* 86–94.

28. **Duncan, C. L. and E. B. Somers.** 1972. Quantitation of *Clostridium perfringens* type A enterotoxin by electroimmunodiffusion. Appl. Microbiol. *24:* 801–804.

29. **Duncan, C. L., D. H. Strong, and M. Sebald.** 1972. Sporulation and enterotoxin production by mutants of *Clostridium perfringens.* J. Bacteriol. *110:* 378–391.

30. **Duncan, C. L.** 1973. Time of enterotoxin formation and release during sporulation of *Clostridium perfringens* type A. J. Bacteriol. *113:* 932–936.

31. **Duncan, C. L.** 1976. *Clostridium perfringens,* pp. 170–197, In: *Food Microbiology: Public Health and Spoilage Aspects,* edited by M. P. de Figueiredo and D. F. Splittstoesser (Ave Publ.: Westport, Conn.).

32. **Eklund, M. and F. Poysky.** 1965. *Clostridium botulinum* type E from marine sediments. Science *149:* 306.

33. **Ellner, P. D.** 1956. A medium promoting rapid quantitative sporulation in *Clostridium perfringens.* J. Bacteriol. *71:* 495–496.

34. **Elter, B.** 1966. Beitrag zum Problem der Lebensmittelvergiftigungen durch aerobe Sporenbildner. Z. Ges. Hyg. Grenzgeb. *12:* 65–69.

35. **Emodi, A. S., and R. V. Lechowich.** 1969. Low Temperature growth of Type E *Clostridium botulinum* spores. 1. Effects of sodium chloride, sodium nitrite and pH. J. Food Sci. *34:* 78–81.

36. **Emswiler, B. S., C. J. Pierson, and A. W. Kotula.** 1977. Comparative study of two methods for detection of *Clostridium perfringens* in ground beef. Appl. Environ. Microbiol. *33:* 735–737.

37. **Fantasia, L. D. and A. P. Duran.** 1969. Incidence of *Clostridium botulinum* Type E in commercially and laboratory dressed white fish chubs. Food Technol. *23:* 793–794.

38. **Foster, E. M., J. Deffner, T. L. Bott, and E. McCoy.** 1965. *Clostridium botulinum* food poisoning. J. Milk & Food Technol. *28:* 86–91.

39. **Geiger, J. C.** 1941. An outbreak of botulism. J. Am. Med. Assoc. *117:* 22.

40. **Genigeorgis, C., G. Sakaguchi, and H. Riemann.** 1973. Assay methods for *Clostridium perfringens* type A enterotoxin. Appl. Microbiol. *26:* 111–115.

41. **Gerwing, J., C. Dolman, M. Reichmann, and H. Bains.** 1964. Purification and molecular weight determination of *Clostridium botulinum* type E toxin. J. Bacteriol. *88:* 216–219.

42. **Gerwing, J., R. W. Morrell, and R. M. Nitz.** 1968. Intracellular synthesis of *Clostridium botulinum* type B toxin. J. Bacteriol. *95:* 22–27.

43. **Gilbert, R. J. and A. J. Taylor.** 1976. *Bacillus cereus* food poisoning, pp. 197–213, In: *Microbiology in Agriculture, Fisheries and Food,* edited by F. A. Skinner and J. G. Carr (Academic Press: N.Y.).

44. **Glatz, B. A. and J. M. Goepfert.** 1973. Extracellular factor synthesized by *Bacillus cereus* which evokes a dermal reaction in guinea pigs. Infect. Immunity *8:* 25–29.

45. **Glatz, B. A. and J. M. Goepfert.** 1976. Defined conditions for synthesis of *Bacillus cereus* enterotoxin by fermenter-grown cultures. Appl. Environ. Microbiol. *32:* 400–404.

46. **Goepfert, J. M., W. M. Spira, and H. U. Kim.** 1972. *Bacillus cereus:* Food poisoning organism. A review. J. Milk Food Technol. *35:* 213–227.

47. **Goepfert, J. M., W. M. Spira, and B. A. Glatz, and H. U. Kim.** 1973. Pathogenicity of *Bacillus cereus,* pp. 69–75, In: *The Microbiological Safety of Food,* edited by B. C. Hobbs and J. H. B. Christian (Academic Press: N.Y.).

48. **Goepfert, J. M. and H. U. Kim.** 1975. Behavior of selected foodborne pathogens in raw ground beef. J. Milk Food Technol. *38:* 449–452.

49. **Graikoski, J. T. and Kempe, L. L.** 1963. Factors affecting the toxin production by *Clostridium botulinum* type E. Bact. Proc., A.S.M., p. 8.

50. **Hall, W. M., J. S. Witzeman, and R. Janes.** 1969. The detection and enumeration of *Clostridium perfringens* in foods. J. Food Sci. *34:* 212–214.

51. **Harmon, S. M., D. A. Kautter, and J. T. Peeler.** 1971. Comparison of media for the enumeration of *Clostridium perfringens.* Appl. Microbiol. *21:* 922–927.

52. **Harmon, S. M. and D. A. Kautter.** 1976. Beneficial effect of catalase treatment on growth of *Clostridium perfringens.* Appl. Environ. Microbiol. *32:* 409–416.

53. **Hauge, S.** 1955. Food poisoning caused by aerobic spore-forming bacilli. J. Appl. Bacteriol. *18:* 591–595.

54. **Hauschild, A. H. W. and R. Hilsheimer.** 1971. Purification and characteristics of the enterotoxin of *Clostridium perfringens* type A. Can. J. Microbiol. *17:* 1425–1433.

55. **Hauschild, A. H. W., L. Niilo, and W. J. Dorward.** 1971. The role of enterotoxin in *Clostridium perfringens* type A enteritis. Can. J. Microbiol. *17:* 987–991.

56. **Hauschild, A. H. W. and R. Hilsheimer.** 1974. Enumeration of foodborne *Clostridium perfringens* in egg yolk-free tryptose-sulfite-cycloserine agar. Appl. Microbiol. *27:* 521–526.

57. **Hobbs, B., M. Smith, C. Oakley, G. Warrack, and J. Cruickshank.** 1953. *Clostridium welchii* food poisoning. J. Hyg. *51:* 75–101.

58. **Hobbs, B. C.** 1962. Staphylococcal and *Clostridium welchii* food poisoning. In: *Food Poisoning.* Royal Society of Health, London, pp. 49–59.

59. **Ivers, J. T. and N. N. Potter.** 1977. Production and stability of hemolysis, phospholipase C, and lethal toxin of *Bacillus cereus* in foods. J. Food Protec. *40:* 17–22.

60. **Johannsen, A.** 1965. *Clostridium botulinum* type E in foods and the environment generally. J. Appl. Bacteriol. *28:* 90–94.

61. **Johnston, R., S. Harmon, and D. Kautter.** 1964. Method to facilitate the isolation of *Clostridium botulinum* type E. J. Bacteriol. *88:* 1521–1522.

62. **Kang, C. K., M. Woodburn, A. Pagenkopf, and R. Cheney.** 1969. Growth, sporulation, and germination of *Clostridium perfringens* in media of controlled water activity. Appl. Microbiol. *18:* 798–805.

63. **Kao, I., D. B. Drachman, and D. L. Price.** 1976. Botulinum toxin: Mechanism of presynaptic blockade. Science *193:* 1256–1258.

64. **Kautter, D. A., S. M. Harmon, R. K. Lynt, Jr., and T. Lilly, Jr.** 1966. Antagonistic effect on *Clostridium botulinum* type E by organisms resembling it. Appl. Microbiol. *14:* 616–622.

65. **Kawabata, T. and G. Sakaguchi.** 1963. The problem of type E botulism in Japan. In: *Microbiological Quality of Foods,* L. W. Slanetz *et al.,* Editors, Academic Press, N.Y., pp. 71–76.

66. **Kempe, L. L.** 1964. Characteristics of various types of *Clostridium botulinum* that lead to different food processing and handling requirements. Assoc. Food & Drug Off. U.S., Quart. *28:* 199–206.

67. **Kindler, S. H., J. Mager, and N. Grossowicz.** 1956. Toxin production by *Clostridium parabotulinum* type A. J. Gen. Microbiol. *15:* 394–403.

68. **Labbe, R. G. and C. L. Duncan.** 1974. Sporulation and enterotoxin production by *Clostridium perfringens* type A under conditions of controlled pH and temperature. Can. J. Microbiol. *20:* 1493–1501.

69. **Labbe, R. G. and C. L. Duncan.** 1977. Evidence for stable messenger ribonucleic acid during sporulation and enterotoxin synthesis by *Clostridium perfringens* type A. J. Bacteriol. *129:* 843–849.

70. **Lamanna, C., D. E. McElroy, and H. W. Eklund.** 1946. Purification and crystallization of *Clostridium botulinum* type A toxin. Science *103:* 613–614.

71. **Lamanna, C., R. A. Hillowalla, and C. C. Alling.** 1967. Buccal exposure to botulinal toxin. J. Inf. Dis. *117:* 327–331.

72. **Lamanna, C., L. Spero, and E. J. Schantz.** 1970. Dependence of time to death on molecular size of botulinum toxin. Infect. Immunity *1:* 423–424.

73. **Lamanna, C. and G. Sakaguchi.** 1971. Botulinal toxins and the problem of nomenclature of simple toxins. Bacteriol. Rev. *35:* 242–249.

74. **Lynt, R. K., D. A. Kautter, and R. B. Read, Jr.** 1975. Botulism in commercially canned foods. J. Milk Food Technol. *38:* 546–550.

75. **Mahony, D. E.** 1977. Stable L-forms of *Clostridium perfringens:* Growth, toxin production, and pathogenicity. Infect. Immunity *15:* 19–25.

76. **Marshall, R., J. Steenbergen, and L. McClung.** 1965. Rapid technique for the enumeration of *Clostridium perfringens*. Appl. Microbiol. *13:* 559–563.

77. **McClung, L.** 1945. Human food poisoning due to growth of *Clostridium perfringens (C. welchii)* in freshly cooked chicken: preliminary note. J. Bacteriol. *50:* 229–231.

78. **McClung, L. S.** 1964. Selected references on botulism, clostridia that produce botulinal toxins, and related topics. In: *Botulism: Proceedings of a Symposium.* U.S. Pub. Hlth. Pub. No. 999, Cincinnati, Ohio, pp. 257–313.

79. **Meyer, K. F. and J. B. Gunnison.** 1929. Botulism due to home canned bartlett pears. J. Inf. Dis. *45:* 135–147.

80. **Midura, T., C. Taclindo, Jr., G. S. Mygaard, H. L. Bodily, and R. M. Wood.** 1968. Use of immunofluorescence and animal tests to detect growth and toxin production by *Clostridium botulinum* type E in food. Appl. Microbiol. *16:* 102–105.

81. **Midura, T., M. Gerber, R. Wood, and A. R. Leonard.** 1970. Outbreak of food poisoning caused by *Bacillus cereus*. Publ. Hlth. Rept. *85:* 45–47.

82. **Midura, T. F., G. S. Nygaard, R. M. Wood, and H. L. Bodily.** 1972. *Clostridium botulinum* type F: Isolation from venison jerky. Appl. Microbiol. *24:* 165–167.

83. **Moller, V. and I. Scheibel.** 1960. Preliminary report on the isolation of an apparently new type of *Cl. botulinum*. Acta Path. Microbiol. Scan. *48:* 80.

84. **Morbidity and Mortality Weekly Report, Center for Disease Control,** *26:* No. 16, 4/22/77.

85. **Mortimer, P. R. and G. McCann.** 1974. Food-poisoning episodes associated with *Bacillus cereus* in fried rice. Lancet *1:* 1043–1045.

86. **Mossell, D. A. A., A. S. DeBruin, H. M. J. VanDiepen, C. M. A. Vendrig, and G. Zoutewelle.** 1956. The enumeration of anaerobic bacteria, and of *Clostridium* species in particular, in foods. J. Appl. Bacteriol. *19:* 142–154.

87. **Mossel, D. A. A.** 1959. Enumeration of sulfite-reducing clostridia occurring in foods. J. Sci. Food Agr. *19:* 662–669.

88. **Mossel, D. A., M. J. Koopman, and E. Jongerius.** 1967. Enumeration of *Bacillus cereus* in foods. Appl. Microbiol. *15:* 650–653.

89. **Mossel, D. A. A.** 1968. Bacterial toxins of uncertain oral pathogenicity. In: *The Safety of Foods.* H. D. Graham *et al.,* Editors, Avi Publishing Company, Westport, Conn., pp. 168–182.

90. **Nakamura, M. and J. A. Schultze.** 1970. *Clostridium perfringens* food poisoning. Ann. Rev. Microbiol. *24:* 359–372.

91. **Niilo, L.** 1971. Mechanism of action of the enteropathogenic factor of *Clostridium perfringens* type A. Infect. Immunity *3:* 100–106.

92. **Nikodemusz, I., M. Bojan, V. Hoch, M. Kiss, and P. Kiss.** 1963. Das vorkommen von *Bac. cereus* in Lebensmitteln. Arch. Lebensmittelhyg. *14:* 172–173.

93. **Ohishi, I. and G. Sakaguchi.** 1975. Molecular construction of *Clostridium botulinum* type F progenitor toxin. Appl. Microbiol. *29:* 444–447.

94. **Ohye, D. F. and W. J. Scott.** 1953. The temperature relations of *Clostridium botulinum* types A and B. Austr. J. Biol. Sci. *6:* 178–189.

95. **Pace, P. J., E. R. Krumbiegel, R. Angelotti, and H. J. Wisniewski.** 1967. Demonstration and isolation of *Clostridium botulinum* types from whitefish chubs collected at fish smoking plants of the Milwaukee area. Appl. Microbiol. *15:* 877–884.

96. **Pederson, H. O.** 1955. On type E botulism. J. Appl. Bacteriol. *18:* 619–629.

97. **Peterson, D., H. Anderson, and R. Detels.** 1966. Three outbreaks of foodborne disease with dual etiology. Pub. Hlth. Repts. *81:* 899–904.

98. **Plazikowski, U.** 1947. Further investigation regarding the cause of food poisonings. Proc., 4th Internat. Cong. Microbiol., Copenhagen, pp. 510–511.

99. **Rey, C. R., H. W. Walker, and P. L. Rohrbaugh.** 1975. The influence of temperature on growth, sporulation, and heat resistance of spores of six strains of *Clostridium perfringens*. J. Milk Food Technol. *38:* 461–465.

100. **Riemann, H.** 1962. Anaerobe toxins. In: *Chemical and Biological Hazards in Food.* J. C. Ayres *et al.,* Editors, Iowa State Univ. Press, pp. 279–302.

101. **Riha, W. E., Jr., and M. Solberg.** 1971. Chemically defined medium for the growth of *Clostridium perfringens*. Appl. Microbiol. *22:* 738–739.

102. **Saylor, G. S., J. D. Nelson, Jr., A. Justice, and R. R. Colwell.** 1976. Incidence of *Salmonella* spp., *Clostridium botulinum,* and *Vibrio parahaemolyticus* in an estuary. Appl. Environ. Microbiol. *31:* 723–730.

103. **Schmidt, C. F., R. V. Lechowich, and J. F. Folinazzo.** 1961. Growth and toxin production by type E *Clostridium botulinum* below 40 F. J. Food Sci. *26:* 626–630.

104. **Schneider, M., N. Grecz, and A. Anellis.** 1963. Sporulation of *Clostridium botulinum* types A, B, and E, *Clostridium perfringens,* and putrefactive anaerobe 3679 in dialysis sacs. J. Bacteriol. *85:* 126–133.

105. **Segner, W. P., C. F. Schmidt, and J. K. Boltz.** 1966. Effect of sodium chloride and pH on the outgrowth of spores of Type E *Clostridium botulinum* at optimal and suboptimal temperatures. Appl. Microbiol. *14:* 49–51.

106. **Skjelkvale, R. and C. L. Duncan.** 1975. Enterotoxin formation by different toxigenic types of *Clostridium perfringens*. Infect. Immunity *11:* 563–575.

107. **Skulberg, A.** 1965. The resistance of *Clostridium botulinum* type E toxin to radiation. J. Appl. Bacteriol. *28:* 139–141.

108. **Smith, L. D. S.** 1963. *Clostridium perfringens* food poisoning. In: *Microbiological Quality of Foods.* L. W. Slanetz *et al.,* Editors, Academic Press, N.Y., pp. 77–83.

109A. **Smith, L. DS.** 1975. Inhibition of *Clostridium botulinum* by strains of *Clostridium perfringens* isolated from soil. Appl. Microbiol. *30:* 319–323.

109B. **Smith, L. DS.** 1977. *Botulism—The Organism, Its Toxins, The Disease.* (Chas. C. Thomas: Springfield, Ill.).

110. **Snudden, B. H. and R. V. Lechowich.** 1967. Growth and sporulation of *Cl. botulinum* type E in chemically defined media. In: *Botulism, 1966.* M. Ingram and T. A. Roberts, Editors, Chapman & Hall, London, pp. 144–149.

111. **Solomon, H. M., R. K. Lynt, Jr., D. A. Kautter, and T. Lilly, Jr.** 1971. Antigenic relationships among the proteolytic and nonproteolytic strains of *Clostridium botulinum*. Appl. Microbiol. *21:* 295–299.

112. **Spira, W. M. and J. M. Goepfert.** 1975. Biological characteristics of an enterotoxin produced by *Bacillus cereus*. Can. J. Microbiol. *21:* 1236–1246.

113. **Stark, R. L. and C. L. Duncan.** 1971. Biological characteristics of *Clostridium perfringens* type A enterotoxin. Infect. Immunity *4:* 89–96.

114. **Sterne, M. and L. M. Wentzel.** 1950. A new method for the large-scale production of high-titre botulinum formol-toxoid types C and D. J. Immunol. *65:* 175–183.

115. **Sterne, M. and W. E. van Heyningen.** 1953. The Clostridia. In: *Bacterial and Mycotic Infections of Man.* 3rd Edition, J. B. Lippincott, Philadelphia, Chapter 14.

116. **Strong, D. H. and J. C. Canada.** 1964. Survival of *Clostridium perfringens* in frozen chicken gravy. J. Food Sci. *29:* 479–482.

117. **Strong, D. H., J. C. Canada, and B. Griffiths.** 1963. Incidence of *Clostridium perfringens* in American foods. Appl. Microbiol. *11:* 42–44.

118. **Strong, D. H., C. L. Duncan, and G. Perna.** 1971. *Clostridium perfringens* type A food poisoning. II. Response of the rabbit ileum as an indication of enteropathogenicity of strains of *Clostridium perfringens* in human beings. Infect. Immunity *3:* 171–178.

119. **Sugii, S. and G. Sakaguchi.** 1976. Molecular construction of *Clostridium botulinum* type A toxins. Infect. Immunity *12:* 1262–1270.

120. **Sugiyama, H. and K. H. Yang.** 1975. Growth potential of *Clostridium botulinum* in fresh mushrooms packaged in semipermeable plastic film. Appl. Microbiol. *30:* 964–969.

121. **Tarr, H. L. A.** 1953. The action of hydroxylamine on bacteria. J. Fish. Res. Bd. Canada *10:* 69–75.

122. **Thatcher, F. S., J. Robinson, and I. Erdman.** 1962. The 'vacuum pack' method of packaging foods in relation to the formation of the botulinum and staphylococcal toxins. J. Appl. Bacteriol. *25:* 120–124.

123. **Tompkin, R. B. and L. N. Christiansen.** 1976. *Clostridium botulinum,* pp. 156–169, In: *Food Microbiology: Public Health and Spoilage Aspects,* edited by M. P. deFigueiredo and D. F. Splittstoesser (Avi Publ.: Westport, Conn.).

124. **Tong, J. L., H. M. Engle, J. S. Cullyford, D. J. Shimp, and C. E. Love.** 1962. Investigation of an outbreak of food poisoning traced to turkey meat. A. J. Pub. Hlth. *52:* 976–990.

125. **Trakulchang, S. P. and A. A. Kraft.** 1977. Survival of *Clostridium perfringens* in refrigerated and frozen meat and poultry items. J. Food Sci. *42:* 518–521.

126. **Walker, H. W.** 1975. Food borne illness from *Clostridium perfringens*. CRC Crit. Rev. Fd. Sci. Nutri. *7:* 71–104.

127. **Ward, B. Q. and B. J. Carroll.** 1967. Production of toxin by *Clostridium botulinum* type E in defined media. Can. J. Microbiol. *13:* 109–110.

128. **Warnecke, M. O., J. A. Carpenter, and R. L. Saffle.** 1967. A study of *Clostridium botulinum* type E, in vacuum packaged meat. Food Technol. *21:* 115A–116A.

129. **Weiss, K. F. and D. H. Strong.** 1967. Some properties of heat-resistant and heat-sensitive strains of *Clostridium perfringens*. I. Heat resistance and toxigenicity. J. Bacteriol. *93:* 21–26.

130. **Wentz, M., R. Scott, and J. Vennes.** 1967. *Clostridium botulinum* type F: Seasonal inhibition by *Bacillus licheniformis*. Science *155:* 89–90.

131. **Williams-Walls, N. J.** 1968. *Clostridium botulinum* type F: Isolation from crabs. Science *162:* 375–376.

132. **Woodburn, M. and C. H. Kim.** 1966. Survival of *Clostridium perfringens* during baking and holding of turkey stuffing. Appl. Microbiol. *14:* 914–920.

133. **Yang, K. H. and H. Sugiyama.** 1975. Purification and properties of *Clostridium botulinum* type F toxin. Appl. Microbiol. *29:* 598–603.

134. **Yotis, W. W. and N. Catsimpoolas.** 1975. Scanning isoelectric focusing and isotachophoresis of *Clostridium perfringens* type A enterotoxin. J. Appl. Bacteriol. *39:* 147–156.

135. **Zeissler, J. and L. Rassfeld-Sternberg.** 1949. Enteritis necroticans due to *Clostridium welchii* type F. Brit. Med. J. *1:* 267–271.

18

Food Poisoning
Caused By
Gram Negative Bacteria

A rather large number of gram negative rods have been reported to cause food-borne gastroenteritis, the most important of which are members of the genus *Salmonella*. In addition to the salmonellae, enteropathogenic *Escherichia coli* and *Vibrio parahaemolyticus* cause food poisoning in man.

In the early bacteriological literature, the organisms now designated *Salmonella* were referred to as *Bacterium* and later as *Eberthella,* the latter in honor of Eberth. The present generic name is in honor of D. E. Salmon, an early American bacteriologist. Contrary to the belief of some, the word "salmonella" bears no relationship to the salmon fish.

SALMONELLA FOOD POISONING

The genus *Salmonella* consists of small, gram negative, nonsporing rods that are indistinguishable from *E. coli* under the microscope, or on ordinary nutrient media. The organisms are widely distributed in nature, with man and animals being their primary reservoirs. *Salmonella* food poisoning results from the ingestion of foods containing appropriate strains of this genus in significant numbers. While the minimum number of salmonellae cells differs from strain to strain, it is considerably lower than is necessary for streptococcal and *B. cereus* food poisoning.

All species and strains of *Salmonella* may be presumed to be pathogenic for man, and the disease syndromes divide themselves into several distinct clinical types. Typhoid fever is caused by *S. typhi* and is the most severe of all diseases caused by this genus. Typhoid fever is a classic example of an enteric fever.

In the same general category with typhoid fever are the paratyphoid fevers caused by *S. paratyphi* A, *S. paratyphi,* B, *S. paratyphi* C, and others. The paratyphoid syndrome tends to be milder than that of typhoid. In the latter, the period of incubation is longer, a higher body temperature is produced, the organisms may be isolated from the blood and sometimes urine, and the mortality rate is higher. Blood cultures are often positive in the paratyphoid syndrome. The etiologic agents of the typhoid and paratyphoid syndromes are specifically pathogenic for man.

The third disease entity caused by *Salmonella* spp. is gastroenteritis. This syndrome differs from the enteric fevers in having an incubation period as short as 8 hr, generally negative blood cultures, and a lack of host specificity among the numerous serotypes capable of causing this type of infection. While some of the food poisoning strains can be identified on the basis of biochemical and cultural characteristics, by far the largest numbers are identified on the basis of antigenic analysis.

The salmonellae may be divided into the following three groups based on host predilections *(10)*. 1. **Primarily adapted to man.** The typhoid and paratyphoid agents are the prime examples of this group. 2. **Primarily adapted to particular animal hosts.** Included in this group are *S. choleraesuis* and serotypes of *S. enteritidis* such as *S. pullorum, S. gallinarum, S. dublin,* and so on. 3. **Unadapted.** This group includes over 1800 serotypes of *S. enteritidis* that attack man and other animals and do not show any host preference. Foodborne salmonellosis is caused by members of this group primarily.

1. Classification of Salmonella.   The classification of these organisms by antigenic analysis is based upon the original work of Kauffmann and White and is often referred to as the Kauffmann-White Scheme. Classification by this scheme makes use of both somatic and flagellar antigens. Somatic antigens are designated O antigens while flagellar antigens are designated H antigens. The K antigens are capsular antigens which lie at the periphery of the cell and prevent access of anti-O agglutinins (antibodies) to their homologous somatic antigens. The K antigen differs from ordinary O antigens in being destroyed by heating for 1 hr at 60 C and by dilute acids and phenol. The use of H, O, and K antigens as the basis of classification of *Salmonella* spp. is based upon the fact that each antigen possesses its own genetically determined specificity.

When classification is made by use of antigenic patterns, species and varieties are placed in Groups designated A, B, C, and so on, according to similarities in content of one or more O antigens. Thus, *S. hirschfeldii, S. choleraesuis, S oranienburg,* and *S. montevideo* are placed in Group C_1 because they all possess O antigens 6 and 7 in common. *S. newport* is placed in Group C_2 due to its possession of O antigens K and 8 (see Table 18-1). For further classification, the flagellar or H antigens are employed. These antigens are of two types: specific phase or phase-1, and group phase or phase-2. Phase-1 antigens are shared with only a few other species or varieties of *Salmonella,* while phase-2 may be more widely distributed among several species. Any given culture of *Salmonella* may consist of organisms in only one phase, or of organisms in both flagellar phases. The H antigens of phase-1 are designated with small letters, and those of phase-2

TABLE 18-1.   Antigenic structure of some of the more common enteric organisms (35).

Group	Type	O Antigens	H Antigens phase 1	phase 2
A	S. paratyphi A	(1), 2, 12	a	—
B	S. schottmuelleri	(1), 4, (5), 12	b	1,2
	S. typhimurium	(1), 4, (5), 12	i	1,2
C₁	S. hirschfeldii	6, 7, (Vi)	c	1,5
	S. choleraesuis	6, 7	c	1,5
	S. oranienburg	6, 7	m,t	—
	S. montevideo	6, 7	g,m,s	—
C₂	S. newport	6, 8	e,h	1,2
D	S. typhi	9, 12, (Vi)	d	—
	S. enteritidis	(1), 9, 12	g,m	—
	S. gallinarum-pullorum	1, 9, 12	—	—
E	S. anatum	3, 10	e,h	1,6

(   ) indicates that antigen may be absent.

are designated by arabic numerals. Thus, the complete antigenic analysis of *S. choleraesuis* is as follows: 6, 7, c, 1, 5, where 6 and 7 refer to O antigens, c to phase-1 flagellar antigens, and 1 and 5 to phase-2 flagellar antigens (see Table 18-1). *Salmonella* subgroups of this type are referred to as serotypes. With a relatively small number of O, phase-1, and phase-2 antigens, a large number of permutations is possible which allows for the possibility of a large number of serotypes. Over 1800 *Salmonella* serotypes are presently known, and the number increases yearly.

The naming of *Salmonella* is now done by international agreement. Under this system, a serotype is named after the place where it was first isolated, e.g., *S. london, S. miami, S. richmond,* and so on. Prior to the adoption of this convention, species and subtypes were named in various ways, e.g., *S. typhimurium* as the cause of typhoid fever in mice. Most foodborne salmonellae are serotypes of *S. enteritidis*.

2. Distribution of Salmonella.   The primary habitat of *Salmonella* spp. is the intestinal tract of animals such as birds, reptiles, farm animals, man, and occasionally insects. Although their primary habitat is the intestinal tract, they may be found in other parts of the body from time to time. As intestinal forms, these organisms are excreted in feces from which they may be transmitted by insects and other living creatures to a large number of places. As intestinal forms, they may also be found in water, especially polluted water. When polluted water and foods that have been contaminated by insects or by other means are consumed by man and other animals, these organisms are once again shed through fecal matter with a continuation of the cycle. The augmentation of this cycle through the international shipment of animal products and feeds is in large part responsible for the present world-wide distribution of salmonellosis and its consequent problems.

While *Salmonella* spp. have been recovered repeatedly from a large number of different animals, their incidence in various parts of animals has been shown to vary. In a study of slaughterhouse pigs, Kampelmacher *(22)* found these organisms in spleen, liver, bile, mesenteric and portal lymph nodes, diaphragm, and pillar, as well as in feces. A higher incidence was found in lymph nodes than in feces. The frequent occurrence of *Salmonella* spp. among susceptible animal populations is due in part to the contamination of *Salmonella*-free animals by animals within the population that are carriers of these organisms or are infected by them. A **carrier** is defined as a person or an animal that repeatedly sheds *Salmonella* spp., usually through feces, without showing any signs or symptoms of the disease. Upon examining poultry at slaughter, Sadler and Corstvet *(44)* found an intestinal carrier rate of 3–5%. During and immediately after slaughter, carcass contamination from fecal matter may be expected to occur. In an examination of the rumen contents of healthy cattle after slaughter, Grau and Brownlie *(16)* found 45% to contain salmonellae. Some 57% of samples taken from the environment of cattle in transit to slaughter were positive for these organisms. From 53.1 to 61.9% of inspected broiler carcasses have been found to be contaminated with *Salmonella* spp.

Equally serious is the contamination of eggs and egg products. Reports from various countries indicate that 2.6–7.0% of eggs are contaminated, mostly with *S. typhimurium (41)*. Duck eggs have been reported to have an even higher contamination rate, reaching 20%. Another common source of these organisms to animal populations is animal feed (Figure 18-1). Some investigators feel that this source is perhaps the most important in terms of the overall control of salmonellosis. In a study of animal feeds in England for the years 1958–1960, Taylor *(48)* isolated *Salmonella* serotypes from meat and bone products 855 times; from mixtures, mashes, and the like 158 times; from fish products 100 times; and from vegetable products 17 times. *S. senftenberg* was isolated most frequently followed by *S. anatum* and *S. cubana*. *S. senftenberg* has been shown to be the most heat-resistant of all *Salmonella* serotypes, and its higher incidence in animal feeds may be due to the fact that most others are destroyed by heat in the processing of these products. A very large number of different serotypes was found in bile, liver, spleen, and lymph glands of slaughter animals by Taylor. Among the many serotypes reported by this author, *S. dublin* was found to be associated with bovines more than with any other animals. The presence of salmonellae in animal feeds has been shown by Clise and Swecker *(9)*, Moyle *(37)*, and Loken *et al. (27)*. In an examination of 1395 bulk protein feed supplements, these workers found 241 or 17% to contain salmonellae. The samples that were most consistently positive were those that had total counts of 100,000/g and above.

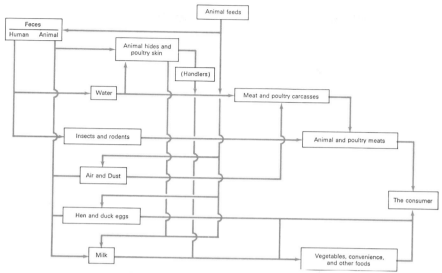

FIGURE 18-1.   Scheme illustrating the way in which *Salmonella* spp. and serotypes are dissiminated.

The most common food vehicles of salmonellosis in man are eggs, poultry, meat and meat products. In a study of 61 outbreaks of *Salmonella* food poisoning for the period 1963–65, Steele and Galton *(46)* found that eggs and egg products accounted for 23, chicken and turkey for 16, beef and pork for 8, ice cream for 3, potato salad for 2, and other miscellaneous foods for 9. The most common food vehicles involved in 12,836 cases of salmonellosis from 37 U.S. states in 1967 were beef, turkey, eggs and egg products, and milk *(52)*. Of 7,907 salmonellae isolations made by CDC in 1966, 70% were from raw and processed food sources. Turkey and chicken sources accounted for 42%. Of the foodborne disease outbreaks with known etiology traced to poultry for the period 1972–1974, 44% were salmonellae *(20)*. In a recent study on the incidence of salmonellae in 69 packs of raw chicken pieces, 34.8% were positive for these organisms and 11 serotypes were represented with *S. muenchen* being the most common *(11)*. Poultry products are important sources of salmonellosis outbreaks, and the wide distribution of these organisms among slaughter animals makes all meats potential sources. Even when such meats are cooked sufficiently to destroy *Salmonella* spp., in the raw state they may still serve as sources of these organisms to other foods such as vegetables, salads, and the like (see also Chapter 5).

In a study of the incidence of *Salmonella* spp. in commercially prepared and packaged foods, Adinarayanan *et al. (1)* found these organisms in 17 of 247 foods examined. Among the contaminated foods were cake mixes, cookie doughs, dinner rolls, corn bread mixes, and others. Foods of this type usually become contaminated from infected eggs, bulk egg products, or by contact with rodents, flies, or even man. *Salmonella* spp. have been found in cocoanut meal, salad dressing, mayonnaise, milk, and many other foods. Not only are the salmonellae to be found world-wide, they may be found throughout the household on many types of foods, and even on pets (especially pet turtles) and pet foods.

Of the various serotypes in food-borne outbreaks of salmonellosis, *S. typhimurium* is invariably the most frequently encountered throughout the world. Of 1713 serotypes isolated from foods during 1963–1965, Steele and Galton *(46)* found the following 5 to be the most common: *S. infantis* (215), *S. oranienberg* (192), *S. typhimurium* (171), *S. montevideo* (151), and *S. heidelberg* (147). In general, the incidence of *Salmonella* serotypes in foods closely parallels their incidence in man and animals. The 7 most frequently isolated strains from man in England and Wales for 1956–1960 were as follows *(48)*: *S. typhimurium* (15,808), *S. heidelberg* (1,013), *S. enteritidis* (734), *S. newport* (667), *S. thompson* (623), *S. saint-paul* (282), and *S. anatum* (279). For the period of April 1962 to April 1963 the 9 most prevalent salmonellae serotypes identified from animal sources in the United States were as follows: *S. typhimurium*, *S. heidelberg*, *S. anatum*, *S. choleraesuis*, *S. infantis*, *S. montevideo*, *S. derby*, *S. saint-paul*, and *S. oranienberg*. The 5 most frequently isolated salmonellae from human sources in 1972 along with percent frequency of isolation were: *S. typhimurium* (25.8), *S. newport* (8.4), *S. enteritidis* (6.5), *S. infantis* (6.3), and *S. heidelberg* (5.6) *(49)*. Human isolations of salmonellae in the U.S. for the period 1966–1974 were between 500–700/week (Fig. 18-2).

3. Characteristics of the Growth and Destruction of Salmonella. *Salmonella* strains are typical of other gram negative bacteria in being able to grow on a large number of culture media and produce visible colonies within 24 hr at about 37 C. These organisms are generally unable to ferment lactose, sucrose, or salicin, although glucose and certain other monosaccharides are fermented, with the production of gas. It should be noted that while lactose fermentation is not usual for these organisms, it is not unknown.

The pH of optimum growth for salmonellae is around neutrality, with values above 9.0 and below 4.0 being bactericidal *(41)*. A minimum growth pH of 4.05 has been recorded for salmonellae (with HCl and citric acids), but depending upon the acid used to lower pH, the minimum may be as high as 5.5 *(8)*. The effect of the acid used to lower pH on minimum growth is presented in Table 18-2. Aeration was found to favor growth at the lower pH values. The parameters of pH, a_w, nutrient content, and temperature are all interrelated for salmonellae as they are for most bacteria *(49)*. For best growth, the salmonellae require pH between 6.6–8.2. The lowest temperatures at which growth has been reported are 5.3 C for *S. heidelberg* and

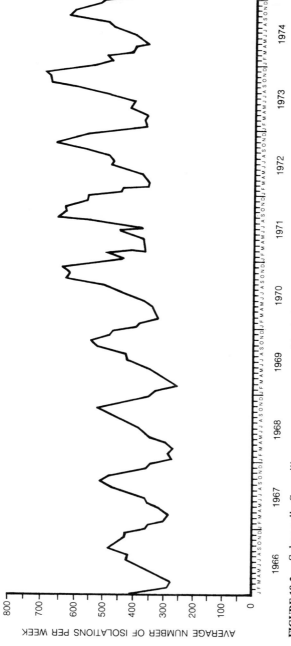

FIGURE 18-2. Salmonella-Surveillance program reported isolations from humans by month, United States, 1966–1974 (source: ref. 16, Chapter 17).

TABLE 18-2. Minimum pH at which Salmonellae would initiate growth under optimum laboratory conditions.[1] (8).

Acid	pH
Hydrochloric	4.05
Citric	4.05
Tartaric	4.10
Gluconic	4.20
Fumaric	4.30
Malic	4.30
Lactic	4.40
Succinic	4.60
Glutaric	4.70
Adipic	5.10
Pimelic	5.10
Acetic	5.40
Propionic	5.50

[1]Tryptone–yeast extract–glucose broth was inoculated with 10^4 cells per milliliter of *Salmonella anatum, S. tennessee* or *S. senftenberg*.

Copyright © 1970, Institute of Food Technologists.

6.2 C for *S. typhimurium (30)*. A low of 6.7 C was reported by Angelotti *et al. (2)*. Temperatures of about 45 C have been reported by several authors to be the upper limit for growth. In regard to available moisture for growth, growth inhibition has been reported for a_w values below 0.94 in media with neutral pH, with higher a_w values being required as pH is decreased towards growth minima. Unlike the staphylococci, the salmonellae are unable to tolerate high salt concentrations. Brine concentrations above 9% have been reported to be bactericidal. Nitrite is effective against these organisms, with the effect being greatest at the lower pH values. This suggests that the inhibitory effect of this compound is referable to the undissociated HNO_2 molecule. The survival of *Salmonella* spp. in mayonnaise was studied by Lerche *(24)* who found that these organisms were destroyed in this product if the pH was below 4.0. This author found that several days may be required for destruction in mayonnaise if the level of contamination was high, but within 24 hr for low numbers of *Salmonella*. *S. thompson* and *S. typhimurium* were found by this worker to be more resistant to acid destruction than *S. senftenberg*.

With respect to heat destruction, all salmonellae are readily destroyed at milk pasteurization temperatures, and thermal D values for the destruction of *S. senftenberg* 775W under various conditions are given in Chapter 12, Table 12-2. Shrimpton *et al. (45)*, reported that *S. senftenberg* 775W required 2.5 min for a 10^4–10^5 reduction in numbers at 64.4 C in liquid whole egg. This strain is the most heat-resistant of all salmonellae serotypes, as previously stated. The above treatment of liquid whole egg has been shown to produce a *Salmonella*-free product and destroy egg alpha-amylase (see Chapter 12 for the heat pasteurization of egg white). It has been suggested by Brooks (7) that the alpha-amylase test may be used as a means of determining the adequacy of heat pasteurization of liquid egg (compare with the pasteurization of milk and the enzyme phosphatase). In a study on the heat resistance of *S. senftenberg* 775W, Ng *et al.* (38) found this strain to be more heat sensitive in the log phase than in the stationary phase of growth. These authors also found that cells grown at 44 C were more heat-resistant than those grown at either 15 or 35 C.

With respect to the destruction of *Salmonella* in baked foods, Beloian and Schlosser (4) found that baked foods reaching a temperature of 160 F or higher in the slowest heating region can be considered *Salmonella*-free. These authors employed *S. senftenberg* 775W at a concentration of 7,000 to 10,000 cells/ml placed into reconstituted dried egg. With respect to the heat destruction of this strain in poultry, it is recommended that internal temperatures of at least 165 F be attained (33). Although *S. senftenberg* 775W has been reported to be 30 times more heat-resistant than *S. typhimurium* (38), the latter organism has been found to be more resistant to dry heat than the former (15). These authors tested dry heat resistance in milk chocolate.

The destruction of *S. pullorum* in turkeys was investigated by Rogers and Gunderson (42), who found that it required 4 hr and 55 min to destroy an initial inoculum of 115,000,000 in 10–11 lb. turkeys with an internal temperature of 160 F, and for 18 lb. turkeys with an initial inoculum of 320,000,000 organisms, 6 hr and 20 min were required for destruction. The salmonellae are quite sensitive to ionizing radiation, with doses of 0.5 to 0.75 Mrads being sufficient to eliminate these organisms from most foods and feed (41). The decimal reduction dose has been reported to range from 0.04–0.07 Mrad for *Salmonella* spp. in frozen eggs. The effect of various foods on the radiosensitivity of salmonellae is shown in a study by Ley *et al.* (25). These investigators found that for frozen whole egg, 0.5 Mrad gave a 10^7 reduction in the numbers of *S. typhimurium*, while 0.65 Mrad was required to give a 10^5 reduction in frozen horsemeat, between 0.5–0.75 Mrad for a 10^5–10^8 reduction in bone meal, and only 0.45 Mrad to give a 10^3 reduction of *S. typhimurium* in desiccated cocoanut.

4. The Salmonella Food Poisoning Syndrome.   This syndrome is caused by the ingestion of foods that contain significant numbers of nonhost specific species or serotypes of the genus *Salmonella*. From the time of ingestion of food, symptoms usually develop between 12–24 hr, although shorter and longer times have been reported. The symptoms consist of nausea, vomiting, abdominal pain (not as severe as with staphylococcal food poisoning), headache, chills, and diarrhea. These symptoms are usually accompanied by prostration, muscular weakness, faintness, moderate fever, restlessness, and drowsiness. Symptoms usually persist for 2–3 days. The average mortality rate is 4.1%, varying from 5.8% during the first year of life, to 2% between the first and fiftieth year, and 15% in persons over fifty *(41)*. Among the different species of *Salmonella, S. choleraesuis* has been reported to produce the highest mortality rate—21%.

While these organisms generally disappear rapidly from the intestinal tract, up to 5% of patients may become carriers of the organisms upon recovery from this disease.

One of the primary differences between salmonellae and staphylococcal food poisoning is the need for the ingestion of viable cells in the case of the former, while only the culture filtrate is necessary in the case of the latter disease. Because viable cells are necessary to cause salmonellosis, this disease is designated as a food infection. Numbers of cells on the order of several million to several billion/g are necessary for salmonellae food poisoning, with strains such as *S. enteritidis* and *S. anatum* being more infective than *S. pullorum.* The ingestion of viable cells seems necessary for this syndrome, although some investigators claim that the ingestion of killed cells will give rise to symptoms. The occurrence of significant numbers of viable salmonellae in the stools of victims tends to favor the role of living cells in the syndromes. The symptoms of the disease are presumed to be caused by endotoxin liberated from the cells by the action of low stomach pH and proteases. Although salmonellae food poisoning outbreaks usually occur from foods in which these organisms have been allowed to grow, there is no evidence at this time that any extracellular toxic substances are produced that might have any relationship to the disease. The relatively short incubation period also tends to support the lack of growth of cells in the intestinal tract as being necessary to cause the disease.

The precise incidence of salmonellae food poisoning is not known, since small outbreaks are often not reported to public health officials. There has been a steady increase in the number of reported cases over the past 17 yr. For the four yr 1951, 1961, 1963, and 1964, respectively, 1,733, 8,500, 18,000 and 21,113 cases were reported. In 1967, a total of 273 outbreaks of food poisoning were reported to the National Communicable Disease Center from 37 states. These outbreaks involved 22,171 cases, with 12,836 cases or 58% having been caused by *Salmonella* spp. *(52)*. Some authorities feel that as many as 2 million human cases of salmonellosis may occur each year in the United States alone. As is the case for staphylococcal food poisoning, the largest outbreaks of salmonellosis usually occur at banquets.

There are no doubt several reasons why salmonellae food poisoning has increased so greatly in the past several years. Kampelmacher *(22)* has offered the following explanations for this increase: (1) The increase in mass food preparation, which favors spread of *Salmonella,* (2) inappropriate methods of storing food which, because of modern living conditions, is sometimes accumulated in excessive amounts, (3) the increasing habit of eating raw or insufficiently heated foods, partly because of overreliance on food inspection, (4) increasing international food trade, and (5) decreased resistance to infection resulting from improved standards of general hygiene.

5. Recovery of Salmonella From Foods.   In view of the federal regulation prohibiting the occurrence of *Salmonella* spp. in foods, the recovery of these organisms from food presents a difficult problem to food microbiologists and food scientists. How can one be certain that a 1,000 lb. lot of powdered eggs is free of *Salmonella* if there might be only 1 organism/100 g? The problem is made all the more difficult by the fact that foods normally contain larger numbers of microorganisms other than *Salmonella,* such as *Proteus, Pseudomonas, Acinetobacter,* and *Alcaligenes* spp., all of which may develop on some of the media employed for the recovery of salmonellae. The problems encountered in the recovery of these organisms from foods are similar to those that one encounters in the recovery of staphylococci from foods, i.e., a generally high ratio of total numbers of other organisms to pathogens. Where the ratio of total flora to salmonellae is rather low, any one of a large number of media may be suitable for recovering these organisms. Special methods are necessary, however, when few salmonellae exist in the presence of a high total count.

In view of the generally low ratio of salmonellae to total flora, it is necessary to inhibit or kill nonsalmonellae types, while the salmonellae are allowed to increase to numbers high enough to better the chances of finding them on plates or in culture. The fluorescent antibody technique has become an important method for identifying salmonellae, and this and other recovery methods are discussed in Chapter 4. For cultural and identification methods, a reference in Table 4-1 should be consulted.

6. The Prevention and Control of Food-Borne Salmonellosis.   An overall view of the primary sources and transmission of *Salmonella* spp. to man is presented in Figure 18-1. As previously stated, the primary source of these organisms is the intestinal tract of man and other animals. From human feces, salmonellae may enter water from which meats, poultry, and other foods may become contaminated when such water is used. Insects and rodents may also become contaminated with these organisms from polluted waters and disseminate them directly to either prepared or raw foods. Animal fecal matter is of greater importance than human, and it may be noted from Figure 18-1 that animal hides and poultry products may become contaminated from this source. *Salmonella* spp. are maintained within an animal population by means of nonsymptomatic animal infections, and by means of animal feeds, which have been shown by many investigators to harbor comparatively large numbers of these organisms. Both of these sources serve to keep slaughter animals reinfected in a cyclical manner.

Secondary contamination is another of the more important means of transmission of salmonellae to man. The presence of these organisms on eggs, meats, and in the air makes their presence in certain foods inevitable through the agency of handlers and direct contact of noncontaminated foods with contaminated foods *(19)*. Prost and Riemann *(41)* have pointed out that this secondary contamination may occur at various stages of food preparation, and is most frequently of animal origin, although human sources can be of some importance.

In view of the world-wide distribution of salmonellosis involving numerous serotypes and species from many different animals, the ultimate control of this problem consists of freeing animals and man of the organisms. This is obviously a difficult task. It is not impossible, however, when it is considered that only 33 of the more than 1800 species and serotypes account for almost 90% of human isolates and approximately 80% of nonhuman isolates *(29)*. The reinfection of animals through animal feeds can be controlled by treatment of feeds so that they are rendered salmonellae-free. One of the more promising ways of achieving the latter consists of heat treatments.

At the consumer level, the *Salmonella* carrier is presumed to play an important role, although just how important is not known. The proper cooking of vehicle foods, their proper handling and subsequent storage at temperatures below the growth range of these organisms will obviously do a great deal to lessen the incidence of salmonellosis at this level. As long as these organisms remain among the animal population, however, the potential hazards of salmonellosis in man will likewise remain.

ESCHERICHIA COLI FOOD POISONING

Reports incriminating *Escherichia coli* as the etiologic agent of food poisoning involving a variety of foods, such as cream pie, mashed potatoes, cream puffs, and creamed fish, date from around 1900 *(47)*. The etiologic status of *E. coli* as a food poisoning agent has been established. It appears that not all strains are equally capable of producing the food-poisoning syndrome but only enteropathogenic (EEC) strains. These strains are differentiated from the more normal *E. coli* strains in being more virulent and by reacting with *E. coli* OB and O antisera. The feeding of volunteers with serotypes 055:B5, 0111:B4, and 0127:B8 at levels of 10^6–10^8 organisms has been shown to produce food-poisoning symptoms *(13, 21, 51)*. In a study of the incidence of EEC serotypes among 219 food handlers, Hall and Hauser *(17)* found 14 (6.4%) who were carriers of a total of 8 different serotypes. Other investigators found the incidence of EEC in the feces of healthy adults and children to range from 1.8–15.1%. This is a higher carriage rate among humans than for the salmonellae.

In spite of the above, EEC strains were not given much attention as foodborne disease agents in the U.S. until an outbreak occurred in 1971 in which 387 persons became ill after consuming imported cheese. The cheese was imported from France and sold under several names (Brie, Camembert, Coulommiers), but all had been made in the same way. The etiologic agent of this outbreak was EEC 0124:B17 (28). The median time for onset of symptoms was 18 hr with a mean duration of 2 days. Common symptoms were: diarrhea, fever, and nausea, in order of decreasing frequency. Less common symptoms included cramps, chills, vomiting, aches, and headaches. Among those who ate the soft ripened cheese, the attack rate was 94%. The outbreak covered 14 U.S. states including the District of Columbia. While some of the cheese contained 10^3–10^5 coliforms/g, *Enterobacteriaceae* at levels of 10^6–10^7/g were found (36).

Enteropathogenic *E. coli* produces at least two enterotoxins. One is designated S(table) T(oxin) and the other L(abile) T(oxin). The ST will withstand 100 C for 15 min while LT is destroyed with less heat (43). LT is labile to mild acid while ST is resistant (12). The toxins may be assayed by the rabbit ligated ileal loop method where differentiations can be made between LT and ST. The latter gives the most immediate response with maximum fluid accumulation occurring between 4–6 hr (12). Maximum ligated ileal loop response for LT requires 16–18 hr. A 6-hr loop test for ST and an 18-hr test for LT enterotoxins have been recommended (32).

Foodborne EEC outbreaks for the years 1969–1972 have been tabulated by Mehlman *et al.* (32). In 1969 there were 5 outbreaks, 396 cases; in 1970 7 outbreaks, 1297 cases; in 1971 2 outbreaks, 395 cases; and in 1972 only 1 outbreak with 39 cases. The foods involved have included meats, fish, poultry, milk and dairy products, vegetables, baked products, rice, and so on. For the years 1969–1972, <2% of all foodborne outbreaks were attributed to EEC (32). In addition to conventional type foodborne outbreaks, these organisms have been implicated as etiologic agents in travelers' diarrhea.

Apparently prompted by the 1971 outbreak involving Camembert cheese, Park *et al.* (40) inoculated *ca.* 100/ml of EEC into pasteurized milk and used this milk to make Camembert cheese. Little growth of the organism occurred until after curd was cut and hooped. Cells reached populations >10^4/g in some cheese. Following storage at 10 C, the organisms died off rapidly in some products but remained in some up to 9 weeks. Attempts to define and isolate antienterotoxic factors from lactic acid bacteria were made recently (34). While some activity against EEC by *Lactobacillus bulgaricus* was noted, the cell-free substance proved difficult to isolate. It had a molecular weight of <10^3 and was unstable. No antienterotoxic factor could be found among other lactic acid bacteria.

The EEC enterotoxins apparently act by stimulating the release of adenyl cyclase activity in the intestinal tract. The increase in cyclic AMP leads to an increase in electrolyte secretion, which in turn leads to fluid accumulation, hence, profuse watery stools are a symptom of this syndrome. In this regard, the EEC syndrome resembles that of cholera.

The recovery of these organisms from foods requires special methods. The rapid lactose fermenters may be pre-enriched in MacConkey broth followed by inoculation into lauryl sulfate tryptose broth and incubation at

44 C. The slow lactose fermenters may be recovered by pre-enrichment in nutrient broth followed by inoculation into EE broth with incubation at 41.5 C *(31)*.

With respect to related organisms, foodborne outbreaks have been attributed to paracolons *(18)* and *Arizona* spp. *(50)*. The paracolons were distinguished by their delayed fermentation of lactose and this group is no longer recognized. It is possible that in all of these instances the etiologic agents were EEC strains.

VIBRIO PARAHAEMOLYTICUS FOOD POISONING

While all other known food poisoning syndromes may be contracted from a variety of foods, *V. parahaemolyticus* food poisoning is contracted almost solely from seafoods. When non-seafoods are involved, they represent cross contamination from seafood products. Another unique feature of this syndrome is the natural habitat of the etiologic agent—the sea.

The identity of this organism as a cause of food poisoning was made first by Fujino in 1951. It has been referred to as *Pasteurella parahaemolytica* and as *Pseudomonas enteritis,* but *V. parahaemolyticus* has been adopted by all authorities. It is a gram negative rod with a single polar flagellum, oxidase and catalase positive, capable of fermenting some sugars without gas, generally halophilic, and lysine and ornithine decarboxylase positive. In its native oceanic habit, the strains are highly variable. Its G + C content of DNA is 40–48%.

This organism grows well over the range 22–42 C with optimum growth temperatures of 35–37 C *(39)*. The lowest temperature at which growth has been recorded is between 3 and 13 C *(5)*. Growth has been demonstrated at 5 C in a laboratory medium while at 9.5–10 C growth has been demonstrated in food products. Maximum growth temperatures are between 42–44 C. Growth occurs over the pH range 5–11 with 7.1–7.7 being optimal and necessary for minimal temperature growth. The organism grows in the presence of 0.5–10% salt and dies off rapidly in distilled water *(5)*. Minimum a_w for growth is around 0.94 depending upon agent employed. Its sensitivity to heat is consistent with the low level found for most gram negative bacteria. $D_{47\ C}$ values ranging from 0.8 to 65.1 min have been reported *(6)*. Cells that are grown at high temperatures and heated in the presence of around 7.0% salt are the most heat resistant.

The *V. parahaemolyticus* syndrome occurs within 2–48 hr (12–24 more commonly) following the consumption of vehicle foods. The symptoms consist of diarrhea, abdominal pain, nausea, pyrexia, vomiting, and chills. The organisms are known to multiply in the gastrointestinal tract.

Not all *V. parahaemolyticus* isolates are capable of producing the food poisoning syndrome. In general, only Kanagawa positive strains are virulent, although some Kanagawa negative strains have been found to cause the syndrome *(3)*. The **Kanagawa phenomenon** is one whereby most strains isolated from patients (diarrheal stools) produce β-hemolysis on human red blood cells in Wagatsuma's agar medium while most marine isolates do not. To determine the Kanagawa reaction of a culture, a suspension of human red blood cells is added to a special agar base and allowed to harden. The culture is surface plated, incubated at 37 C for 18–24 hr, and read for the presence of β-hemolysis. Kanagawa positive cultures produce a heat-stable hemolysin which is toxic, but whether it is responsible for the food poisoning syndrome is not clear at this time. This hemolysin is a protein with a molecular weight of 45,000. A heat-labile hemolysin has been recovered also. Toxic culture filtrates are lethal to mice, induce a cutaneous response in guinea pigs, and cause fluid accumulation in ligated ileal loops of rabbits. Just why pathogenesis is associated with Kanagawa positive cultures is not entirely clear but a greater adherance of these strains to intestinal walls of victims may be partly responsible for their pathogenesis.

As noted above, the primary vehicle foods for *V. parahaemolyticus* food poisoning are seafoods such as oysters, shrimps, crabs, lobsters, clams, and so forth. While its incidence is quite low in the U.S. and some European countries, it is the leading cause of food poisoning in Japan *(14)*. Canada, Australia, and some European countries have remained free of the syndrome. From Table 17-2, it can be seen that for the yr 1973–1975, 3 outbreaks and 224 cases were reported to CDC.

The organism is common in oceanic and coastal waters. Its detection in ocean waters appears to be related to water temperature, with numbers of organisms being undetectable until water temperature rises to around 19–20 C. A study of the Rhode River area of the U.S. Chesapeake Bay showed that the organisms survive in sediment during the winter and later are released into the water column where they associate with the zooplankton from April to early June *(23)*. Their numbers become detectable in open waters with water temperatures of 19–20 C. In ocean waters, they tend to be associated more with shellfish than with other forms *(26)*.

Identification of *V. parahaemolyticus* by serological typing employing O and K antisera has been attempted with some success. Like other foodborne microorganisms, they are subject to environmental stresses and undergo metabolic injuries which must be dealt with in their recovery from foods (see Chapter 4).

The emergence of *V. parahaemolyticus* over the past 25 yr as the leading cause of food poisoning in at least one country and as an organism of concern in many others raises many questions. Where was it prior to the 1950s? Is its emergence a reflection of changes in food processing methods and/or our eating habits? Why are not marine isolates capable of inducing the syndrome in man? Are Kanagawa positive strains examples of host-induced changes in otherwise avirulent strains, or do they represent host-selected, naturally occurring virulent strains? While there is much speculation on these and other similar questions, few concrete answers are available at this time.

REFERENCES

1. **Adinarayanan, N., V. D. Foltz, and F. McKinley.** 1965. Incidence of Salmonellae in prepared and packaged foods. J. Inf. Dis. *115:* 19–26.

2. **Angelotti, R., M. J. Foter, and K. H. Lewis.** 1961. Time-temperature effects on salmonellae and staphylococci in foods. I. Behavior in refrigerated foods. II. Behavior at warm holding temperatures. Am. J. Pub. Hlth. *51:* 76–88.

3. **Barrow, G. I. and D. C. Miller.** 1976. *Vibrio parahaemolyticus* and seafoods, pp. 181–195, In: *Microbiology in Agriculture, Fisheries and Food,* edited by F. A. Skinner and J. G. Carr (Academic Press: N.Y.).

4. **Beloian, A. and G. C. Schlosser.** 1963. Adequacy of cooking procedures for the destruction of salmonellae. Am. J. Pub. Hlth. *53:* 782–791.

5. **Beuchat, L. R.** 1975. Environmental factors affecting survival and growth of *Vibrio parahaemolyticus.* A review. J. Milk Food Technol. *38:* 476–480.

6. **Beuchat, L. R. and R. E. Worthington.** 1976. Relationships between heat resistance and phospholipid fatty acid composition of *Vibrio parahaemolyticus.* Appl. Environ. Microbiol. *31:* 389–394.

7. **Brooks, J.** 1962. Alpha amylase in whole eggs and its sensitivity to pasteurization temperatures. J. Hyg. *60:* 145–151.

8. **Chung, K. C. and J. M. Goepfert.** 1970. Growth of *Salmonella* at low pH. J. Food Sci. *35:* 326–328.

9. **Clise, J. D. and E. E. Swecker.** 1965. Salmonellae from animal by-products. Pub. Hlth. Repts. *80:* 899–905.

10. **Committee on Salmonella.** 1969. *An Evaluation of the Salmonella Problem.* Pub. #1683, Nat'l. Acad. Sci. (Washington, D.C.).

11. **Duitschaever, C. L.** 1977. Incidence of *Salmonella* in retailed raw cut-up chicken. J. Food Protec. *40:* 191–192.

12. **Evans, D. G., D. J. Evans, Jr., and N. F. Pierce.** 1973. Differences in the response of rabbit small intestine to heat-labile and heat-stable enterotoxins of *Escherichia coli.* Infect. Immunity *7:* 873–880.

13. **Ferguson, W. W. and R. C. June.** 1952. Experiments on feeding adult volunteers with *Escherichia coli* 111:B4, a coliform organism associated with infant diarrhea. Am. J. Hyg. *55:* 155–169.

14. **Fujino, T., G. Sakaguchi, R. Sakazaki, and Y. Takeda.** 1974. *International Symposium on Vibrio parahaemolyticus.* (Saikon Publ. Co.: Tokyo).

15. **Goepfert, J. M. and R. A. Biggie.** 1968. Heat resistance of *Salmonella typhimurium* and *Salmonella senftenberg* 775W in milk chocolate. Appl. Microbiol. *16:* 1939–1940.

16. **Grau, F. H. and L. E. Brownlie.** 1968. Effect of some pre-slaughter treatments on the *Salmonella* population in the bovine rumen and feces. J. Appl. Bacteriol. *31:* 157–163.

17. **Hall, H. E. and G. H. Hauser.** 1966. Examination of feces from food handlers for salmonellae, shigellae, enteropathogenic *Escherichia coli,* and *Clostridium perfringens.* Appl. Microbiol. *14:* 928–933.

18. **Hobbs, B. C., M. E. M. Thomas, and J. Taylor.** 1949. School outbreak of gastroenteritis associated with pathogenic paracolon bacillus. Lancet *2:* 530–532.

19. **Hobbs, B. C.** 1961. Public health significance of *Salmonella* carriers in livestock and birds. J. Appl. Bacteriol. *24:* 340–352.

20. **Horwitz, M. A. and E. J. Gangarosa.** 1976. Foodborne disease outbreaks traced to poultry, United States, 1966–1974. J. Milk Food Technol. *39:* 859–863.

21. **June, R. C., W. W. Ferguson, and M. T. Waifel.** 1953. Experiments in feeding adult volunteers with *Escherichia coli* 055:B5, a coliform organism associated with infant diarrhea. Am. J. Hyg. *57:* 222–236.

22. **Kampelmacher, E. H.** 1963. The role of salmonellae in foodborne diseases. In: *Microbiological Quality of Foods.* L. W. Slanetz *et al.,* Editors, Academic Press, N.Y., pp. 84–101.

23. **Kaneko, T. and R. R. Colwell.** 1973. Ecology of *Vibrio parahaemolyticus* in Chesapeake Bay. J. Bacteriol. *113:* 24–32.

24. **Lerche, M.** 1961. Zur Lebenfahigkeit von Salmonellabakterien in Mayonnaise und Fleischsalat. Wien. tierarztl. Mschr. *6:* 348–361.

25. **Ley, F. J., B. M. Freeman, and B. C. Hobbs.** 1963. The use of gamma radiation for the elimination of salmonellae from various foods. J. Hyg. *61:* 515–529.

26. **Liston, J.** 1973. *Vibrio parahaemolyticus,* pp. 203–213, In: *Microbial Safety of Fishery Products,* edited by C. O. Chichester and H. D. Graham (Academic Press: N.Y.).

27. **Loken, K. I., K. H. Culbert, R. E. Solee, and B. S. Pomeroy.** 1968. Microbiological quality of protein feed supplements produced by rendering plants. Appl. Microbiol. *16:* 1002–1005.

28. **Marier, R., J. G. Wells, R. C. Swanson, W. Callahan, and I. J. Mehlman.** 1973. An outbreak of enteropathogenic *Escherichia coli* foodborne disease traced to imported French cheese. Lancet *2:* 1376–1378.

29. **Martin, W. J. and W. H. Ewing.** 1969. Prevalence of serotypes of *Salmonella.* Appl. Microbiol. *17:* 111–117.

30. **Matches, J. R. and J. Liston.** 1968. Low temperature growth of *Salmonella.* J. Food Sci. *33:* 641–645.

31. **Mehlman, I. J., N. T. Simon, A. C. Sanders, and J. C. Olson, Jr.** 1974. Problems in the recovery and identification of enteropathogenic *Escherichia coli* from foods. J. Milk Food Technol. *37:* 350–356.

32. **Mehlman, I. J., M. Fishbein, S. L. Gorbach, A. C. Sanders, E. L. Eide, and J. C. Olson, Jr.** 1976. Pathogenicity of *Escherichia coli* recovered from food. J. Assoc. Off. Anal. Chem. *59:* 67–80.

33. **Milone, N. A. and J. A. Watson.** 1970. Thermal inactivation of *Salmonella senftenberg* 775W in poultry meat. Hlth. Lab. Sci. *7:* 199–225.

34. **Mitchell, I. De G. and R. Kenworthy.** 1976. Investigations on a metabolite from *Lactobacillus bulgaricus* which neutralizes the effect of enterotoxin from *Escherichia coli* pathogenic for pigs. J. Appl. Bacteriol. *41:* 163–174.

35. **Morgan, H. R.** 1958. The *Salmonella.* In: *Bacterial and Mycotic Infections of Man.* R. J. Dubos, Editor, J. B. Lippincott & Company, Philadelphia, Chapter 16.

36. **Mossel, D. A. A.** 1974. Bacteriological safety of foods. Lancet *1:* 173.

37. **Moyle, A. I.** 1966. Salmonellae in rendering plant by-products. J. Am. Vet. Med. Assoc. *149:* 1172–1176.

38. **Ng, H., H. G. Bayne, and J. A. Garibaldi.** 1969. Heat resistance of *Salmonella:* The uniqueness of *Salmonella senftenberg* 775W. Appl. Microbiol. *17:* 78–82.

39. **Nickelson, R. and C. Vanderzant.** 1971. *Vibrio parahaemolyticus*—a review. J. Milk Food Technol. *34:* 447–452.

40. **Park, H. S., E. H. Marth, and N. F. Olson.** 1973. Fate of enteropathogenic strains of *Escherichia coli* during the manufacture and ripening of Camembert cheese. J. Milk Food Technol. *36:* 543–546.

41. **Prost, E. and H. Riemann.** 1967. Food-borne salmonellosis. Ann. Rev. Microbiol. *21:* 495–528.

42. **Rogers, R. E. and M. F. Gunderson.** 1958. Roasting of frozen stuffed turkeys. I. Survival of *Salmonella pullorum* in inoculated stuffing. Food Res. *23:* 87–95.

43. **Sack, R. B.** 1975. Human diarrheal disease caused by enterotoxigenic *Escherichia coli.* Ann. Rev. Microbiol. *29:* 333–353.

44. **Sadler, W. W. and R. E. Corstvet.** 1965. Second survey of market poultry for *Salmonella* infections. Appl. Microbiol. *13:* 348–351.

45. **Shrimpton, D. H., J. B. Monsey, B. C. Hobbs, and M. E. Smith.** 1962. A laboratory determination of the destruction of alpha amylase and salmonellae in whole egg by heat pasteurization. J. Hyg. *60:* 153–162.

46. **Steele, J. H. and M. M. Galton.** 1967. Epidemiology of foodborne salmonellosis. Hlth. Lab. Sci. *4:* 207–212.

47. **Tanner, F. W. and L. P. Tanner.** 1953. *Food-borne Infections and Intoxications.* The Garrard Press, Champaign, Ill., Chapter 14.

48. **Taylor, J.** 1962. *Salmonella* and salmonellosis In: *Food Poisoning.* Royal Soc. of Hlth. Publication, Great Britain, pp. 15–32.

49. **Troller, J. A.** 1976. *Salmonella* and *Shigella,* pp. 129–155, In: *Food Microbiology: Public Health and Spoilage Aspects,* edited by M. P. deFigueiredo and D. F. Splittstoesser (Avi Publ.: Westport, Conn.).

50. **Verder, E.** 1961. Bacteriological and serological studies of organisms of the *Arizona* group associated with a food-borne outbreak of gastroenteritis. J. Food Sci. *26:* 618–621.

51. **Wentworth, F. H., D. W. Broek, C. S. Stulberg, and R. H. Page.** 1956. Clinical bacteriological and serological observations of two human volunteers following ingestion of *E. coli* 0127:B8. Proc. Soc. Exptl. Biol. Med. *91:* 586–588.

52. **Woodward, W. E., R. A. Armstrong, and E. L. Gangarosa.** 1968. Food-borne disease surveillance in the United States—1967. Proc., 98th Ann. Meeting, Am. Pub. Hlth. Assoc., p. 170.

19

Biological Hazards in Foods Other Than Food Poisoning Bacteria

In addition to the food poisoning bacteria covered in the last three chapters, foods may also transmit to man other harmful organisms and toxins, among which are bacteria, viruses, fungi (and/or their toxic products), some parasitic protozoa, and various parasitic flat and round worms. Other biological hazards sometimes present in foods consist of various poisonous plants and animals which are sometimes eaten directly as food by man (e.g., Moray eel poisoning) or indirectly by consuming the flesh of animals that have in turn consumed the poisonous organisms, (e.g., paralytic shellfish poisoning). Some of these groups of organisms are discussed in the following sections.

FOOD-BORNE BACTERIA

A large number of bacterial infections may be contracted through various foods as may be noted from Table 19-1. Among these diseases are anthrax, brucellosis, tuberculosis, listeriosis, leptospirosis, pasteurellosis, typhoid and paratyphoid fevers, shigellosis, and erysipeloid. All of these infections differ from the food-poisoning syndromes presented in the three previous chapters in that they also may be contracted by means other than through the ingestion of foods. Since most are treated at length in standard textbooks of medical microbiology, only erysipeloid will be discussed here.

Erysipeloid is caused by *Erysipelothrix rhusiopathiae (insidiosa)*. They exist in decomposing animal matter and cause infections in man and other animals. These organisms are gram positive, nonsporulating rods that are catalase negative and nonmotile. While they are micro-aerophilic, they can grow under both aerobic and anaerobic conditions. They ferment glucose and lactose with the production of acid, but do not ferment maltose, sucrose, or mannitol. They have been reported to be killed by holding at 55 C for 15 min. They survive salting, pickling, and smoking and can survive in such treated meats for 1–3 mon *(54)*. These organisms have also been reported to

TABLE 19-1.   Animal diseases transmissible to man through various foods (56).*

Dolman's (1957) listing of animal diseases transmissible to man through meat

Zoonoses acquired occasionally through the intestinal tract
  Pasteurellosis
    Tularemia
    Pseudotuberculosis
    P. multocida infection
  Leptospirosis
  Erysipelas (Swine)
  Listerellosis
  Miscellaneous
    Foot and Mouth Disease
    Q Fever
    Ornithosis
    V. foetus infection
Commoner zoonoses acquired occupationally by meat handlers
  Anthrax
  Bovine Tuberculosis
  Brucellosis

Zoonoses acquired chiefly through the intestinal tract
  Salmonellosis
  Shigellosis
Meat-borne helminthic zoonoses
  Trichinosis
  Cysticercosis (Taeniasis)
  Echinococcosis[a]
Rare meat-borne zoonoses, possibly acquired by ingestion
  Toxoplasmosis
  Sarcosporidiosis
  Intestinal Myiasis

Kaplan et al. (1962) listing of animal diseases transmissible to man through milk

Anthrax[b]
Brucellosis
Coli infections (pathogenic strains of *E. coli*)
Foot and mouth disease
Leptospirosis[b]
Listeriosis[b]
Paratyphoid fever
Q Fever
Salmonellosis (other than typhoid and paratyphoid fevers)
Staphylococcal enterotoxic gastro enteritis
Streptococcal infections
Tick-borne encephalitis
Toxoplasmosis[b]
Tuberculosis

Galton and Arnstein' (1959) listing of animal diseases transmissible to man through poultry meat or eggs

Arthropod-borne encephalitides[c]
Brucellosis[c]
Erysipelas (swine)
Listeriosis[c]
Newcastle disease
Paracolon infections
Pasteurella multocida infection[c]

Pseudotuberculosis[c]
Psittacosis-ornithosis
Salmonellosis
Staphylococcal infections or intoxications
Tuberculosis
Toxoplasmosis[c]

[a]Indirect transmission only.
[b]Not conclusively incriminated as milk-borne, but epidemiologically probable or suspect.
[c]No proved transmission.
*Reproduced from *The Safety of Foods,* ed. H. D. Graham, with permission of The Avi Publishing Co., Inc., Westport, Conn.

survive from 4–5 days in drinking water and from 12–14 days in sewage. Their relevancy to food microbiology is due to their natural infections in at least 20 species of animals including man, swine, sheep, mice, cattle, horses, fish, and various species of fowl *(45)*. The infections they cause in man are of the localized cutaneous variety, known as Rosenbach's erysipeloid. The disease is primarily an occupational disease of workers processing or handling meat of animals susceptible to the disease *(56)*. The disease has been known in food handlers since the turn of the century. It may be contracted by workers from handling crabs, other shellfish, fish, poultry, and meats. The disease has also been seen in cooks, butchers, cheese handlers, and bone-button factory workers. It commonly appears as a mild cutaneous infection 1–7 days after inoculation, usually in a finger or a thumb. Although the infection is normally localized and generally mild, Klauder *et al. (38)* have described the case of a butcher who cut his finger while working and died 6 mon later from erysipeloid. There have been few authenticated reports of infection through the ingestion of infected meats.

An organism that may prove to be of significance as a foodborne pathogen is *Yersinia enterocolitica*. While the first human disease shown to be caused by this bacterium was recorded in 1939, it is now an important human pathogen having been shown to cause organ abscesses, septicemia, peritonitis, mesenteric lymphadenitis, and gastroenteritis. The latter is the most frequently caused syndrome of this organism in man *(62)*. Although it was implicated in intra-family gastroenteritis cases in the U.S. during the early 1970s, the first documented foodborne outbreak occurred during the fall of 1976 among school children in Oneida County, New York. The vehicle food was chocolate milk which was prepared by adding chocolate syrup to pasteurized milk in an open vat. The chocolate milk was packaged in small cartons without subsequent heat treatment. *Y. enterocolitica* serotype 8 was recovered from an unopened 8-oz carton of chocolate milk from one school and from stools of some victims. The symptoms consisted of abdominal pain, fever, and in some, diarrhea. A total of 218 children were affected among five county schools *(50)*.

This organism has been found in beef, oysters, mussels, and in river waters. Recently, *Y. enterocolitca*-like organisms were isolated from vacuum-packaged beef and lamb samples stored at 1–3 C. Ten samples of vacuum-packed beef and 2 of lamb yielded these organisms at 1–3 C after 21–35 days, but none of those incubated for 0, 7, or 14 days was positive *(27)*.

This organism is a psychrotroph, and is unusual as a pathogen in that low-temperature grown cells are more virulent than those grown at high temperatures. It grows slowly and produces very small colonies on media selective for the *Enterobacteriaceae* of which it is a member. The finding of Hanna *et al. (27)* noted above suggests that it cannot compete with other bacteria that are common on meats. Mice are susceptible to *Y. enterocolitica* where fatal infections may occur. Its growth curve in mice has been found to be similar to that of *S. enteritidis (7)*.

L-FORMS AND MYCOPLASMAS

Bacterial L-forms are varient forms of normal or classical bacteria which have lost most of their cell wall. They can be induced by exposing classical gram positive bacteria to antibiotics, such as penicillin, in the presence of osmotic stabilizers. The expressions "cell wall deficient form" and "spheroplasts" are used by some investigators synomously with L-form even though fine distinctions do exist (46). In the present context, "L-form" is used to designate all varients that lack substantial parts of their cell wall, require osmotic stabilizers for growth, display resistance to antibiotics to which the parent forms are sensitive, and produce a "fried egg" type colony on agar. A vast amount of L-form research has been done, and it appears that all bacterial genera and species may be induced to the L phase under appropriate cultural conditions.

For several reasons, the existence of L-forms in foods should be expected. In spite of this, there is a paucity of reports on their presence in foods. Apparently the first report was that of Brueckner and Sherman (4) who examined aseptically drawn milk from healthy cows and found the number of L-form colonies to reach 10^{12}/g (!) from a majority of the healthy cows examined. More recently, it was shown that L-phase enterotoxigenic S. aureus strains produced A, B, and C enterotoxins under experimental conditions. The L-form colonies were induced from classical strains of S. aureus by use of penicillin and methicillin (14). L-phase induction of S. aureus appears to be easier to effect than for S. epidermidis, as Dalton et al. (15) found that while 203 of 256 clinical isolates of S. aureus produced L-forms, none could be induced from 132 strains of S. epidermidis. Stable enterotoxin-producing L-forms of C. perfringens have been produced by employing penicillin (41). Since many gram negative bacteria are known to produce endotoxins in their L-phases (63), it appears that these forms of food poisoning bacteria may, indeed, exist in foods and not be detected by normal procedures of culturing for classical bacteria.

The mycoplasmas belong to the order Mycoplasmatales and four genera are recognized: Mycoplasma, Acholeplasma, Ureoplasma, and Thermoplasma. These are the smallest known living cells that can exist free of other living organisms. Formerly referred to as PPLO (pleuropneumonia-like organisms), the mycoplasmas are common in cattle and all other warm-blooded meat animals. While they lack a cell wall, some are known to live free in soils, and some are thermophiles. They are gram negative, and can produce extracellular products such as neurotoxins, hemolysins, and exoenzymes. They can withstand freezing and tolerate bile salts and methylene blue (42). Only one human disease is known to be caused by mycoplasmas—primary atypical pneumonia—and at least one species is pathogenic for poultry. The mycoplasmas are considered along with L-forms because of the difficulty of separating the two upon isolation and because of the similarity of growth of both types. A typical isolation medium for mycoplasmas is heart infusion agar supplemented with horse serum, yeast extract, DNA, and RNA. Typical *Mycoplasma* colonies display the fried-egg appearance with a central dense core embedded in the agar and a lighter peripheral zone on the agar surface. These organisms can pass a 220 nm filter and are resistant to some antibiotics *(61)*.

Due to the fact that neither L-forms nor mycoplasmas develop on the standard plating media employing standard plate count procedures, it is possible that one or both of these types is responsible for part of the 60% of food poisoning outbreaks for which classical etiologic agents are not found. Because of their gram negative properties, large numbers of mycoplasmas in foods could lead to gastroenteritis outbreaks.

FOOD-BORNE VIRUSES

The precise role that foods play in the transmission of viruses to man is poorly understood at this time. While a large number of viruses that may be transmitted by foods is known, the degree of occurrence or significance of such transmissions is not well understood. The lack of more information on the role of foods in the transmission of viruses is due in large part to the fact that the isolation and culturing of viruses from foods is much more difficult and time consuming than for bacteria. While proof is lacking, it may be assumed that viruses are at least as common in foods as are intestinal bacteria. The enteric virus family contains over 94 types consisting of the polio, Coxsackie groups A and B, ECHO, adeno, and infectious hepatitis viruses. Joseph *(36)* has pointed out that all of these groups possess the following characteristics which make them likely candidates for food dissemination: (1) they invade the body via the oral route, (2) they establish in the intestinal tract for short periods of time, (3) they are excreted in tremendous quantities in the feces, (4) they possess the fecal-oral dissemination pattern, (5) carriers and asymptomatic cases involving these viruses far exceed clinical cases, (6) they possess high resistance to physical and chemical environmental exposures, and (7) poor personal hygiene and poor sanitary conditions favor their dissemination. In spite of the fact that our knowledge of viruses as related to foods is not equal to that of food-borne bacteria, foods have been known for some time to serve as vehicles for at least two virus diseases of man—poliomyelitis and infectious hepatitis.

Ten outbreaks of polio of food origin which occurred between 1914 and 1949 have been summarized by Cliver *(11)*. Of these outbreaks, involving around 161 persons, the food vehicle appeared to be milk 8 times (pasteurized milk in 2 instances), and lemonade and cream-filled pastries one time each. Since the polio viruses are destroyed in the pasteurization of milk, the occurrence of these viruses in pasteurized milk may be taken to indicate either improper pasteurization or post-pasteurization contamination.

The virus disease which has been studied most in regard to food transmission is infectious hepatitis. In spite of the fact that the incubation period for the onset of symptoms is long (about 4 weeks), the characteristic symptoms of the disease allow for some epidemiological assessments of its occurrence and incidence. In examining reports covering a period of 20 yr, Cliver noted over 3,000 cases of food-borne infectious hepatitis involving a variety of foods and conditions. When shellfish were not involved, the vehicle foods consisted of raw milk, potato salad, cold meat cuts, custard, sandwiches, and so on, and in most instances the source of the virus was a food handler who either had infectious hepatitis at the time or had recently recovered from the disease. In the case of shellfish-borne hepatitis, the victims reported having eaten raw shellfish. These animals obtain the virus from their polluted water environment. One such outbreak involving oysters was reported by Mason and McLean *(44)*. At least 45 foodborne outbreaks of hepatitis have been confirmed *(12)*. In addition to the hepatitis virus, Metcalf and Stiles *(49)* isolated a Coxsackie and an ECHO virus from oysters taken from estuary waters. These viruses could be isolated from oysters up to 4 miles from the nearest raw sewage outlet. These authors also demonstrated the survival of viruses in oyster tissues stored at 5 C for at least 28 days. The capacity of oysters to take up and retain the polio virus was demonstrated by Hedstrom and Lycke *(29)*. These authors found the virus to be more stable in oyster tissue than in surrounding sea water. While oysters normally cleanse themselves of bacteria within 5 hr when transferred to fresh uninfected water, these investigators reported that the polio virus was still present in oyster tissues after 5 hr in uninfected water.

In addition to the viruses listed above, the virus of foot-and-mouth disease has been shown to survive for longer than 2 mon in infected cattle tissues in both cured and uncured meat *(13)*. The Newcastle disease virus causes mild infections in man as well as in chickens and turkeys. Poultry plant workers have been reported to contract the disease by splash infections of the eye *(28)*. Poultry plant workers have also been reported to contract a rickettsial infection, ornithosis (parrot fever), while handling infected turkeys *(17)*. Other virus diseases of animals transmissible to man through foods are presented in Table 19-1.

The true incidence of food-borne virus infections may be considerably higher than the data now indicate. The intestinal viruses described above are shed from the body in feces in the same way as coliforms, salmonellae, and *C. perfringens.* Although many apparently cannot survive for long periods in water and sewage or withstand adverse environmental conditions, some can and do. Their origin in foods would presumably be the same as that for enteric bacteria (see Chapter 5 for further discussion of viruses in foods).

MYCOTOXINS

The mycotoxins are poisonous substances produced by fungi. A survey of the literature reveals a long list of compounds produced by a large number of fungi toxic for one or more animal species. In lieu of an exhaustive list of all known mycotoxins, only the most important ones will be discussed in any detail.

1. Aflatoxins.   Knowledge of the aflatoxins dates from 1960 when more than 100,000 turkey poults died in England after eating peanut meal imported from Africa and South America. From the poisonous feed were isolated *Aspergillus flavus* and a toxin produced by this organism that was designated aflatoxin (*Aspergillus flavus* toxin—A-fla-toxin). Studies on the nature of the toxic substances revealed the 4 components presented below.

It was later determined that *A. parasiticus* also produces aflatoxins. The aflatoxins are highly substituted coumarins, and at least 8 closely related toxins are known at this time. In addition to the 4 noted above, M_1, M_2, B_{2a}, and G_{2a} are known. When chromatograms of aflatoxin-containing extracts are viewed under ultraviolet light, the individual toxins fluoresce as follows:

B_1 and B_2—blue
G_1 —green
G_2 —green-blue
M_1 —blue-violet
M_2 —violet

Toxins M_1 and M_2 are hydroxylated derivatives of B_1 and B_2 and are excreted in milk, urine, and feces as metabolic products (20). With respect to toxicity, B_1 is the most toxic with M_1 being almost as toxic. These are followed in decreasing order by G_1, B_2, M_2, and G_2 (1).

With respect to temperature of growth and aflatoxin production, a recent study by Schindler (58) employing 25 isolates of *A. flavus/parasiticus* on wort agar showed that no aflatoxins were produced at temperatures of 2, 7, 41, or 46 C within 8 days, or none <7.5 C or >40 C even under otherwise favorable conditions. The optimum temperature for aflatoxin production appears to be around 24–28 C. Several investigators have found that G_1 is produced at lower growth temperatures than B_1. While some have found more B_1 production than G_1 at around 30 C, others have found equal production in this temperature range. In fermented sausage, 10 times more G_1 as B_1 was found (39). Aflatoxin production has been demonstrated on fresh beef, ham, and bacon inoculated with toxigenic cultures and stored at 15, 20, and 30 C (2), and on country cured hams during aging when temperatures approached 30 C, but not at temperatures <15 C and r. h. <75% (3). At 25 C, 160 and 426 ppm of G_1 were produced in 10 and 18 days, respectively, in fermented sausage (39). Aflatoxin production has been shown to occur in whole-rye and whole-wheat breads, in tilsit cheese, and in apple juice at 22 C (24). Production has been demonstrated in the upper layer of 3-mon-old Cheddar cheese held at room temperature (40), and on brick cheese at 12.8 C by A. parasiticus after one week but not by A. flavus (60).

With respect to moisture, optimal a_w for aflatoxin production is 0.93–0.98 with limiting values being 0.71–0.94 (43). Limiting r. h. for aflatoxin production on peanuts was found to be 83% or higher at 30 C (18). Lower limiting temperature for peanuts was 11–12 C, while the upper was 40.5 C at 98% r. h. (18). Minimum pH for aflatoxin production is difficult to define. Production at pH 4.0 has been demonstrated and production no doubt occurs at lower pH values. It appears that aflatoxin production can occur under any and all conditions that allow for good fungal growth. Several chemically defined media have been developed that allow for toxin production, and these have been reviewed and discussed elsewhere (43). Levels of toxin up to 110 mg/liter on defined media and up to 350 μg/g on complex media have been produced.

The distribution of aflatoxin-producing fungi is widespread. A. flavus is one of the storage fungi (A. flavus-oryzae group) that develop on a wide variety of stored grains such as wheat, peanuts, soybeans, corn, and so forth. Toxin production has been demonstrated on an endless number of food products in addition to those noted above. Under optimal conditions of growth, some toxin can be detected within 24 hr, otherwise within 4–10 days (8).

Hesseltine et al. (30) have demonstrated the capacity of aflatoxin-producing fungi to grow and form toxins in rice, wheat, corn, oats, and on soybeans, while Wogan (67) reports that barley, peas, beans, cowpeas, cassava, sweet potatoes, and others have been found to contain these toxic compounds. In addition, aflatoxins have been found in nature in cottonseed meal (31). With respect to the growth and production of aflatoxin-producing organisms in peanuts, Hesseltine (31) has made the following observations: (1) Growth of the mold and the formation of aflatoxin occur mostly during the curing of the peanuts after removal from soil, (2) in a toxic lot of peanuts, only a comparatively few kernels contain toxin. Attempts to detect aflatoxin, therefore, depend on collecting a relatively large sample for assay, such as 1 kg. (3) The toxin will vary greatly in amount even within a single kernel, and (4) the two most important factors affecting aflatoxin formation are moisture and temperature. Products with moisture levels above 16% are capable of supporting *A. flavus* growth.

While unequivocable proof of aflatoxicosis in man is still wanting, circum-stantial evidence is strongly in favor. Among human conditions believed to result from aflatoxins is the EFDV syndrome of Thailand, Reye's syndrome of New Zealand *(5),* and an acute hepatoma of a Ugandan child. In the latter, a fatal case of acute hepatis disease revealed histological changes in the liver identical to those observed in monkeys treated with aflatoxins. According to Serck-Hanssen *(59),* an aflatoxin etiology was strongly suggested by the find-ings. Recently, two men who worked with purified aflatoxin developed colon carcinoma *(16).*

With respect to mode of action, B_1 is known to suppress DNA synthesis, cause mitotic arrest, and inhibit RNA polymerase *(20).* The carcinogenesis of these compounds is believed to result from their effects on nucleic acids in ways not yet fully understood.

The toxicity of aflatoxins has been shown for several different animal species (see Table 19-2). Young ducklings are among the most sensitive to the lethal effect of these compounds, followed by rats and others. Most species of susceptible animals die within 3 days after administration of toxins and show gross liver damage, which upon postmortem examination reveals the aflatoxins to be hepatocarcinogenic *(66).*

TABLE 19-2.   Comparative lethality of single doses of aflatoxin B_1 *(66).*

Animal	Age (or weight)	Sex	Route*	LD_{50}
				mg/kg
Duckling	1 day	M	PO	0.37
	1 day	M	PO	0.56
Rat	1 day	M-F	PO	1.0
	21 days	M	PO	5.5
	21 days	F	PO	7.4
	100 g	M	PO	7.2
	100 g	M	ip	6.0
	150 g	F	PO	17.9
Hamster	30 days	M	PO	10.2
Guinea pig	Adult	M	ip	ca. 1
Rabbit	Weanling	M-F	ip	ca. 0.5
Dog	Adult	M-F	ip	ca. 1
	Adult	M-F	PO	ca. 0.5
Trout	100 g	M-F	PO	ca. 0.5

*PO = oral; ip = intraperitoneal.

The toxicity of the aflatoxins is higher for young animals and males than for older animals and females. The toxic effects are enhanced by low protein or cirrhogenic diets.

Due to their chemical nature, aflatoxins are not destroyed by boiling or by many other simple means. Autoclaving for 4 hr has been reported to reduce but not destroy toxicity. Fishbach and Campbell *(23)* have recommended the use of 5% NaOCl, an oxidizing agent, as a means of disposing of contaminated materials and laboratory equipment. After screening about 1,000 microorganisms on their capacities to detoxify aflatoxins, Ciegler *et al. (10)* found that only *Flavobacterium aurantiacum* (NRRL B-184) could detoxify B_1 completely.

2. Other Mycotoxins.   According to Mossel *(52)*, at least 150 mycotoxigenic molds were known by mid 1974. Thirty-nine of these were *Aspergillus*, 37 *Penicillium*, and 15 *Fusarium* spp.

Some of the many known mycotoxins are carcinogenic as are the aflatoxins. Among these are sterigmatocystin (produced by *A. nidulans, A. versicolor,* and *Bipolaris* sp.), luteoskyrin (produced by *P. islandicum*), and patulin (produced by a large number of molds including *P. expansum* and *A. clavatus*). The latter toxin has been shown to be produced in experimentally infected McIntosh apples held at 25 C by 60 isolates of *P. expansum (65)*.

Among the many other mycotoxins either known or thought to be toxic to man are the fusarial toxins (produced by *F. poae, F. sporotrichoiodes*), stachybotryotoxin (by *Stachybotrys alternans* and others), trichothecenes (by *Myrothecium* and *Fusarium* spp.), and so on. These and others have been discussed more fully elsewhere *(8, 9, 35, 43, 53)*.

CESTODES

Of the flatworms that are parasitic for man, 3 are obtained by eating the flesh of pork, beef, and fish. The beef tapeworm, *Taenia saginata,* lives as an adult in the human intestine and in larval form in the muscles of bovines. The form of the organism found in bovine and porcine flesh is referred to as a **cysticerus.** They have been reported to occur in skeletal muscles, as well as in the muscles of the tongue, neck, jaws, heart, diaphragm, and esophagus where they may remain viable for as long as a yr *(33)*. Upon the ingestion of cysticercus-containing beef by man, the scolex of the parasite evaginates and attaches to the intestine where growth of the tapeworm begins. Development to adult stage requires from 8–10 weeks at which time **proglottids** may appear in the feces.

The pork tapeworm, *T. solium,* has a life cycle similar to that of *T. saginata,* but unlike the latter, man can serve as the intermediate host as well. The larval stage of *T. solium* is *Cysticercus cellulosae.* Human infestations from this form are referred to as **cysticercosis** *(51)*. Autoinfection in man with eggs from the adult worm reaching the upper intestines is not unknown.

Both the beef and pork tapeworms are distributed world-wide with the incidence being highest where raw or improperly cooked pork is eaten. In the United States the pork tapeworm has been reported to be rarer than the beef tapeworm.

Prevention of tapeworm infections in man requires the rejection of tapeworm-infested meats, which is one of the primary functions of federal and local meat inspection laws. If present in meats, the cysts of these parasites can be destroyed by cooking to a temperature of at least 60 C (33). These forms are destroyed upon the freezing of meats to at least −10 C with holding for 10–15 days, or by immersion in concentrated salt solutions for up to 3 weeks. The epidemiologic control of these infestations involves breaking the host cycle.

The third tapeworm which man acquires from food is the broad fish tapeworm, *Diphyllobothrium latum*. Man obtains this organism by ingesting fish such as pike, river perch, trout, and other fish which become infested upon the ingestion of **copepods,** in which the eggs of the parasite develop into **procercoid larva** measuring around 0.5 mm in length. The adult stage of this parasite develops in the intestinal tract of man from which eggs are shed through feces. Once in water, the eggs hatch and the first larval stage is taken up by certain copepods with a continuation of the cycle. In addition to man, fish-eating mammals such as dogs, bears, and others may become infested with this tapeworm. Like the other tapeworms, cooking for at least 10 min at 50–55 C is adequate to destroy the larval forms of this organism. Freezing to at least −10 C has also been reported to destroy these forms.

Hydatidosis is an infestation in man caused by ingesting the larval stage (or hydatid cyst) of two species of cestodes, *Echinococcus granulosis* and *E. multilocularis*. Unlike the tapeworm diseases, man becomes infested by these forms as a result of ingesting infective ova which are shed by dogs and certain other carnivores. The natural host of the adult parasite is the dog. Problems arise here only where dogs are allowed access to uncooked slaughter-house viscera, or by lack of proper hygiene of persons in close contact with dogs.

NEMATODES

Of the several species of nematodes pathogenic for man, the genus *Trichinella,* especially *T. spiralis,* is perhaps of greatest importance from the standpoint of foods. *T. spiralis* causes trichinosis in man, and Jacobs (33) has stated that this disease has been known for 300 yr. Unlike the tapeworm infections discussed above, both larval and adult stages of *T. spiralis* are passed in the same host. Upon the ingestion of encysted larvae in muscle tissue, adults develop in the intestine, discharge larvae into the blood, and these larvae again settle in striated muscles and again become encysted. The cycle is repeated if another animal consumes these muscles containing encysted larvae. In addition to man and swine, other animals such as dogs, cats, and bear may be involved. Man usually contracts trichinosis by eating infected pork. The disease can be prevented by the thorough cooking of pork. A temperature of 58.3 C in the center of the product has been recommended as minimum treatment for the destruction of these forms. In a study by Carlin *et al. (6)* on the heat destruction of trichina larvae in pork roasts, all roasts cooked to an internal temperature of 140 F or higher were subsequently found to be free of these organisms. These authors found larvae in all roasts cooked at 130 F or lower and in some roasts cooked at 135 F. The encysted forms may be destroyed by freezing, but freezing times and temperatures are dependent upon the thickness of the product and range from -15 C for 20 days, -23.3 C for 10 days, or -28.9 C for 6 days for pieces 6 in or less in thickness (U.S. Dept. of Agriculture Regulation of 1947). Progressively longer holding times are needed for larger pieces of meat. It has been shown, however, that the destruction of trichina larvae in infected pork can be achieved in shorter holding times as related to temperature of freezing and freezing time as follows (Gould and Kaasa, *26,* courtesy of The American Journal of Epidemiology):

Freezing temperatures		Freezing time
°F	°C	
-16.6	-27	36 hours
-22	-30	24 hours
-27.4	-33	10 hours
-31	-35	40 minutes
-34.6	-37	2 minutes

The encysted larvae have been reported to be destroyed by irradiation somewhere between $15,000-20,000\,r$.

The effect of curing and smoking on the viability of trichina in pork hams and shoulders was investigated by Gammon *et al. (25)*. These investigators employed the meat of hogs experimentally infested with *T. spiralis* as weanling pigs. After curing, the meat was hung for 30 days followed by smoking for approximately 24 hr at 90–100 F with subsequent aging. Live trichinae were found in both hams and shoulders 3 weeks after smoking, but none could be detected after 4 weeks.

Trichinosis can be controlled by avoiding the feeding of infected meat scraps to swine and preventing the consumption of infested tissues by other animals. The feeding of uncooked garbage to swine helps to perpetuate this disease. Where only cooked garbage is fed to pigs, the incidence of trichinosis has been shown to fall sharply. The decreasing incidence of trichina larvae in pork products may be seen from data obtained by Zimmerman *et al. (68)* from commercial pork available in Iowa. In 1944–46, 12.4% of bulk link sausage and 11.7% of processed link pork sausage contained this parasite; in the period 1953–1960, the incidence had fallen to 1, 2.4, and 0.2%, respectively, for fresh bulk, fresh link and processed link sausage samples. The latter findings were based upon 8,402, 1,432, and 861 samples of fresh bulk, fresh link, and processed link sausage samples, respectively.

The incidence of trichinosis in the U.S. has declined since 1947 when the disease first became reportable to an annual mean of 133 cases for the period 1965–1974. In 1975, however, 284 cases were reported with pork from domestic swine accounting for 73 percent of the cases with identified sources. Of the 161 cases involving pork, sausage was responsible for 149 while meat from wild animals accounted for 33 and ground beef which had been adulterated with pork for 34 cases. Studies on the presence of pork in retail ground beef have revealed that from 3 to 38 percent of beef samples contained pork. The presence of pork in ground beef may be deliberate on the part of some stores, or the result of using the same grinder for both products. The increasing consumption of ground beef along with the fact that it is often not well cooked offers the potential for a continuing increase in the incidence of trichinosis unless the adulteration of beef with pork is reduced.

TREMATODES

A large number of trematodes or flukes cause disease in man but only one species will be mentioned here, *Clonorchis sinensis*. This organism, which is an important pathogen in Asia, carries out its adult stage in the bile ducts and gall bladder of mammals including man; its larval stage is found in some 40 species of fresh-water fish. Man contracts the organisms by eating one of the susceptible fish. Fish become reinfected when human feces or those of dogs, cats, hogs, and others are deposited in water. Like the cestode infestations, the encysted larval forms in fish flesh can be destroyed by heating. A temperature of at least 50 C for 15 min has been recommended to destroy these forms. Unlike the cestode infestations, these forms have been reported to resist drying and salting *(33)*.

PROTOZOA

The only protozoan disease of any known importance in food microbiology is **toxoplasmosis,** caused by *Toxoplasma gondii*. This organism has gained importance over the past 25 yr, during which time it has been shown to cause a variety of diseases in man. The organism has been reported to multiply in many different cells of the human body. In chronic infections, it is found encysted in the brain and other tissues, especially the skeletal muscles. The cyst consists of many parasites packed together within a cyst wall. The disease is somewhat similar to rubella (German measles) in that during pregnancy it may cause serious damage to the fetus. In adults, symptoms consist of fever with rash, headache, muscle aches and pain, and swelling of the lymph nodes. The muscle pain, which is rather severe, may last up to a month or more. While the incidence is not known, estimates place the number of worldwide cases at about 1 million. The disease is known to be transmitted by congenital means, and evidence suggests the role of certain nematode ova in its transmission *(47)*. Cats that are fed mice chronically infected with toxoplasmosis have been shown to excrete *T. gondii* in their feces for 7 days or longer, and up to 21–24 days after eating infected cat feces *(19)*.

The importance of this disease to food microbiology is based upon the belief that infected meat serves as a source of human toxoplasmosis. Serologic studies show that a number of individuals have antibodies to the organism. In their study of 1,191 individuals, Feldman and Miller *(21)* found that 426 or 36% had toxoplasma antibodies and 23% or 801 animals had antibodies to this organism. In another study of U.S. Army recruits ranging in age from 17–26, Feldman *(22)* found that 13% were positive for toxoplasma antibodies. Pork and mutton have been reported to show a high rate of infection by this organism with beef being rare although reported. The first demonstration of this organism in slaughter animals was made by Jacobs *et al. (32)* who showed that 24, 1.7, and 9.3% of diaphragm muscles of swine, beef cattle, and sheep, respectively, were infected with toxoplasma cysts. Jacobs *(33)* reports that it is similar to the *Trichinella* larvae with respect to heat, freezing, and irradiation destruction. The encysted parasites are scattered and too small to be seen with the unaided eye, thus making their detection by visual inspection all but impossible. For further information on this parasite relative to foods, see Jacobs *(34)*.

PARALYTIC SHELLFISH POISONING

This condition is contracted by eating oysters, mussels, or clams. Symptoms usually develop within 30 min after the ingestion of susceptible mollusks and death within 3–12 hr. The symptoms are characterized by paresthesia (tingling, numbness, or burning) which begins about the mouth, lips, and tongue and later spreads over the face, scalp, neck, and to the finger tips and toes. The disease is contracted from eating mussels, clams, or oysters which have fed upon certain dinoflagellates of which *Gonyaulax catenella* is representative of the U.S. Pacific Coast flora. Along the north Atlantic coast of the U.S. and over to northern Europe *G. tamarensis* is found, and its poison is more toxic than that of *G. catenella*. Along the coast of British Columbia *G. acatenella* is found. The toxin has been assigned an empirical formula of $C_{10}H_{17}N_7O_4 \cdot 2HCl$, and exerts its effect in man through cardiovascular collapse and respiratory failure *(64)*. A fatal dose of the poison for man, which may be obtained from a single serving of highly toxic shellfish, is 0.54–0.9 mg *(57)*. The toxin acts by blocking the propagation of nerve impulses without depolarization *(57)*. There is no known effective antidote. From 1793 to 1958, some 792 cases were recorded with 173 (22%) deaths *(48)*. Paralytic shellfish or "mussel poisoning" in man has a mortality rate variously reported to range from 1–22%. The dinoflagellates from which mollusks and other organisms obtain the toxic principle of paralytic shellfish poisoning have a wide distribution in marine waters. Masses or blooms of toxic dinoflagellates give rise to **"red tide"** or other related conditions of seas. Prevention of this type of food hazard consists of avoiding seafoods from waters laden with toxic dinoflagellates. The toxin can be reduced by heating above 100 C. Thorough cooking may effect a reduction of 70% of the poison in meat.

CIGUATERA POISONING

According to Wills *(64),* ciguatera poisoning was first reported from the West Indies in 1555. It is contracted by man upon the ingestion of any one of approximately 300 species of fishes such as barracudas, groupers, sea basses, jacks, sharks, eels, and others *(55).* Almost all fishes involved in ciguatera poisoning are reef or shore species which become toxic by feeding upon herbivorous fishes, which in turn feed upon toxic algae or other toxophoric matter present in coralline reefs or from related areas. Russell has pointed out that the flesh of source fishes is less toxic than the viscera, with the liver being the most poisonous part of the fish. The toxin has been assigned an empirical formula of $C_{35}H_{65}NO_8$, and acts by causing respiratory paralysis in man. The mortality rate has been reported to vary from 2–7%. Symptoms of ciguatera poisoning develop within 4 hr following ingestion of the offending fish and consist of nausea, paresthesia about the mouth, tongue, throat, and sometimes over the face and distal parts of the fingers and toes. Weakness, abdominal pain, vomiting, diarrhea, and chills are often experienced *(55).*

OTHER POISONOUS FISHES

In addition to ciguatera poisoning, numerous other types of fish poisoning have been reported dating back to the Fifth Dynasty of the ancient Egyptians *(55).* Of these many types, only 2 others will be mentioned—**puffer fish** and **Moray eel poisoning.** Puffer fish (tetraodon, fugu) poisoning may be obtained from at least 50 of the approximately 100 species of puffer fish scattered among 10 genera. The puffers are widely distributed and contain the most lethal of all fish poisons, having a mortality rate of 60–70% associated with their ingestion by man. The responsible toxin has been purified and assigned an empirical formula of $C_{11}H_{17}N_3O_8$. Except in Asia, these fishes are rarely used for food. Further information may be obtained from Kao *(37)* and Wills *(64).*

Moray eel poisoning is caused by eating the Moray eel. This poisoning syndrome is much like ciguatera poisoning. It has a mortality rate of 10% and is restricted to those parts of the world where the Moray eel is eaten for food. Death is thought to be due to paralysis of the diaphragm.

REFERENCES

1. **Ayres, J. C.** 1973. Aflatoxins as contaminants of foods, fish, and foods, pp. 261–272. In: *Microbial Safety of Fishery Products,* edited by C. O. Chichester and H. D. Graham (Academic Press: N.Y.).

2. **Bullerman, L. B., P. A. Hartman, and J. C. Ayres.** 1969. Aflatoxin production in meats. I. Stored meats. Appl. Microbiol. *18:* 714–717.

3. **Bullerman, L. B., P. A. Hartman, and J. C. Ayres.** 1969. Aflatoxin production in meats. II. Aged dry salamis and aged country cured hams. Appl. Microbiol. *18:* 718–722.

4. **Brueckner, H. J. and J. M. Sherman.** 1932. Primitive or filtrable forms of bacteria and their occurrence in aseptic-milk. J. Inf. Dis. *51:* 1–16.

5. **Butler, W. H.** 1974. Aflatoxin, pp. 1–28, In: *Mycotoxins,* edited by I. F. H. Purchase (Elsevier: N.Y.).

6. **Carlin, A. F., C. Mott, D. Cash, and W. Zimmerman.** 1969. Destruction of trichina larvae in cooked pork roasts. J. Food Sci. *34:* 210–212.

7. **Carter, P. B. and F. M. Collins.** 1974. Experimental *Yersinia enterocolitica* infection in mice: Kinetics of growth. Infect. Immunity *9:* 851–857.

8. **Christensen, C. M.** 1971. Mycotoxins. CRC Crit. Rev. Environ. Cont. *2:* 57–80.

9. **Christensen, C. M.** 1975. *Molds, Mushrooms, and Mycotoxins* (U. Minn. Press).

10. **Ciegler, A., E. B. Lillehob, R. E. Peterson, and H. H. Hall.** 1966. Microbial detoxification of aflatoxin. Appl. Microbiol. *14:* 934–939.

11. **Cliver, D. O.** 1967. Food-associated viruses. Hlth. Lab. Sci. *4:* 213–221.

12. **Cliver, D. O.** 1976. Viruses, pp. 257–270, In: *Food Microbiology: Public Health and Spoilage Aspects,* edited by M. P. deFigueiredo and D. F. Splittstoesser (Avi Publ.: Westport, Conn.).

13. **Cox, B. F., G. E. Cottral, and D. E. Baldwin.** 1961. Further studies on the survival of foot-and-mouth disease virus in meat. Am. J. Vet. Res. *22:* 224–226.

14. **Czop, J. K. and M. S. Bergdoll.** 1970. Synthesis of enterotoxin by L-forms of *Staphylococcus aureus.* Infect. Immunity *1:* 169–173.

15. **Dalton, H. P., M. R. Escobar, and M. J. Allison.** 1971. Correlation of staphylococcal bacteriophage types to L-form colony production. Infect. Immunity *3:* 774–776.

16. **Deger, G. E.** 1976. Aflatoxin—human colon carcinogenesis? Ann. Intern. Med. *85:* 204.

17. **Delaplane, J. P.** 1958. Ornithosis in domestic fowl: Newer findings in turkeys. Ann. N.Y. Acad. Sci. *70:* 495–500.

18. **Diener, U. L. and N. D. Davis.** 1969. Production of aflatoxin on peanuts under controlled environments. J. Stored Prod. Res. *5:* 251–258.

19. **Dubey, J. P., N. L. Miller, and J. K. Frenkel.** 1970. Characterization of the new fecal form of *Toxoplasma gondii.* J. Parasitol. *56:* 447–456.

20. **Enomoto, M. and M. Saito.** 1972. Carcinogens produced by fungi. Ann. Rev. Microbiol. *26:* 279–312.

21. **Feldman, H. A. and L. T. Miller.** 1956. Serological study of toxoplasmosis prevalence. Am. J. Hyg. *64:* 320–335.

22. **Feldman, H. A.** 1963. A nationwide serum survey of United States military recruits, 1962. *Toxoplasma* antibodies. Am. J. Epidemiol. *81:* 385–391.

23. **Fischbach, H. and A. D. Campbell.** 1965. Detoxification of the alfatoxins. J. Assoc. Off. Agric. Chemists *48:* 28–30.

24. **Frank, H. K.** 1968. Diffusion of aflatoxins in foodstuffs. J. Food Sci. *33:* 98–100.

25. **Gammon, D. L., J. D. Kemp, J. M. Edney, and W. Y. Varney.** 1968. Salt, moisture and aging time effects on the viability of *Trichinella spiralis* in pork hams and shoulders. J. Food Sci. *33:* 417–419.

26. **Gould, S. E. and L. J. Kaasa.** 1949. Low temperature treatment of pork: Effect of certain low temperatures on viability of trichina larvae. Am. J. Hyg. *49:* 17–24.

27. **Hanna, M. O., D. L. Zink, Z. L. Carpenter, and C. Vanderzant.** 1976. *Yersinia enterocolitica*-like organisms from vacuum-packaged beef and lamb. J. Food Sci. *41:* 1254–1256.

28. **Hanson, R. P. and C. A. Brandly.** 1958. Newcastle disease. Ann. N.Y. Acad. Sci. *70:* 585–597.

29. **Hedstrom, C. E. and E. Lycke.** 1964. An experimental study of oysters as virus carriers. Am. J. Hyg. *79:* 134–142.

30. **Hesseltine, C. W., O. L. Shotwell, J. J. Ellis, and R. D. Stubblefield.** 1966. Investigation of aflatoxin formation by *Aspergillus flavus*. Bacteriol. Rev. *30:* 795–805.

31. **Hesseltine, C. W.** 1967. Aflatoxins and other mycotoxins. Hlth. Lab. Sci. *4:* 222–228.

32. **Jacobs, L., J. S. Remington, and M. L. Melton.** 1960. A survey of meat samples from swine, cattle, and sheep for the presence of encysted *Toxoplasma*. J. Parasitol. *46:* 23–28.

33. **Jacobs, L.** 1962. Parasites in food. In: *Chemical and Biological Hazards in Food*. J. C. Ayres *et al.,* Editors, Iowa State Univ. Press, pp. 248–266.

34. **Jacobs, L.** 1967. *Toxoplasma* and toxoplasmosis. Adv. in Parasitol. *5:* 1–45.

35. **Jarvis, B.** 1976. Mycotoxins in food, pp. 251–267, In: *Microbiology in Agriculture, Fisheries and Food,* edited by F. A. Skinner and J. G. Carr (Academic Press: N.Y.).

36. **Joseph, J. M.** 1965. Virus diseases transmitted through foods. Assoc. Food & Drug Off. U.S., Quarterly Bull. *29:* 10–15.

37. **Kao, C. Y.** 1966. Tetrodotoxin, sanitoxin, and their significance in the study of excitation phenomena. Pharmacol. Rev. *18:* 997–1049.

38. **Klauder, J. V., D. W. Kramer, and L. Nicholas.** 1943. *Erysipelothrix rhusiopathiae* septicemia: Diagnosis and treatment. J. Am. Med. Assoc. *122:* 938–943.

39. **Leistner, L. and F. Tauchmann.** 1970. Aflatoxinbildung in Rohwurst durch verschiedene *Aspergillus flavus*-Stämme und einer *Aspergillus parasiticus*-Stamm. Fleischwirtschaft *50:* 965–966.

40. **Lie, J. L. and E. H. Marth.** 1967. Formation of aflatoxin in Cheddar cheese by *Aspergillus flavus* and *Aspergillus parasiticus*. J. Dairy Sci. *50:* 1708–1710.

41. **Mahony, D. E.** 1977. Stable L-forms of *Clostridium perfringens:* Growth, toxin production, and pathogenicity. Infect. Immunity *15:* 19–25.

42. **Maniloff, J. and H. J. Morowitz.** 1972. Cell biology of the mycoplasmas. Bacteriol. Rev. *36:* 263–290.

43. **Marth, E. H. and B. G. Calanog.** 1976. Toxigenic fungi, pp. 210–256, In: *Food Microbiology: Public Health and Spoilage Aspects,* edited by M. P. de-Figueiredo and D. F. Splittstoesser (Avi Publ.: Westport, Conn.).

44. **Mason, J. O. and W. R. McLean.** 1962. Infections hepatitis traced to the consumption of raw oysters. An epidemiologic study. Am. J. Hyg. *75:* 90–111.

45. **Mattman, L. H.** 1968. Personal communication.

46. **Mattman, L. H.** 1974. *Cell Wall Deficient Forms,* Chapter 2 (CRC Press: Ohio).

47. **McCulloch, W. F., R. P. Crawford, and R. L. Hoff.** 1968. Ecologic aspects of toxoplasmosis. Proc., 98th Ann. Meeting, Am. Pub. Hlth. Assoc., p. 238.

48. **McFarren, E. F., M. L. Schafer, J. E. Campbell, K. H. Lewis, E. T. Jensen, and E. J. Schantz.** 1960. Public health significance of paralytic shellfish poison. Adv. Food Res. *10:* 135–179.

49. **Metcalf, T. G. and W. C. Stiles.** 1965. The accumulation of enteric viruses by the oyster, *Crassostrea virginica.* J. Inf. Dis. *115:* 68–76.

50. **Morbidity and Mortality Weekly Report.** 1977. *Yersinia enterocolitica* outbreak—New York. *26:* No. 7, 2/18/77 (U.S. Dept. Hlth. Educ. Wlfre).

51. **Morgan, P. M.** 1968. Meat animal parasites and the importance of an effective and standardized meat inspection system. In: *The Safety of Foods.* H. D. Graham *et al.,* Editors, Avi Publishing Co., Westport, Conn., Chapter 16.

52. **Mossel, D. A. A.** 1975. Occurrence, prevention, and monitoring of microbial quality loss of foods and dairy products. CRC Crit. Rev. Environ. Cont. *5:* 1–140.

53. **Purchase, I. F. H., editor.** 1974. *Mycotoxins* (Elsevier: N.Y.).

54. **Reed, R. W.** 1958. *Listeria* and *Erysipelothrix.* In: *Bacterial and Mycotic Infections of Man,* 3rd Edition. R. J. Dubos, Editor, J. B. Lippincott Co., Philadelphia, Chapter 20.

55. **Russell, F. E.** 1968. Poisonous marine animals. In: *The Safety of Foods.* H. D. Graham *et al.,* Editors, Avi Publishing Co., Westport, Conn., Chapter 14.

56. **Sadler, W. W.** 1968. Food-borne diseases of animal origin. In: *The Safety of Foods.* H. D. Graham *et al.,* Editors, Avi Publishing Co., Westport, Conn., Chapter 25.

57. **Schantz, E. J.** 1973. Some toxins occurring naturally in marine organisms, pp. 151–162, In: *Microbial Safety of Fishery Products,* edited by C. O. Chichester and H. D. Graham (Academic Press: N.Y.).

58. **Schindler, A. F.** 1977. Temperature limits for production of aflatoxin by twenty-five isolates of *Aspergillus flavus* and *Aspergillus parasiticus.* J. Food Protec. *40:* 39–40.

59. **Serck-Hanssen, A.** 1970. Aflatoxin-induced fatal hepatitis? Arch. Environ. Hlth. *20:* 729–731.

60. **Shih, C. N. and E. H. Marth.** 1972. Experimental production of aflatoxin on brick cheese. J. Milk Food Technol. *35:* 585–587.

61. **Stanbridge, E. J.** 1976. A reevaluation of the role of mycoplasmas in human disease. Ann. Rev. Microbiol. *30:* 169–187.

62. **Toma, S. and L. LaFleur.** 1974. Survey of the incidence of *Yersinia enterocolitica* infection in Canada. Appl. Microbiol. *28:* 469–473.

63. **Weibull, C., W. D. Bickel, W. T. Haskins, K. C. Milner, and E. Ribi.** 1967. Chemical, biological, and structural properties of stable *Proteus* L forms and their parent bacteria. J. Bacteriol. *93:* 1143–1159.

64. **Wills, J. H., Jr.** 1966. Seafood toxins. In: *Toxicants Occurring Naturally in Foods.* Pub. # 1354, Nat'l. Acad. Sci., Washington, D.C., pp. 147–163.

65. **Wilson, D. M. and G. J. Nuovo.** 1973. Patulin production in apples decayed by *Penicillium expansum.* Appl. Microbiol. *26:* 124–125.

66. **Wogan, G. N.** 1966. Chemical nature and biological effects of the aflatoxins. Bacteriol. Rev. *30:* 460–470.

67. **Wogan, G. N.** 1968. Aflatoxin risks and control measures. Fed. Proc. *27:* 932–938.

68. **Zimmermann, W. J., L. H. Schwarte, and H. E. Biester.** 1961. On the occurrence of *Trichinella spiralis* in pork sausage available in Iowa (1953–1960). J. Parasitol. *47:* 429–432.

20

Characteristics and Growth of Psychrophilic/Psychrotrophic Microorganisms

Psychrophilic microorganisms have been defined as those organisms capable of growing at temperatures between 0–7 C and producing visible colonies within 7 days. This definition conforms with that suggested by Ingraham and Stokes *(27)* and Stokes *(56)*. The former report suggested that growth at 0 C within 2 weeks be employed to define psychrophiles, while the latter report has suggested growth at this temperature within 1 week. Although most all microorganisms that grow at and below this temperature range are commonly referred to as psychrophiles, some authors employ the use of optimum temperatures of growth as the basis for this classification. Among the optimum temperature ranges suggested by various authors for psychrophiles are the following: 10–20, 12–18, 15–25, 18–20, and 20–40 C. While many authors have suggested temperature optima below 20 C, only recently have organisms been isolated and studied that actually have optimum growth temperatures this low. In an effort to give more precision to the term psychrophile, Eddy *(10)* and Mossel and Zwart *(42)* have suggested that the term "psychrotrophic" (*psychros,* cold, and *trephein* to nourish upon or to develop) be applied to those organisms able to grow at 5 C and below, without regard to their optimum temperatures. These authors further suggested that the word psychrophile be defined according to optimum growth temperature without reference to minimum temperatures. In spite of the apparent soundness of this proposal, most investigators continue to describe low-temperature growing microorganisms as psychrophiles regardless of optimum temperature.

Recently, Morita *(41)* suggested that the term psychrophile be applied only to those organisms whose optimum and maximum temperatures of growth are about 15 and 20 C respectively. Those organisms that have optimum and maximum growth temperatures above 20 C, but which can grow at refrigerator temperatures, are psychrotrophs after Eddy as noted above. Most investigators have continued to employ the term psychrophile even though the organisms under study probably fit Eddy's definition of a psychrotroph better than that of Morita's for a psychrophile. By these definitions, it is unlikely that one would encounter psychrophiles in meats, poultry, and vegetable products. The arctic and oceanic environments are the primary sources of psychrophiles even though psychrotrophs as well as mesophiles exist in the same environments.

Because of the difficulty of determining whether organisms employed by some workers were psychrotrophs or psychrophiles, the terminology employed by each author is employed in the remainder of this chapter.

With respect to the distribution of psychrophilic bacteria among the various genera, Farrell and Rose *(13)* have called attention to the fact that the preponderant types are gram-negative rods, largely pseudomonads. These authors listed the following genera in which psychrophilic strains are fairly common: *Pseudomonas, Flavobacterium, Alcaligenes,* and *Arthrobacter.* Less frequently, but nevertheless regularly isolated, are psychrophilic strains of the genera *Escherichia, Aeromonas, Serratia, Proteus, Acinetobacter, Enterobacter, Chromobacterium, Vibrio, Clostridium, Citrobacter, Salmonella, Shigella, Hafnia,* and *Bacillus (13, 42, 67).* Occasional reports of a less definitive nature on the isolation of psychrophilic strains of *Corynebacterium, Lactobacillus,* and *Micrococcus* have been made. As far as low temperature food preservation is concerned, the genus *Pseudomonas* represents by far the most important psychrophilic bacteria. Psychrophilic strains of yeasts have been reported from the genera *Candida, Cryptococcus, Rhodotorula,* and *Torulopsis,* with *Candida* strains being the more common.

Since psychrophilic microorganisms are so important in the low-temperature preservation of foods, it is highly desirable to understand the basic mechanism by which these organisms are able to grow at such low temperatures. This information could lead to the preservation of foods by specifically altering or controlling the molecular mechanisms that allow for low temperature growth and activity. While the basic mechanisms that underlie psychrophilism are not yet well understood, most of what is presently known about this phenomenon can be grouped into three broad categories: (1) Temperature-induced changes in the production of metabolic end products, (2) temperature effects on physiologic mechanisms, and (3) the low heat resistance of psychrophilic microorganisms. Each of these categories is treated below.

TEMPERATURE-INDUCED CHANGES

There are at least 4 temperature-induced changes known to occur in psychrophilic microorganisms that have been studied.

1. There is a greater increase in the proportion of unsaturated fatty acid residues in the lipids of psychrophiles than in those of mesophiles. The usual lipid content of most bacteria is between 2–5%, most or all of which is in the cell membrane. Bacterial fats are glycerol esters of two types: (a) neutral lipids in which all 3 or only 1 or 2 of the —OH groups of glycerol are esterified with long-chain fatty acids, and (b) phospholipids in which one of the —OH groups is linked through a phosphodiester bond to choline, ethanolamine, glycerol, inositol, or serine, and the other two —OH groups are esterified with long-chain fatty acids *(50)*. It has been shown by various investigators that most psychrophiles synthesize neutral lipids and phospholipids containing an increased proportion of unsaturated fatty acids when grown at low temperatures as compared with growth at higher temperatures. As much as a 50% increase in content of unsaturated bonds in fatty acids from mesophilic and psychrophilic *Candida* spp. grown at 10 compared to 25 C was reported by Kates and Baxter *(33)* with no effect on the phospholipid composition of the yeasts. Studies on a psychrophilic (psychrotrophic) *Candida* by McMurrough and Rose *(39)* substantiated the findings of others that unsaturated fatty acids increased with decreasing growth temperatures (Table 20-1). Linolenic acid increased at the expense of oleic acid at the lower temperatures.

The widespread occurrence of low-temperature-induced changes in fatty acid composition suggests that they are associated with physiological mechanisms of the cell. It is known that an increase in the degree of unsaturation of fatty acids in lipids causes a decrease in lipid melting point. It has been suggested that increased synthesis of unsaturated fatty acids at low temperatures has the function of maintaining the lipid in a liquid and mobile state, thereby allowing membrane activity to continue to function. This concept is referred to as the **"lipid solidification"** theory and was first proposed by Gaughran *(16)* and Allen *(3)*. It has been shown by Byrne and Chapman *(8)* that the melting point of fatty acid side chains in lipids is more important than the entire lipid structure.

Although full support for the lipid solidification idea is wanting, there is circumstantial evidence available such as the phenomenon of **"cold shock,"** which is the dying off of many cells of mesophilic bacteria upon the sudden chilling of a suspension of viable cells grown at mesophilic temperatures. It

TABLE 20-1.   **Effects of incubation temperature on the fatty acid composition of stationary cultures of Candida utilis *(39)*.**

Incubation temperature (C)	Cell concn (mg/ml)	Fatty acid composition[a]				
		16:0	16:1	18:1	18:2	18:3
30	2.0	18.9	4.6	39.1	34.3	2.1
20	2.0	20.3	11.4	31.6	27.7	6.1
10	2.0	27.4	20.6	20.7	17.6	10.7
5	1.7	19.2	15.9	18.2	16.3	27.3

[a]Values quoted are expressed as percentages of the total fatty acids. Fatty acids are designated x:y, where x is the number of carbon atoms and y is the number of double bonds per molecule.

has been shown for a large number of gram negative bacteria including *E. coli* and is generally a property of gram negative bacteria and not of gram positives. Cold shock has been shown to be accompanied by the release of certain low molecular weight cell constituents, an effect which presumably occurs by virtue of damage to the plasma membrane. According to Rose *(50)*, cold shock seems to result from a sudden release of cell constituents from bacteria following the "freezing" of certain membrane lipids after sudden chilling with consequent development of "holes" in the membrane. To support this hypothesis, Farrell and Rose *(14)* grew a mesophilic strain of *Ps. aeruginosa* at 30 C and showed that the cells were susceptable to cold shock while the same strain grown at 10 C was not susceptible.

2. Psychrophiles display a greater synthesis of polysaccharides than do mesophiles. Well-known examples of this effect include the production of ropy milk and ropy dough, both of which are favored by low temperatures. The production of extracellular dextrans by *Leuconostoc* and *Pediococcus* spp. are known also to be favored at temperatures below the growth optima of these organisms. The greater production of dextran at lower temperatures is apparently due to the fact that dextransucrase is very rapidly inactivated at temperatures in excess of 30 C *(45)*. A temperature-sensitive dextransucrase synthesizing system has been shown also for a *Lactobacillus* spp. *(9)*.

From a practical standpoint, increased polysaccharide synthesis at low temperatures manifests itself in the characteristic appearance of low-temperature spoiled meats. As discussed in Chapter 7, slime formation is characteristic of the bacterial spoilage of frankfurters, fresh poultry, and ground beef. The coalescence of surface colonies leads to the sliminess of such meats and no doubt contributes to the increased hydrating capacity that accompanies low-temperature meat spoilage.

3. Pigment-producing microorganisms produce more pigment under psychrophilic conditions than under mesophilic. This effect seems to be confined to those organisms that synthesize phenazine and carotenoid pigments. The best documented example of this phenomenon involves pigment production by *S. marcescens*. According to Williams *et al.* *(66)*, this organism possesses an abnormally heat-sensitive enzyme which catalyzes the coupling of a monopyrrole and a bipyrolle precursor to give prodigiosin (the red pigment). The increased production of pigments at sub-optimum temperatures has been reported by other authors *(62, 66)*. It is interesting that a very large number of marine psychrophiles are pigmented. This is true for bacteria as well as yeasts. On the other hand, none of the more commonly studied thermophiles are pigmented.

4. Under psychrophilic conditions, some organisms show a differential attack rate on certain metabolizable substrates. Greene and Jezeski *(18)* reported that sugar fermentation at temperatures below 30 C gave rise to both acid and gas, while above 30 C only acid was produced. Similarly, Upadhyay and Stokes *(64)* studies a psychrophile which fermented glucose and other sugars with the formation of acid and gas at 20 C and lower, but produced only acid at higher temperatures. This difference was ascribed to a temperature-sensitive formic hydrogenase system. These investigators *(65)* studied a similar effect and attributed the difference to a temperature-sensitive hydrogenase synthesizing system of the cell. Beef spoilage bacteria have been shown to liquefy gelatin and utilize water-soluble beef proteins more at 5 C than at 30 C *(31)*, but whether this effect is due to temperature-sensitive enzymes is not yet known.

THE EFFECT OF LOW TEMPERATURES ON MICROBIAL PHYSIOLOGIC MECHANISMS

Of the effects that low incubation temperatures have on the growth and activity of food-borne microorganisms, the 5 that have received the most study are presented below.

1. Psychrophiles have a slower metabolic rate than mesophiles. This is one of the best known effects of low temperatures on microorganisms, and as pointed out in Chapter 11, it is the rationale behind the low-temperature preservation of foods. The effect of temperature on the generation time of a psychrophile (pseudomonad) and a mesophile *(E. coli)* is presented in Figure 20-1, where the curve of the psychrophile is shifted to a lower range by 10

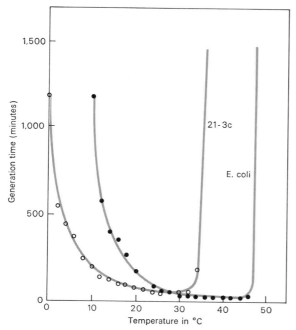

FIGURE 20-1.   The effect of temperature on generation time of a psychrophilic pseudomonad (21-3c) and the mesophile *E. coli*, K-12 *(28)*.

degrees or so. An Arrhenius plot of these two organisms (Figure 20-2) shows the general shape of the two curves to be similar *(28)*. The striking difference between the two curves, however, is the difference in the slope of the linear region where the **temperature characteristic** (μ) of growth rate is about 28,000 cal/mole for the mesophile but only 18,000 cal/mole for the psychrophile. It should be noted here that an Arrhenius plot relates rate of enzyme reaction on a logarithmic scale to the reciprocal of the absolute temperature, while μ (and Q_{10}) are measures of the rate of decrease or rate of a given process with temperature. In general, as incubation temperatures are lowered, μ increases.

Although the practice of employing μ values as a means of differentiating between mesophiles and thermophiles is long standing, Shaw *(52)*, Hanus and Morita *(22)*, and others have questioned the appropriateness of this practice. Shaw has stated that psychrophiles cannot be distinguished from mesophiles on the basis of μ values which he found to be about the same for both types of organisms—about 12,000 cal/mole. Hanus and Morita *(22)* have agreed with this notion and have stated that it is difficult to say that psychrophiles have a lower μ value than do mesophiles. In the more traditional sense, the deviation of the psychrophile curve from linearity as seen in Figure 20-2 is common for this group of organisms, and though rather difficult to interpret in kinetic terms, may be explained in part by the various

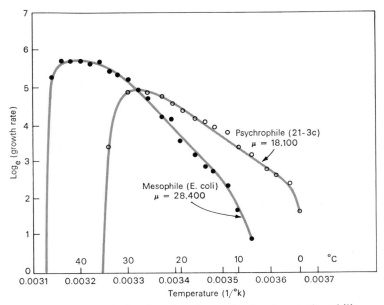

FIGURE 20-2.   An Arrhenius plot of growth rate of a psychrophilic pseudomonad (21-3c) as compared with that of the mesophile *E. coli*, K-12 (after Ingraham, *28*, with corrected μ as suggested by Hanus and Morita, *22*).

parameters that affect growth. Other investigators have produced results that vary relative to the utility of μ and these have been reviewed and discussed elsewhere (30, 41, 51).

The precise reasons why metabolic rates are slowed at low temperatures are not fully understood. As noted above, psychrophilic growth decreases more slowly than that of mesophilic with decreasing temperatures. The temperature coefficients (Q_{10} values) for various substrates such as acetate and glucose have been shown by several investigators to be lower for growing psychrophiles than for mesophiles. The end products of mesophilic and psychrophilic metabolism of glucose were shown by Brown (5) and Ingraham and Bailey (26) to be the same, with the differences largely disappearing when the cells were broken. In other words, the temperature coefficients are about the same for psychrophiles and mesophiles when cell-free extracts are employed.

As temperature is decreased, the rate of protein synthesis is known to decrease, and this occurs in the absence of changes in the amount of cellular DNA. One reason for this may be the increase in intramolecular hydrogen bonding that occurs at low temperatures, leading to increased folding of enzymes with losses in catalytic activity (34). On the other hand, the decrease in protein synthesis appears to be related to a decreased synthesis of individual enzymes at low growth temperatures. Although the precise mechanism of reduced protein synthesis is not yet well understood, Marr et al. (37) have suggested that low temperatures affect the synthesis of a repressor protein, and that the repressor protein itself is thermolabile (61). Several investigators have also suggested that low temperatures may influence the fidelity of the translation of mRNA during protein synthesis. For example, in their studies on E. coli, Goldstein et al. (17) showed that a leucine-starved auxotroph of this mesophile incorporated radioactive leucine into protein at 0 C. These investigators suggested that at this temperature all essential steps in protein synthesis apparently go on, and involve a wide variety of proteins. The rate of synthesis at 0 C was estimated to be about 350 times slower than at 37 C for this organism.

Whatever the specific mechanism of lowered metabolic activity of microorganisms as growth temperature is decreased, psychrophiles growing at low temperature have been shown to possess good enzymatic activity, since motility, endospore formation, and endospore germination all occur at 0 C (57). Alford and Pierce (2) have shown that Ps. fragi along with several organisms produced lipases within 2–4 days at −7 C, within 7 days at −18 C, and within 3 weeks at −29 C. Stokes (57) has pointed out that minimum growth temperatures may be determined by the structure of the enzymes and cell membrane as well as by enzyme synthesis. Lack of production of enzymes at high temperatures by psychrophiles, on the other hand, is due apparently to the inactive nature of enzyme-synthesizing reactions rather than to enzyme inactivation (57), although the latter is known to occur (see below). With respect to individual groups of enzymes, Peterson and Gunderson *(47)* showed that yields of endocellular proteolytic enzymes were greater in *Ps. fluorescens* grown at 10 than at either 20 or 35 C, while Nashif and Nelson *(43, 44)* have shown that *Ps. fragi* preferentially produces lipase at low temperatures, with none being produced at 30 C or higher. Alford and Elliott *(1)* found that *Ps. fluorescens* produced just as much lipase at 5 C as at 20 C, but produced only a slight amount at 30 C. On the other hand, Hurley *et al. (25)* found that a proteolytic enzyme system of *Ps. fluorescens* showed more activity on egg white and hemoglobin at 25 C than at 15 and 5 C.

2. Psychrophiles display a greater transport of solutes across the cell membrane than do mesophiles. Several investigators have shown that upon lowering the growth temperature of mesophiles, within the psychrophilic range, solute uptake is decreased. Studies by Baxter and Gibbons *(4)* indicate that minimum growth temperature of mesophiles is determined by the temperature at which transport permeases are inactivated. Farrell and Rose *(13)* have offered three basic mechanisms by which low temperature could affect solute uptake: (a) inactivation of individual permease-proteins at low temperature as a result of low-temperature induced conformational changes which have been shown to occur in some proteins, (b) changes in the molecular architecture of the cytoplasmic membrane which prevent permease action, and (c) a shortage of energy required for the active transport of solutes. Although it is not clear at this time as to what the precise mechanism of reduced uptake of solutes at low temperatures is, (b) above seems most likely, according to Farrell and Rose.

On the other hand, psychrophiles appear to be more efficient in transporting solutes across their cytoplasmic membranes. As stated above, psychrophiles tend to possess in their membranes lipids that decrease the melting point. The greater mobility of the psychrophile membranes may be expected to facilitate membrane transport at low temperatures. In addition, the transport permeases of psychrophiles are apparently more operative under these conditions than those of mesophiles. Whatever the specific mechanism of increased transport might be, it has been demonstrated that psychrophiles are more efficient than mesophiles in the uptake of solutes at low temperatures. Baxter and Gibbons *(4)* showed that a psychrophilic *Candida* sp. incorporated glucosamine more rapidly than a mesophilic *Candida*. The psychrophile transported glucosamine at 0 C while scarcely any was transported by the mesophile at this temperature or even at 10 C.

3. Some microorganisms produce larger cells when growing under psychrophilic conditions. Yeasts and fungi have been reported to produce larger cell sizes when growing under psychrophilic conditions than when growing under mesophilic. This observation has not been reported for bacteria. Employing *Candida utilis,* Rose *(50)* found that it increased in cell size upon the lowering of growth temperature, and he suggested that the increase in size was a consequence of the lowering of growth temperature, which might be correlated with increases in RNA and protein content of the cells.

Low-temperature induced synthesis of additional RNA has been reported by several investigators (6, 23, 60). However, Frank *et al.* (15) found no increase in the amount of RNA at 2 C when *Pseudomonas* 92 cells were grown at 2 and 30 C at equivalent physiological stages. Also, no increases in cell size, protein content, or catalase activity could be found by these workers.

4. Psychrophiles are more efficient producers of flagella than mesophiles. It has been reported that flagella production is often favored at low temperatures though not at higher temperatures. Examples of this phenomenon include *E. coli, B. inconstans, S. paratyphi B,* and others, including some psychrophiles. The opposite effect is also known for some psychrophiles.

5. Psychrophiles are more favorably affected by aeration than mesophiles. The effect of aeration on the generation time of *Ps. fluorescens* at temperatures from 4 to 32 C employing three different carbon sources is presented in Table 20-2. It may be noted that the greatest effect of aeration (shaking) occurred at 4 and 10 C while at 32 C the aerated cultures produced a longer generation time (46). The significance of this effect is not clear at this time. In a study of facultatively anaerobic psychrophiles under anaerobic conditions, Upadhyay and Stokes (63) showed that these organisms grew more slowly, survived longer, died more rapidly at higher temperatures, and produced lower maximal cell yields under anaerobic conditions than under aerobic. It has been commonly observed that plate counts on many foods are higher when incubated at low temperatures than at temperatures of 30 C and above. The generally higher cell yields that are obtained at lower temperatures are due to the increased solubility and, consequently, availability of O_2 (53). The latter authors have found that equally high cell yields can be obtained at both low and high incubation temperatures when O_2 is not limiting. This greater availability of O_2 in refrigerated foods undoubtedly exerts selectivity on the spoilage flora of such foods. The vast majority of psychrophilic bacteria studied are aerobes or facultative anaerobes, and these are the types that are associated with the spoilage of

TABLE 20-2. Effect of growth temperature, carbon source, and aeration on generation times (hr) of *Pseudomonas fluorescens* (46).

Growth medium*	Culture	Growth temp					
		4°C	10°C	15°C	20°C	25°C	32°C
Glucose	Stationary	8.20	3.52	2.02	1.47	0.97	1.19
	Aerated	5.54	2.61	2.00	1.46	0.93	1.51
Citrate	Stationary	8.20	3.46	2.00	1.43	1.01	1.24
	Aerated	6.68	2.95	2.02	1.26	0.98	1.45
Casamino Acids	Stationary	7.55	3.06	1.78	1.36	1.12	0.95
	Aerated	4.17	2.57	1.56	1.12	0.87	1.10

*Basal salts + 0.02% yeast extract + the carbon source indicated.

refrigerated foods. Few anaerobic psychrophiles have been isolated and studied. One of the first to isolate and study such organisms was McBryde *(38)* who studied *Clostridium putrefaciens.* More recently, Sinclair and Stokes *(54)* and Roberts and Hobbs *(49)* have reported the isolation of psychrophilic clostridia.

NATURE OF THE LOW HEAT RESISTANCE OF PSYCHROPHILES

It has been known for many years that psychrophilic microorganisms are generally unable to grow much above 30–35 C. Among the first to suggest reasons for this limitation of growth were Edwards and Rettger *(11),* who concluded that the maximum growth temperatures of bacteria may bear a definite relationship to the minimum temperatures of destruction of respiratory enzymes. The conclusion of these authors has been borne out by results from a large number of investigators over the past few years. It has been shown that many respiratory enzymes are inactivated at the temperatures of maximum growth of various psychrophilic types (Table 20-3). Thus, the thermal sensitivity of certain enzymes of psychrophiles is at least one of the factors that limit the growth of these organisms to low temperatures.

It has also been shown that when some psychrophiles are subjected to temperatures above their growth maxima, cell death is accompanied by the leakage of various intracellular constituents *(20, 21, 59).* The leakage substances have been shown to consist of proteins, DNA, RNA, free amino acids, and lipid phosphorus. The latter substance was thought by Hagen *et al.* to represent phosphorus of the cytoplasmic membrane. While the specific reasons for the release of cell constituents is not fully understood, it would appear to involve rupture of the cell membrane. These events appear to follow those of enzyme inactivation cited above.

Whatever is the true mechanism of psychrophile death at temperatures a few degrees above their growth maxima, their destruction at these relatively low temperatures is characteristic of this group of organisms. This is especially true of those that have optimum growth temperatures at and below 20 C. Reports by several investigators on psychrophiles isolated and studied over the past 9 yr reveal that all are capable of growing at 0 C with growth optima at either 15 or between 20–25 C, and growth maxima from 20–35 C *(20, 21, 35, 54, 55, 58).* These organisms include gram negative rods, gram positive aerobic and anaerobic rods, sporeformers and nonsporeformers, gram positive cocci, vibrios, and yeasts. One of these, *Vibrio fisheri (marinus),* was shown by Morita and Albright *(40)* to have an optimum growth temperature at 15 C and a generation time of 80.7 min at this temperature. In almost all cases, the growth maxima of these organisms was only 5–10 degrees above the growth optima. The *Enterobacteriaceae* isolated by Mossel and Zwart *(42)* appear to be exceptions to the foregoing.

TABLE 20-3. Some heat-labile enzymes of psychrophilic microorganisms (as reported by various investigators).

Enzyme	Organism	Temp. of maximum growth, C	Temp. of enzyme inactivation, C	References
Extracellular lipase*	Ps. fragi		30	(44)
α-oxoglutarate synthesizing enzymes and others	Cryptococcus	~28	30	(19)
Alcohol dehydrogenase	Candida sp.	<30		(4)
Formic hydrogenlyase	Psychrophile 82	35	45	(64)
Hydrogenase	Psychrophile 82	35	>20	(65)
Malic dehydrogenase	Marine Vibrio	30	30	(7)
Pyruvate dehydrogenase	Candida sp.	~20	25	(12)
Isocitrate dehydrogenase	Arthrobacter sp.	~35	37	(12)
Fermentative enzymes	Candida sp. P16	~25	35	(55)
Reduced NAD oxidase	Psychrophile 82	35	46	(48)
Cytochrome c reductase	Psychrophile 82	35	46	(48)
Lactic and glycerol dehydrogenase	Psychrophile 82	35	46	(48)
Pyruvate clastic enzymes	Psychrophile 82	35	46	(48)
Protein and RNA synthesizing	Micrococcus cryophilus	25	30	(36)

*Enzyme-forming system inactivated.

425

These organisms were shown to be capable of growing at 6 C and were designated by these authors as psychrotrophs. They differed from the more classical psychrophilic types in being able to grow at 41–42 C.

SUMMARY

The precise mechanisms of psychrophilism/psychrotrophism are yet to be completely understood. It appears that the ability to grow at low temperatures is under genetic control. The conversion of mesophiles to psychrophiles/psychrotrophs through transduction and ultraviolet light has been effected as well as the reverse process by use of UV light *(41)*. Psychrophilic strains or species of genera that also include mesophiles may then be viewed as having adapted themselves to growth under such conditions through mutations and low temperature selection. From information available at this time, the most basic differences between mesophiles and psychrophiles are to be found in the membrane lipids, in permease-proteins that affect cellular transport, and the general lability of psychrophile enzymes 5–10 degrees above the growth optimum. Also, the temperature characteristic (μ) of growth rate is generally thought to be different for psychrophiles and mesophiles although this notion has recently been questioned. Various authors have made suggestions as to what determines the minimum growth temperature of microorganisms. The findings of Harder and Veldkamp *(24)* and Strange and Shon *(59)*, employing obligately psychrophilic *Pseudomonas* spp., suggest that cessation of RNA synthesis may be the controlling factor. Jezeski and Olsen *(32)* have suggested that there are preformed elements in microbial cells grown at any temperature which are selectively temperature sensitive. Ingraham and Maaløe *(29)* have pointed out that the ability of bacteria to grow at different temperatures probably gives them a selective advantage. These authors further believe that bacteria cease to grow at a certain low temperature because of excessive sensitivity in one or several control mechanisms, the effectors of which cannot be supplied in the growth medium. They further indicated that interaction between effector molecules and the corresponding allosteric proteins may be expected to be a strong function of temperature.

Overall, psychrophiles may be characterized as having enzymes and enzyme-forming systems that differ from other microorganisms in being able to function at low temperatures and being simultaneously susceptible to heat inactivation at relatively moderate temperatures; in having cell membranes that permit the transport of substrates at low temperatures; and in possessing lipids that do not solidify at the lowest temperatures of growth.

REFERENCES

1. **Alford, J. A. and L. E. Elliott.** 1960. Lipolytic activity of microorganisms at low and intermediate temperatures. I. Action of *Pseudomonas fluorescens* on lard. Food Res. *25:* 296–303.

2. **Alford, J. A. and D. A. Pierce.** 1961. Lipolytic activity of microorganisms at low and intermediate temperatures. III. Activity of microbial lipases at temperatures below 0°C. J. Food Sci. *26:* 518–524.

3. **Allen, M. B.** 1953. The thermophilic aerobic sporeforming bacteria. Bacteriol. Rev. *17:* 125–173.

4. **Baxter, R. M. and N. E. Gibbons.** 1962. Observations on the physiology of psychrophilism in a yeast. Can. J. Microbiol. *8:* 511–517.

5. **Brown, A. D.** 1957. Some general properties of a psychrophilic pseudomonad: The effect of temperature on some of these properties and the utilization of glucose by this organism and *Pseudomonas aeruginosa*. J. Gen. Microbiol. *17:* 640–648.

6. **Brown, C. M. and A. H. Rose.** 1969. Effects of temperature on composition and cell volume of *Candida utilis*. J. Bacteriol. *97:* 261–272.

7. **Burton, S. D. and R. Y. Morita.** 1963. Denaturation and renaturation of malic dehydrogenase in a cell-free extract from a marine psychrophile. J. Bacteriol. *86:* 1019–1024.

8. **Byrne, P. and D. Chapman.** 1964. Liquid crystalline nature of phospholipids. Nature *202:* 987–988.

9. **Dunican, L. K. and H. W. Seeley.** 1963. Temperature-sensitive dextransucrease synthesis by a lactobacillus. J. Bacteriol. *86:* 1079–1083.

10. **Eddy, B. P.** 1960. The use and meaning of the term "psychrophilic." J. Appl. Bacteriol. *23:* 189–190.

11. **Edwards, O. F. and L. F. Rettger.** 1937. The relation of certain respiratory enzymes to the maximum growth temperatures of bacteria. J. Bacteriol. *34:* 489–515.

12. **Evison, L. M. and A. H. Rose.** 1965. A comparative study on the biochemical bases of the maximum temperatures for growth of three psychrophilic microorganisms. J. Gen. Microbiol. *40:* 349–364.

13. **Farrell, J. and A. Rose.** 1967. Temperature effects on micro-organisms. Ann. Rev. Microbiol. *21:* 101–120.

14. **Farrell, J. and A. H. Rose.** 1968. Cold shock in mesophilic and psychophilic pseudomonads. J. Gen. Microbiol. *50:* 429–439.

15. **Frank, H. A., A. Reid, L. M. Santo, N. A. Lum, and S. T. Sandler.** 1972. Similarity in several properties of psychrophilic bacteria grown at low and moderate temperatures. Appl. Microbiol. *24:* 571–574.

16. **Gaughran, E. R. L.** 1947. The thermophilic micro-organisms. Bacteriol. Rev. *11:* 189–225.

17. **Goldstein, A., D. B. Goldstein, and L. I. Lowney.** 1964. Protein synthesis at O C in *Escherichia coli,*. J. Mol. Biol. *9:* 213–235.

18. **Greene, V. W. and J. J. Jezeski.** 1954. The influence of temperature on the development of several psychrophilic bacteria of dairy origin. Appl. Microbiol. *2:* 110–117.

19. **Hagen, P.-O. and A. H. Rose.** 1962. Studies on the biochemical basis of low maximum temperature in a psychophilic *Cryptococcus*. J. Gen. Microbiol. *27:* 89–99.

20. **Hagen, P.-O., D. J. Kushner, and N. E. Gibbons.** 1964. Temperature-induced death and lysis in a psychrophilic bacterium. Can. J. Microbiol. *10:* 813–322.

21. **Haight, R. D. and R. Y. Morita.** 1966. Thermally induced leakage from *Vibrio marinus,* an obligately psychrophilic marine bacterium. J. Bacteriol. *92:* 1388–1393.

22. **Hanus, F. J. and R. Y. Morita.** 1968. Significance of the temperature characteristic of growth. J. Bacteriol. *95:* 736–737.

23. **Harder, W. and H. Veldkamp.** 1967. A continuous culture study of an obligately psychrophilic *Pseudomonas* species. Arch. Mikrobiol. *59:* 123–130.

24. **Harder, W. and H. Veldkamp.** 1968. Physiology of an obligately psychrophilic marine *Pseudomonas* species. J. Appl. Bact. *31:* 12–23.

25. **Hurley, W. C., F. A. Gardner, and C. Vanderzant.** 1963. Some characteristics of a proteolytic enzyme system of *Pseudomonas fluorescens.* J. Food Sci. *28:* 47–54.

26. **Ingraham, J. L. and G. F. Bailey.** 1959. Comparative study of effect of temperature on metabolism of psychrophilic and mesophilic bacteria. J. Bacteriol. *77:* 609–613.

27. **Ingraham, J. L. and J. L. Stokes.** 1959. Psychrophilic bacteria. Bacteriol. Rev. *23:* 97–108.

28. **Ingraham, J. L.** 1962. Newer concepts of psychrophilic bacteria. In: *Proceedings Low Temperature Microbiology Symposium—1961.* Campbell Soup Co., Camden, N.J., pp. 41–62.

29. **Ingraham, J. L. and O. Maalϕe.** 1967. Cold-sensitive mutants and the minimum temperature of growth of bacteria. In: *Molecular Mechanisms of Temperature Adaptation.* C. L. Prosser, Editor., Pub. # 84, A.A.A.S., pp. 297–309.

30. **Inniss, W. E.** 1975. Interaction of temperature and psychrophilic microorganisms. Ann. Rev. Microbiol. *29:* 445–465.

31. **Jay, J. M.** 1967. Nature, characteristics, and proteolytic properties of beef spoilage bacteria at low and high temperatures. Appl. Microbiol. *15:* 943–944.

32. **Jezeski, J. J. and R. H. Olsen.** 1962. The activity of enzymes at low temperatures. In: *Proceedings Low Temperature Microbiology Symposium—1961.* Campbell Soup Co., Camden, N.J., pp. 139–155.

33. **Kates, M. and R. M. Baxter.** 1962. Lipid comparison of mesophilic and psychrophilic yeasts (*Candida* species) as influenced by environmental temperature. Can. J. Biochem. Physiol. *40:* 1213–1227.

34. **Kavanau, J. L.** 1950. Enzyme kinetics and the rate of biological processes. J. Gen. Physiol. *34:* 193–209.

35. **Larkin, J. M. and J. L. Stokes.** 1966. Isolation of psychrophilic species of *Bacillus.* J. Bacteriol. *91:* 1667–1671.

36. **Malcolm, N. L.** 1968. Synthesis of protein and ribonucleic acid in a psychrophile at normal and restrictive growth temperatures. J. Bacteriol. *95:* 1388–1399.

37. **Marr, A. G., J. L. Ingraham, and C. L. Squires.** 1964. Effect of the temperature of growth of *Escherichia coli* on the formation of B-galactosidase. J. Bacteriol. *87:* 356–362.

38. **McBryde, C. N.** 1911. A bacteriological study of ham souring. U.S. Bureau of Animal Industry, Bull. No. 132.

39. **McMurrough, I. and A. H. Rose.** 1973. Effects of temperature variation on the fatty acid composition of a psychrophilic *Candida* species. J. Bacteriol. *114:* 451–452.

40. **Morita, R. Y. and L. J. Albright.** 1965. Cell yields of *Vibrio marinus*, an obligate psychrophile, at low temperatures. Can. J. Microbiol. *11:* 221–227.

41. **Morita, R. Y.** 1975. Psychrophilic bacteria. Bacteriol. Rev. *39:* 144–167.

42. **Mossel, D. A. A. and H. Zwart.** 1960. The rapid tentative recognition of psychrotrophic types among *Enterbacteriaceae* isolated from foods. J. Appl. Bacteriol. *23:* 185–188.

43. **Nashif, S. A. and F. E. Nelson.** 1953a. The lipase of *Pseudomonas fragi*. I. Characterization of the enzyme. J. Dairy Sci. *36:* 459–470.

44. **Nashif, S. A. and F. E. Nelson.** 1953b. The lipase of *Pseudomonas fragi*. II. Factors affecting lipase production. J. Dairy Sci. *36:* 471–480.

45. **Neely, W. B.** 1960. Dextran: structure and synthesis. Adv. Carbohy. Chem. *15:* 341–369.

46. **Olsen, R. H. and J. J. Jezeski.** 1963. Some effects of carbon source, aeration, and temperature on growth of a psychrophilic strain of *Pseudomonas fluorescens*. J. Bacteriol. *86:* 429–433.

47. **Peterson, A. C. and M. F. Gunderson.** 1960. Some Characteristics of proteolytic enzymes from *Pseudomonas fluorescens*. Appl. Microbiol. *8:* 98–104.

48. **Purohit, K. and J. L. Stokes.** 1967. Heat-labile enzymes in a psychrophilic bacterium. J. Bacteriol. *93:* 199–206.

49. **Roberts, T. A. and G. Hobbs.** 1968. Low temperature growth characteristics of Clostridia. J. Appl. Bacteriol. *31:* 75–88.

50. **Rose, A. H.** 1968. Physiology of microorganisms at low temperatures. J. Appl. Bacteriol. *31:* 1–11.

51. **Rouf, M. A. and M. M. Rigney.** 1971. Growth temperatures and temperature characteristics of *Aeromonas*. Appl. Microbiol. *22:* 503–506.

52. **Shaw, M. K.** 1967. Effect of abrupt temperature shift on the growth of mesophilic and psychrophilic yeasts. J. Bacteriol. *93:* 1332–1336.

53. **Sinclair, N. A. and J. L. Stokes.** 1963. Role of oxygen in the high cell yields of psychrophiles and mesophiles at low temperatures. J. Bacteriol. *85:* 164–167.

54. **Sinclair, N. A. and J. L. Stokes.** 1964. Isolation of obligately anaerobic psychrophilic bacteria. J. Bacteriol. *87:* 562–565.

55. **Sinclair, N. A. and J. L. Stokes.** 1965. Obligately psychrophilic yeasts from the polar regions. Can. J. Microbiol. *11:* 259–269.

56. **Stokes, J. L.** 1963. General biology and nomenclature of psychrophilic microorganisms. In: *Recent Progress in Microbiology,* Symp. Intern. Congr. Microbiol., 8th, Montreal, 1962. N. E. Gibbons, Editor, Univ. of Toronto Press, pp. 187–192.

57. **Stokes, J. L.** 1967. Heat-sensitive enzymes and enzyme synthesis in psychrophilic microorganisms. In: *Molecular Mechanisms of Temperature Adaptation.* C. L. Prosser, Editor, Pub. #84, A.A.A.S., pp. 311–323.

58. **Straka, R. P. and J. L. Stokes.** 1960. Psychrophilic bacteria from Antarctica. J. Bacteriol. *80:* 622–625.

59. **Strange, R. E. and M. Shon.** 1964. Effect of thermal stress on the viability and ribonucleic acid of *Aerobacter aerogenes* in aqueous suspensions. J. Gen. Microbiol. *34:* 99–114.

60. **Tempest, D. W. and J. R. Hunter.** 1965. The influence of temperature and pH value on the macromolecular composition of magnesium-limited and glycerol-limited *Aerobacter aerogenes* growing in a chemostat. J. Gen. Microbiol. *41:* 267–273.

61. **Udaka, S. and T. Horiuchi.** 1965. Mutants of *Escherichia coli* having a temperature sensitive regulatory mechanisms in the formation of arginine biosynthetic enzymes. Biochem. Biophys. Res. Commun. *19:* 156–160.

62. **Uffen, R. L. and E. Canale-Parola.** 1966. Temperature-dependent pigment production by *Bacillus cereus* var. *alesti.* Can. J. Microbiol. *12:* 590–593.

63. **Upadhyay, J. and J. L. Stokes.** 1962. Anaerobic growth of psychrophilic bacteria. J. Bacteriol. *83:* 270–275.

64. **Upadhyay, J. and J. L. Stokes.** 1963a. Temperature-sensitive formic hydrogenlyase in a psychrophilic bacterium. J. Bacteriol. *85:* 177–185.

65. **Upadhyay, J. and J. L. Stokes.** 1963b. Temperature-sensitive hydrogenase and hydrogenase synthesis in a psychrophilic bacterium. J. Bacteriol. *86:* 992–998.

66. **Williams, R. P., M. E. Goldschmidt, and C. L. Gott.** 1965. Inhibition by temperature of the terminal step in biosynthesis of prodigiosin. Biochem. Biophys. Res. Commun. *19:* 177–181.

67. **Witter, L. D.** 1961. Psychrophilic bacteria—a review. J. Dairy Sci. *44:* 983–1015.

21
Characteristics and Growth of Thermophilic Microorganisms

On the basis of growth temperature, thermophilic bacteria may be characterized as organisms with a minimum at 30 C, an optimum between 50–60 C, and maximum growth temperatures between 70–80 C *(35)*. Thermophilic algae, *Streptomyces*, and fungi have all been described *(13, 14, 37)*. The thermophilic eubacteria, however, are by far the most important from the standpoint of food microbiology. The most important genera of eubacteria that contain thermophiles are *Bacillus* and *Clostridium*, with *Actinomycetes* and lactobacilli being of less importance in foods.

Thermophilic growth may be characterized as follows: (1) The lag phase is short and sometimes difficult to measure. Spores germinate and grow rapidly. (2) The logarithmic phase of growth is of short duration. Some thermophiles have been reported to have generation times as short as 10 min when growing at high temperatures. (3) The rate of death or "die off" is rapid; and (4) Loss of viability or "autosterilization" below the thermophilic growth range is characteristic of organisms of this type. The growth curves of a bacterium at 55, 37, and 20 C are compared in Figure 21-1.

The most basic question to be asked about this group of organisms is, why do they require such high temperatures for growth? As is the case with psychrophilic organisms, the precise mechanisms of growth of these forms is not too well understood, but research over the past ten yr or so has greatly illuminated the underlying mechanisms and pointed to the following as being important to the welfare of these organisms.

THERMOSTABILITY

Thermophilic microorganisms possess thermostable components which permit a high minimum growth temperature. The three groups of thermostable components that have been most studied are enzymes, ribosomes, and flagella.

431

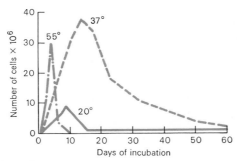

FIGURE 21-1.   Growth curves of bacterial
strain incubated at 20, 37, and 55 C *(36)*.

1. Thermostable enzymes.   The enzymes of thermophiles may be dividied
into three groups: (1) Those stable at the temperature of production but
requiring slightly higher temperatures for inactivation. Examples of this type
are malic dehydrogenase, ATPase, inorganic pyrophosphatase, aldolase,
and certain peptidases. (2) Enzymes that are inactivated at the temperature
of production in the absence of specific substrates. These include asparagine
deamidase, catalase, pyruvic acid oxidase, isocitrate lyase, and certain
membrane-bound enzymes. (3) Highly heat-resistant enzymes and proteins
among which are alpha-amylase, protease, glyceraldehyde-3-phosphate
dehydrogenase, amino acid activating enzymes, flagella proteins, esterases,
and thermolysin. The heat stability of some enzymes from mesophilic and
thermophilic bacteria is presented in Table 21-1. It should be noted that the
alpha-amylase of *B. stearothermophilus* remained completely active after
being heated at 70 C for 24 hr. However, in a study by Pfueller and Elliott
(32) the alpha-amylase of *B. stearothermophilus* was found to have a
molecular weight of approximately 53,000 instead of 15,500 and to be heat
sensitive in the purified state although it hydrolyzed starch at 70 C and
above. These authors reported that the enzyme was protected from thermal
denaturation above 55 C by metal ions, particularly Ca^{++}, and by protein.
The alpha-amylase from a facultative thermophile grown at 37 C was shown
by Campbell *(9)* to be heat-labile whereas that produced at 55 C was heat
stable. Studies on the heat-stable alpha-amylase by Campbell and Manning
(10) and Manning and Campbell *(26)* showed that the enzyme contained a
high content of proline, which resulted in a large negative optical rotation. It
may be that since the enzyme exists in an unfolded configuration in the native state, heat has relatively little effect upon its denaturation. Koffler and
Gale *(23)* studied the heat stability of cytoplasmic proteins isolated from 4
thermophiles and 4 mesophiles and showed that the enzymes from
thermophiles were more heat-stable than similar preparations from mesophiles. These investigators further found that over 50% of the mesophile proteins coagulated when heated at 60 C and pH 6 for 8 min, while hardly any of
the thermophile proteins coagulated under similar conditions. Ascione and

TABLE 21-1. Comparison of thermostability and other properties of enzymes from mesophilic and thermophilic bacteria (29). *

Species	Enzyme	Heat stability[a] (%)		Half-cystine (mole/ mole of protein)	Molecular weight	Metal required for stability
B. subtilis	Subtilisin BPN'	45	(50°C, 30 min)	0	28,000	Yes
B. subtilis	Neutral protease	50	(60°C, 15 min)	0	44,700	Yes
Ps. aeruginosa	Alkaline protease	80	(60°C, 10 min)	0	48,400	Yes
Ps. aeruginosa	Elastase	{ 86	(70°C, 10 min)	4.6	39,500	Yes
		10	(75°C, 10 min)			
Group A streptococci	Streptococcal protease	0	(70°C, 30 min)	1	32,000	
Cl. histolyticum	Collagenase	1.5	(50°C, 20 min)	0	90,000	
St. griseus	Pronase	60	(60°C, 10 min)			Yes
B. thermoproteolyticus	Thermolysin	{ 95	(60°C, 120 min)	0	42,700	Yes
		50	(80°C, 60 min)			
B. subtilis	α-Amylase	55	(65°C, 20 min)	0	50,000	Yes
B. stearothermophilus	α-Amylase	100	(70°C, 24 hr)	4	15,500	

[a] Activity remaining after heat treatment shown in parentheses.

Fresco *(3)* demonstrated the heat-resistance of amino acid activating enzymes from thermophilic algae. These authors found that enzymes from algae grown at 70 C withstood boiling for 5 min. The thermostable aldolase of *Thermus aquaticus* has been found to have optimal activity at 95 C with little activity <60 C *(17)*.

Just why thermophilic enzymes and proteins are so heat resistant is not clear at this time. Among the promising leads are findings that thermophiles have higher levels of hydrophobic amino acids in their proteins than do mesophiles. The amino acid sequences of clostridial ferredoxin from a mesophilic and a thermophilic strain reveal a somewhat higher number of charged amino acids in the thermophilic protein than in that of the mesophile. It is conceivable that this could lead to a tighter binding or a more hydrophobic character of the thermophile protein. A hydrophobic protein would presumably be more heat resistant than one with a more hydrophilic character. Thermophilic proteins have been found to undergo a type of conformational change around 55–60 C and this might have the effect of "melting" the hydrophobic cluster within the protein molecule. Many or most thermophilic proteins may owe their high stability to their binding of metal ions such as Mg^{2+}. This arrangement would make for a more tightly bound structure and consequently one more refractive to heat. The structural integrity of the membrane of *B. stearothermophilus* protoplasts has been shown to be affected by divalent cations *(39)*. The proteins of thermophiles are similar in molecular weight, amino acid composition, allosteric effectors, subunit composition, and primary sequences to their mesophile counterparts. These and other aspects of thermostable proteins have been discussed by Singleton and Amelunxen *(34)*.

2. Thermostable ribosomes. It has been reported that ribosomes from thermophilic bacteria are more heat-stable than those from mesophiles such as *E. coli (25)*. Furthermore, the thermal stability of ribosomes corresponds well with the maximal growth temperatures of organisms (Table 21-2). Heat-resistant ribosomal RNA has been reported *(16)*, but not DNA *(27)*. Friedman *et al. (19)* have reported that thermophile ribosomes have a higher melting point than their corresponding rRNA (ribosomal RNA), whereas this effect was not observed with *E. coli*. A study by Saunders and Campbell *(33)* of the amino acid composition of the ribosomes of *B. stearothermophilus* and *E. coli* showed them to be quite similar. As to why ribosomes of *B. stearothermophilus* are thermally stable, Saunders and Campbell suggested that they may reflect either an unusual packing arrangement of the protein to the RNA, or differences in the primary structure of the ribosomal proteins. Ansely *et al. (1)* were unable to find any unusual chemical features of *B. stearothemophilus* ribosomal proteins which could explain the thermal stability of the respective ribosomes, while Bassel and Campbell *(4)* found no significant differences in either the size or the arrangement of surface filaments of *B. stearothermophilus* and *E. coli* ribosomes.

TABLE 21-2.   Ribosome melting and maximal growth temperatures of 19 selected organisms *(31).*

Organism and strain no.	Max. growth temp (°C)	Ribosome T_m, (°C)
(1) *V. marinus* (15381)	18	69
(2) 7E-3	20	69
(3) 1–1	28	74
(4) *V. marinus* (15382)	30	71
(5) 2–1	35	70
(6) *D. desulfuricans (cholinicus)*[1]	40	73
(7) *D. vulgaris* (8303)[1]	40	73
(8) *E. coli* (B)	45	72
(9) *E. coli* (Q13)	45	72
(10) *S. itersonii* (SI−1)[2]	45	73
(11) *B. megaterium* (Paris)	45	75
(12) *B. subtilis* (SB−19)	50	74
(13) *B. coagulans* (43P)	60	74
(14) *D. nigrificans* (8351)[3]	60	75
(15) Thermophile 194	73	78
(16) *B. stearothermophilus* (T−107)	73	78
(17) *B. stearothermophilus* (1503R)	73	79
(18) Thermophile (Tecce)	73	79
(19) *B. stearothermophilus* (10)	73	79

[1]*Desulfovibrio* [2]*Spirillum* [3]*Desulfotomaculum*

The thermal stability of bacterial ribosomes is affected by rRNA base composition. For example, in a study of 19 organisms, Pace and Campbell *(31)* found that with a few exceptions, the G-C (guanine-cytosine) content of rRNA molecules increased and the A-U (adenine-uracil) content decreased with increasing maximal growth temperature. An increased G-C content makes for a more stable structure through more extensive hydrogen bonding. The 16 S and 23 S fractions of thermophile rRNA have been shown by Saunders and Campbell to be more heat-stable than the corresponding components from *E. coli.* Friedman *(20)* has pointed out that the thermostability of ribosomes could be the limiting factor in determining the upper growth temperature for organisms. The thermal stability of soluble RNA from thermophiles and mesophiles appears to be the same *(2, 19).* The latter authors have shown that ribosomes and soluble components from *B. stearothermophilus* are capable of incorporating some amino acids at temperatures between 30–70 C.

3. Thermophilic flagella.   In ᴀ series of studies by Koffler and his associates, the flagella of thermophilic bacteria were shown to be more thermostable than these structures from mesophilic organisms. Koffler *(22)* and Koffler *et al.* *(23)* have reported that flagella of thermophiles remain intact at temperatures as high as 70 C, while those of mesophiles are disintegrated at

50 C. Thermophilic flagella were also found by these investigators to be more resistant to urea and acetamide than mesophile flagella, suggesting that more effective hydrogen bonding occurs in those of thermophiles. Thermophile flagella were further shown to possess only one-half as many titratable or basic and acidic groups as did flagellin from mesophiles. Other than having more overall intermolecular stability, the precise reasons why flagella from thermophiles are more heat-stable than those from mesophiles is not yet well understood.

In addition to the above thermostable components of thermophiles, other aspects of thermophilic microorganisms and thermophilic growth have received the attention of various investigators, and some of these are discussed below.

OTHER ASPECTS OF THERMOPHILIC MICROORGANISMS

1. Nutrient Requirements.   Thermophiles generally have a higher nutrient requirement than mesophiles when growing at thermophilic temperatures. This is illustrated for 2 strains of *B. coagulans* and three of *B. stearothermophilus* in Table 21-3. It may be noted that one strain of *B. coagulans* showed no differences in growth requirements regardless of incubation temperature. The obligately thermophilic strain of *B. stearothermophilus* showed one additional requirement as incubation temperature increased, while one temperature facultative strain of this organism showed additional requirements as the incubation temperature was lowered. Although this aspect of thermophilism has not received much study, changes in nutrient requirements as incubation temperature is raised may be due to a general lack of

TABLE 21-3.   **Effect of incubation temperature on the nutritional requirements of thermophilic bacteria** *(8).*

Organism and strain no.	Nutritional requirements at		
	36°	*45°*	*55°*
B. coagulans 2 (F)*	His, thi, bio, fol	His, thi, bio, fol	His, thi, bio, fol
B. coagulans 1039 (F)	Thi, bio, fol	Thi, bio, fol	His, met, thi, nic, bio, fol
B. stearothermophilus 3690 (F)	Met, leu, thi, nic, bio, fol	Met, thi, bio, fol	Met, thi, bio, fol
B. stearothermophilus 4259 (F)	Bio, fol	Met, his, nic, bio, fol	Met, his, nic, bio, fol
B. stearothermophilus 1373b (O)	No growth	Glu, his, met, leu, bio	Glu, his, met, leu, bio, rib

*F, facultative; O, obligate thermophile.
His, histidine; thi, thiamine; bio, biotin; fol, folic acid; met, methionine; leu, leucine; nic, nicotinic acid; glu, glutamic acid; rib, riboflavine.

efficiency on the part of the metabolic complex. Certain enzyme systems might well be affected by the increased temperature of incubation as well as the overall process of enzyme synthesis.

2. Oxygen Tension.   Thermophilic growth is affected by oxygen tension. As the temperature of incubation is increased, the growth rate of microorganisms increases, thereby increasing the oxygen demand on the culture medium, while at the same time the solubility of oxygen is reduced. This is thought by some investigators to be one of the most important limiting factors of thermophilic growth in culture media. Downey *(15)* has shown that thermophilic growth is optimal at or near the oxygen concentration normally available in the mesophilic range of temperatures—143 to 240 μM. Although it is conceivable that thermophiles are capable of high temperature growth due to their ability to consume and conserve oxygen at high temperatures, a capacity that mesophiles and psychrophiles lack, further data in support of this notion are wanting.

3. Cellular Lipids.   The state of cellular lipids affects thermophilic growth. Since an increase in degree of unsaturation of cellular lipids is associated with psychrophilic growth, it is reasonable to assume that a reverse effect occurs in the case of thermophilic growth. This idea finds support in the investigations of many authors. Gaughran *(21)* showed that mesophiles growing above their maximum range showed decreases in lipid content and more lipid saturation. According to this author, cells cannot grow at temperatures below the solidification point of their lipids. Marr and Ingraham *(28)* showed a progressive increase in saturated fatty acids and a corresponding decrease in unsaturated fatty acids in *E. coli* as the temperature of growth increased. The general decrease in the proportion of unsaturated fatty acids as growth temperatures increase has been found to occur in a large variety of animals and plants *(12)*.

There seems to be little doubt that thermophiles display a preference for saturated fatty acids. More recent work reveals that branched fatty acids are common also among these organisms. Weerkamp and Heinen *(38)* found a preferential synthesis of branched heptadecanoic acid and the total elimination of unsaturated fatty acids by two thermophilic *Bacillus* spp.

4. Cellular Membranes.   The nature of cellular membranes affect thermophilic growth. Brock *(6)* reported that the molecular mechanism of thermophilism is more likely to be related to the function and stability of cellular membranes than to the properties of specific macromolecules. This investigator has pointed out that there is no evidence that organisms are killed by heat because of the inactivation of proteins or other macromolecules, a view that is widely held. According to Brock, an analysis of thermal-death curves of various microorganisms shows that this is a first-order process compatible with an effect of heat on some large structure such as the cell membrane, since a single hole in the membrane could result in leakage of cell constituents and subsequent death. Brock has also pointed out that thermal killing due to the inactivation of heat-sensitive enzymes, or heat-sensitive ribosomes, of which there are many copies in the cell, should not result in simple first-order kinetics. The leakage of ultraviolet light-absorbing and other material from cells undergoing "cold shock" would tend to implicate the membrane in high temperature death. Since most animals die when body temperatures reach between 40–45 C and most psychrophilic bacteria are killed at about this temperature range, the suggestion that lethal injury is due to the melting of lipid constituents of the cell or cell membrane is not only plausible; it has been supported by the findings of various investigators *(34)*. The unit cell membrane is thought to consist of layers of lipid surrounded by layers of protein and to depend upon the lipid layers for its biological functions. The disruption of this structure would be expected to cause cell damage and perhaps death. In view of the changes in cellular lipid saturation noted above, the cell membrane appears to be critical to growth and survival at thermophilic temperatures.

5. Effect of Temperature.   Brock *(6)* has also called attention to the fact that thermophiles apparently do not grow as fast at their optimum temperatures as one would predict or is commonly believed. Arrhenius plots of thermophile growth compared to *E. coli* over a range of temperatures indicated that overall, the mesophilic types were more efficient. This author believes that thermophile enzymes are inherently less efficient than mesophiles because of thermal stability, that is, the thermophiles have had to discard growth efficiency in order to survive at all. Somewhat in support of the latter notion, Brock and Brock *(7)* have shown that the optimum temperature for glucose incorporation was similar to the environmental temperature for thermophilic bacteria occurring at various temperatures along a thermal gradient of a hot spring in Yellowstone Park. This is a good example of the adaptation of microorganisms to temperature changes. As was suggested for psychrophiles in the previous chapter, the underlying control of thermophilism appears to be under genetic control.

6. Genetic Bases.   A significant discovery towards an understanding of the genetic bases of thermophilism was made by McDonald and Matney *(30)*. These investigators effected the transformation of thermophilism in *B. subtilis* by growing cells of a strain that could not grow above 50 C in the presence of DNA extracted from one that could grow at 55 C. The more heat-sensitive strain was transformed at a frequency of 10^{-4}. These authors also noted that only 10–20% of the transformants retained the high-level streptomycin resistance of the recipient, which indicated that the genetic loci for streptomycin resistance and that for growth at 55 C were closely linked.

Although much has been learned about the basic mechanisms of thermophilism in microorganisms, the precise mechanisms underlying this high-temperature phenomenon still remain a mystery. The facultative thermophiles such as some *B. coagulans* strains present a picture as puzzling as the obligate thermophiles. The facultative thermophiles display both mesophilic and thermophilic types of metabolism. In their studies of these types from the genus *Bacillus,* which grew well at both 37 and 55 C, Bausum and Matney *(5)* reported that the organisms appeared to shift from mesophilism to thermophilism between 44–52 C.

SUMMARY

There are essentially two views regarding the underlying mechanisms of thermophilism in microorganisms. The first is that of Brock *(6),* which relates this phenomenon more to the function and stability of cellular membranes than to the thermal properties of specific macromolecules. The second is that the reason for thermophilic growth is the greater thermostability of essential cell components than those of their mesophilic counterparts. Friedman and Weinstein *(18)* have suggested that thermophilism may be explained by the following: (1) The rapid resynthesis of heat-damaged components, (2) the presence of stabilizing cofactors, or (3) the synthesis of inherently thermostable macromolecules. These authors have emphasized the role of Mg^{++} as a stabilizer in thermophiles and that these organisms may maintain high levels of this ion or polycations such as polyamines.

REFERENCES

1. **Ansley, S. B., L. L. Campbell, and P. S. Sypherd.** 1969. Isolation and amino acid composition of ribosomal proteins from *Bacillus stearothermophilus.* J. Bacteriol. *98:* 568–572.

2. **Arca, M., C. Calvori, L. Frontali, and G. Tecce.** 1964. The enzymic synthesis of aminoacyl derivatives of soluble ribonucleic acid from *Bacillus stearothermophilus.* Biochim. Biophys. Acta *87:* 440–448.

3. **Ascione, R. and J. R. Fresco.** 1964. Heat-stable amino activating enzymes from a thermophile. Fed. Proc. *23:* 163.

4. **Bassel, A. and L. L. Campbell.** 1969. Surface structure of *Bacillus stearothermophilus* ribosomes. J. Bacteriol. *98:* 811–815.

5. **Bausum, H. T. and T. S. Matney.** 1965. Boundary between bacterial mesophilism and thermophilism. J. Bacteriol. *90:* 50–53.

6. **Brock, T. D.** 1967. Life at high temperatures. Science *158:* 1012–1019.

7. **Brock, T. D. and M. L. Brock.** 1968. Relationship between environmental temperature of bacteria along a hot spring thermal gradient. J. Appl. Bacteriol. *31:* 54–58.

8. **Campbell, L. L. and O. B. Williams.** 1953. The effect of temperature on the nutritional requirements of facultative and obligate thermophilic bacteria. J. Bacteriol. *65:* 141–145.

9. **Campbell, L. L.** 1955. Purification and properties of an alpha-amylase from facultative thermophilic bacteria. Arch. Biochem. Biophys. *54:* 154–161.

10. **Campbell, L. L. and G. B. Manning.** 1961. Thermostable α-amylase of *Bacillus stearothermophilus*. III. Amino acid composition. J. Biol. Chem. *236:* 2962–2965.

11. **Campbell, L. L. and B. Pace.** 1968. Physiology of growth at high temperatures. J. Appl. Bacteriol. *31:* 24–35.

12. **Chapman, D.** 1967. The effect of heat on membranes and membrane constituents. In: *Thermobacteriology.* A. H. Rose, Editor, Academic Press, N.Y., pp. 123–146.

13. **Cooney, D. G. and R. Emerson.** 1964. *Thermophilic Fungi.* W. H. Freeman & Company, San Francisco.

14. **Cross, T.** 1968. Thermophilic actinomycetes. J. Appl. Bacteriol. *31:* 36–53.

15. **Downey, R. J.** 1966. Nitrate reductase and respiratory adaptation in *Bacillus stearothermophilus*. J. Bacteriol. *91:* 634–641.

16. **Farrell, J. and A. Rose.** 1967. Temperature effects on microorganisms. Ann. Rev. Microbiol. *21:* 101–120.

17. **Freeze, R. and T. D. Brock.** 1970. Thermostable aldolase from *Thermus aquaticus*. J. Bacteriol. *101:* 541–550.

18. **Friedman, S. M. and I. B. Weinstein.** 1966. Protein synthesis in a subcellular system from *Bacillus stearothermophilus*. Biochim. Biophys. Acta *114:* 593–605.

19. **Friedman, S. M., R. Axel, and I. B. Weinstein.** 1967. Stability of ribosomes and ribosomal ribonucleic acid from *Bacillus stearothermophilus*. J. Bacteriol. *93:* 1521–1526.

20. **Friedman, S. M.** 1968. Protein-synthesizing machinery of thermophilic bacteria. Bacteriol. Rev. *32:* 27–38.

21. **Gaughran, E. R. L.** 1947. The saturation of bacterial lipids as a function of temperature. J. Bacteriol. *53:* 506.

22. **Koffler, H.** 1957. Protoplasmic differences between mesophiles and thermophiles. Bacteriol. Rev. *21:* 227–240.

23. **Koffler, H. and G. O. Gale.** 1957. The relative thermostability of cytoplasmic proteins from thermophilic bacteria. Arch. Biochem. and Biophys. *67:* 249–251.

24. **Koffler, H., G. E. Mallett, and J. Adye.** 1957. Molecular basis of biological stability to high temperatures. Proc. Nat'l. Acad. Sci. U.S. *43:* 464–477.

25. **Mangiantini, M. T., G. Tecce, G. Toschi, and A. Trentalance.** 1965. A study of ribosomes and of ribonucleic acid from a thermophilic organism. Biochim. Biophys. Acta *103:* 252–274.

26. **Manning, G. B. and L. L. Campbell.** 1961. Thermostable α-amylase of *Bacillus stearothermophilus*. I. Crystallization and some general properties. J. Biol. Chem. *236:* 2952–2957.

27. **Marmur, J.** 1960. Thermal denaturation of deoxyribosenucleic acid isolated from a thermophile. Biochim. Biophys. Acta *38:* 342–343.

28. **Marr, A. G. and J. L. Ingraham.** 1962. Effect of temperature on the composition of fatty acids in *Escherichia coli*. J. Bacteriol. *84:* 1260–1267.

29. **Matsubara, H.** 1967. Some properties of thermolysin. In: *Molecular Mechanisms of Temperature Adaptation.* C. L. Prosser, Editor, Pub. #84, Am. Assoc. Adv. Sci., Washington, D.C., pp. 283–294.

30. **McDonald, W. C. and T. S. Matney.** 1963. Genetic transfer of the ability to grow at 55 C in *Bacillus subtilis*. J. Bacteriol. *85:* 218–220.

31. **Pace, B. and L. L. Campbell.** 1967. Correlation of maximal growth temperature and ribosome heat stability. Proc. Nat'l. Acad. Sci. U.S. *57:* 1110–1116.

32. **Pfueller, S. L. and W. H. Elliott.** 1969. The extracellular α-amylase of *Bacillus stearothermophilus*. J. Biol. Chem. *244:* 48–54.

33. **Saunders, G. F. and L. L. Campbell.** 1966. Ribonucleic acid and ribosomes of *Bacillus stearothermophilus*. J. Bacteriol. *91:* 332–339.

34. **Singleton, R., Jr. and R. E. Amelunxen.** 1973. Proteins from thermophilic microorganisms. Bacteriol. Rev. *37:* 320–342.

35. **Stokes, J. L.** 1967. Heat-sensitive enzymes and enzyme synthesis in psychrophilic microorganisms. In: *Molecular Mechanism of Temperature Adaption.* C. L. Prosser, Editor, Pub. #84, Am. Assoc. Adv. Sci. Washington, D.C. pp. 311–323.

36. **Tanner, F. W. and G. I. Wallace.** 1925. Relation of temperature to the growth of thermophilic bacteria. J. Bacteriol. *10:* 421–437.

37. **Tendler, M. D. and P. R. Burkholder.** 1961. Studies on the thermophilic *Actinomycetes*. I. Methods of cultivation. Appl. Microbiol. *9:* 394–399.

38. **Weerkamp, A. and W. Heinen.** 1972. Effect of temperature on the fatty acid composition of the extreme thermophiles, *Bacillus caldolyticus* and *Bacillus caldotenax*. J. Bacteriol. *109:* 443–446.

39. **Wisdom, C. and N. E. Welker.** 1973. Membranes of *Bacillus stearothermophilus:* Factors affecting protoplast stability and thermostability of alkaline phosphatase and reduced nicotinamide adenine dinucleotide oxidase. J. Bacteriol. *114:* 1336–1345.

22

Nature of
Radiation Resistance
In Microorganisms

Prior to 1956, the most radiation-resistant microorganisms known were the sporeforming bacteria, especially *Clostridium botulinum* strains. That sporeforming bacteria should be especially resistant to radiation is perhaps not surprising in view of their high levels of resistance to heat, drying, chemicals, and other environmental conditions. In 1956, Anderson *et al. (2)* announced the discovery of a nonsporeforming gram positive coccus which was later named *Micrococcus radiodurans (3)*. This organism was shown to be more radioresistant than any bacterium previously known. It has variously been referred to as a *Sarcina* sp. and as *M. rubens*. A second organism with a similar high degree of radioresistance was reported by Murray and Robinow *(32)*. This isolate was later determined to be *M. radiodurans* and has been designated the Sark strain. A third radiation-resistant coccus was isolated by Davis *et al. (8)* from irradiated haddock and appears to be a variant of the original isolate by Anderson and co-workers, but differs from it in several ways as noted later in this chapter. More recently, another highly resistant coccus has been isolated from Bombay duck *(28)*. This isolate was subsequently named *Micrococcus radiophilus*. It differs from *M. radiodurans* in being smaller in size, salt tolerant, and more radioresistant.

Both *M. radiodurans* and *M. radiophilus* are more resistant to radiations than sporeformers. With the discovery of these organisms, a great deal of attention has been focused on the mechanisms of their high level of radiation resistance. There are several reasons for the relatively intense interest in these organisms. First, as food-borne organisms, an understanding of the mechanism of their radiation resistance would possibly enable one to prevent the accumulation of such types in foods through mutations or other means. This accumulation of radiation-resistant organisms would reduce the effectiveness of radiations when applied to foods, or cause the radiation dosage to be increased to levels that would lead to undesirable changes in irradiated foods. Second, the doses of ionizing radiation necessary to effect sterilization of foods (4–5 Mrads) cause undesirable changes in some foods. The addition of suitable radiation-sensitizing agents would allow for a reduction in the dose of radiation and thereby lessen the development of undesirable changes. Research of this nature would also facilitate the search for or the design of radiation-sensitizing agents for food use. Third, since the advent of atomic energy, numerous unsuccessful attempts have been made to uncover a nontoxic radiation-protective agent for human and animal uses. It is not inconceivable that such compounds may be developed from a better understanding of the nature of the high level of radioresistance in microorganisms.

Presented below are characteristics of these highly radiation-resistant microorganisms in addition to other aspects of radiation resistance as determined by studies involving other organisms.

THE MICROBIOLOGY OF *M. RADIODURANS* STRAINS

1. Morphologic and Cultural Characteristics.   All 3 of the original isolates of this organism are gram positive cocci that occur principally as tetrads and occasionally in pairs or singly. On culture media, they produce pigmented colonies that range from flesh-colored to pink to brownish-red and to bright red, depending upon age of culture and type of medium employed. The pigments have been identified as carotenoids *(22)*. In the case of the original isolate of Anderson and co-workers, the pigment is water-soluble while the strain isolated by Davis and co-workers is not. The temperature growth range for *M. radiodurans* was reported by Anderson *et al.* to be 5–40 C with an optimum around 25–30 C. No growth occurred at 45 C. Visible colonies are produced within 48 hr at 35 C. Anderson and co-workers reported that *M. radiodurans* resembled *M. roseus* and *M. rubens* in its cultural and morphologic characteristics. The strain isolated by Davis and co-workers differs from the original isolate by its inability to reduce nitrate to nitrite, its inability to produce gelatinase, and its smaller cell size. Although facultative in nature, the growth of this organism was reported by Anderson and co-workers to be accelerated by aeration.

In regard to its nutrition, the R_1 strain (described below) grows on a synthetic medium containing 4 B-vitamins, the amino acids glutamate and methionine, glucose as an energy source, and minerals. Methionine is the only essential amino acid and it is incorporated rapidly into the cells *(33)*. Several monosaccharides are readily utilized by this organism, and it appears that they are oxidized through the tricarboxylic acid cycle.

2. Relative Radiation Resistance.   The R_1 strain of *M. radiodurans* was obtained by Anderson *et al.* from ground beef and pork which had been exposed to 2–3 Mrads of gamma radiation. A strain isolated from unirradiated meat was designated U_1. A variant of the R_1 strain having different colony form, pigmentation, and smaller cell size was designated R_4R, while the R_w strain is a variant of R_1 that possesses less color *(11)*. Some of these strains survived 6 Mrads on agar slants. The isolate of Davis and co-workers was obtained from irradiated haddock and was shown to consist of smooth and rough strains and to be more resistant than the R_2 isolate of Anderson *et al.*, especially in buffer (see Figure 22-1). Of the smooth and rough strains of Davis and co-workers, the former was reported to be more resistant than the latter. The relative resistance of the 4 strains of *M. radiodurans* isolated by Anderson and co-workers is shown in Figure 22-2. This organism has been shown to possess very high resistance to ultraviolet (UV) radiation as well as to ionizing radiations *(10, 37)*.

The radiation survival of the R_1 strain was shown to be greater in raw beef and raw chicken than in raw fish and cooked beef *(11)*. In beef, this strain was reduced by a factor of about 10^{-5} by 3 Mrads and by a factor of 10^{-9} by 4 Mrads, making it more resistant than sporeformers in this environment. The original isolate of R_1 required 4.8 Mrads for a 10^{-7} reduction. The effect of various treatments on the radiation-resistance of the R_1 strain was studied by Duggan *et al. (12)*. These investigators found that freezing in raw puréed beef did not significantly affect its sensitivity. Irradiation of the organism in menstra at temperatures between 40–50 C reduced its resistance as did pre-irradiation heat treatments. No effect upon radiation resistance was noted when the organism was irradicated at pH values of 5, 7, or 9 in buffer. The

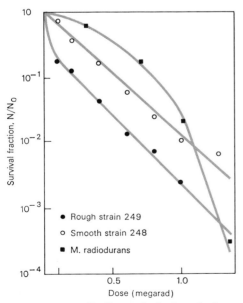

FIGURE 22-1. Radiosurvival of washed cells of the rough and smooth variants of newly isolated coccus and *Micrococcus radiodurans (8)*.

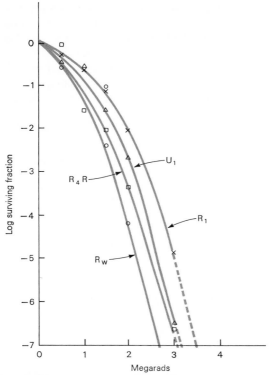

FIGURE 22-2.   Survival curves of four cultures of radiation-resistant bacteria in beef *(11)*.

response of this organism under similar conditions was found to be essentially the same to UV light *(10)*. The latter authors *(12)* also studied the effect of oxidizing and reducing conditions upon resistance to radiation of this organism in phosphate buffer (see Table 22-1). The flushing of buffer suspensions with nitrogen or O_2 had no significant effect on radiation sensitivity when compared to the control. Likewise, the presence of 100 ppm of H_2O_2 had no significant effect upon sensitivity. Treatment with cysteine rendered the cells less sensitive while ascorbate increased their sensitivity. The effect of N-ethylmaleimide (NEM) and iodoacetic acid (IAA) upon resistance was investigated by Lee *et al. (27)*. It was found that IAA reduced resistance but not NEM when tested at nontoxic levels. These compounds responded in the same way in the presence and absence of O_2.

A study of Krabbenhoft *et al. (25)* showed that growth of *M. radiodurans* in media supplemented with NZ-case caused significant reductions in radio resistance. The LD_{50} for this organism was found to be 700 krads when grown on plate count agar supplemented with DL-methionine, but was approximately one-half as resistant when grown on this medium supplemented with 0.5% NZ-case. Its resistance to UV light was shown also to change in a similar manner. The growth patterns of the organism on the two media were reported to be similar. In addition to being 10 times more sensitive on the NZ-case supplemented medium, the cultures were also less pigmented.

The carotenoid pigments of this organism have been shown to be radiosensitive and apparently do not contribute to its radioresistance (22). The carotenoid pigments of *M. radiophilus* play no role in the resistance of this organism to radiations (29).

Radiation death rate curves for *M. radiodurans* have been reported by various authors to be sigmoidal in nature. This can be seen from Figure 22-1 for the R_1 strain. Some possible explanations for this are presented later in this chapter.

3. Thermal Resistance.   Studies by Duggan *et al.* (13) on strain R_1 showed its thermal death rate to approximate exponential form, unlike its radiation death rate. In beef, these investigators found this strain to possess a D_{140} value of 0.75 and a z value of 10.65 min. Unlike its radiation resistance, it is rather sensitive to heat and will not survive milk pasteurization temperatures.

4. Distribution in Nature.   The distribution of these organisms was studied by Krabbenhoft *et al.* (24). These authors were able to isolate this organism from ground beef, pork sausage, the hides of animals, and from creek water. All isolates were shown to possess high levels of radiation resistance. None could be found in soil, hay, or fecal matter samples. The finding by Davis and co-workers of this organism in haddock along with the above suggests its rather widespread distribution in nature. Its apparent low incidence may be due to its lack of competition with other members of its habitat (24).

TABLE 22-1.   Effects of oxidizing and reducing conditions on resistance to radiation of *Micrococcus radiodurans* (table of means)* (12).

Condition	Log of surviving fraction†
Buffer, unmodified	−3.11542
Oxygen flushed	−3.89762
Nitrogen flushed	−2.29335
H_2O_2 (100 ppm)	−3.47710
Thioglycolate (0.01 M)	−1.98455
Cysteine (0.1 M)	−0.81880
Ascorbate (0.1 M)	−5.3605

*Determined by count reduction after exposure to 1 megarad of gamma radiation in 0.05 M phosphate buffer. LSD: P = 0.05 (1.98116); P = 0.01 (2.61533).
†Averages of four replicates.

5. Sensitivity to Other Inhibitors.   The sensitivity of *M. radiodurans* to 10 antibiotics was determined by Hawiger and Jeljaszewicz *(17)*. These authors found this organism to be sensitive to all 10 antibiotics and thus to resemble *Staphylococcus aureus* in its sensitivity to these compounds. The sensitivity to ethylene oxide was investigated by Gammon *et al. (15)*. In a mixture of ethylene oxide and Freon 12 at a level of 500 mg/liter, 50% R. H., and at a temperature of 130 F, the D value of *M. radiodurans* and several sporeformers on nonporous surfaces were found to be as follows:

M. radiodurans	5.4 min.
B. globigii	6.0 min.
C. sporogenes	3.8 min.
B. stearothermophilus	2.8 min.

6. Nature of Cell Walls.   In general, the cell walls of this organism are atypical for gram positive cocci. They have been shown to contain lipoproteins as well as mucopeptides in which L-ornithine is the principal diamino acid instead of the more usual diaminopimelic acid or lysine *(43, 44)*. The cell-wall lipids consist of even and odd-numbered straight-chain saturated and unsaturated fatty acids *(23)*. Polysaccharides of the cell wall were found by Work and Griffiths to contain mannose and rhamnose but not heptose. One of the more unusual features of the cell wall of this organism is its thickness and the existence of at least 3–4 layers, with one of these layers consisting of hexagonally arranged subunits or holes *(40, 41, 44)*. The existence of a multilayered cell wall in bacteria was previously unknown and its significance relative to the radiation resistance of this organism is not clear at this time. Its cytoplasm and nuclear structures appear to be normal *(41)*.

7. Mechanism of Radiation Resistance.   Upon the irradiation of complex substrates such as foods, a series of chemical changes begins immediately, the extent and duration of which depend upon the dosage of radiation applied, nature of the substrate, and so forth (see Chapter 10). An overall view of some of the events that take place when radiation is applied may be seen from Figure 22-3. When water-containing samples are irradiated, the radiolysis of water is one of the consequences that leads to the formation of free radicals, peroxides, and so forth. Peroxides are produced by both UV and ionizing radiations *(21)*. It has been known for some time that some of the damaging effects of irradiation can be minimized or halted by application of certain radioprotective compounds prior to irradiation. The compounds most effective for this purpose are those that contain —SH groups. According to Serianni and Bruce *(35)*, the most effective chemical radioprotective agents found so far are amino-thiols of the following general formula: R_2N—$(CH_3)_n$—SH, where R = H or NH_2 and n is not greater than 3. Since such compounds possess the capacity of increasing radioresistance in microorganisms, most of the earlier work on the mechanism of

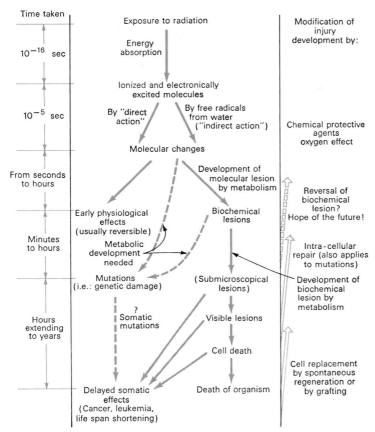

Time taken

Exposure to radiation

Modification of
injury
development by:

10^{-16} sec

Energy
absorption

Ionized and electronically
excited molecules

10^{-5} sec

By "direct
action"

By free radicals
from water
("indirect action")

Chemical protective
agents
oxygen effect

Molecular changes

From seconds
to hours

Development of
molecular lesion
by metabolism

Reversal of
biochemical
lesion?
Hope of the future!

Early physiological
effects
(usually reversible)

Biochemical
lesions

Minutes
to hours

Metabolic
development
needed

Intra-cellular
repair (also applies
to mutations)

Mutations
(i.e.: genetic damage)

(Submicroscopical
lesions)

Development of
biochemical
lesion by
metabolism

Hours
extending
to years

?
Somatic
mutations

Visible lesions

Cell death

Cell replacement
by spontaneous
regeneration or
by grafting

Delayed somatic
effects
(Cancer, leukemia,
life span shortening)

Death of organism

FIGURE 22-3. Summary of radiation and post-irradiation effects in organic matter (Bacq and Alexander, *4*, reprinted with permission of the authors, *Fundamentals of Radiobiology*, copyright 1961 Pergamon Press).

radioresistance in *M. radiodurans* sought to uncover and identify such compounds in this organism.

An important finding in the studies on radioresistance in this organism was that of Bruce *(6)*, who was able to confer radioresistance upon *Escherichia coli* by employing extracts from *M. radiodurans*. Similar extracts from *Sarcina lutea*, a relatively radiosensitive organism, were shown to sensitize rather than protect *E. coli*. This investigator stated that the amount of protective substance in the *M. radiodurans* extracts was sufficient to explain the high radioresistance of the organism. Additional information on this protective action was reported by Serianni and Bruce *(35)*. These workers found that extracts from *M. radiodurans* cultured in chemically defined media had very little free —SH activity but, nevertheless, exhibited great protective activity. These authors further reported that extracts from stationary phase cultures, while highly protective, were apparently devoid of components containing sulfur, thus suggesting that the radioprotective component or components in the extract were devoid of —SH groups. No evidence for the presence of either mercaptoethylamine (MEA) or mercaptopropylamine (MPA) was found. The protective substance was reported to be of low molecular weight and to be produced within the cell.

8. Genetic Aspects of Radiation Resistance.   Kaplan and Zavarine *(20)* reported that the sensitivity of microorganisms to ionizing radiations appeared to increase with an increase in G-C content of their DNA. Various investigators have since studied DNA of *M. radiodurans* in an effort to determine its role in the high radioresistance of this organism. Setlow and Duggan *(37)* found that DNA from this organism was not resistant to UV light. These investigators postulated that the organism must have a very efficient DNA repair mechanism in view of its high UV resistance and the low resistance of its DNA. The G + C/A + T ratio of *M. radiodurans* was found by these workers to be 1.6 compared to 1.0 for *E. coli*. Moseley and Schein *(30)* found a high G-C content (67%) in the DNA of this organism and noted that this level was the same as that for *Pseudomonas* spp., which represent some of the most radiosensitive bacteria. *M. radiodurans* does not fall into the pattern of those organisms studied by Kaplan and Zavarine *(20)*, in which radiation resistance is correlated with a low G-C content. Moseley and Schein *(30)* found the quantity of DNA in *M. radiodurans* to be about the same as that for other bacteria.

In view of the evidence that the radioresistance of this organism is apparently not due to the radioresistance of its DNA, several investigators have postulated a rapid repair of DNA damage by this organism *(31, 36, 37)*. Moseley and Laser have shown that ionizing radiation damage in *M. radiodurans* can be repaired enzymatically and is similar to that operative in UV dark repair. Setlow *(36)* found that the UV irradiation of DNA results in dimerization of adjacent thymines in the same polynucleotide chain, and that these dimers can be removed by the cell in the form of oligonucleotides. According to this author, formation of the dimers blocks DNA synthesis.

The resistance of this organism to X rays can be reduced by around 90% if irradiation of cells is carried out in the presence of $10^{-3} M$ iodoacetamide *(9)*. These workers showed that this radiosensitizer must be present during irradiation to be effective. This compound was not effective against sensitive *E. coli* and *Ps. fluorescens*, but was effective against a radiation-resistant strain of *E. coli*, strain B/r. Moseley and Laser *(31)* have suggested that iodoacetamide acts by inhibiting the cell's DNA repair mechanism.

There is general agreement among researchers that the radiation resistance of these organisms is due in large part to their possession of an excellent pyrimidine dimer excision repair system. Studies with UV-sensitive mutants have shown that they possess reduced rates of pyrimidine dimer excision. A recent study by Lavin *et al.* *(26)* employing *M. radiophilus* revealed that this organism also possesses an efficient excision repair system which these authors believe is at least in part responsible for its extreme radioresistance.

THE GENERAL NATURE OF RADIORESISTANCE IN OTHER MICROORGANISMS

1. Role of Radioprotective Compounds.   Much attention has been devoted to the finding of radioprotective substances in radioresistant microorganisms other than *M. radiodurans*. The finding by Raj *et al.* *(33)* that methionine was the only amino acid required by *M. radiodurans* in a synthetic medium provided some support for the notion of intracellular radioprotective substances since this amino acid contains a sulfur group. More direct support came from studies on the radioresistance of bacterial endospores, structures known to be far more resistant than their corresponding vegetative cells. Romig and Wyss *(34)* studied UV resistance in sporulating *B. cereus* and found that UV resistance increased about 2 hr before heat resistance as vegetative cells went into spore production. The same general phenomenon was shown by Vinter *(42)* for X-ray resistance. The latter author further demonstrated that spores contained higher levels of cysteine and cystine sulfur than corresponding vegetative cells. The increase in sulfur was found to coincide with an increase in radiation resistance of sporulating cells. Vinter *(42)* showed further that changes in cystine sulfur and radioresistance were distinct from the changes in Ca^{++} and dipicolinic acid. The cystine-rich substance occurred in spore coats and was not released into the medium upon spore germination as with dipicolinic acid. These studies were supported by the finding of Bott and Lundgren *(5)* that radioresistance in *B. cereus* increased with progressive spore maturity and that the spores had extremely low, or no, —SH content. Upon treating the spores of *B. cereus* with thioglycolic acid which ruptured from 10–30% of the spore disulfide bonds to thiol groups, they were found to retain their resistance to gamma radiation and to heat *(19)*.

2. Role of Catalase.   Among the various characteristic features of *M. radiodurans* is the large quantity of catalase it produces *(2)*. This property is apparently shared by all strains of this organism. The possible relationship of catalase production to radioresistance stems from the fact that H_2O_2 is known to be produced by both UV and ionizing radiations. This compound is, of course, destroyed by catalase. As noted above, the irradiation of *M. radiodurans* in the presence of H_2O_2 does not affect its resistance. In a study of the role of catalase and H_2O_2 on the radiation sensitivity of *Escherichia*, Engel and Adler *(1)* found that there was no positive correlation between catalase activity and sensitivity to ionizing radiation. It thus appears that the amount of catalase produced by an organism bears no relationship to its radioresistance. The clostridia are among the more radioresistant of all bacteria and yet they are anaerobic (produce no catalase).

3. Role of Oxygen.   The sensitivity of microorganisms is generally higher when irradiation occurs in the presence of O_2. Conversely, the absence of O_2 during irradiation decreases sensitivity. Stapleton *et al.* *(38)* showed that O_2 removal protected *E. coli* against X-ray inactivation. These investigators suggested that O_2 removal by either cellular enzyme action or autooxidation could also serve to protect this organism against X rays. Cromroy and Adler *(7)* showed that *E. coli* could be protected by MEA more effectively than by O_2 removal. MEA apparently acts to remove this gas from the cell's environment.

4. Genetic Aspects.   Attempts to explain the sigmoidal death rate curve of *M. radiodurans* have caused various authors to suggest the role of packet or tetrad formation by this organism as well as the existence of a multinuclear state. The latter was found by Gunter and Kohn *(16)* to produce multihit curves when diploid yeasts were subjected to X rays. In a study of X-ray resistance in *E. coli,* Stapleton and Engel *(39)* found that the resistance was not due to ploidy. The irradiation of the more sensitive *Sarcina* sp. has shown that tetrad formation *per se* does not give rise to the multihit-type death rate curve.

Penicillin-resistant strains of the *Achromobacter-Alcaligenes* group have been reported to be more sensitive to radiation than their penicillin-sensitive variants *(40)*. This finding may indicate some role of cell walls in radiation resistance, since penicillin affects these structures in bacterial cells. It also suggests that some property controlling penicillin resistance is at least partly responsible for radiation sensitivity *(40)*. On the other hand, *M. radiodurans* is sensitive to penicillin *(17)*.

Adler *(1)* has shown that radiation sensitivity in *E. coli* is under the control of genes that can be genetically transferred. In *E. coli* K12, one of these genes occupies a locus on the linkage map between those controlling the ability to utilize lactose and galactose. Both UV and ionizing radiation sensitivity are affected similarly. In studies with *B. subtilis*, Zamenhof *et al.* *(45)* were able to transform UV sensitivity to a more resistant strain. This feat has not been reported for *M. radiodurans*.

With respect to the role of DNA in the radiation resistance of microorganisms other than *M. radiodurans*. Hill and Simson *(18)* in their studies with a sensitive and a resistant strain of *E. coli* found no variation in the number of nuclei or in the content of DNA or RNA between these strains. DNA was shown by Zamenhof *et al.* *(45)* not to be responsible for radiation resistance in *B. subtilis*. Adler *(1)* has stated that genes affecting radiation sensitivity fall into two broad categories: (1) those that control systems concerned with repair of radiation-damaged DNA, and (2) those that seem to exert their influence on processes somewhat removed from DNA itself, such as cell division.

SUMMARY

Micrococcus radiodurans and *M. radiophilus* are the most radioresistant bacteria known to date. Attempts to associate this high level of resistance with pigment production, tetrad formation, catalase production, DNA resistance, and DNA composition have all been unsuccessful. Extracts have been shown to confer radioresistance upon sensitive organisms. The radioresistant substance or substances have been shown to be of low molecular weight and essentially devoid of —SH activity. Studies on the nature of radioresistance in other organisms have shown some similarities in their modes of resistance to that of *M. radiodurans*. The radiation resistance of *M. radiodurans* and *M. radiophilus* is due in large part to the ability of these organisms to repair damaged DNA.

REFERENCES

1. **Adler, H. I.** 1966. The genetic control of radiation sensitivity in microorganisms. Adv. Radiat. Biol. *2:* 167–191.

2. **Anderson, A. W., H. C. Nordan, R. F. Cain, G. Parrish, and D. Duggan.** 1956. Studies on a radio-resistant *Micrococcus*. I. Isolation, morphology, cultural characteristics, and resistance to gamma radiation. Food Technol. *10:* 575–578.

3. **Anderson, A. W., K. E. Rash, and P. R. Elliker.** 1961. Taxonomy of a recently isolated radio-resistant *Micrococcus*. Bacteriol. Proc., p. 56.

4. **Bacq, Z. M. and P. Alexander.** 1961. *Fundamentals of Radiobiology,* 2nd ed. Pergamon Press, Oxford.

5. **Bott, K. F. and D. G. Lundgren.** 1964. The relationship of sulfhydryl and disulfide constituents of *Bacillus cereus* to radioresistance. Radiat. Res. *21:* 195–211.

6. **Bruce, A. K.** 1964. Extraction of the radioresistant factor of *Micrococcus radiodurans*. Radiat. Res. *22:* 155–164.

7. **Cromroy, H. L. and H. I. Adler.** 1962. Influences of β-mercaptoethylamine and oxygen removal on the X-ray sensitivity of four strains of *Escherichia coil*. J. Gen. Microbiol. *28:* 431–435.

8. **Davis, N. S., G. J. Silverman, and E. B. Masurovsky.** 1963. Radiation-resistant, pigmented coccus isolated from haddock tissue. J. Bacteriol. *86:* 294–298.

9. **Dean, C. J. and P. Alexander.** 1962. Senitization of radio-resistant bacteria to x-rays by iodocetamide. Nature *196:* 1324–1325.

10. **Duggan, D. E., A. W. Anderson, P. R. Elliker, and R. F. Can.** 1959. Ultraviolet exposure studies on a gamma radiation resistant *Micrococcus* isolated from food. Food Res. *24:* 376–382.

11. **Duggan, D. E., A. W. Anderson, and P. R. Elliker.** 1963a. Inactivation of the radiation-resistant spoilage bacterium *Micrococcus radiodurans*. I. Radiation inactivation rates in three meat substrates and in buffer. Appl. Microbiol. *11:* 398–403.

12. **Duggan, D. E., A. W. Anderson, and P. R. Elliker.** 1963b. Inactivation of the radiation-resistant spoilage bacterium *Micrococcus radiodurans*. II. Radiation inactivation rates as influenced by menstrum temperature, preirradition heat treatment, and certain reducing agents. Appl. Microbiol. *11:* 413–417.

13. **Duggan, D. E., A. W. Anderson, and P. R. Elliker.** 1963c. Inactivation rate studies on a radiation-resistant spoilage microorganism. III. Thermal inactivation rates in beef. J. Food Sci. *28:* 130–134.

14. **Engel, M. S. and H. I. Adler.** 1961. Catalase activity, sensitivity to hydrogen peroxide, and radiation response in the genus *Escherichia*. Radiat. Res. *15:* 269–275.

15. **Gammon, R. A., K. Kereluk, and R. S. Lloyd.** 1968. Microbial resistance to ethylene oxide. Bacteriol. Proc., p. 16.

16. **Gunter, S. E. and H. T. Kohn.** 1956. The effect of X-ray on the survival of bacteria and yeast. I. J. Bacteriol. *71:* 571–581.

17. **Hawiger, J. and J. Jeljaszewicz.** 1967. Antibiotic sensitivity of *Micrococcus radiodurans*. Appl. Microbiol. *15:* 304–306.

18. **Hill, R. F. and E. Simson.** 1961. A study of radiosensitive and radioresistant mutants of *Escherichia coli* strain B. J. Gen. Microbiol. *24:* 1–14.

19. **Hitchins, A. D., W. L. King, and G. W. Gould.** 1966. Role of disulphide bonds in the resistance of *Bacillus cereus* spores to gamma irradiation and heat. J. Appl. Bacteriol. *29:* 505–511.

20. **Kaplan, H. S. and R. Zavarine.** 1962. Correlation of bacterial radiosensitivity and DNA base composition. Biochem. Biophys. Res. Commun. *8:* 432–436.

21. **Kelner, A., W. D. Bellamy, G. E. Stapleton, and M. R. Zelle.** 1955. Symposium on radiation effects on cells and bacteria. Bacteriol. Rev. *19:* 22–44.

22. **Kilburn, R. E., W. D. Bellamy, and S. A. Terni.** 1958. Studies on a radiation-resistant pigmented *Sarcina* sp. Radiat. Res. *9:* 207–215.

23. **Knivett, V. A., J. Cullen, and M. J. Jackson.** 1965. Odd-numbered fatty acids in *Micrococcus radiodurans*. Biochem. J. *96:* 2–3C.

24. **Krabbenhoft, K. L., A. W. Anderson, and P. R. Elliker.** 1965. Ecology of *Micrococcus radiodurans*. Appl. Microbiol. *13:* 1030–1037.

25. **Krabbenhoft, K. L., A. W. Anderson, and P. R. Elliker.** 1967. Influence of culture media on the radiation resistance of *Micrococcus radiodurans*. Appl. Microbiol. *15:* 178–185.

26. **Lavin, M. F., A. Jenkins, and C. Kidson.** 1976. Repair of ultraviolet light-induced damage in *Micrococcus radiophilus,* an extremely resistant microorganism. J. Bacteriol. *126:* 587–592.

27. **Lee, J. S., A. W. Anderson, and P. R. Elliker.** 1963. The radiation-sensitizing effects of N-ethylmaleimide and iodoacetic acid on a radiation-resistant *Micrococcus*. Radiat. Res. *19:* 593–598.

28. **Lewis, N. F.** 1971. Studies on a radio-resistant coccus isolated from Bombay duck *(Harpodon nehereus)* J. Gen. Microbiol. *66:* 29–35.

29. **Lewis, N. F., D. A. Madhavesh, and U. S. Kumta.** 1974. Role of carotenoid pigments in radio-resistant micrococci. Can. J. Microbiol. *20:* 455–459.

30. **Moseley, B. E. B. and A. H. Schein.** 1964. Radiation resistance and deoxyribonucleic acid base composition in *Micrococcus radiodurans*. Nature *203:* 1298–1299.

31. **Moseley, B. E. B. and H. Laser.** 1965. Similarity of repair of ionizing and ultraviolet radiation damage in *Micrococcus radiodurans*. Nature *206:* 373–375.

32. **Murray, R. G. E. and C. F. Robinow.** 1958. Cytological studies of a tetrad-forming coccus. Intern. Congr. Microbiol., 7th, Stockholm, pp. 427–428.

33. **Raj, H. D., F. L. Duryee, A. M. Deeney, C. H. Wang, A. W. Anderson, and P. R. Elliker.** 1960. Utilization of carbohydrates and amino acids by *Micrococcus radiodurans*. Can. J. Microbiol. *6:* 289–298.

34. **Romig, W. R. and O. Wyss.** 1957. Some effects of ultraviolet radiation on sporulating cultures of *Bacillus cereus*. J. Bacteriol. *74:* 386–391.

35. **Serianni, R. W. and A. K. Bruce.** 1968. Role of sulphur in radioprotective extracts of *Micrococcus radiodurans*. Nature *218:* 485–486.

36. **Setlow, J. K.** 1964. Physical changes and mutagenesis. J. Cell. and Comp. Physiol. *64:* Sup. 1, 51–68.

37. **Setlow, J. K. and D. E. Duggan.** 1964. The resistance of *Micrococcus radiodurans* to ultraviolet radiation. I. Ultraviolet-induced lesions in the cell's DNA. Biochim. Biophys. Acta *87:* 664–668.

38. **Stapleton, G. E., D. Billen, and A. Hollaender.** 1952. The role of enzymatic oxygen removal in chemical protection against x-ray inactivation of bacteria. J. Bacteriol. *63:* 805–811.

39. **Stapleton, G. E. and M. S. Engel.** 1960. Cultural conditions as determinants of sensitivity of *Escherichia coli* to damaging agents. J. Bacteriol. *80:* 544–551.

40. **Thornley, M. J.** 1963. Radiation resistance among bacteria. J. Appl. Bacteriol. *26:* 334–345.

41. **Thornley, M. J., R. W. Horne, and A. M. Glauert.** 1965. The fine structure of *Micrococcus radiodurans*. Arch. Mikrobiol. *51:* 267–289.

42. **Vinter, V.** 1962. Spores of micro-organisms. IX. Gradual development of the resistant structure of bacterial endospores. Folia Microbiol. *7:* 115–120.

43. **Work, E.** 1964. Amino-acids of walls of *Micrococcus radiodurans*. Nature *201:* 1107–1109.

44. **Work, E. and H. Griffiths.** 1968. Morphology and chemistry of cell walls of *Micrococcus radiodurans*. J. Bacteriol. *95:* 641–657.

45. **Zamenhof, S., H. Bursztyn, T. K. R. Reddy, and P. J. Zamenhof.** 1965. Genetic factors in radiation resistance of *Bacillus subtilis*. J. Bacteriol. *90:* 108–115.

Appendix

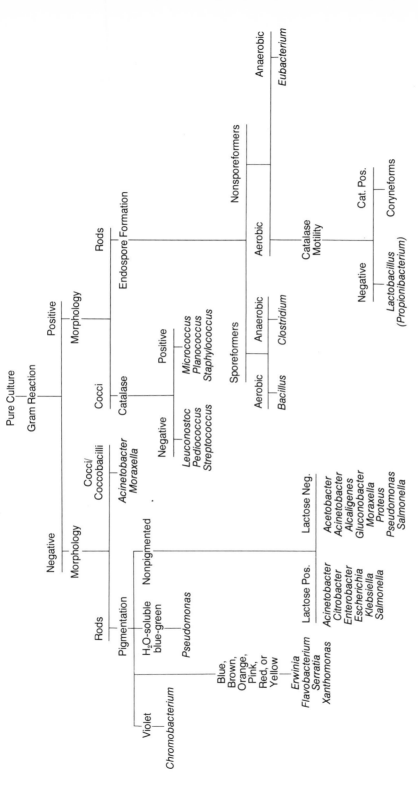

A. Simplified schematic showing the relationship of common food-borne bacterial genera to each other (for details, see ref. 1 and sections B and C below).

B. Generic differentiations among some of the genera in schematic A.

The Gram Positive, Catalase Positive Cocci

	Micrococcus	Staphylococcus	Planococcus
Cell arrangement	Irregular clusters, tetrads	Irregular clusters	Singly, pairs, tetrads
Motility	−	−	+
Ferment glucose anaerobically	−	+	−
Coagulase production	−	+ (some)	−
Pigmentation	Yellow or red	white, gold, yellow to orange	Yellow-brown
Metabolism	Respiratory	Respiratory/ fermentative	Respiratory
Lysostaphin sensitivity	−	+	−
Moles %G + C	66–75	30–40	39–52

The Gram positive, Catalase Negative Cocci

	Streptococcus	Leuconostoc	Pediococcus
Cell arrangement	1-plane division, pairs & chains	1-plane division, pairs & chains	2-plane division, pairs & tetrads
Glucose fermentation	Homofermentatively to lactic acid	Heterofermentatively to lactic acid, CO_2, ethanol and/or acetate	Homofermentatively to lactic acid
Moles %G + C	33–42	38–42 (43–44)	34–44

The Gram Negative Cocci/Coccobacilli (1, 2, 3).

	Moraxella	Acinetobacter
Morphology	Plump rods, pairs, or short chains	Young cells are rods, older cells spherical
Oxidase Reaction	+	−
Penicillin sensitivity	Sensitive	Resistant
Lactose	−	+ or −
Moles %G + C	40–46	39–47

The Gram Negative Pigmented Rods

	Flavobacterium	Xanthomonas	Serratia	Erwinia
Pigmentation	Yellow to orange	Yellow	Pink to red	Blue or yellow
Motility	+ or −	+	+	+ (most)
DNase			+	− (most)
Acid from glucose	Rare	Most		+
Nitrate reduction	+ or −	−		
Indole production	+ or −	−		
Oxidase		− (or weak)		−

The Gram Positive, Nonsporeforming, Aerobic Rods

| | Lactobacillus | The Coryneforms[1] | | | Propionibacterium |
		Corynebacterium	Arthrobacter	Kurthia	
Motility	–	–	– (most)	+	–
Catalase production	–	+	+	+	+ (most)
Glucose fermentation	A or AG				A
Pigmentation					White, gray, pink, red, orange, or yellow
Distinctive character			Log phase cells are rods; older cells are coccoids		
Moles %G + C	35–53	48–70	60–72		59–66

[1]These bacteria are characterized by pleomorphic cells during log-phase growth in suitable culture media and by V- and palisade-arrangements resulting from snapping post-fission movements. The genera *Brevibacterium* and *Microbacterium* are coryneforms that are closely related to the arthrobacters. All members and close relatives of the coryneform group are highly variable not only morphologically but culturally and biochemically as well (1, 6).

The Gram Negative, Lactose Positive Rods

	Escherichia	Enterobacter	Citrobacter	Klebsiella	Salmonella
MR/VP Reactions	MR+ VP−	MR− VP+	MR+ VP−	MR− VP+	MR+ VP−
Motility	+	+	+	−	+ (most)
Indole production	+	−	+ or −	var.	−
H₂S production	−	−	var.	−	+
Ornithine decarboxylase	var.	+	var.	−	+
Utilize citrate as sole C source	−	+	+	var.	+
Sucrose fermentation	var.	+	var.	+	−
Growth on KCN	−	+	+	+	var.
Moles %G + C	50–51	52–59		52–56	50–53

The Gram Negative, Lactose Negative Rods

	Proteus	Gluconobacter	Acetobacter	Pseudomonas	Alcaligenes
Motility	+	+ or −	+ or −	+	+
Gelatinase production		− (or weak)	−	+ or −	−
Growth at pH 4.5		+	+	−	
Acetic or lactic acid oxidized to CO₂		−	+		
Water-soluble brown pigment produced			some		
Urea hydrolyzed	Most	some			
Moles %G + C	39–42 (50)	60–64	55–64	58–70	58–70

C. A grouping of the gram negative asporogenous rods, polar-flagellate, oxidase positive and not sensitive to 2.5 i. u. penicillin, on the results of four other tests *(5)*.

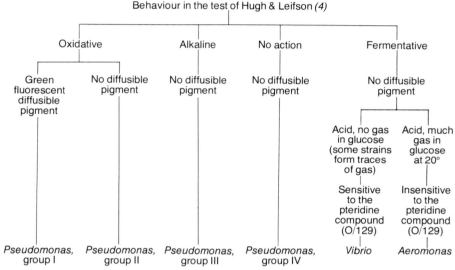

Copyright © 1953, Society for Applied Bacteriology.

REFERENCES

1. **Bergey's Manual of Determinative Bacteriology.** 1974. Edited by R. E. Buchanan and N. E. Gibbons, 8th Edition (Williams & Wilkins Co.: Baltimore).

2. **Henriksen, S. D.** 1973. *Moraxella, Acinetobacter,* and the *Mimeae.* Bacteriol. Rev. *37:* 522–561.

3. **Henriksen, S. D.** 1976. *Moraxella, Neisseria, Branhamella,* and *Acinetobacter.* Ann. Rev. Microbiol. *30:* 63–83.

4. **Hugh, R. and E. Leifson.** 1953. The taxonomic significance of fermentative versus oxidative metabolism of carbohydrates by various gram-negative bacteria. J. Bacteriol. *66:* 24–00.

5. **Shewan, J. M., G. Hobbs, and W. Hodgkiss.** 1960. A determinative scheme for the identification of certain genera of gram-negative bacteria, with special reference to the *Pseudomonadaceae.* J. Appl. Bacteriol. *23:* 379–390.

6. **Veldkamp, H.** 1970. Saprophytic coryneform bacteria. Ann. Rev. Microbiol. *24:* 209–240.

Index*

462